REFLECTIONS ON BIOCHEMISTRY

Published with a special grant from the
Spanish Association against Cancer

Reflections on Biochemistry

IN HONOUR OF SEVERO OCHOA

Editors

A. KORNBERG

Department of Biochemistry, Stanford University Medical School, California

B. L. HORECKER

Department of Biochemistry Roche Institute of Molecular Biology, New Jersey

L. CORNUDELLA

Departamento de Química Macromolecular, C.S.I.C. Universidad Politécnica, Barcelona

J. ORO

Departments of Biophysical Sciences and Chemistry, University of Houston, and Instituto de Biofisica y Neurobiologia, C.S.I.C., Barcelona

PERGAMON PRESS

OXFORD · NEW YORK · TORONTO · SYDNEY · PARIS · FRANKFURT

U.K.	Pergamon Press Ltd., Headington Hill Hall, Oxford OX3 0BW, England
U. S. A.	Pergamon Press Inc., Maxwell House, Fairview Park, Elmsford, New York 10523, U.S.A.
CANADA	Pergamon of Canada Ltd., 75 The East Mall, Toronto, Ontario, Canada
AUSTRALIA	Pergamon Press (Aust.) Pty. Ltd., 19a Boundary Street, Rushcutters Bay, N.S.W. 2011, Australia
FRANCE	Pergamon Press SARL, 24 rue des Ecoles, 75240 Paris, Cedex 05, France
WEST GERMANY	Pergamon Press GmbH, 6242 Kronberg/Taunus, Pferdstrasse 1, Frankfurt-am-Main, West Germany

First edition 1976

Library of Congress Cataloging in Publication Data

International Symposium on Enzymatic Mechanisms in Biosynthesis and Cell Function, Barcelona and Madrid, 1975.
Reflections on biochemistry.
"Organized by the Autonomous Universities of Barcelona and Madrid and the University of Houston, with the collaboration of Stanford University and the Roche Institute of Molecular Biology."
1. Biological chemistry—Congresses. 2. Ochoa de Albornoz, Severo, 1905– I. Ochoa de Albornoz, Severo, 1905– II. Kornberg, Arthur, 1918– III. Universidad Autónoma de Barcelona. IV. Universidad Autónoma de Madrid. V. Title. [DNLM: 1. Biochemistry—History—Congresses. QU11.1 I61r 1975]
QP501.I485 1975 574.1'92'0904 76–24886
ISBN 0–08–021011–2
ISBN 0–08–021010–4 pbk

0 08 021011 2 h.c
ISBN 0 08 021010 4 f.c

PRINTED IN NORTHERN IRELAND AT THE UNIVERSITIES PRESS (BELFAST) LTD.

CONTENTS

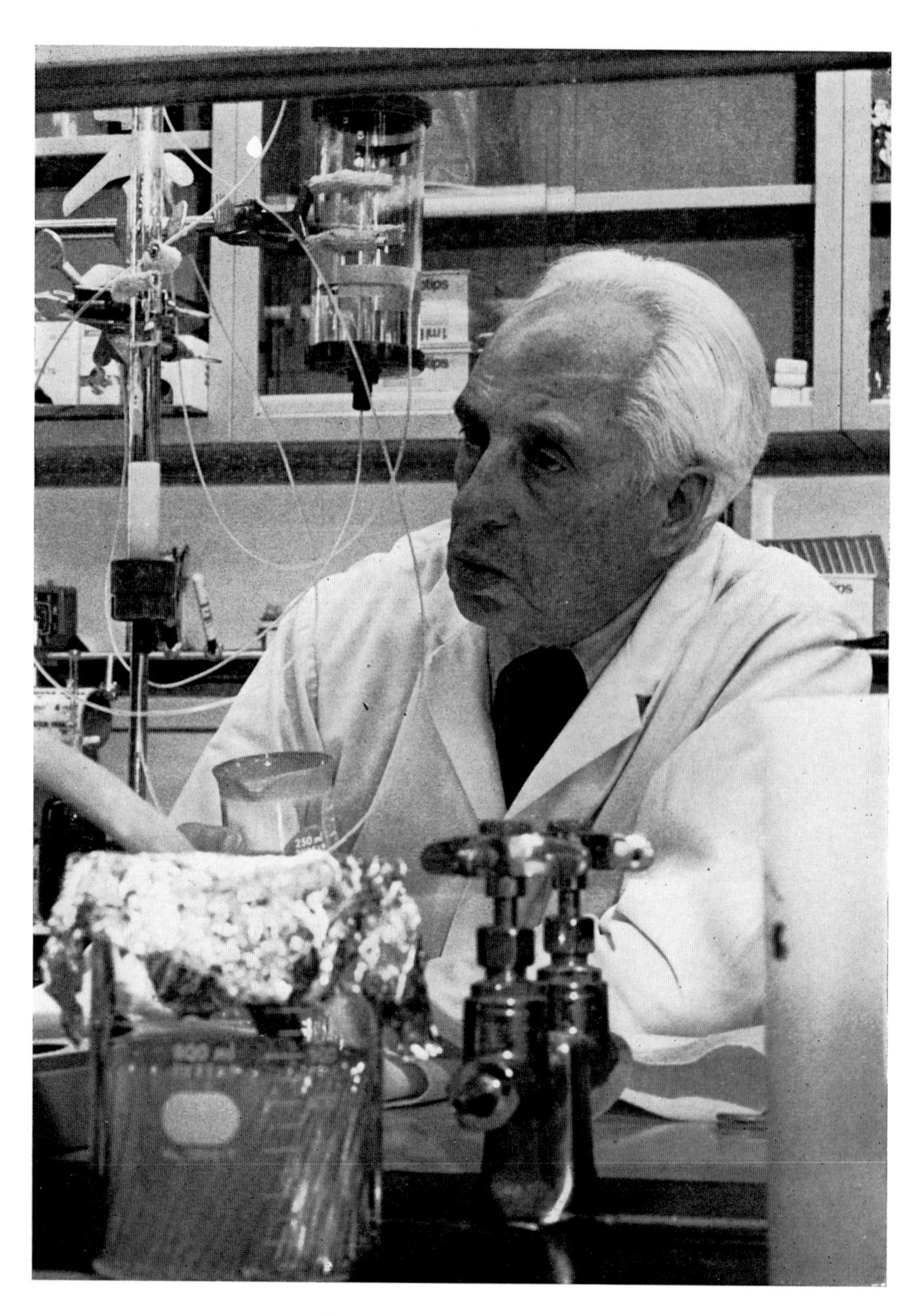

SEVERO OCHOA

PREFACE

This collection of essays by some of Severo Ochoa's students and colleagues celebrates his 70th birthday and his outstanding contributions to biochemistry spanning nearly half a century. These students and colleagues assembled at an International Symposium held in Barcelona and Madrid in September 1975, to honour Ochoa with scientific papers in six colloquia covering fields to which he had made his contributions: energy metabolism, lipids and saccharides, regulation, nucleic acids and the genetic code, protein biosynthesis and cell biology.

It would have been conventional and relatively simple to prepare a *Festschrift* of these papers. Instead the participants attempted something more demanding and difficult. They wrote essays reflecting on the development of a subject, a concept, or an approach to biochemistry. As editors we were struck by the reluctance of scientists who are in the habit of writing numerous papers and reviews to engage in such an exercise and especially to give it an autobiographical flavour. This may explain the paucity of such historical essays by active scientists and account for the rather wide divergence in tone and style in this collection. Still we are pleased that this volume contains so many accounts about the development and developers of biochemistry which are original and scholarly and enrich our appreciation of the subject.

ACKNOWLEDGEMENTS

We wish to express our gratitude to all the contributors to this volume, as well as to all the persons and institutions who, by their support or participation, made possible the celebration of the "International Symposium on Enzymatic Mechanisms in Biosynthesis and Cell Function"* and other activities honouring Severo Ochoa on his 70th birthday, which were a prelude to the publication of this book. The Symposium was organized by the Autonomous Universities of Barcelona and Madrid and the University of Houston with the collaboration of Stanford University and the Roche Institute of Molecular Biology. It was held under the auspices of:

The Spanish Ministry of Education and Science
Higher Council for Scientific Research
Provincial Council of Barcelona
City Hall of Barcelona
Spanish Society of Biochemistry
Juan March Foundation
General Mediterranean Foundation

The publication of the book was made possible, in part, by a special grant received from the Scientific Foundation of the Spanish Association Against Cancer and by the enthusiastic collaboration of the publisher and staff of Pergamon Press, which we gratefully acknowledge. It is also a great pleasure to recognize the generous collaboration of Don Salvador Dalí who created a painting for this occasion as a gallant gesture for the promotion of biochemical science. The painting, significantly reduced, has been printed on the cover of the Symposium program and on the jacket of this book. At the end of the essays we are enclosing a translation of what the painter dictated to one of the editors, a statement which may be a key for the interpretation of this painting.

* Abstracts of the papers presented in Madrid and Barcelona, and biographical sketches of the contributors (in English), and a biography of Severo Ochoa (in Spanish) were included in a bound programme entitled "Homenaje al Profesor Severo Ochoa en su 70 aniversario—International Symposium on Enzymatic Mechanisms in Biosynthesis and Cell Function" (J. Oró and L. Cornudella, eds., Socitra, Salvadors, 22 Barcelona-1).

SEVERO OCHOA AND THE DEVELOPMENT OF BIOCHEMISTRY

FRANCISCO GRANDE AND CARLOS ASENSIO

Wir lernen die Menschen nicht kennen, wenn sie zu uns kommen;
wir müssen zu ihnen gehen, um zu erfahren, wie es mit ihnen steht.

You cannot know a man if he comes to you;
you must go to him to discover his inner self. J. W. Goethe, *Die Wahlverwandtschaften*

Severo Ochoa's scientific biography condenses the history of contemporary biochemistry and connects events which have significantly affected the development of this science.

Ochoa's scientific activity began in the late 1920s. It embraced the golden age of European Biochemistry, which was centered in Germany in the 1920s, shifted to Great Britain in the 1930s, and finally surrendered leadership to the United States since 1940. Ochoa appears as an exceptional participant in the places where the scientific activity reached its greatest intensity: Berlin, Heidelberg, Oxford, Saint Louis, New York. His connection with Biochemistry's development over time and geography is deeply rooted. Indeed Ochoa is one of the very few contemporary scientists who has maintained a position of leadership in this explosively developing and versatile science over such a long interval.

Severo Ochoa was born on 24 September 1905 in the village of Luarca of the province of Asturias, in Northern Spain.

He was the youngest of seven children born to Severo Ochoa, a lawyer and businessman, and Carmen de Albornoz. His father died when Severo was 7. At this time the family went to live in Málaga in Southern Spain. There Severo had his initial education, first in a Jesuit elementary school and then at the State High School, where his enthusiasm and dedication to the subjects of human anatomy and physiology were notable.

Ochoa entered Madrid University in 1922 to study Medicine. Although he was not a student of Cajal, the famous neuroanatomist, who had retired the previous year, Ochoa in his early demonstration of enthusiasm, industry, and determination seems to have been an embodiment of the "heroic young research worker", characterized by Cajal. A book that had a lasting influence on Ochoa's development was: *La Physiologie, Resultats, methodes, hypotheses*, by the Swiss physiologist M. Arthus.

Ochoa's first publication in biochemistry (1929) was based on work in Juan Negrin's physiology laboratory. It was a method for the determination of total creatine in muscle. He had earlier completed a study on the effect of guanidine on the melanophores of the skin of the frog in the physiology laboratory of Noel Paton in Glasgow.

From 1929 to 1931, Ochoa worked in Otto Meyerhof's laboratory in Heidelberg. To quote Ochoa: "Meyerhof was the teacher who most contributed towards my formation, and the most influential in directing my life's work".

He worked on the sources of energy for muscular contraction, a continuation of his first interest in muscle creatine. Phosphagen had been recently discovered. Einar Lundsgaard had just shown that muscle poisoned with iodoacetic acid was able to perform a certain amount of work without liberation of lactic acid. Using muscles with low carbohydrate content, Ochoa was able to demonstrate the muscle's ability to perform work using sources of energy different from those then known. It was an important piece of work on a topic of great interest at the time.

Upon returning to Madrid from Germany in 1931, Ochoa married Carmen Cobian. He resumed his work at the Physiology Laboratory as an Associate Professor. He devoted most of his time to work on the chemistry and energetics of muscle contraction. He found proof for the existence of a fraction of combined creatine which differed from phosphagen. He also continued studies on chemical changes associated with adrenal insufficiency.

In 1932 Ochoa decided to extend his training in enzymology and went to work with H. W. Dudley at the National Institute for Medical Research in London. He discovered the antiglyoxalase effect of pancreatic extracts and their influence on the glycolysis of muscle extracts. He also reported on chemical changes associated with muscular contraction in adrenalectomized animals, and on the influence of toxemia on carbohydrate metabolism.

Back in Madrid in 1934, Ochoa defended his doctoral thesis on the chemistry of muscular contraction in adrenalectomized animals. The chemistry of muscle continued to attract Ochoa's attention in 1935. In that year he presented to the Congress of Physiology in Leningrad a paper on the levels of pyrophosphate in the contracting muscle and initiated with F. Grande a study on the formation of lactic acid by the mammalian heart.

The beginning of the Civil War in 1936 ended all possibilities of scientific work. The Institute was located in the buildings of the new university, soon to become a battlefield.

Ochoa and his wife Carmen decided to leave Spain. Negrin, then a leader of the Republican Government, helped them to get the necessary papers, and the couple went to Heidelberg where Ochoa worked again in Meyerhof's laboratory, concluding some of the work initiated in Spain on the production of lactic acid by the mammalian heart. He collaborated in an investigation on cozymase, his first work on a coenzyme.

After a year in Meyerhof's laboratory, Ochoa, through the help of the British biophysicist A. V. Hill, moved as a research fellow to the Marine Biological Laboratory at Plymouth, England, where he studied enzymatic phosphorylation in invertebrate muscle, and in collaboration with his wife, Carmen, a study of its cozymase content.

In 1938 Ochoa joined the biochemical laboratory of Oxford University directed by Rudolph A. Peters. Studies on the role of vitamin B_1 in pyruvate metabolism led him to some of the biochemical properties of this vitamin and coenzyme and to a singularly important study of oxidative phosphorylation.

The coenzyme of carboxylase (thiamine pyrophosphate) was shown to be the active form of vitamin B_1, a method for its estimation was developed, its enzymatic synthesis in animal tissues was demonstrated. Perhaps the most important part of Ochoa's work at Oxford was the discovery of the coupling of phosphorylation with oxidation of pyruvic acid in brain. Here he demonstrated the obligatory relationship between these two processes, a contribution which came almost at the same time as the observations of Herman Kalckar in Denmark and Vladimir A. Belitzer in the Soviet Union. Ochoa was the first to show the formation of three phosphate bonds for each oxygen atom used in the process.

The beginning of World War II confronted the Ochoas with a decision similar to the one they had faced in Spain in 1936. The war situation in England made it impossible to pursue purely scientific work. Ochoa therefore decided to accept the invitation from Carl and Gerty Cori to join them at Washington University in St. Louis.

When Ochoa arrived, the laboratory was in a period of frantic activity. The crucial experiments leading to the discovery of phosphorylase and the synthesis of glycogen *in vitro* by this enzyme had been performed the previous year. In the Cori laboratory were Herman Kalckar, Sidney Colowick, Earl Sutherland and others.

During a sojourn in St. Louis of about a year and a half, Ochoa extended previous studies on carboxylase and worked on the relation between glycolysis and phosphorylation. Also, in collaboration with the Coris, he isolated and characterized fructose-1-phosphate and inorganic pyrophosphate in liver dispersions. In St. Louis he was imbued with the importance and the techniques of isolating and characterizing enzymes.

The Coris were, in fact, Ochoa's last formal teachers. St. Louis was the penultimate leg of the Ochoas' long peregrination. Their destination was New York. There Ochoa found a place as research associate in the Department of Medicine at the New York University School of Medicine.

Ochoa's academic advancement in New York was slow, his initial appointment in 1945 being that of Assistant Professor of Biochemistry. The following year, however, he was appointed Professor of Pharmacology. In 1954 he succeeded to the chairmanship of the Department of Biochemistry. The delays in Ochoa's academic promotion were to some measure due to his own indifference. When Ochoa was offered the chair of Pharmacology his colleague, Ephraim Racker, recalls long discussions in which Ochoa kept saying: "Why do I need a professorship? I can do my work where I am now; will the research work not suffer if I become a department chairman?" What finally led him to accept the chair was the prospect of getting some well-equipped laboratory space which the previous chairman, James A. Shannon, had developed.

STUDIES ON INTERMEDIARY METABOLISM

In Oxford under Rudolph Peters, Ochoa was introduced to the oxidation of pyruvic acid. The experience gained with the Coris made him familiar with the first steps of glycolysis and gave him valuable experience in dealing with enzymes. In New York, Ochoa began to explore the energetic efficiency of glucose utilization in its total combustion to CO_2 and H_2O. He determined that thirty-six phosphates were esterified to ATP. This 1943 calculation is remarkably close to the value of thirty-eight accepted today.

Ochoa's attention was next concentrated upon one of the central issues of that period: the mechanisms of oxidative degradation of pyruvic acid to CO_2 and H_2O. In the previous decades, Torsten Thunberg had discovered the oxidative capacity of tissue extracts for a number of dicarboxylic acids. Albert Szent-Györgyi had observed the catalytic effect of succinic, fumaric, malic and oxaloacetic acids in the respiratory process and related these observations to the metabolism of carbohydrates. The oxidative process implied the formation of CO_2 and hydrogen, the latter being eventually accepted by respiratory oxygen to form water with the participation of cytochrome.

Hans Krebs, then recently arrived in England as a refugee from Germany, had added citric and α-ketoglutaric acids to the earlier list and, in 1937, demonstrated that the rapid oxidation of pyruvic acid by muscle extracts was catalyzed by these carboxylic acids. It was a cyclic process. Pyruvic acid, a C_3 acid, entered the cycle by condensing with oxaloacetic acid to produce a hypothetical C_7 compound. From this, the C_6 citric acid was formed and then, by successive loss of CO_2 units and H atoms, all the other intermediates were formed. Finally oxaloacetic was regenerated and able to take another pyruvic acid molecule through the cycle.

In the middle 1940s, when Ochoa launched his study of the mechanisms involved in the intermediary steps of the citric acid cycle, he adopted a clearly enzymological approach. Since each step of the process should be catalyzed by a specific enzyme, one must isolate each enzyme in order to clarify the mechanism. Some of the enzymes were already known, but none had been isolated to a sufficient degree of purity.

Figure 1 gives a simplified version of the cycle. The thick arrows denote the steps investigated by Ochoa and his group. Included also are the connections with other metabolic routes later investigated in his laboratory.

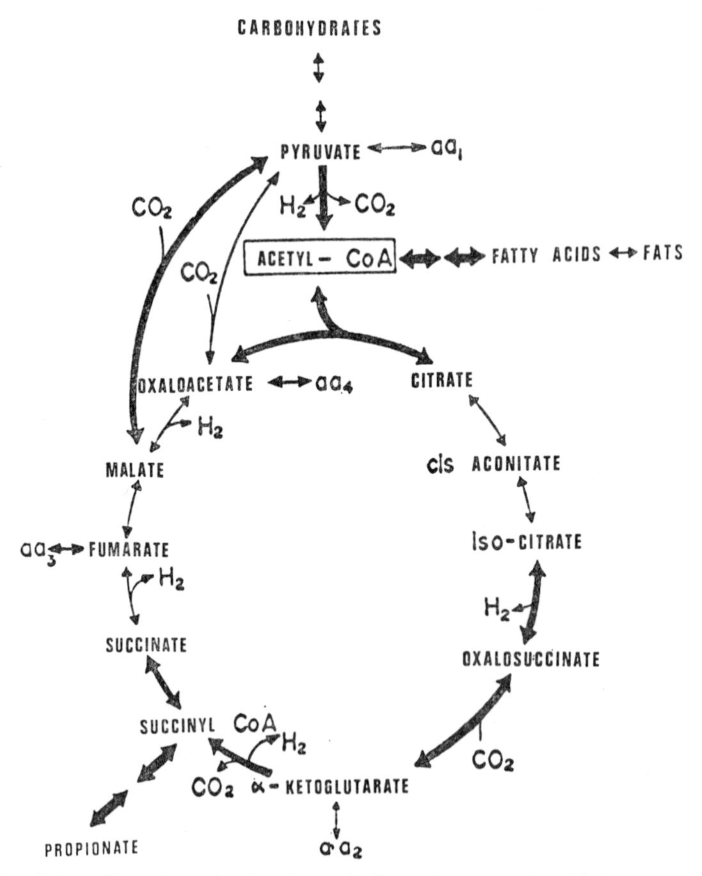

FIG. 1. The citric acid cycle and related metabolic pathways. The thick arrows indicated the steps clarified by Ochoa.

(a) Pyruvic acid oxidation

Pyruvic acid has a central position in metabolism; aa_{1-4} represent amino acids connecting the cycle with protein synthesis (Fig. 1).

The oxidation of pyruvic acid to form acetyl CoA was for many years the elusive "active acetate". It was the target of active research in many laboratories since Peters demonstrated in the 1930s that cocarboxylase (thiamine pyrophosphate) is necessary for the process. The reaction implies an "oxidative decarboxylation" by which pyruvic acid loses CO_2 and 2H to form acetyl CoA, and requires several enzymes and coenzymes (thiamine pyrophosphate, coenzyme A, NAD, lipoic acid and Mg^{2+}).

A key advance took place in 1950 when Fritz Lipmann discovered CoA and Ochoa demonstrated that CoA and NAD were required, thus resolving the system into several soluble fractions. Shortly thereafter Lynen isolated acetyl-CoA and Ochoa demonstrated it to be the "active acetate" in its direct reaction with oxaloacetic acid to form citric acid. Neither oxygen nor other compounds such as acetyl phosphate previously found by Lipmann in bacterial systems were needed. Ochoa's most distinguished coworker in this work was Seymour Korkes, tragically killed in an accident a few years later.

(b) The condensing enzyme

The search for the C_7 compound formed by condensation of pyruvic and oxaloacetic acids, originally postulated by Krebs, had been fruitless for more than 10 years. In 1949 Ochoa with the collaboration of Joseph Stern isolated an enzyme from heart muscle capable of synthesizing stoichiometric amounts of citric acid from ATP, acetate, CoA and oxaloacetate. First named "condensing enzyme", it is now known as citrate synthetase. It was obtained by the Ochoa group in crystalline form, the first enzyme of the cycle to be crystallized. With the pure enzyme in hand, it was clear not only that the product was not a C_7 compound, but that other alternatives to citrate, namely, *cis*-aconitate and isocitrate, could also be rejected. The citrate synthetase condensation takes place between the methyl group of acetyl-CoA and the keto group of oxaloacetic acid; the reaction is reversible.

Using the crystalline enzyme and a precious sample of acetyl-CoA sent by Lynen, Ochoa was able to show in 1951 that the condensing enzyme required only this compound and oxaloacetic acid to produce citric acid. The key reaction of the Krebs cycle was thus formulated:

$$\text{Acetyl-CoA} + \text{Oxaloacetate} \xrightleftharpoons{\text{condensing enzyme}} \text{Citrate} + \text{CoA}$$

(c) Formation of oxalosuccinate and α-ketoglutarate

In 1948 Ochoa demonstrated the existence of an enzyme, isocitric dehydrogenase, which catalyzed the oxidation of isocitric acid and required NADP. He was, however, unable to demonstrate the formation of the expected product: oxalosuccinic acid (Fig. 1). The existence of this acid as an intermediate in the Krebs cycle had been postulated by Carl Martius. Ochoa was able to prepare the compound by chemical synthesis and showed that cell extracts catalyzed the decarboxylation of this very unstable β-keto acid to α-ketoglutaric acid. Similar results were simultaneously obtained by Lynen in Germany.

Thus the enzyme had two functions: dehydrogenation of isocitric acid, to oxalosuccinic acid, followed by decarboxylation to α-ketoglutaric acid. The second stage of the process could be studied separately because of its dependence on Mn^{2+}. Ochoa was able to show the reversibility of the process, namely CO_2 fixation by α-ketoglutarate and subsequent reduction of the presumed oxalosuccinic acid intermediate.

(d) Oxidation of α-ketoglutarate

With the collaboration of Seymour Kaufman, Ochoa obtained in 1952 a soluble preparation from heart muscle which showed a direct coupling between decarboxylative oxidation of α-ketoglutarate and ATP formation. This confirmed Ochoa's earlier results with particulate fractions. NAD and CoA were needed in the soluble system. Formation of succinyl-CoA as an intermediate, analogous to acetyl-CoA, was postulated, and the presence of this compound was quickly demonstrated in both his and David Green's laboratories. The mechanism of this reaction proved to be similar to that of the oxidation of pyruvic acid.

(e) The malic enzyme

The connection between the citric acid cycle and other biochemical activities in the cell, among them the synthesis and degradation of proteins, sugars, and fatty acids, has already been noted.

In 1936, Harland Wood and Chester Werkman made the fundamental discovery that non-photosynthetic, heterotrophic organisms utilize CO_2 for the synthesis of compounds related to those in the Krebs cycle. CO_2 fixation was later observed with extracts of bacterial and animal cells.

The mechanism of carboxylation of pyruvic acid (the "Wood and Werkman reaction") remained obscure for many years. In 1948 Ochoa, with Alan Mehler and Arthur Kornberg, discovered the first enzyme system which fixed CO_2 to pyruvic acid. It produced malic acid. The enzyme, first obtained from liver extracts, was called "malic enzyme" (see Fig. 1). The reaction proved to be reversible, producing pyruvic from malic acid. The enzyme serves, as do similar reactions discovered since then, to replenish the citric acid cycle intermediates ("anaplerotic" reactions). In so doing it is of vital importance in the intermediary metabolism of animal and bacterial cells.

(f) Fatty acid metabolism

With the recognition of acetyl-CoA and its importance in metabolism, Ochoa's laboratory turned to important questions of fatty acid metabolism, among them the identification of the enzymes, crotonase and acetyl-CoA transferase. The latter is closely related to Ochoa's most outstanding contribution in this area: the metabolism of propionic acid, a compound produced by oxidation of odd-numbered fatty acids and certain amino acids.

It had been known that propionic acid could give rise, in the presence of liver cell fractions, to citric acid cycle compounds and eventually carbohydrates.

Ochoa and his coworkers, particularly Martin Flavin and Yoshito Kaziro, were able to clarify all the steps of the process up to the production of succinyl-CoA. In the first step, propionic acid is activated to form propionyl-CoA, by acetate thiokinase and ATP. Next a fixation of CO_2 (simultaneously discovered in 1955 by Ochoa's group and by Joseph Katz and I. L. Chaikoff) by propionyl-CoA carboxylase using ATP and bicarbonate produces D-methylmalonyl-CoA. This in turn is racemized to the L-form by a specific enzyme. Finally, a mutase transforms L-methylmalonyl-CoA to succinyl-CoA in a vitamin B_{12}-dependent reaction. All the steps of the system are reversible and are therefore able to produce propionic acid from the citric acid cycle intermediates. The enzymes were obtained in highly purified form and the biotin-dependent carboxylase was crystallized.

(g) Studies on photosynthesis

Since photosynthesis is, in very broad terms, a reversal of respiration, Ochoa's interest flowed naturally to the study of photosynthetic mechanisms. His work dealt with: (i) correlation of the photosynthetic and respiratory processes at stages of CO_2 fixation and evolution, (ii) mechanism of formation of NADPH under the influence of light, and (iii) enzymatic studies of the production of intermediates of the pentose cycle.

Ochoa's main contribution in this area was the discovery, with Wolf Vishniac in 1951, that spinach grana have the ability to reduce NAD and NADP under the influence of light. This was the first demonstration of hydrogen fixation by physiological acceptors catalyzed by radiant energy. Leonard Tolmach and Daniel Arnon also obtained similar results. The formation of NADPH was coupled under the influence of light, pyruvic acid and CO_2 in the presence of the malic enzyme, with formation of malic acid. Later on, coupling was demonstrated with a variety of purified enzymes.

SYNTHESIS OF RIBONUCLEIC ACID

By 1955 the biosynthetic pathways of the basic units of DNA and RNA, the nucleotides, were largely clarified, but the biological mechanisms for the production of the nucleic acids were totally unknown. That year, Ochoa and Marianne Grunberg-Manago announced the discovery of a new enzyme system capable of synthesizing RNA in the test tube, starting from simple nucleotide units. This extraordinary finding gained for Ochoa the 1959 Nobel prize in Physiology and Medicine, a prize he shared with his former pupil Arthur Kornberg, who shortly after Ochoa's discovery, had achieved the *in vitro* enzymatic synthesis of DNA.

(a) Discovery of polynucleotide phosphorylase

Ochoa's own description of how this enzyme was discovered (Base molecular de la expresión del mensaje genético, pp. 110–111, Madrid, Ed. Moneda y Crédito, 1969) is as follows:

> We were interested in the mechanisms of oxidative phosphorylation and I thought it would be of interest to study the process in the extracts of *Azotobacter vinelandii*, a bacterium in which oxidative processes are very intense. The experimental plan was very simple. Bacterial extracts were incubated with ATP and inorganic P tagged with ^{32}P to measure the incorporation of ^{32}P into ATP. Very soon we obtained positive results showing extensive incorporation of radioactivity.
>
> We had been using amorphous ATP, but one day a sample of crystalline ATP arrived, which when

tested showed no incorporation of radioactivity. The amorphous ATP was then purified by chromatography yielding a very active fraction which was simply ADP. Other nucleoside diphosphates were then tested (GDP, CDP and UDP) and all of them proved to be active. The radioactivity became incorporated into the internal P (for instance in the case of ADP: A-ribose-P*—P).

It was soon found that when the incubation was made with relatively high concentrations of diphosphates, inorganic P was produced. The formation of inorganic P from ADP served as a method for measuring the enzyme's activity. We thought the other product of the reaction to be AMP (ADP → AMP + P_i), in a simple hydrolytic process. However, when the incubation mixture was treated with trichloroacetic acid, AMP, which is soluble in acid, was not found in the supernatant because the product was in fact an acid-insoluble polynucleotide.

(b) The properties of the enzyme

Polynucleotide phosphorylase requires only nucleoside diphosphates and Mg^{2+}. The reaction can thus be formulated:

$$n\ \text{N—R—P—P} \overset{Mg^{2+}}{\rightleftarrows} (\text{N—R—P})_n + n\text{P}.$$

N stands for any of the known bases in RNA, R for ribose and P for phosphate; N—R—P represents a nucleotide.

A remarkable property of the enzyme is that it not only catalyzes the synthesis of RNA from a mixture of the four known nucleoside diphosphates, but also from a single NDP or from uncommon or unnatural NDPs. Another characteristic of the enzyme is that the incorporation of the precursors into the macromolecule is proportional to their concentration in the reaction.

(c) Structure of the synthetic polynucleotides

With the fruitful collaboration of Leon Heppel, a polynucleotide was prepared from the four NDPs including α-^{32}P-labeled ADP (Fig. 2). Hydrolysis by snake venom phosphodiesterase breaks the C—P bond indicated by the arrows in Fig. 2. The labeled products obtained from the synthetic polynucleotide were exclusively ^{32}P-AMP, indicating the same 3′-5′ phosphodiester bond as in natural RNA. Spleen diesterase, which cleaves the phosphodiester bonds as shown in Fig. 3, produced ^{32}P-labeled UMP, CMP, GMP and AMP. Thus there was a transfer of label ("nearest-neighbor") as would be found in RNA.

Confirmatory evidence of macromolecular size was obtained by measurements of molecular weight and viscosity of the solutions. A few years later the polynucleotides synthesized *in vitro* proved to be active as messengers in the biosynthesis of proteins, as is natural messenger RNA.

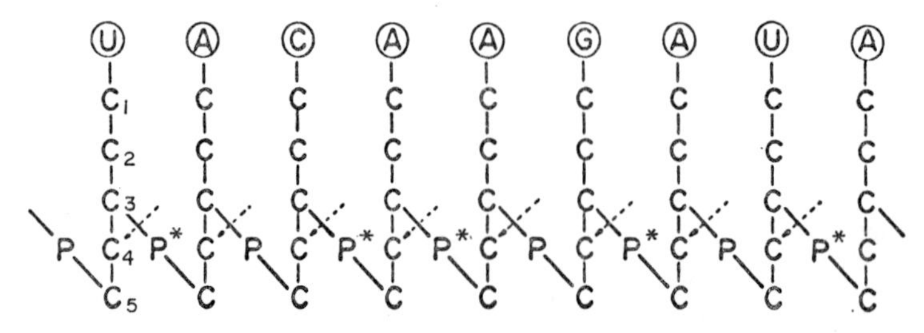

FIG. 2. Degradation of a synthetic polyribonucleotide by snake venom phosphodiesterase. Arrows indicate the bonds broken by the enzyme.

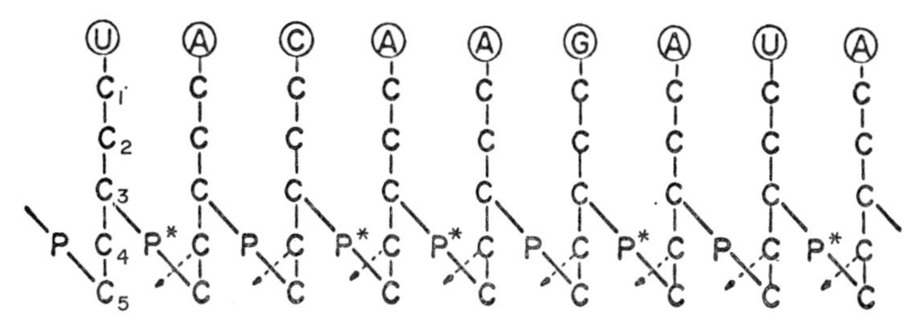

FIG. 3. Degradation of a synthetic polyribonucleotide using the spleen phosphodiesterase.

(d) Biological significance

The precise biological significance of polynucleotide phosphorylase is still unknown. It has been detected in several bacteria besides *Azotobacter vinelandii*, and in plant tissues, but not so far in animal tissues. It is likely to have a degradative rather than biosynthetic role. Highly purified preparations of the enzyme require an oligonucleotide to prime the reaction; but the enzyme does not require or copy a template. After the isolation in 1960 of an *E. coli* enzyme (RNA polymerase) which synthesizes RNA upon directions from a DNA template, Ochoa isolated a similar enzyme from *Azotobacter vinelandii* and showed it to be distinct from polynucleotide phosphorylase.

STUDIES ON THE GENETIC CODE

1. Breaking the genetic code

Ochoa's interest in this field began in 1958. The problem had been defined by James D. Watson and Francis Crick in 1953 in their presentation of the structure of DNA and possible mechanisms of its replication and expression. However, no experimental approach was at hand to explain the translation of nucleic acids into proteins.

The question of how the four-letter language of DNA and RNA is translated into the 20-letter language of protein engaged mathematicians and biologists. What was the genetic code and how was it expressed?

Ochoa's work on the genetic code has been described by Carlos Basilio, a witness and participant in this exciting period of discovery, as follows:

> Sooner or later Ochoa had to tackle the problem of the genetic code. For several years he had in his hands the tool, that is, the polynucleotide phosphorylase, which made it possible to synthesize a variety of ribonucleic acids of known composition and, eventually, to decipher the code. Two more things were needed however; the concept of messenger RNA, introduced by François Jacob and Jacques Monod in 1961, and Sydney Brenner's observation in the same year regarding the restricted specificity of bacterial ribosomes in the translation of RNA messengers of different origin.
>
> It was this last observation that opened the door to the experimental approach. The idea, as a perfect example of scientific dialectics, was translated into clear words by Ochoa: The observation of Brenner *et al*. . . . suggested the use of synthetic polyribonucleotides with known nucleotide sequences as messengers, and a possible method to approach the nucleotide code experimentally. The simplest possible sequences are those of the homopolynucleotides which, if they were additive, should prescribe the formation of homopolypeptide chains. Some members of our department were pessimistic not so Peter Lengyel, but the experiments were completed in a short time and the results appeared in a series of papers in the Proceedings of the National Academy of Sciences, U.S.A., under the title: Synthetic Polynucleotides and the Amino Acid Code. The papers described the codons for the different amino acids and related observations.

A surprising fact, that can only happen in the U.S.A., was the variety of nationalities of the members of the laboratory who directly collaborated with Ochoa in this work. Peter Lengyel was a chemical engineer from Hungary, who was working on his Ph.D. thesis, Joseph Speyer was a young biochemist of German descent, and finally, I myself was a physician from Chile.

No sooner had we started our work than Marshall Nirenberg announced at the International Congress of Biochemistry in Moscow that polyuridylic acid promoted the synthesis of polyphenylalanine. Some scientists attending the Congress said that the audience was electrified. We became petrified when a member of our department told us the news upon his return from Moscow.

One month later Ochoa and Nirenberg presented their results at a meeting in the New York Academy of Medicine. This time the audience received an electric shock when Ochoa showed the codons for 11 amino acids. Next day, Nirenberg came to visit us and we had an exchange of ideas during tea. From this time on, the teams in New York and Bethesda started an academic race that lasted several years.

The codons containing uracil were defined in the first 5 months of work, but codons without uracil could not be detected at the time. Thus the emerging code appeared to be very rich in uracil, which was in disagreement with the composition of the DNA of *E. coli*, the microorganism we were using in our study. This threw serious doubts upon the validity of our results and a distinguished biochemist said in public that he did not bother to learn results that he would soon have to forget. This problem was overcome thanks to the sagacity of Robert S. Gardner, a medical student who was in our laboratory for a period of training. Gardner had read that polylysine, which was later demonstrated to be codified by polyadenylic acid, was soluble in trichloroacetic acid but not in tungstic acid. Therefore by changing the precipitating agent we were able to define the adenine rich codons. A similar technique allowed us to find the cytosine rich codons. Work progressed so fast that 18 months later most of the fundamental information had been discovered.

Ochoa led the group with wisdom and calm. The precision and joy in the work of the team was the consequence of the two most remarkable characteristics of Ochoa's scientific personality: his high ability to attack the problems in the most objective way, and his inborn gifts to lead and to inspire a group. These two qualities were happily blended with his ability to enjoy the good things of life. Ochoa stimulated discussions, particularly in the good restaurants. All of us were aware of the importance of our work and could hardly hide our feelings of pride. But we were unable to detect any change in Ochoa's behavior. He gave the impression of being used to this kind of notoriety. He was not present when the press and TV made the first report.

The property of polynucleotide phosphorylase to synthesize a polyribonucleotide with a composition proportional to the amounts of NDP's present was crucial to its use in elucidating the genetic code. When UDP alone was present, the homopolymer poly U was obtained, the very homopolymer used by Nirenberg and Heinrich Matthei in their classic experiment. Having obtained polyphenylalanine as the polypeptide product of translation, they concluded that U—U—U would code for the amino acid phenylalanine, were the genetic code based on nucleotide triplets.

Ochoa and his coworkers did not limit themselves to the use of homopolymers inasmuch as these could yield triplets for only four amino acids. They resorted to the use of polyribonucleotide copolymers, synthesized by phosphorylase, containing two or more different nucleotides. By calculating the statistical composition of triplets in the heteropolymers it was possible to deduce their correspondence to most of the twenty amino acids. For example, among the three possible triplets of U_2G, one codes for cysteine and another for valine; for the three possible triplets of UG_2, one codes for glycine and another for tryptophan. Similar results were obtained by Nirenberg's group in Bethesda using the same approach. Polynucleotide phosphorylase served to produce a modern "Rosetta stone" for deciphering the genetic code.

In the course of these studies, the important discovery of the "degeneracy" of the code emerged. This pioneer work by Ochoa and Nirenberg was extended and refined by H. Gobind Khorana, Charles Yanofsky and others. By 1966 the composition of the genetic code was almost completely elucidated, a major conquest in the history of science.

2. Genetic expression of the RNA viruses

Around 1962 Ochoa started a systematic investigation of the expression of the genetic message of nucleic acids. This work remains the focus of his laboratory program to this day. It has two main lines: one relates to the replication of the RNA viruses, and the other, to the expression of the nucleic acid message in protein synthesis in bacterial and in developing eukaryotic systems.

Ochoa's studies centered on the RNA bacterial viruses discovered by Norton Zinder in 1962. His main collaborator in this work was Charles Weissmann. Their main contributions were:

(i) Discovery of RNA synthetase (1963) which catalyzes the multiplication of viral RNA in infected cells. This enzyme, after partial purification, requires the four NTPs but does not require the addition *in vitro* of an RNA template because the viral RNA remains firmly bound to the enzyme.

(ii) Characterization of the "replicating form" of the viral RNA, similar to that obtained in animal cells. The genome of the RNA viruses is a single strand which is converted to a duplex by RNA synthetase at the beginning of the infective process. This duplex is a template for the synthesis of RNA chains identical to the original infecting molecule. Many of these chains direct the synthesis of the virus-specific proteins inside the cell; these proteins then become associated with the remaining chains, producing a number of complete viruses identical to the one initiating the process. A "replicating form" of the tobacco mosaic virus was also isolated by the Ochoa group.

(iii) Clarification in collaboration with Eladio Viñuela of the sequential synthesis of specific proteins from the viral message. By inhibiting the synthesis of host proteins, isolation was facilitated of the three principal viral proteins: capsid, "maturation" and RNA synthetase.

3. Mechanisms of protein biosynthesis

(a) *Direction of translation of the genetic message.* Proteins were synthesized *in vitro* using as mRNA, polynucleotide products of polynucleotide phosphorylase; the initial and final triplets of the mRNA were defined. Sequential analysis of the protein then identified the terminal NH_2- and COOH-amino acids. Message reading in the $5' \rightarrow 3'$ direction was thus established.

(b) *Demonstration that UAA is a termination codon.* Genetic evidence had suggested in 1966 that UAG and UAA, the so-called "amber" and "ochre" triplets, are the codons for terminating the reading of the RNA message. The following year the Ochoa group provided the first direct proof that one of them, UAA, was indeed a codon for termination. Synthetic polymers of the type AUGUUUUAAAAA . . . AAA ($AUGU_4A_n$) when used as messengers directed the formation of only the soluble dipeptide formylmethionyl phenylalanine, corresponding to reading of the triplets AUG and UUU which precede the codon UAA in the messenger.

(c) *Initiation factors for protein biosynthesis.* In 1966 Ochoa and his coworkers demonstrated in prokaryotes the existence of proteins required for initiation of polypeptide synthesis. Characterization of these proteins and their functions has been Ochoa's main line of work to date.

Starting with the observation that *in vitro* systems of washed ribosomes translate certain synthetic polyribonucleotides but not natural messengers (e.g. viral RNA), the Ochoa group identified factors for natural mRNA translation in the solutions used for washing the ribosomes. Three proteins were characterized as initiation factors. Two of them, IF-1 and IF-2, are needed for forming the complex initiating the polypeptide chain, that is, for fixing f-Met-tRNA to the 30S ribosomal subunit, using as messenger the trinucleoside diphosphate ApUpG (AUG); IF-3 is needed for forming the initial complex with natural messengers.

Further work, summarized in Fig. 4, describes a cycle of dissociation and reassociation of the two ribosomal subunits (30S and 50S) during protein biosynthesis, a cycle dependent on the recycling of the initiation factors discovered by Ochoa.

Ochoa's current work is directed toward further elucidating the mechanisms of protein biosynthesis, in prokaryotic and in the intriguingly complex brine shrimp, *Artemia salina*, which provides an interesting biological model to approach the study of biological differentiation.

The scientific biography has for the most part deleted personal commentaries (see Acknowledgements), but it would be remiss to exclude one brief note. Representative of the feelings of the large number of his students and colleagues is the epitomizing remark made by one of the most distinguished of this group: "Ochoa is the biochemist of the biochemists and the ideal of what the true scientist should be".

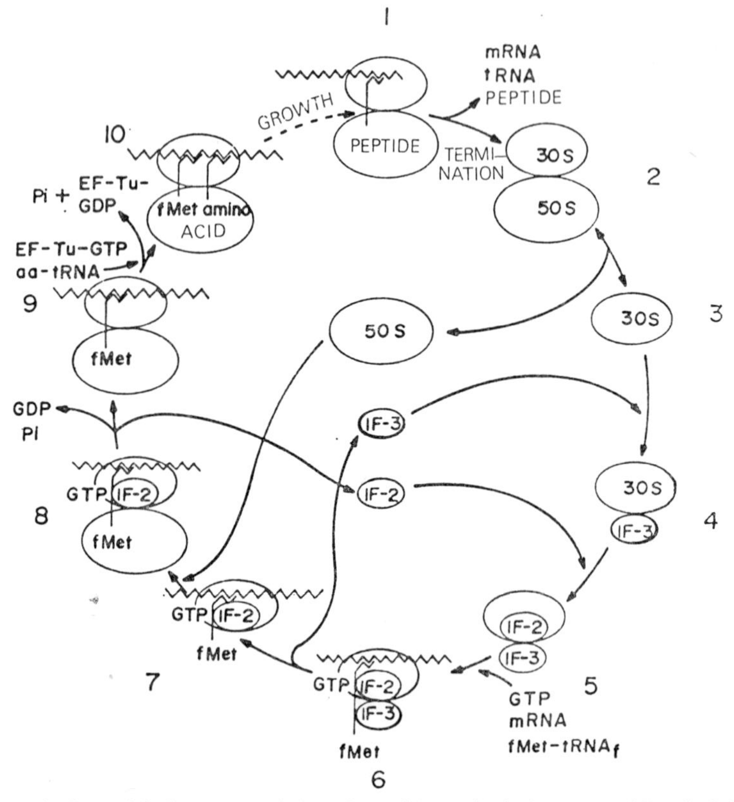

FIG. 4. The bacterial ribosome cycle in polypeptide synthesis (courtesy of Dr. S. Ochoa).

ACKNOWLEDGEMENTS

We wish to thank all of those who have helped us in the preparation of this paper, beginning with Dr Severo Ochoa himself. Valuable information was provided by Doctors C. Basilio, C. F. Cori, C. Gilvarg, S. Grisolia, A. Kornberg, L. F. Leloir, D. Nachmansohn and E. Racker. We wish to thank Dr. Alberto Sols for his suggestions regarding the preparation of the manuscript and Amalia Montes for technical help.

An important part of the information contained in this biography, in particular that related to Ochoa's early years has been taken from the following publications:

F. GRANDE, "Severo Ochoa", *ICSU Review of World Science*, **5**, 147–158 (1963).
J. M. GARCÍA-VALDECASAS (Editor), "El Dr Severo Ochoa, Nobel Prize of Medicine". Mexico, D.F., Mexico (1962).

An extended version including more personal commentaries has been printed in Spanish in the program for Homenaje a Profesor Severo Ochoa en su 70 Aniversario (1975) Barcelona, Socitra-Salvadors, 22.

A list of selected papers from Dr. Severo Ochoa

Up to 1974 Ochoa published 246 papers.

Figure 5 describes the number of papers he published during successive 2-year periods

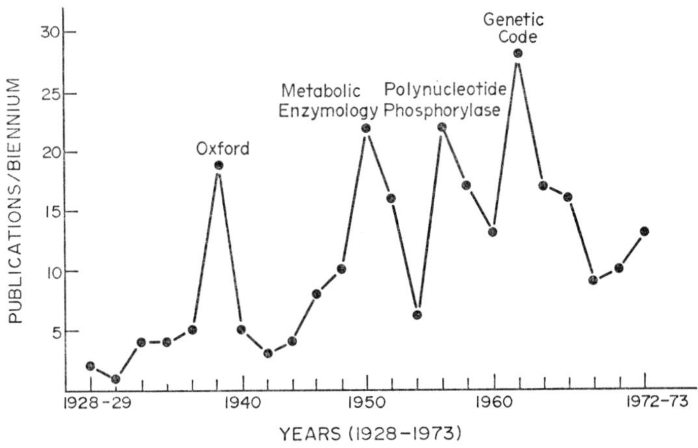

FIG. 5. Papers published by Severo Ochoa from 1928 through 1973, grouped in 2-year periods.

(taken from '*Collected Papers of Severo Ochoa, 1928–1975*', published by the Ministry of Education and Science, edited by A. Sols and C. Estevez).

The following list includes his first two papers and reviews in which more detailed bibliographies can be found.

REFERENCES

OCHOA, S., The action of guanidine on the melanophores of the skin of the frog. *Proc. Roy. Soc.* B **102,** 256 (1928).

OCHOA, S. and VALDECASAS, J. G., A micro-method for the estimation of creatine in muscle. *J. Biol. Chem.* **81,** 351 (1929).

OCHOA, S., Cocarboxylase. In *A Symposium on the Biological Action of Vitamins* edited by E. A. EVANS, Jr., p. 17, University of Chicago Press, 1942.

OCHOA, S., Biological mechanisms of carboxylation and decarboxylation. *Physiol. Rev.* **31,** 56 (1951).
OCHOA, S., Enzymatic mechanism of carbon dioxide fixation. *The Enzymes*, vol. 11, p. 929, Academic Press, 1952.
OCHOA, S., Enzyme studies in biological oxidations and synthesis. *The Harvey Lectures*, Series XLVI, 153, 1950–51.
OCHOA, S. and VISHNIAC, W., Carboxylation reactions and photosynthesis. *Science*, **115,** 297 (1952).
VISHNIAC, W. and OCHOA, S., Reduction of pyridine nucleotides in photosynthesis. In *Phosphorus Metabolism*, vol. 11, pp. 467–490, Johns Hopkins Press, 1952.
COON, M. J., STERN, J. R., DEL CAMPILLO, A. and OCHOA, S., Role of coenzyme A in the enzymatic synthesis and breakdown of acetoacetate. *Federation Proc.* **12,** 191 (1953).
OCHOA, S., Enzymatic mechanisms in the citric acid cycle. *Advances in Enzymology*, XV, 183 (1954).
VISHNIAC, W., HORECKER, B. L. and OCHOA, S., Enzymic aspects of photosynthesis. *Advances in Enzymology*, **19,** 1 (1957).
KAZIRO, Y. and OCHOA, S., The metabolism of propionic acid. *Advances in Enzymology*, **26,** 283 (1964).
OCHOA, S. and HEPPEL, L. A., Polynucleotide synthesis. In MCELROY, W. D. and GLASS, B., *Chemical Basis of Heredity*, p. 615, Baltimore, 1957.
OCHOA, S., Enzymatic synthesis of ribonucleic acid. *Les Prix Nobel en 1959*, p. 146 (1960).
WEISSMANN, C., SIMON, L., BORST, P. and OCHOA, S., Induction of RNA synthetase in *E. coli* after infection by the RNA phage, MS2. *Cold Spring Harbor Symposia on Quantitative Biology*, **28,** 99 (1963).
OCHOA, S., Replication of viral RNA. *Proc. Robert A. Welch Foundation Conferences of Chemical Research*, VIII. Selected Topics, November 1964, p. 9.
OCHOA, S., Synthetic polynucleotides and the genetic code. *Federation Proc.* **22,** 62 (1963).
OCHOA, S., Synthetic polynucleotides and the genetic code. *Symposium on Informational Macromolecules*, p. 437, Academic Press, Inc., New York, N.Y., 1963.
SPEYER, J. F., LENGYEL, P., BASILIO, C., WAHBA, A. J., GARDNER, R. S. and OCHOA, S., Synthetic polynucleotides and the amino acid code. *Cold Spring Harbor Symposia on Quantitative Biology*, **28,** 559 (1963).
OCHOA, S., Chemical basis of heredity. The genetic code. *Experientia*, **20,** 57 (1964).
OCHOA, S. and MAZUMDER, R., Polypeptide chain initiation. In *The Enzymes*, vol. X, 3rd ed., pp. 1–51, Academic Press, 1974.

I. ENERGY METABOLISM, PHOTOSYNTHESIS, FERMENTATION

THE ROLE OF LACTIC ACID IN THE DEVELOPMENT OF BIOCHEMISTRY

CARL F. CORI

The Enzyme Research Laboratory, Massachusetts General Hospital and
The Department of Biochemistry, Harvard Medical School,
Boston, Massachusetts

The incredibly fast advance in many areas of biochemistry and molecular biology makes for equally rapid obsolescence of previous findings. Even the basic observation on which a new advance is based is rapidly forgotten because it has become common knowledge. What may be irretrievably lost in this natural course of events is really something else. It is the passion, the art, the very flavor which characterizes a particular scientific period that quickly sinks into oblivion together with the men and women who were the participants. Whereas one need not know too much about the lives of artists to appreciate their work, this is not so with scientists. Here one wants to know what education they had, who inspired them and under what conditions they worked.

With such thoughts in mind, I have tried my hand at a little history of biochemistry and have chosen lactic acid as a theme. Ochoa himself has contributed much to this subject. Some of his early work was on enzymatic lactate formation in heart muscle and brain and on the role of DPN and diphosphothiamin as coenzymes, all published between 1937 and 1942.

Lactic acid was recognized early—nearly 200 years ago—as one of the important endproducts of bacterial and animal metabolism. The study of the mechanism of its formation led to great advances in biochemistry and physiology. Some of the milestones along this path were the recognition of the role of ATP in the transfer of energy, the role of the pyridine nucleotides in oxidation-reduction reactions and not least, the isolation and detailed study of individual enzymes responsible for the step-wise degradation of the glycogen or glucose molecule to lactic acid. The problem of the mechanism of muscular contraction was greatly influenced by these biochemical investigations. On comparing lactic and alcoholic fermentation it became clear that there is a basic plan in nature common to many different organisms.

It is of interest to go back to the sources of these modern developments. Table 1 shows in chronological order some of the early work on lactic acid in three parts; first, the chemistry; second, the search for the source of energy for muscular work which is intimately connected with the lactic acid problem; and third, what is called here the classical period for want of a better name. Only the first two parts will be covered in this presentation.

TABLE 1. HIGHLIGHTS OF WORK ON LACTIC ACID

Year	Author	Early chemical findings
1780	Scheele (1742–1786)	Isolation from sour milk
1807	Berzelius (1779–1848)	Detection in muscle
1832	Liebig (1803–1873)	Elementary analysis yields $(CH_2O)_x$
1847	Liebig	Two Zn salts of lactic acid
1869	Wislicenus (1835–1902)	Optically active lactic acid
		Fuel for muscular work
1840	Liebig	Protein
1865	Fick (1829–1901)	
	Wislicenus	Carbohydrate
1866	Pettenkofer (1818–1901)	
	Voit (1831–1908)	"Explosive" substance
1877	Hoppe-Seyler (1825–1895)	Glycogen → lactic acid
1877	Cl. Bernard (1813–1878)	*Idem*—(glycogen discovered in 1856)
		The classical period
1905	Harden and Young	Esterification of phosphate
1907	Fletcher and Hopkins	Quantitation of lactate formation
1913	A. V. Hill	Myothermic measurements
1919	Meyerhof	Role of lactate in contraction process
1923	Warburg	Aerobic glycolysis of tumors
1927	Parnas	Ammonia formation from AMP
1927–8	Eggleton, Fiske, Lohmann	Phosphocreatine and ATP
1930	Lundsgaard	Alactic contraction
1933	Embden	Glycolytic scheme
1936	Cori and Cori	Glycogen + P_i → glucose-1-P → glucose-6-P
1939	Warburg	DPN, oxidation and phosphorylation

CHEMISTRY

The well-known Swedish chemist and apothecary, Karl Wilhelm Scheele, demonstrated that the acidity of sour milk was due to an acid which he was able to isolate and which was later named appropriately milk or lactic acid. Scheele was unequalled in his time as the discoverer of new substances in spite of lack of funds and laboratory facilities. Although his most important discoveries were in the field of inorganic chemistry—among others, he isolated the element chlorine—he made many important contributions to biochemistry, particularly in the field of organic acids. He isolated citric, oxalic, malic and tannic acid from plants; uric acid from a bladder stone; and mucic acid from milk sugar heated with nitric acid. He is also the discoverer of glycerol which he isolated from olive oil as a viscid fluid with sweet taste. It seems clear that the isolation of lactic acid from sour milk was no accidental discovery but was based on Scheele's knowledge of the chemistry of organic acids. He introduced the method of precipitating these acids as calcium or lead salts and of decomposing these salts with mineral acids.

Next in order in Table 1 is another Swedish chemist, Jöns Jacob Berzelius, who in the beginning of his scientific career was mainly interested in biochemical problems. He studied medicine and became Professor of Medicine and Pharmacy in Stockholm. His influence on contemporary chemistry was enormous and he is chiefly known for his investigations of chemical proportions, making possible the use of rational chemical formulas in inorganic and organic chemistry. He recognized the essential property of a "catalyst" as a substance which in small amounts can accelerate the decomposition or formation of another substance without participating in the reaction and, in fact, it was he who coined the word. This contribution which ultimately became of the greatest importance to biochemistry would seem to overshadow his detection of lactic acid in muscle, but here again he started an important new development. The story goes that he followed the hunters with a piece of litmus paper in his pocket and that he inserted it in the muscle of a stag killed after a long chase. He isolated the acid but met considerable opposition, even 20 years later, when he claimed that it was lactic acid. Nearly 30 years were to pass before this was firmly established. The German physiologist Du Bois-Reymond (1818–1892) is generally credited with the observation made in 1859 that the neutral reaction of normal muscle becomes acid on stimulation or in rigor mortis owing to the formation of lactic acid. A pupil of Du Bois-Reymond, Johannes Ranke (1836–1916) concluded in 1865 that lactate was a fatigue substance which accumulated to a maximum during tetanic stimulation of muscle.

We come now to a man of great capacity—Justus von Liebig—who is often referred to as the father of biochemistry. He and Mitcherlich produced the first elementary analysis of lactic acid which was later confirmed by Gay-Lussac.* The formula was that of a carbohydrate which was a bit confusing since the substance also had strong acidic properties. Some 15 years later Liebig isolated the zinc salt of muscle lactic acid and found that it crystallized with $2H_2O$, whereas the zinc salt of lactic acid isolated from sour milk crystallized with $3H_2O$.† The reason for this behavior was explained by Johannes Wislicenus, a German chemist, who showed that muscle lactic acid was optically active, whereas fermentation lactic acid was optically inactive. On the basis of this work he expressed the opinion that molecules with identical chemical structure but differing physical properties must have different position of their atoms in space. Thus lactic acid became one of the model substances for the Le Bel–van't Hoff theory about stereoisomerism.

UTILIZATION OF LACTATE

Configurationally muscle lactic acid belongs to the L-series, that is, it has the same configuration on carbon 2 as the reference substance L-glyceraldehyde. However, free L-lactic acid is dextrorotatory, whereas its Zn salt is levorotatory, and this has caused considerable confusion in the literature. To bring matters up to date, the reason why only one of the two stereoisomers of lactic acid is formed enzymatically in animal tissues is explained by the demonstration of Westheimer and Vennesland in 1953 that the enzymatic hydrogen transfer from DPNH to the double bond of pyruvate and the reverse reaction is stereospecific.

The utilization of lactate in the animal body must overcome the very unfavorable equilibrium of lactate dehydrogenase in the direction of formation of pyruvate. Nonetheless, utilization by oxidation and glycogen formation in the liver can be quite rapid as the following data (published over 40 years ago) show (Table 2). For nearly equal amounts

* Liebig, as a young student, had gone to Paris because of student unrest in Southern Germany. There he entered the laboratory of Gay-Lussac where he learned the methods of elementary analysis.

† This may seem trivial, but until 1910 isolation as the Zn salt was the standard method for the quantitative determination of lactic acid in muscle. Fletcher and Hopkins in their classical study on the formation of lactate in amphibian muscle still used this method. Fürth and Charnas in 1910 introduced the method of oxidizing lactic acid to acetaldehyde with permanganate.

TABLE 2. UTILIZATION OF LACTATE
Values are given in mg per 100 g rat per 3 hours
(Data from *J. Biol. Chem.* **81**, 389 (1929))

Type of lactate fed	Absorbed	Excreted	Liver glycogen
—	—	—	4
D(−)	124	36.5	10
R(0)	108	1.6	26
L(+)	111	0.5	44
L(+)	120*	0.9	33

* Injected subcutaneously.

absorbed from the intestine, the unnatural or D(−) form is rapidly excreted in the urine and forms little if any liver glycogen. The racemic lactate is intermediate, while of the L(+) form as much as 35% of the amount absorbed is retained as liver glycogen. Lactate injected subcutaneously also forms liver glycogen. These and other experiments led us to propose a cycle, whereby blood glucose going to muscle glycogen is returned as lactate going to liver glycogen. Recent measurements with isotopes have shown that in man, from 10 to 30% of the hepatic glucose output recycles by way of lactate, about one-half of the lactate coming from muscle and one-half from brain.

The reconversion of lactate to glucose or glycogen passes over oxaloacetate as intermediate. The necessary enzymes are present in liver and kidney, the chief sites of gluconeogenesis, but they are weak or absent in skeletal muscle. Muscle can oxidize lactate but cannot convert it back to glycogen to any significant extent. Meyerhof in 1920 stimulated whole hind legs of frogs and allowed them to recover for 23 hours in a moist chamber with oxygen. He found that twice as much lactate disappeared than glycogen was reformed and concluded that part of the former had been reconverted to the latter. We did similar experiments on frog sartorius muscle some 17 years later and included hexose-6-phosphate in our analyses (Table 3). In the first experiment we could not tell which was the source of the glycogen that was reformed during a 2-hour rest period after the stimulation. In the second experiment we allowed some of the lactate formed during stimulation to be carried away by the blood-stream before isolating the muscle. Here it is clear that the principal source of the reformed glycogen was the hexose-6-phosphate which had accumulated during the preceding stimulation.

TABLE 3. CHANGES IN GLYCOGEN, HEXOSE-6-PHOSPHATE AND LACTATE IN ISOLATED FROG SARTORIUS MUSCLE 2 HOURS AFTER RECOVERY FROM STIMULATION

One of a pair of simultaneously stimulated muscles was fixed immediately or after 5 minutes with the circulation intact, while the other was fixed after 2 hours of rest in oxygen.
(Data from *J. Biol. Chem.* **120**, 193 (1937))

Glycogen	H-6-P	Lactate
mg per 100 g muscle		
+97	−72	−98
+90	−80	−38*

* Circulation intact for 5 minutes after stimulation.

SOURCE OF ENERGY FOR MUSCULAR WORK

The main topic I would like to discuss is the long debate about the source of energy for muscular work. This was only one of the great controversies that were going on at that time. In fact, most of the prominent scientists of this period were engaged in heated discussions with each other.

Liebig for nearly 30 years put all his authority behind the idea that muscle protein, or organized protein as he called it, was broken down during muscular activity and served as the principal fuel for energy production. Although this idea could not be sustained in the end, the debate brought out many interesting points and so served a useful purpose. In other respects Liebig did not come off too badly;* his contributions in organic chemistry, agricultural chemistry, nutrition, to name only a few, were of far-reaching importance. What kind of a man he was, his principal concerns and why he had such a strong influence on his contemporaries may be illustrated by a letter he wrote to Wöhler in 1862.† To quote: "Our advances in art and science do not improve the conditions of the existence of man, and even if a small fraction of human society gains in material wellbeing, the sum of misery among the large masses of people remains the same". He then goes on to say that through an improvement in agriculture and food production the worst of human misery might be mitigated. In fact, by creating the scientific basis for the proper use of fertilizer and for agriculture in general, he was responsible for an unprecedented increase in agricultural yields. He also published a most influential book entitled "Chemistry in its application to agriculture and physiology", which first appeared in 1840 and was revised in 1862.

To return to the problem under discussion, Liebig distinguished between nitrogen-containing foods which were essential for the formation of blood and tissues—he called them plastic nutrients—and the nitrogen-free respiratory nutrients such as carbohydrates and fat which served the purpose of producing heat. Although Liebig claimed originally that the food proteins were not used directly for metabolism (but only after they had been reorganized as tissue protein), he later modified his ideas and admitted that after consumption of an excess of food proteins they could also be metabolized directly. It should be remembered here that Liebig and his contemporaries, for example, Claude Bernard (1813–1878) firmly believed that oxygen consumption and CO_2 production took place in the blood itself.‡

The replacement of tissue protein by food protein did not seem possible if the latter were metabolized and excreted as urea, uric acid and CO_2. Although some amino acids had been isolated, little was known about protein structure at that time. It was only in 1902 that Emil Fischer and Franz Hofmeister independently demonstrated the occurrence of the peptide linkage in protein.

Another strong impetus for Liebig's ideas came from his investigation of meat and

* He became professor at the University of Giessen at 21 years of age and in 1852, when he moved to Munich, he was not obliged to teach.

† Wöhler (1800–1882), who is best known for his synthesis of urea in 1828, was a life-long friend and collaborator of Liebig.

‡ Between 1873 and 1877, i.e. after Liebig's death, Pflueger (1829–1910) championed the idea that it is the living cell which consumes oxygen and which regulates the magnitude of oxygen uptake. The matter seemed to be finally clinched by the famous experiment with the "salt frog" in 1877. A frog which had all its blood replaced by salt solution did not show much change in oxygen consumption. Nevertheless, the idea of respiration in the blood lingered on for another 20 years. Claude Bernard, in the end, agreed with Pflueger.

its nitrogenous extractives. In a paper in 1847 entitled "The Components of the Muscle Fluid" he describes the preparation of what was later to become the renowned Liebig's meat extract. The name became official in 1862.

Liebig in many ways had the outlook of an organic chemist. For example, he described a series of sixteen different derivatives of uric acid, among them alloxan. In a paper with Wöhler, which came 9 years after the synthesis of urea by Wöhler, the two authors predict that it will be possible in the future to synthesize in the laboratory most of the organic compounds occurring in nature. In his work on muscle extractives he describes the conversion of creatine to creatinine and to sarcosine. He gives directions for obtaining creatine from muscle extract in semi-quantitative yield as the crystalline hydrate and he calls attention to the fact that the muscles of a fox killed after a chase yielded ten times as much creatine hydrate as the muscles of a fox held in captivity. He then says: "creatine has unusual chemical properties not found in any other known compound and suggests that it may play a role in the energy production of muscle". This seems prophetic but what he had in mind was that creatine derived from protein served as substrate for muscular work. He also analyzed the inorganic constituents of meat and speculated about the preponderance of potassium salts in muscle as compared to blood. Here he again encountered lactic acid and recorded the observation that much more lactic acid was present in extracts of aged as compared to fresh meat.

Liebig thought that his meat extract had considerable nutritive value and that when added to bread or to a vegetable diet, it could substitute for meat. This was vigorously opposed by Voit, a former student of Liebig, who said that it could not even spare protein intake, and to make matters worse, he prepared (in 1870, a few years before Liebig's death) his own meat extract using pressure but without adding any water. Naturally, he could claim that his more concentrated extract had greater nutritive value than Liebig's. Another teacher of Voit—Pettenkofer—tried to make peace. Ironically enough, after the death of Liebig and Pettenkofer, the control of the preparation of the meat extract passed on to Voit. Meat extracts live on to this day in many forms; e.g. Bovril in England and Maggi's in Europe—but they are mainly used as condiments to stimulate the appetite.

Liebig's ability to popularize his ideas had a strong influence on dietary habits, especially in continental Europe, where the main meal always began with beef broth—referred to as "Kraftbrühe" or brew of strength. Only a few years back I saw a menu in a German restaurant which featured a "double" brew of strength. How strongly this idea about the nutritive value of soup was embedded in people's minds is shown by the story of the suppen-Kaspar from *Struwwelpeter*, a book first published at about that time.*

Opposition to Liebig's views came from several quarters. A number of authors measured nitrogen excretion during work, but found only an insignificant increase. Moritz Traube (1826–1894), in particular, denied that muscle proteins were used up during work and he opposed the idea that there was any correlation between nitrogen excretion and work performed. Of decisive importance was an experiment carried out by Fick and Wislicenus in 1865. We have already encountered Wislicenus who was primarily an organic chemist. Fick was professor of physiology in Zürich. These two climbed the Faulhorn, a mountain

* In this story a boy, strong and healthy, suddenly refuses to eat his soup, probably a case of anorexia nervosa. The next day and the next day and so on when the soup was brought in he did the same thing and got thinner and thinner until he was buried with a soup tureen on his grave, no doubt as a warning to other boys.

TABLE 4. THE CLIMB OF THE FAULHORN (2745 m) BY FICK AND WISLICENUS IN 1865

Protein-free diet 17 hours before and during climb. Nitrogen excreted during climb equivalent to 37 g of protein or 150 Cal.

Lift 76 Kg 2000 m	150,000 Kg meters of work
150 Cal from protein	64,000 Kg meters of work

of about 9000 feet opposite the Berner Oberland. They had not eaten any nitrogenous food for 17 hours before the climb and they also abstained during the climb, but they did not abstain in other ways, since their protocols show they drank a lot of wine. The gist of their experiments is summarized in Table 4. The data shown are for Fick—he excreted nitrogen equivalent to 37 g of protein or to 150 Cal.*

For the sake of argument the authors assumed that all of the protein calories were available for climbing. This implied 100% efficiency of the muscle machine, whereas between 20–30% efficiency is the present-day assumption. They concluded that protein could not be the sole source of energy for muscular work and that most of the work was done at the expense of carbohydrate and fat. The nitrogen excretion in the urine merely indicated the wear and tear of the machine.

They were violently attacked by Pettenkofer and Voit for these (to us) rather obvious conclusions, especially the idea that a machine would be worn down at the same rate whether it worked or not. Furthermore, they said, the calculation of Fick and Wislicenus was meaningless, unless they could prove that there was no reserve involved, that the muscles actually expended without delay all the energy which they had produced. In order to understand what they meant, we have to review some of their work.

Voit (1866) had done careful experiments on a dog in nitrogen equilibrium and could not demonstrate an increase in nitrogen excretion during considerable work. Pettenkofer and Voit obtained similar results in experiments on man. One would have thought that these results would be incompatible with Liebig's ideas. Actually, Pettenkofer and Voit presented their own hypothesis according to which (to quote them): "through oxygen uptake and through protein degradation at a steady rate there accumulates a 'Spannkraft' (literally a tension) which is slowly used up also during rest and which is convertible at will into mechanical work". What they had in mind was a combination of oxygen with a tissue component which thereby acquired a high chemical potential and they were obviously still under the influence of Liebig since they were thinking of protein in this connection. Although Voit 2 years later gave up the idea of the formation of an "explosive" substance as it was then called, it had a strong influence on contemporaries and lingered on until 1903, when Verworn revived it with his hypothesis of the formation of "biogen" molecules. The reason was partly to be found in the work of two physiologists. L. Hermann had observed in 1867 that a frog muscle even when completely freed of oxygen was still capable of a limited amount of work. This capability he said was due to the presence of an energy and oxygen-rich substance formed from muscle protein which could sustain work for a while in the absence of oxygen until it was used up. Pflueger enlarged upon these ideas and following Liebig emphasized the role of living protoplasm as the source of a reactive substance.

* Actually they assumed 6.72 Cal per g of protein instead of 4.2, since the exact heat of combustion of protein and of the nitrogenous excretory products was not known at that time.

It is beyond the present discussion to explain how this *idée fixe* about the preeminence of protein* was finally given up. Other lines of work eventually led to the realization that the anaerobic breakdown of carbohydrate could provide the energy for muscular contraction. Claude Bernard had discovered glycogen in the liver in 1856 and had recognized it as the source of blood glucose. Ten years earlier Helmholtz (1812–1894) had reported that extracts of fatigued frog muscle contained less water-soluble and more alcohol-soluble substances than extracts of resting muscle. This has been confirmed by many subsequent investigators, among others by Ranke and Otto Nasse (1839–1904). The latter found that muscle glycogen decreased during contraction and that this accounted for the decrease in water-soluble substances reported by Helmholtz. The increase in alcohol-soluble substances was presumably due to the formation of lactic acid. Hermann who found that acid production in muscle occurred mainly anaerobically has already been mentioned.

In the well-known textbook of physiological chemistry published by Hoppe-Seyler in 1877 we find the following summary:

1. During muscular work there occurs an increase in oxygen consumption and CO_2 production with an approach of the respiratory quotient to unity. This indicates increased combustion of carbohydrate.
2. The results of Helmholtz are explained by a decrease in glycogen and increase in lactic acid during contraction.
3. The source of energy comes from chemical transformations in which oxygen is not initially involved.
4. Such a chemical transformation is the splitting of glycogen possibly first to glucose or perhaps directly to lactic acid.

The idea that muscle glycogen could be transformed to lactic acid was also expressed by Claude Bernard in the *Leçons sur le Diabete* (1877). He speaks of "la matière glycogène qui subit très rapidement une fermentation lactique".

We have reached the classical period and with it the end of my discussion.

My aim has been to show how the path was cleared for a truly remarkable development—for one of the great periods of biological science. Within a few years were made many of the fundamental discoveries which form the basis of modern biochemistry. The momentum continues to the present day. Apart from clearing the path there were many other factors which contributed to, or indeed made possible, this rapid development. To describe these would require a much longer exposition.

It seems to me that one of the purposes of studying the past history of biochemistry, or of any other science, is to maintain a feeling of continuity, to provide depth, to learn from past experience and not the least to contribute to a philosophy of science. Finally, science, always under attack, will be in a better position to defend itself if it can explain the present in the light of the past.

As a little momento for biologists, I would like to quote Goethe (1749–1832). Mephisto, impersonating Faust, is talking to a student who seeks advice:

They who try to comprehend what's live
First the spirit seek away to drive.
They then hold sundry parts in hand
Gone—alas—the unifying band.
When chemists knowledge thus avow
They mock themselves and know not how.

* The name protein was coined by Mulder in 1840, derived from the Greek verb πρωτευω—I occupy the first place. Berzelius apparently suggested the name to Mulder according to H. B. Vickery.

More than 100 years later the same problem, in a less poetical vein, was presented to F. G. Hopkins. "Has the biochemist, in analyzing the organism into parts, so departed from reality that his studies no longer have biological meaning?" Hopkins answers: "So long as his analysis involves the isolation of events and not merely of substances, he is not in danger of such departure". At the time these questions were asked, vitalism still had a strong influence on biological thinking and for all we know, it will raise its hydra-head again, because the idea that life processes can be entirely explained by known laws of physics and chemistry is hard to accept by some. *Chance and Necessity*, the brilliant book by Jaques Monod, has set the stage anew for such a discussion.

NOTES ON BIBLIOGRAPHY

A complete bibliography was beyond the scope of this article, but early references have been dated by year in order to facilitate their retrieval. For more complete references see: *Die Geschichte der physiologischen Chemie* by Fritz Lieben (Franz Deuticke), Vienna, 1935.

POSTSCRIPT

The Faulhorn is accessible on the north side from Interlaken which is the route that Fick and Wislicenus took on their famous climb, or it can be climbed from the south starting in Grindelwald. In June 1958 Carmen and Severo Ochoa and I happened to be at the same time in Switzerland for meetings and we decided to use this opportunity for a short vacation in the mountains. We chose Grindelwald where we had been before and favored by good weather we went up the next day by chairlift to the First. From there one can walk up to the Faulhorn in about 3 hours. The path leads over mountain meadows in full view of the giants of the Berner Oberland—Jungfrau, Mönch and Eiger. There was still much snow, and since the Ochoa's were not as well equipped as I, they stayed behind near a little lake below the peak of the Faulhorn. There were very few people there that day.

I had climbed this mountain as a sort of scientific pilgrimage on previous occasions and have climbed it since, so there was no way of holding me. The going was hard through the deep snow, but I got up without difficulty and finding myself alone in a sheltered spot, I took a little nap.

When I woke up, the sun was low and the mountain completely deserted. The Ochoas had left to catch the last chairlift down. I then had a most exhilarating experience watching the mountain marmots cavorting in the snow. They must have been just up from their hibernation and they were the most playful animals imaginable. They gave me a real feeling of *joie de vivre*. Running down the mountain, I arrived in Grindelwald after dark, having spent a most memorable day.

In the meantime Ochoa, having gotten worried about my late return, had sent out some Swiss students to look for me. As I came out of the woods, I ran into them and they carried me off to a beer party where a contest was in progress over how many chairs one could jump. Fortunately, I was not obliged to participate in this sport, since I was well tired out by that time.

On the previous night in Bern I had nearly been arrested with the students. They had decided to re-enact a famous episode of their history and I was chosen to be it. This obliged me to climb a ladder leaning against an imaginary city wall. At the top I was to extend my hat which would be filled from the other side by a woman with hot soup. I was then to descend the ladder without spilling any soup. This would have been awkward, but before it came to that, a policeman who had misinterpreted the proceedings wanted to arrest us and was only with difficulty persuaded not to do so. It is gratifying to report that students are much the same everywhere.

ORIGINS OF THE CONCEPT OF OXIDATIVE PHOSPHORYLATION: RETROSPECTIVES AND PERSPECTIVES

HERMAN M. KALCKAR

Department of Biological Chemistry, Harvard Medical School and the John Collins Warren Laboratories of the Huntington Memorial Hospital of Harvard University at the Massachusetts General Hospital, Boston, Massachusetts 02114

For this volume, I have selected the topic "Origins of the Concept of Oxidative Phosphorylation" not merely on account of my early encounters in this field but also because Severo Ochoa made such important contributions to its further development. Our friendship stems back to the Cori laboratory of 1940. Besides, Ochoa's early friendship with my two mentors in Copenhagen, Fritz Lipmann and the late Ejnar Lundsgaard, 10 years earlier, may have shaped the later approach towards oxidative phosphorylation.

In 1930 Lundsgaard discovered that glycolysis is totally arrested by iodoacetic acid (IAA) and yet he found the "alactacid" muscles able to perform a considerable amount of work before they became exhausted in a state of rigor which also was alactacid. The 30-year-young physiologist published within 1 year a series of most creative articles.

The succeeding text includes excerpts from an article in *Molecular and Cellular Biochemistry*[1] reprinted from a symposium on the history of biochemistry at the American Academy of Arts and Sciences.

It is interesting to note that Lundsgaard by early 1930, 2 to 3 years before Lipmann arrived in Copenhagen, pursued the phenomenon of alactacid muscle contraction along such fundamental biological lines. Unlike the Kaiser Wilhelm Institute in Berlin where the Meyerhof and Warburg schools already by 1930 had developed refined ideas and techniques for the study of cell metabolism, Lundsgaard was practically alone in this field at the University of Copenhagen, perhaps with the exception of the plant physiologist Boysen-Jensen.

At that time, the Meyerhof-Hill dogma about lactic acid formation and muscle contraction had conquered practically all of the intellectuals. This ingenious formulation, which is well known, dominated thinking in physiology during the 1920s until Fiske and Subbarow's discoveries of phospho-creatine and ATP in 1927 and 1929 attracted increasing attention. One of the cardinal points of the Meyerhof bioenergetics cycle was the recovery balance, that one-fifth of the lactic acid formed during a contraction period was burned to CO_2 and water, harnessing energy for the conversion of the four-fifths of the lactic acid back to glycogen.

Lundsgaard studied Meyerhof's work with excitement but it inevitably followed from the phenomenon of the alactacid contractions that a fundamental new revision was needed. Lundsgaard unseated the theory of lactic acid formation as the indispensible chemical energy for muscular contraction and recovery.

Since iodoacetate-treated muscles can carry out a considerable amount of mechanical work anaerobically without a trace of lactic acid being formed, lactic acid formation could not be counted among the primary energy sources. However, since the splitting of phospho-creatine (also called phosphagen) in IAA poisoned muscles proceeded at a high rate driven by "phosphate bond energy" (as Lipmann later called it) this must be counted as a possible primary energy source for mechanical work and contraction in general. This was fully corroborated by Lundsgaard himself as a Rockefeller Research Fellow in the completely new Meyerhof Institute in Heidelberg, a satisfactory beginning for this new laboratory. The linear relationship between mechanical work performed under a variety of conditions and the amount of phosphate released from phospho-creatine was demonstrated unambiguously.[2]

Lundsgaard's presentation did not pretend to prove a 100% efficiency in coupling since he was quoting Meyerhof's ΔH values (not ΔF values) and Hill's myothermic data. Lipmann's later computations of the ΔF values of these processes assessed the efficiency as being of the order of 70%.[3]

A second important idea advanced by Lundsgaard a few months after his original discovery relates to respiratory recovery, possibly by-passing lactic acid metabolism. Lundsgaard found namely that IAA, while arresting glycolysis or alcoholic fermentation completely, did not affect respiration in muscle or yeast.[4] He, therefore, advanced the idea that the prolonged mechanical work demonstrable in IAA muscles in oxygen and the more moderate phospho-creatine splitting was due to a direct aerobic rephosphorylation. Ochoa, newly arrived in the Meyerhof Institute, also encouraged support of Lundsgaard's idea of emancipation from lactic acid control during aerobic recovery.[5] Ochoa found that frog muscles almost exhausted in their carbohydrate storage after insulin treatment of the frog (promoting insulin convulsions) performed normal mechanical work under aerobic conditions, and showed the usual marked increase in oxygen consumption and recovery in spite of the lack of, or only minimal formation of lactic acid. Fuel other than carbohydrate or carbohydrate split products must have been used for contraction and recovery.[5] Essentially what Lundsgaard formulated was a direct forerunner of the concept of oxidative phosphorylation.

Engelhardt[6] had encountered aerobic phosphorylation in mammalian and avian intact erythrocytes in the presence of redox dyes, with glucose as a metabolite.

In studies in 1928 shortly following Warburg's work on tumor metabolism and its monumental methodology, György and co-workers[7] had compared Q_{O_2} data of various tissues as tissue slices. They sharply distinguished between kidney cortex which contains the tubules operating the active transport of sugars, amino acids and ions and kidney medulla which contain the broader tubules involved in excretion. They found that slices of kidney cortex showed the highest Q_{O_2} seen in any mammalian tissue and the lowest glycolysis. I, therefore, considered respiration as a possible critical factor when it came to demonstration of active phosphorylation in minced kidney or in kidney extracts. In my first experiments, I used to compare respiration and phosphorylation of briefly dialyzed kidney extracts with or without added glucose. Fluoride was always present. Since I found respiration as well as phosphorylation proceeding rapidly,[8] I varied also the nature of the electron donors added, since I was more or less in the dark with respect to the nature of the metabolites used in the respiration. However, Krebs experiments of 1935[9] had shown that glutamic acid increased the Q_{O_2} of kidney slices from a variety of species, especially rabbit kidney where the response was a doubling of Q_{O_2} or more. This encouraged me to add glutamic acid, dicarboxylic and tricarboxylic acids to the minced kidney or to kidney extracts. Q_{O_2} as well as phosphorylation went way up.

In one way, I was fortunate in having started my efforts on kidney cortex "brei" or extracts. The high respiration helped drive the phosphorylation. Moreover, the phosphatase activity of the crude kidney preparations was readily inhibited by fluoride. The price which I eventually had to pay was probably the presence of phosphatases which were not completely inhibited. The P/O ratios varied between 0.5 and 1.5[10]. Without oxygen consumption no phosphorylation was discernible.

There was no sign that glucose or glucose phosphoric esters formed in kidney extracts were used as electron donors. Evidently, glucose was merely phosphorylated and something else present in the kidney homogenate served as electron donors.[11,12]

I was particularly interested in the possibility of a coupling between oxidation of dicarboxylic acids and phosphorylation, and more specifically, the step succinate to fumarate. This was a step which, according to the work of Szent Györgyi and his group, involved something else than nucleotides, namely cytochrome c. Was this step able to furnish energy for phosphorylation or not? I did not manage to obtain unambiguous proof that succinate oxidation could be coupled with phosphorylation. However, in the cases where the extra oxygen uptake was accompanied by additional phosphorylation, 5 to 6 moles of extra oxygen consumed showed 2.5 to 4 μmoles P of additional phosphoric ester formation. The P/O ratio varied between 0.4 and 0.5.[10]

A couple of years later, Colowick, Welch and Cori[13] resolved the succinate dehydrogenase system from that of malic acid by prolonged dialysis. They showed beyond doubt that the succinate–fumarate step can be coupled with phosphorylation.

I was able to obtain malate oxidation either in dialyzed yet concentrated kidney cortex extracts or by a very reproducible technique with kidney "brei" washed 3 to 4 times with distilled water and then extracted with phosphate. The phosphate extract was incubated with malate in the presence of fluoride. Glucose was not added. Malate underwent oxidation and phosphopyruvate accumulated.[11,12]

This conversion of malate to phospho-enolpyruvate (PEP) turned out to be of much interest later, especially thanks to work from Harland Wood's group. It is probably one of the main pathways in gluconeogenesis from pyruvate and CO_2 and Utter and Kurahashi described precisely the coupling between carboxylation–decarboxylation reactions of the pyruvate system and phosphorylation of guanosine diphosphate to triphosphate.[14]

THE P/O RATIOS IN VARIOUS TISSUES

I mentioned that in kidney cortex P/O ratios, varying from 0.5 to 1.5, were observed (in most cases the P/O ratios were close to 1, in a few cases close to 2[11]).

Belitser[15] discovered that oxidative phosphorylation in heart muscle and pigeon breast muscle, minced or homogenized, clearly gave P/O ratios of the order of magnitude of 5. In a series of beautiful experiments, the formation of phospho-creatine or adenylic acid was compared with the stimulation of phosphorylation and oxygen consumption, brought about by addition of di- or tricarboxylic acids to homogenized pigeon muscle depleted of endogenous substrate. I quote from V. A. Belitser and E. T. Tsybakova:[15] "Perhaps the explanation of the high coefficients of "respiratory" phosphorylation must be based on the possibility that not only primary, but also some intermediate oxidation-reductions are coupled with phosphorylation".

On the same page, coupling of phosphorylation to the "Fumaric system" is briefly discussed. Their comments to the additional phosphorylation (phosphagen synthesis) in the presence of succinic acid as compared to malic acid concludes as follows: "The difference, although not great, is nevertheless real. The surplus synthesis in the presence of succinic acid should be regarded as relating to the first step of oxidation". Further substantiation of this type of coupling was shown in experiments using arsenous acid.

About the same time, Ochoa also found high P/O ratios in incubates of pigeon brain homogenates. In the first publication, a P/O ratio of about 3 was presented and independently many of the same arguments as Belitser's were discussed.[16]

Ochoa's paper of 1940 is the first paper which used the term "oxidative phosphorylation" in the title as well. His emphasis on the involvement of several electron carrier systems for obtaining the high P/O ratios which he has observed, was expressed in a lucid style.

"*ENERGIZED STATE*" OF MEMBRANES

Two new developments helped us understand better the "topography" of such a coupling between electron carrier systems and formation of "energy-rich phosphate bonds" or "∼P".[3] The first development was Kennedy and Lehninger's realization that the seat of Ox ∼ P in higher organisms is the mitochondrial membrane. The second development was Peter Mitchell's formulation of proton motive force and the separation of charges in the mitochondrial membrane during respiration due to vectorial oxidation-reductions.

I realize that I will have to skip the chapter on the mitochondria and oxidative phosphorylation which has been developed to such a high degree by Racker and others. This chapter is well known to everybody. Its further interpretation may have been facilitated by the finding that the bacterial membrane (inner membrane) functions in many ways like the mitochondrial membrane and that it can perform active transport. Mitchell's electrochemical model of oxidative phosphorylation has been greatly strengthened by studies on active transport of nutrients in bacterial strains, some of which have defective membrane ATPases. Heppel and his coworkers have provided us with some important guidelines based on experiments carried out by themselves as well as by other laboratories.

A particularly good illustration of the vectorial proton motive force in action was provided by studies on a gram positive micro-organism *Streptococcus lactis* by T. H. Wilson and his associates.[17,18] *S. lactis* is able to store considerable amounts of K^+; if it is starved for carbon sources, subsequently, it is still able to retain K^+ but the ATP breaks down to a low level. Addition of glucose will restore resynthesis of ATP from phosphate and ADP generated by the glycolytic pathway. However, a sudden, albeit short-lived, burst of ATP synthesis in the cell can also be generated by adding a specific K^+ ionophore, i.e. valinomycin which is soluble in the membrane. This makes the membrane specifically permeable to K^+. If the medium contains only the usual traces of K^+ found in minimum media, a diffusion potential of 150 to 200 mV may well be generated (initial proton motive force) prior to the "in rush" of protons to neutralize the excess negative charges. The initial proton motive force has also been shown to be effective in active transport of methyl-thiogalactoside creating a short-lived 20- to 25-fold accumulation of the sugar analogue in the cells.[18] The K^+ membrane potential is eliminated in the presence of inhibitors of oxidative phosphorylation. In *E. coli* mutants of lac permease, the coupling between membrane potentials and transport is absent.

The membrane potential might well serve as a useful illustration of the so-called "energized state" of the membrane, also called "~", a notion which is important for accounting for the bioenergetics of the transport of certain amino acids as well as for the motoric apparatus in chemotaxis.

TABLE 1

Functions	Structures	Examples of substrates or attractants	Energy coupling	Reference
1a. Active transport	Membrane translocase	Proline	~	(19)
		Methyl thiogalactoside	~	(20)
1b. "Scavenging" & transport	Periplasmic binding proteins+membrane translocase	Glutamine	ATP	(19)
		Galactose	(~) ATP	(21)
2a. Swimming	Flagella (counter-clockwise rotation)		~	(22) (23)
2b. Tumbling	Flagella (clockwise rotation)		≁ Adenosyl methionine?	(22,24)
2c. Tumbling regulation	?		Adenosyl methionine?	(22,25)
3a. Chemoreceptors A		Serine		
		Fructose	?	(22)
		Mannose		
3b. Chemoreceptors B	Periplasmic binding proteins	Galactose		(22,26)
		Ribose	0	(25)
		Maltose		(27)

Key:
~ Energized state of membrane.
≁ Uncoupling of energized state of membrane

Notwithstanding the exciting demonstrations of the energized state in membrane vesicles, I shall here confine myself to the functions of the intact cell.

High affinity receptor protein located in the so-called periplasmic space between the inner membrane and the cell wall participates in active transport of nutrients or their analogues. Certain types of sugar chemotaxis are also initiated in the periplasmic space, especially galactose chemotaxis. Table 1 summarizes various types of transport and other functions of the bacterial cell in terms of bioenergetics (see also references).

After the introduction of the concept of "~" membrane, one might debate whether phosphorylation is optional or obligatory. I have discussed this at another occasion[28] and I will merely add here that it would be difficult to rule out phosphorylation as an obligatory step. Mutants defective in membrane ATPase have become of much interest lately (see review by Berger and Heppel[19]) and further studies on these types of mutants should help to provide us with orientation in this direction.

REFERENCES

1. KALCKAR, H. M. (1974) Origins of the concept oxidative phosphorylation. *Molec. Cell. Biochem.* **5**, 55–60.
2. LUNDSGAARD, E. (1932) The significance of the phenomenon: alactacid muscle contractions. H. M. KALCKAR'S *Biological Phosphorylation*, Prentice-Hall (1969). Translation from Danish.
3. LIPMANN, F. (1941) Metabolic generation and utilization of phosphate bond energy. *Adv. Enzymol.* **1**, 99–162, Interscience Publ.
4. LUNDSGAARD, E. (1930) Über die Einwirkung der Monojodessigsäure auf der Spaltungs- und Oxydationsstaffwechsel. *Biochem. Z.* **220**, 8–18.
5. OCHOA, S. (1930) Über den Tätigkeitsstoffwechsel kohlenhydratarmer Kaltblütermuskeln. *Biochem. Z.* **227**, 116–134.
6. ENGELHARDT, W. A. (1930) Ortho- und Pyrophosphat im aeroben und anaeroben Stoffwechsel der Blutzellen. *Biochem. Z.* **227**, 16–38.
7. GYÖRGY, P., KELLER, W. and BREHME, TH. (1928) Nierenstoffwechsel und Nierenentwicklung. *Biochem. Z.* **200**, 356–366.
8. KALCKAR, H. M. (1937) Phosphorylations in kidney tissue. *Enzymol.* **2**, 47–52.
9. KREBS, H. A. (1935) Metabolism of amino acids. *Biochem. J.* **29**, 1951–1969.
10. KALCKAR, H. M. (1938) Fosforyleringsprocesser i Dyrisk Vaev. Nyt Nordisk Forlag, København.
11. KALCKAR, H. M. (1939) Phosphorylations in kidney extracts. *Enzymol.* **5**, 365–371.
12. KALCKAR, H. M. (1939) The nature of phosphoric esters formed in kidney extracts. *Biochem. J.* **33**, 631–641.
13. COLOWICK, S. P., WELCH, M. and CORI, C. F. (1940) Phosphorylation of glucose in kidney extracts. *J. Biol. Chem.* **133**, 359–373.
14. UTTER, M. F. and KURAHASHI, K. (1954) Mechanism of action of oxalacetic carboxylase. *J. Biol. Chem.* **207**, 821–841.
15. BELITSER, V. A. and TSYBAKOVA, E. T. (1939) The mechanism of phosphorylation associated with respiration. *Biokhimiya*, **4**, 516–534, translated into English in H. M. KALCKAR'S *Biological Phosphorylations*, pp. 211–227, Prentice-Hall (1969). Translation from Russian.
16. OCHOA, S. (1940) Nature of oxidative phosphorylation in brain tissue. *Nature*, **146**, 267.
17. MALONEY, P. C., KASHKET, E. R. and WILSON, T. H. (1974) A protonmotive force drives ATP synthesis in bacteria. *Proc. Nat. Acad. Sci., U.S.A.* **71**, 3896–3900.
18. KASHKET, E. R. and WILSON, T. H. (1974) Proton-coupled accumulation of galactoside in *Streptococcus lactis* 7962. *Proc. Nat. Acad. Sci., U.S.A.* **70**, 2866–2869.
19. BERGER, E. A. and HEPPEL, L. A. (1974) Different mechanisms of energy coupling for the shock-sensitive and shock-resistant amino acid permeases of *Escherichia coli*. *J. Biol. Chem.* **249**, 7747–7755.
20. HAROLD, F. M. (1972) Conservation and transformation of energy by bacterial membranes. *Bact. Rev.* **36**, 172–230.
21. PARNES, J. R., BOOS, W. and KALCKAR, H. M. (1974) Forms of energy coupling of active transport in *Escherichia coli*, *Dynamics of Energy-transducing Membranes* (ERNSTER, ESTABROOK and SLATER, eds.) pp. 389–403, Elsevier Scientific Publ. Co., Amsterdam.
22. ADLER, J. (1974) Chemotaxis in bacteria. *Biochemistry of Sensory Functions*, ed. L. JAENICKE, pp. 107–131, Springer-Verlag, Berlin.
23. BERG, H. C. (1974) Dynamic properties of bacterial flagella motors. *Nature*, **249**, 77–79.
24. ORDAL, W. G. and GOLDMAN, D. J. (1975) Chemotaxis away from uncouplers of oxidative phosphorylation in *Bacillus subtilis*, *Science* (in press).
25. KOSHLAND, D. E., JR. (1974) The chemotactic response in bacteria. *Biochemistry of Sensory Functions* (L. JAENICKE, ed.), pp. 133–160, Springer-Verlag, Berlin.
26. SILHAVY, T. J., BOOS, W. and KALCKAR, H. M. (1975) The role of the *Escherichia coli* galactose-binding protein in galactose transport and chemotaxis. *Biochemistry of Sensory Functions* (L. JAENICKE, ed.), pp. 1–30, Springer-Verlag, Berlin.
27. HAZELBAUER, G. L. (1975) Maltose chemoreceptor of *Escherichia coli*. *J. Bact.* **122**, 206–214.
28. KALCKAR, H. M. (1966) High energy phosphate bonds: optional or obligatory? *Phage and the Origins of Molecular Biology* (J. CAIRNS, G. S. STENT and J. D. WATSON, eds.), pp. 43–49, Cold Spring Harbor Laboratory of Quantitative Biology, Cold Spring Harbor, N.Y.

REFLECTIONS ON THE EVOLUTIONARY TRANSITION FROM PROKARYOTES TO EUKARYOTES

FRITZ LIPMANN
The Rockefeller University, New York, 10021

For some time I have felt unsure whether one can consider the development in bacteria as evolution in the sense applied to multicellular organisms. What we see in bacteria, I feel, is not a true evolution but it might rather be called a diversification. On the metabolic side, the bacterial world displays a startling inventiveness; in it, the base line is developed on which evolution can take off, but take off only after a rather profound cellular reorganization. It is only recently that we have become used, accordingly, to dividing all living organisms into two classes: (1) the prokaryotes, or bacteria and blue-green algae, and (2) the eukaryotes, destined to develop into multicellular organisms, animals and plants.

In their marvellous book entitled "The Microbial World",[1]* Stanier, Doudoroff and Adelberg added, in the second edition of 1963, a new chapter dealing with the gap in the transition from prokaryotes to eukaryotes, and in a concluding paragraph say:

> The numerous and fundamental differences between the eucaryotic and the procaryotic cell . . . have been fully recognized only in the past few years . . . this basic divergence in cellular structure, *which separates the bacteria and blue-green algae from all other cellular organisms*, probably represents the greatest single evolutionary discontinuity to be found in the present-day living world. It is not too unreasonable to consider that the bacteria and blue-green algae represent *vestiges of a stage in the evolution of the cell* which, once it achieved a eucaryotic structure in the ancestors of the present-day higher protists, did not undergo any further fundamental changes through the entire subsequent course of biological evolution.

The reconstruction of the events that bridged this gap has been rather heatedly discussed during the last 10 years. Since it appeared to me to be more satisfying to approach the early history of organismic development by attempting to go backwards from the present to the past, my interest in this phase was aroused through recent work on the comparison of prokaryote and eukaryote mechanisms of protein synthesis.

Before becoming involved in the background discussion, however, I will present results which seem to have originated through the construction of the eukaryote cell, namely, a logarithmic increase in the rate of organismic evolution. This, one might say, began very slowly and increased dramatically during the last $1.0–0.5 \times 10^9$ years. A chart may be drawn by taking as abscissa the time clock of evolution presented in a scheme found in a paper by Schopf[2] (Fig. 1). The ordinate represents the rate of evolution over the time elapsed in years since the existence of a viable Earth, about 5 billion years ago. An early rise up to minus $4.5–3.0 \times 10^9$ years represents the transition from prokaryotes to eukaryotes, or of the bacteria and blue-green algae to nucleated, mitotically dividing pro-eukaryotic, the monocellular protists. Since evolution in the stationary plants was

* Reprinted by permission of the publishers.

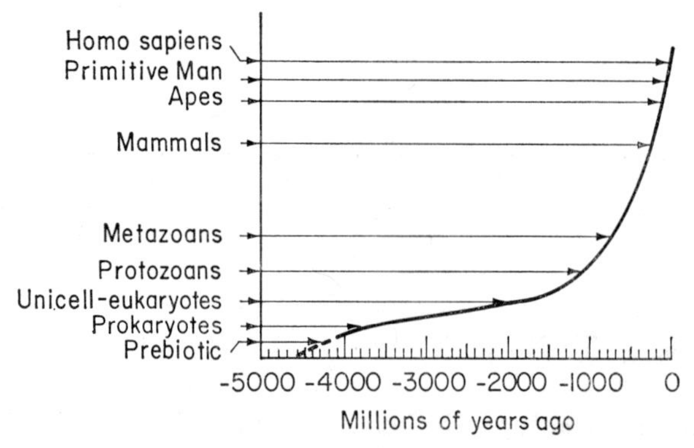

FIG. 1. Plot of evolutionary change against the age of the Earth.[2]

relatively slow, only animal evolution is considered in the following discussion. After minus 2×10^9 years, eukaryotes began to aggregate to multicellular organisms, multiplying by mitosis followed by meiosis and fusion of male–female type gametes.

Then one begins to recognize a logarithmic rate increase which ends in the fast transition from ape to man during the last 200 million years. During this period, the large increase in brain volume and structuration had to be paralleled with the change to a larynx usable for speaking and to a hand adapted to the use of tools. This eventual fast acceleration after a long induction period must be rooted in the drastic transformation of a prokaryote into a eukaryote cell type.

This discussion will now deal with a necessarily rather sketchy attempt to reconstruct backwards how the gap between the two cell types may have been bridged, and will be largely based on the heterogeneity between the two energy-transducing organelles and the body of the eukaryote cells within which the mitochondria and chloroplasts impress one as being islands of prokaryotic structure. Such a reconstruction lies within experimentally accessible domains since both the prokaryotes and the eukaryotes are easily available for study. Yet only recently has evidence built up to enable one to begin to understand the insularity of the energy-transducing organelles within the eukaryote cell.

My own appreciation of the importance of the gap was due to work in the laboratory on the discrepancies between the mechanism of protein synthesis in mitochondria and that in eukaryotic protoplasm. Earlier, when studying homogenates of *E. coli* and liver, Ochoa and Rendi[3] and Nathans and Lipmann[4] simultaneously showed that the respective ribosomes were not cross-complemented by their supernatant factors (Table 1). However,

TABLE 1. RIBOSOME SPECIFICITY OF TRANSFER FACTOR

Ribosomes	Supernatant factor	^{14}C-Leu transferred (cpm)
E. coli	None	90
E. coli	Liver	80
E. coli	*E. coli*	665
Liver	None	21
Liver	*E. coli*	19
Liver	Liver	244

For details see Nathans and Lipmann.[4]

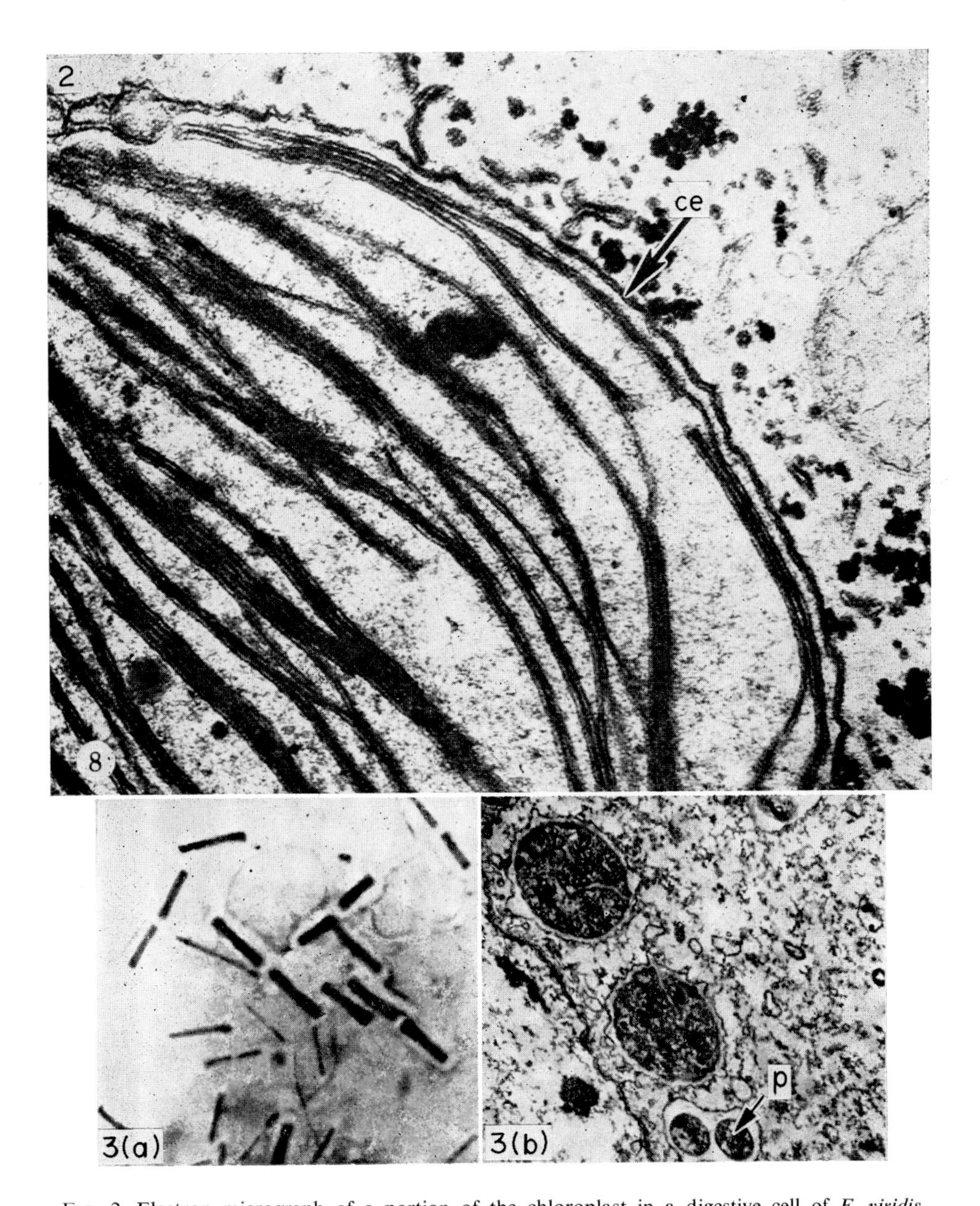

FIG. 2. Electron micrograph of a portion of the chloroplast in a digestive cell of *E. viridis*, bounded only by the chloroplast envelope indicated by the arrow.[11]

FIG. 3. (a) Electron micrograph of the symbiotic bacteria from the homogenate of *Pelomyxa palustris*, magnification 16,000×. Most of the thin rods condense by doubling up to the heavier ones. (b) Border of a nucleus of the amoeba showing cross-sections of the bacteria, which are lined up against the nuclear membrane.[14]

TABLE 2. INTERCHANGEABILITY OF MITOCHONDRIAL AND BACTERIAL PEPTIDE-SYNTHESIZING SYSTEM

Supernatant factors	Ribosomes	
	Mitochondrial	Bacterial
	Phe polymerized ($\mu\mu$moles/mg protein)	
Mitochondrial	7.1	22.1
E. coli	18.1	421.0
Cytoplasmic	0.5	0.8

For details see Richter and Lipmann.[7]

several recent reports had shown that complementation could be found with extracts of eukaryotic cells.[5] The best studied case seemed to be that of yeast, where Albrecht, Prenzel and Richter[6] showed such cross-complementation. Parallels between bacteria, and mitochondrial as well as chloroplast protein synthesis were hinted at by observations on their response to chloramphenicol, which is bacteria-specific, and their resistance to cycloheximide, which is a cytoplasm-specific poison of protein synthesis. Therefore, when Richter joined this laboratory, we carefully separated yeast mitochondria and cytoplasm, and were indeed able to show that the former did contain bacterial supernatant factors and the latter cytoplasmic (i.e. eukaryote) factors.[7] This is shown in Table 2.

In the meantime, a lively discussion began,[8] arguing about what seemed, to me, increasingly convincing evidence that mitochondria were descended from bacteria and chloroplasts from the prokaryotic blue-green algae. This has been promoted into an endosymbiotic theory of a prokaryote–eukaryote transition by Margulis,[9] and has more recently been discussed soberly by Taylor.[10] Symbiotic fusion of different lines of organisms at a crucial stage of evolution would be an unexpected addition to chance mutation, and to judge from present-day examples, be attributable to choice rather than to chance.

SOME RELEVANT EXAMPLES OF PRESENT-DAY SYMBIOSIS

There is great use of photosynthesizing implementation of eukaryotes, predominantly in the fungi and represented by the great variety of lichens, and in aquatic polycellulars; the latter generally develop in families called *Viridans*. For example, varieties of *Hydra viridans* associate with monocellular algae, nucleated or blue-green. Of particular interest is the mollusk *Elysia atroviridis*[11] which stores the chloroplasts of *Codium fragile* in the cells of its digestive diverticulum after digestion of the body of the algae. The chloroplasts left intact are then obviously enclosed by pinocytosis into the cells of the digestive tract. A segment of a chloroplast surrounded by a double membrane in such a cell can be seen in an electron micrograph in Fig. 2 taken from Trench *et al.*[11] It shows an intact chloroplast structure which is retained for up to 3 months with essentially unchanged functions; it supplies its products to the host although it is incapable of division.

A rather thorough integration of the presumptive chloroplast precursors, the blue-green algae, when associated with a host cell, has been reported and the associated "organelle" has been called plastoid or cyanophors. In our laboratory, we tested ribosomes of *Chlamydomonas reinhardtii* chloroplasts,[12] compared them with their cytoplasmic ribosomes, and found them positive for a specific function characteristic of a prokaryote ribosome, in contrast to their cytoplasmic counterparts.

BACTERIAL ASSOCIATION WITH EUKARYOTE CELLS

Bacteria are not as frequently found in symbiosis with eukaryote cells. However, one outstanding and very successful example is available in the combination between the root cells of legumes and a variety of Gram-positive bacteria, the root nodule-forming bacteria.[13] After entering the root cell, these organisms change shape and lose replicability, but only in symbiosis develop the ability to fix nitrogen. On the other hand, they induce the root cells to form a hemoglobin, the leg-hemoglobin, which is deposited near the bacteroid and delivers the oxygen essential for nitrogen fixation. The combination creates the food plants that are richest in protein.

The growing interest in a descent of mitochondria from aerobic bacteria has focused attention on a symbiosis of an anaerobic polynucleated amoeba-like organism, *Pelomyxa palustris*, which is lacking in mitochondria, with a respiring rod-shaped bacterium.[14] In the electron micrograph (Fig. 3a), the isolated rod-shaped bacteria, which contain the only respiratory activity in the cell homogenates, are shown; the cytoplasmic fraction is free of respiration but has a rather potent glycolytic activity. Figure 3b shows cuts through bacteria lining the border of the nuclei; high levels of lactic acid are not well tolerated but are removed by bacterial respiration and thus convert the anaerobic amoeba into a viable aerobic organism.

A most persistent objection to an identification of mitochondria with aerobic bacteria has been that they contain very different electron transport systems.[8] Recently, however, Smith *et al.*[15] have found that *Micrococcus denitrificans* contains an electron transfer system quite analogous to that of mitochondria, including a very similar cytochrome *c* and cytochrome aa_3. This observation has been amplified in a careful study by John and Whatley,[16] directed specifically towards a comparison of the electron transport and ATP-generating systems in mitochondria, using the same micrococcus now renamed *Paracoccus denitrificans*. They applied their results to mapping the events that transform bacteria into mitochondria, using as a starting point the mentioned symbiosis between lactic acid-forming *Pelomyxa palustris* and aerobic bacteria. They assume a similar combination to have formed between anaerobic proto-eukaryotic host cells and a (micro)-paracoccus-like aerobe. The latter, in the course of prolonged coexistence, instead of removing the product of the energy-delivering glycolysis, replaced the glycolysis by becoming permeable to aerobically formed ATP. In the process, the bacterial outer wall was lost but the internal plasma membrane-linked electron transport and ATP-generating systems were retained. Thereby, the bacterium was converted to a Heppel-type protoplast known to be permeable to nucleotides and nucleic acids.

Actually, the eventual conversion of the free-living bacterium into an integrated part of the eukaryote cells involves, as we know, a transfer to the nucleus of most templates necessary for the synthesis of respiratory enzymes. Such transfer of original functions drastically cuts down the mitochondrial DNA to a mere vestige of the ancestral bacterial DNA. Although the remaining DNA is distinctly of bacterial type, it contains space only to template for a few proteins and some mitochondria-specific RNAs.

CONCLUSION

Retracing the transition from prokaryotes to eukaryotes: an abrupt change produced a cell structure which, after a slow beginning during the period of protists, led to the construction of multicellular organisms. From then on, at least in the eyes of the observing

Homo sapiens, evolution accelerated and, in the age of mammals, made animals increasingly able to dominate the environment. This culminated in the development of man's capacity to extrovert evolution, so to say, by adding to the intrinsic genetic memory the ability to transmit to future generations, orally or in writing, the experience acquired during an individual's lifetime.

I dedicate this paper to Severo Ochoa with my best wishes. Through a great part of our scientific lives, our paths have intersected pleasantly; some of these occasions I recall more vividly. We first met in Meyerhof's laboratory in 1929–30, in the last years of innocence before the Second World War. There I remember him in the Heidelberg period, gaily driving his sports car, enjoying the driving as he still does, and often taking us along. Soon afterwards, we both had to leave our countries; he went to England and I to Denmark, and we found each other again in the United States where we both settled.

Eventually we both came to stay in New York and there developed a close contact. From that time I remember with particular pleasure his exciting Saturday morning seminars with overflow audiences of students, faculty, and guests, after which a small group would assemble for luncheon at a nice Italian restaurant at the corner of 30th Street and Second Avenue. These were the years one happily recalls as the period when biochemistry was one of the most highly respected disciplines in medical schools, and when those aspiring to a chairmanship, particularly in internal medicine but also in surgery and other clinical fields, were generally well served by spending a few years in the biochemical laboratory.

REFERENCES

1. STANIER, R. Y., DOUDOROFF, M. and ADELBERG, E. A. (1963) The evolutionary significance of eucaryotic and procaryotic cellular structure. *The Microbial World*, pp. 84–85, Prentice Hall, Englewood Cliffs, N.J.
2. SCHOPF, J. W. (1970) Precambrian micro-organisms and evolutionary events prior to the origin of vascular plants. *Biol. Rev.* **45**. 319–352. cited in Lipmann, F. (1975) The Roots of Bioenergetics. In *Energy Transformation in Biological Systems*, Ciba Foundation Symposium 31 (new series), Associated Scientific Publishers, Amsterdam, pp. 3–22.
3. OCHOA, S. and RENDI, R. (1962) Species specificity in activation and transfer of leucine from carrier ribonucleic acid to ribosomes. *J. Biol. Chem.* **237**, 3707–3710.
4. NATHANS, D. and LIPMANN, F. (1961) Amino acid transfer from aminoacyl ribonucleic acids to protein on ribosomes of *Escherichia coli*. *Proc. Nat. Acad. Sci. U.S.A.* **47**, 497–504.
5. CANNING, L. and GRIFFIN, A. C. (1965) Specificity in the transfer of aminoacyl-s-ribonucleic acid to microbial, liver, and tumor ribosomes. *Biochim. Biophys. Acta*, **103**, 522–525.
6. ALBRECHT, U., PRENZEL, K. and RICHTER, D. (1970) Amino acid transfer factors from yeast. III. Relationships between transfer factors and functionally similar protein fractions. *Biochemistry*, **9**, 361–368.
7. RICHTER, D. and LIPMANN, F. (1970) Separation of mitochondrial and cytoplasmic peptide chain elongation factors from yeast. *Biochemistry*, **8**, 5065–5070.
8. BOGORAD, L. (1975) Evolution of organelles and eukaryotic genomes. *Science*, **188**, 891–898.
9. MARGULIS, L. (1971) Symbiosis and evolution. *Sci. Am.* **225**, 48–57.
10. TAYLOR, F. J. R. (1974) Implications and extensions of the serial endosymbiosis theory of the origin of eukaryotes. *Taxon*, **23**, 229–258.
11. TRENCH, R. K., BOYLE, J. E. and SMITH, D. C. (1973) The association between chloroplasts of *Codium fragile* and the mollusc *Elysia viridis*. I. Characteristics of isolated *Codium* chloroplasts. II. Chloroplast ultrastructure and photosynthetic carbon fixation in *E. viridis*. *Proc. R. Soc. Lond.* B, **184**, 51–61, 63–81.
12. SY, J., CHUA, N.-H., OGAWA, Y. and LIPMANN, F. (1974) Ribosome specificity for the formation of guanosine polyphosphates. *Biochem. Biophys. Res. Commun.* **56**, 611–616.
13. BERGERSEN, F. J. (1974) Formation and function of bacteroids. In *The Biology of Nitrogen Fixation* (A. QUISPEL, ed.), pp. 473–498, North Holland Pub. Co., Amsterdam.

14. LEINER, M., SCHWEIKHARDT, F., BLASCHKE, G., KÖNIG, K. and FISCHER, M. (1968) Die Gärung und Atmung von *Pelomyxa palustris* Greeff. *Biol. Zbl.* **87**, 567–591.
15. SMITH, L., NEWTON, N. and SCHOLES, P. (1966) Reaction of cytochrome *c* oxidases of beef heart and *Micrococcus denitrificans* with mammalian and bacterial cytochromes *c*. In *Hemes and Hemoproteins* (B. CHANCE, R. W. ESTABROOK and T. YONETANI, eds.), pp. 395–403, Academic Press, New York.
16. JOHN, P. and WHATLEY, F. R. (1975) *Paracoccus denitrificans* and the evolutionary origin of the mitochondrion. *Nature*, **254**, 495–498.
WHATLEY, F. R. (1975) Discussion remark. In *Energy Transformation in Biological Systems*, Ciba Foundation Symposium 31 (new series), p. 64, Associated Scientific Publishers, Amsterdam.

A PATHWAY FROM PSYCHIATRY TO CANCER

EFRAIM RACKER
Section of Biochemistry, Molecular and Cell Biology, Cornell University, Wing Hall, Ithaca, N.Y. 14853

When I started as a medical student in 1932 in Vienna, I knew I was going to become a psychiatrist. The famous Viennese psychiatrist Wagner-Jauregg, who received the Nobel Prize for his pioneering work on the treatment of general paresis, warned medical students that the major difference between the inmates and the doctor in a state mental hospital is that the latter has a key. In the first lecture to the medical students Wagner-Jauregg explained that there are two kinds of psychiatrists, those who always wanted to be psychiatrists and those who got into the profession either because they did not want to deliver babies in the middle of the night or because they were offered an easy job in a state mental hospital. "I want you to know", he said, "that I belong to the second category".

I am afraid I belonged to the first category. I felt drawn to mental patients and wanted to understand their disease. I became fascinated with an article written by J. H. Quastel in 1936[1] on biochemistry and mental disorders. Shortly after Hitler invaded Vienna, I left for England and went to work with Dr. Quastel at the Cardiff City Mental Hospital, where I did my first experimental work in biochemistry. In 1941 I immigrated to the United States and after some brief excursions into clinical medicine, I settled for research and teaching in the department of bacteriology at New York University School of Medicine under the leadership of Colin Macleod. There I investigated the brain metabolism of mice infected with poliomyelitis. During these studies I noticed that brain tissue homogenized with distilled water did not glycolyze. This observation was a turning point in my career because it induced me to seek out the advice of a young assistant professor at New York University in the Department of Psychiatry. His name was Severo Ochoa. Soon after our first conversation we started to meet regularly for lunch and coffee breaks. There is nothing in my life which had a greater influence on my scientific development than these daily contacts with Severo and his bright young associates. Science was shared, it was exciting and it was fun.

Geiger[2] and Ochoa[3] had reported that rat brains that were homogenized with distilled water did not glycolyze because of the release of a glycolytic inhibitor. With the help of Ochoa I soon discovered that this "inhibitor" was a DPNase. Mann and Quastel[4] had reported that inactivation of DPN by brain homogenates was prevented by nicotinamide. Indeed, on addition of nicotinamide, hexose diphosphate was rapidly converted to lactate in water-treated brain tissue preparations. However, there was still no fermentation of glucose. Then I found that Na^+, which was present, somehow interfered with glucose phosphorylation.[5] When I realized that this was caused by a rapid hydrolysis of ATP, I introduced what I believe was the first ATP regenerating system. With phosphocreatine and creatine kinase present, the ATP concentration in the brain homogenates was maintained and glucose phosphorylation was sustained. The ATP-regenerating system soon became a favorite tool in our and many other laboratories. However, in a sense it was unfortunate that I had discovered this symptomatic "cure" because it prevented me from searching into the basic cause of the Na^+ induced breakdown of ATP. Had I pursued this phenomenon in depth, it would have led me to the responsible enzyme, namely the Na^+K^+ ATPase of brain, and I would have discovered this important enzyme 10 years before Skou.[6] This is only one of several important discoveries which I have missed by following one road instead of another.

Since I was supported by a grant from the National Foundation for Infantile Paralysis, I resumed my work on the effect of poliomyelitis on brain glycolysis and I began to learn the hard way how to analyze the complex multi-enzyme systems of glycolysis. We could not make collect calls to a Sigma company and I had to prepare DPN, ATP and even synthesize chemically glucose-6-phosphate and phosphocreatine. However, these chores had the advantage that they brought me closer to Severo Ochoa who had similar needs and interests. We had fun together preparing ATP from rabbit muscle, laughing at the mistakes we made. I am sure he still remembers how amazed we were watching our ATP preparation producing bubbles because we had mistakenly taken an aluminum pot which did not respond too well to the presence of HCl. Cheerfully, we started the preparation all over again.

In brains infected with poliomyelitis, the utilization of glucose was impaired to a much greater extent than that of hexose diphosphate. My first attention was therefore focused on the enzymes hexokinase and phosphofructokinase. I developed methods that allowed me to measure these enzymes in crude tissue extracts and I fell in love with spectrophotometric analyses. Although this was just methodology, it led me later to interesting discoveries, as you will see. I first ruled out hexokinase as the site of the lesion in infected brain and turned to phosphofructokinase. When I added the purified phosphofructokinase to the infected brain homogenate there was no effect on glycolysis, but the crude rabbit muscle extract which did not glycolyze by itself, completely restored activity. Luckily, Cori *et al.*[7] had just described the crystallization of glyceraldehyde-3-phosphate dehydrogenase from rabbit muscle and we prepared this enzyme as described. It cured the glycolytic defect in poliomyelitis infected brain! This then led me into the investigations of the mechanism of action of glyceraldehyde-3-phosphate dehydrogenase. We made one observation which proved most useful for future work on purification of enzymes. Cori *et al.*[7] reported that the enzyme was rather inactive unless cysteine was added. For our studies we needed an enzyme which did not require an external reducing agent. On a hunch that metals may be responsible for an oxidation of the SH group of the enzyme, we added KCN during its purification. It worked beautifully on the enzyme, but not on my collaborator, Mr. Krimsky, who contracted severe headaches from cyanide. In deference to his grumbling, I made a literature search for other metal chelators and came across the chelating properties of ethylene diamine tetraacetate which had just been described in the chemical literature. EDTA worked better for us than any other chelator and we thus introduced this useful reagent to biochemists.[8] This was perhaps the beginning of the philosophy which I developed later that headaches in research are good for you (particularly when your assistant gets them!).

When glyceraldehyde-3-phosphate dehydrogenase was added to infected brain homogenates, it was slowly inactivated. We suspected that we were dealing with a proteolytic process. We therefore added various peptides for protection and found that glutathione had a rather specific protective effect for glyceraldehyde-3-phosphate dehydrogenase.

Moreover, there was some glutathione present in the crystalline enzyme and we proposed that glutathione is a prosthetic group of the enzyme.[8] This turned out, later, to be incorrect. But the idea led me into a side-path which proved very productive. In searching for a sensitive assay for glutathione I turned to glyoxalase which had a specific requirement for this peptide. The manometric assay for glyoxalase was not sensitive enough for my purposes. Since methyl glyoxal, the substrate for glyoxalase, exhibited an absorption at 240 mμ and lactate, the end product, did not, I decided to explore the possibility of a spectrophotometric assay. On addition of an active yeast preparation of glyoxalase to methyl glyoxal and glutathione, I fully expected to see a disappearance of the 240 mμ absorption but, instead, observed a rapid increase. In pursuing this clue I separated two enzymes which participate in the transformation of methyl glyoxal into lactate. Since I did not know what they did, I gave them the "imaginative" names, glyoxalase I and glyoxalase II.[9] I then isolated the product formed by glyoxalase I and identified it as the thiolester lactoyl glutathione which absorbed at 240 mμ. It was hydrolyzed by glyoxalase II to lactate and glutathione. Lactoyl glutathione was the first example of a biological thiolester. It occurred to me that a thiolester may be an intermediate in the mechanism of action of glyceraldehyde-3-phosphate dehydrogenase and undergo phosphorylative cleavage to 1,3-diphosphoglycerate. I proposed this possibility to many biochemists and chemists but encountered a lot of opposition. How could a simple thiolester be a precursor of a high-energy acyl phosphate? There were two persons who encouraged me to persevere: Severo Ochoa and Henry Lardy. I therefore included in the paper on the mode of action of glyoxalase the proposal that a thiolester may be an intermediate in glyceraldehyde-3-phosphate oxidation.[9] Just when we started to obtain experimental evidence for such an enzyme-bound thiolester intermediate, the exciting report on the isolation of acetyl-coenzyme A was published by Lynen.[10] This important discovery changed the climate for thiolesters and our formulation of the mechanism of glyceraldehyde-3-phosphate dehydrogenase became accepted even before we had solid kinetic evidence that the thiolester formation was rapid enough for a true intermediate.[11]

The daily lunch meetings with Ochoa continued. One of his post-doctoral fellows was Bill Slater who worked on oxidative phosphorylation. One day he proposed that mitochondrial phosphorylation may operate by a mechanism similar to that of glyceraldehyde-3-phosphate dehydrogenase. At the time I was already interested in oxidative phosphorylation because in collaboration with Giff Pinchot, we had embarked on the problem of oxidative phosphorylation in bacteria. Officially I was still a bacteriologist and was indeed impressed with the possibilities offered by microorganisms in the study of biochemical mechanisms. We published only a brief note on our work on oxidative phosphorylation in *E. coli*[12] and I decided to switch to mitochondria while Giff Pinchot continued with bacteria. A few months ago Arthur Kornberg asked me whether I think, looking back, that I had made a mistake leaving bacteria. Obviously he thought I had. Perhaps, he is right and if nothing else I would have been spared some of the controversies that have plagued the mitochondrial field. I remember quite clearly the reasons for my decision to abandon bacteria. I felt that the approach which we needed most was resolution and reconstitution of oxidative phosphorylation into individual components. One of the key requirements for such an approach is the availability of large and stable quantities of starting material. In the early fifties, large-scale growth of bacteria with the facilities we had available then was simply not practical.

After a short interval of a few busy years on the elucidation of the pentose phosphate cycle, I returned to oxidative phosphorylation. By that time David Green's group had developed large-scale methods for the isolation of bovine heart mitochondria. Soon we were able to store a few 100 g of mitochondria in the deep freeze and could start working on resolution and reconstitution. In 1958, in collaboration with Pullman and Penefsky,[13] we isolated and characterized the first coupling factor (F_1) and identified it as mitochondrial ATPase. By systematical degradation and resolution of the mitochondrial membrane utilizing a variety of procedures, we obtained evidence for the participation of additional coupling factors which we named F_2, F_3 and F_4.[14] We could not tell what they were doing but we did know that each was required at all three sites of phosphorylation. Meanwhile, work on oxidative phosphorylation, similar to ours, was conducted by David Green and his collaborators. However, they reported that they had coupling factors that were site-specific and they published data on the isolation of the site-specific, high-energy intermediates. There was so little resemblance between their work and ours that I entitled the seminar I gave at the time: "Oxidative phosphorylation in New York City". After a while people were so confused that the entire field was under a cloud.

There is a time in the life of virtually every scientist when he has some scientific disagreement with a colleague. Some controversies are fruitful and stimulate work of greater sophistication. But this particular controversy seemed fruitless, since we disagreed on basic facts and I proposed to David Green that we should get together and straighten the matter out. David immediately and graciously invited me to come to Wisconsin to demonstrate our experiments and offered to do likewise. We travelled in triplicate to Madison. Jeff Schatz represented site I of oxidative phosphorylation, I was site II and June Fessenden-Raden was site III. Our counterparts were Archie Smith, Robert Beyer and George Webster. As a result of this visit David Green withdrew publicly at the next Federation Meeting their claims on the intermediates and factors required for oxidative phosphorylation. This lifted only partly the fog of confusion. The system became increasingly complex and difficult to analyze. F_4 was actually a mixture of F_2, F_3 and of a new factor F_5. MacLennan and Tzagoloff,[15] starting with F_4, isolated from it a protein which they called OSCP (oligomycin-sensitivity conferral protein) which replaced F_3 and F_5 in most (but not in all) systems. We therefore abandoned for the time being further work on F_3 and F_5 and substituted OSCP. Meanwhile Sanadi and his collaborators had started to work on coupling factors which he called factors A, B, C, etc.[16] This did not help to diminish the confusion. Once again I proposed collaboration to Dr. Sanadi which resulted in a joint publication on the identity of F_2 and factor B.[17] By that time it had become apparent that factor A and F_1 had identical functions.

Looking back, I wonder how we could have avoided some of the confusion that was caused by these multiple factors. In the history of vitamins we find similar confusions and complications. There were vitamins B_1, B_2, B_3 to B_{12} and more, but only a few have retained their individuality. Perhaps we should expect that in complex multifactorial systems which are being studied by different approaches and assay methods, confusion will prevail in the early stages. On the other hand, I feel we could have done better by maintaining a better communication system with our colleagues and by seeking more active collaborations.

Although we have now four fairly well-characterized protein coupling factors—F_1, F_2, OSCP and F_6[18]—we still do not know their individual functions but the evidence is mounting that they participate in association with the membranous hydrophobic proteins, in the assembly of the mitochondrial proton pump.

This brings me to the latest chapter in the field of oxidative phosphorylation and the current controversies on the mechanism of oxidative phosphorylation. Slater's formulations based on the mode of glyceraldehyde-3-phosphate dehydrogenase action became known as the chemical hypothesis.[19] In 1961 Mitchell proposed a chemiosmotic hypothesis.[20] I admit that when I first read his proposition, I was not impressed. In 1965 I published a book on "Mechanisms in Bioenergetics" and did not even mention the chemiosmotic hypothesis.[21] Phil Handler who wrote a generous review of my book in *Science* objected to my failure to discuss Mitchell's hypothesis. By the time his review appeared I knew that his criticism was justified because Peter Mitchell had visited me in New York in 1965. This was another important event in my scientific life. Not that I really understood most of what Mitchell said during these days of intensive discussions, but I opened my mind to a new way of thinking. I am now convinced that the basic formulations of his chemiosmotic hypothesis are correct, namely that the function of the respiratory chain is to translocate protons and that the return of those protons via the oligomycin-sensitive ATPase is responsible for ATP formation. Thus the problem of the mechanism of coupling of oxidation and phosphorylation is basically solved.

However, I began to realize a few years ago that this only solves the first half of the problem. The second equally important question to be answered is: how does a proton (or any other) pump utilize an ion gradient to make ATP? How does ATP generate an ion gradient in the reverse direction?

It was at this stage that I once again made a decision which changed the course of our work.[22] The complexity of the mitochondrial proton pump with its multiple coupling factors seemed a formidable barrier to an understanding of its mechanism. Two other pumps appeared to be simpler and more approachable. The Ca^{++} pump of sarcoplasmic reticulum and the Na^+K^+ pump of the plasma membrane. About 4 years ago we started working on these systems and I have recently formulated a molecular mechanism of the operation of the Ca^{++} pump.[23]

Curiously enough our work on the mechanism of pump action has now merged with another research project that has fascinated me for many years. About 20 years ago I started working on the phenomenon of the high aerobic glycolysis of tumor cells described by Warburg.[24] In collaboration with R. Wu and S. Gatt, we examined the rate-limiting factors that control glycolysis in tumor cells and in reconstituted systems.[25,26] We obtained evidence that the rate-limiting reaction of glycolysis is the hydrolysis of ATP to ADP and P_i. How are these two essential cofactors generated? During the past few years we obtained good evidence that the responsible catalysts in some tumor cells is the Na^+K^+ ATPase of the plasma membrane, in other cells the mitochondrial ATPase and in a third group an unidentified ATPase activity. We have also obtained evidence that it is not an excess of pumps, e.g. number of Na^+K^+ ATPase molecules which is responsible for the increased ATPase activity. It appears that the tumor cells pump Na^+ and K^+ inefficiently so that several ATP molecules are cleaved for each K^+ that is translocated.[27] We have found that naturally occurring plant products called bioflavonoids somehow repair this lesion and increase the efficiency of pumping. We are therefore actively engaged at present in the elucidation of the mechanism and the control of pump action.

Whatever we have contributed to biochemistry was an outgrowth of the basic approach of enzymology: isolation of the individual catalysts and components of a pathway in pure form and exploration of their function. This is the thread I see in common in all of my scientific endeavors and I am proud to have been guided in this approach by my friend Severo Ochoa.

REFERENCES

1. QUASTEL, J. H. (1936) Biochemistry and mental disorder. In *Perspectives in Biochemistry* (J. NEEDHAM and D. E. GREEN, eds.), pp. 269. Cambridge Univ. Press.
2. GEIGER, A. (1940) Glucolytic activity of brain. *Biochem. J.* **34,** 465.
3. OCHOA, S. (1941) Glycolysis and phosphorylation in brain extracts. *J. Biol. Chem.* **41,** 245.
4. MANN, P. J. G. and QUASTEL, J. H. (1941) Nicotinamide, cozymase and tissue metabolism. *Biochem. J.* **35,** 502.
5. RACKER, E. and KRIMSKY, I. (1945) Effect of nicotinic acid amide and sodium on glycolysis and oxygen uptake in brain homogenates. *J. Biol. Chem.* **161,** 453.
6. SKOU, J. C. (1957) The influence of some cations on an ATPase from peripheral nerve. *Biochim. Biophys. Acta,* **23,** 394.
7. CORI, G. T., SLEIN, M. W. and CORI, C. F. (1948) Crystalline D-glyceraldehyde-3-phosphate dehydrogenase from rabbit muscle. *J. Biol. Chem.* **173,** 605.
8. KRIMSKY, I. and RACKER, E. (1952) Glutathione, a prosthetic group of glyceraldehyde-3-phosphate dehydrogenases. *J. Biol. Chem.* **198,** 721.
9. RACKER, E. (1951) The mechanism of action of glyoxalase. *J. Biol. Chem.* **190,** 685.
10. LYNEN, F., REICHERT, E. and RUEFF, L. (1951) Isolation and chemical nature of activated acetate. *Anal. Chem.* **574,** 1.
11. RACKER, E. and KRIMSKY, I. (1952). The mechanism of oxidation of aldehydes by glyceraldehyde-3-phosphate dehydrogenase. *J. Biol. Chem.* **198,** 731.
12. PINCHOT, G. B. and RACKER, E. (1951). Ethyl alcohol oxidation and phosphorylation in extracts of *E. coli.* In *Phosphorus Metabolism* (W. D. MCELROY and B. GLASS, eds.), Vol. I, p. 366. The Johns Hopkins Press, Baltimore.
13. PULLMAN, M. E., PENEFSKY, H. and RACKER, E. (1958) A soluble protein fraction required for coupling phosphorylation to oxidation in submitochondrial fragments of beef heart mitochondria. *Arch. Biochem. Biophys.* **76,** 227.
14. RACKER, E. and CONOVER, T. E. (1963) Multiple coupling factors in oxidative phosphorylation. *Fed. Proc.* **22,** 1088.
15. MACLENNAN, D. H. and TZAGOLOFF, A. (1968) Purification and characterization of the oligomycin sensitivity conferring protein. *Biochemistry,* **7,** 1603.
16. ANDREOLI, T. E., LAM, K. W. and SANADI, D. R. (1965) A coupling enzyme which activates reversed electron transfer. *J. Biol. Chem.* **240,** 2644.
17. RACKER, E., FESSENDEN-RADEN, J. M., KANDRACH, M. A., LAM, D. W. and SANADI, D. R. (1970) Identity of coupling factor 2 and factor B. *Biochim. Biophys. Res. Commun.* **41,** 1474.
18. RACKER, E. (1974) Oxidative phosphorylation. In *Molecular Oxygen in Biology* (O. HAYAISHI, ed.), p. 339. North-Holland Publishing Co.
19. SLATER, E. C. (1953) Mechanisms of phosphorylation in the respiratory chain. *Nature* (*Lond.*) **172,** 975.
20. MITCHELL, P. (1961) Coupling of phosphorylation to electron and H transfer by a chemiosmotic type of mechanism. *Nature,* **191,** 144.
21. RACKER, E. (1965) *Mechanisms in Bioenergetics,* Academic Press Inc., New York.
22. RACKER, E. (1974) Mechanism of ATP formation in mitochondria and ion pumps. In *Dynamics of Energy-transducing Membranes* (ERNSTER, ESTABROOK and SLATER, eds.), p. 269. Elsevier Scientific Publishing Co., Amsterdam, The Netherlands.
23. RACKER, E. (1975) Reconstitution, mechanism of action and control of ion pumps. Hopkins Lecture. In *Biochem. Soc. Trans.* **3,** 785.
24. WARBURG, O. (1926) *Uber den Stoffwechsel der Tumoren,* Springer, Verlag, Berlin.
25. GATT, S. and RACKER, E. (1959) Regulatory mechanisms in carbohydrate metabolism. II. Pasteur effect in reconstructed systems. *J. Biol. Chem.* **234,** 1024.
26. WU, R. and RACKER, E. (1959) Regulatory mechanisms in carbohydrate metabolism. IV. Pasteur effect and crabtree effect in ascites tumor cells. *J. Biol. Chem.* **234,** 1036.
27. SUOLINNA, E-M., LANG, D. R. and RACKER, E. (1974) Quercetin, an artificial regulator of the high aerobic glycolysis of tumor cells. *J. Nat. Cancer Inst.* **53,** 1515.

BIOENERGETICS: PAST, PRESENT AND FUTURE

E. C. SLATER

Laboratory of Biochemistry, B.C.P. Jansen Institute, University of Amsterdam
Amsterdam, The Netherlands

PAST

Severo Ochoa began his career as a biochemist not long after the discovery of the importance of the pyrophosphate bond in energy transduction. Lohmann, working in the laboratory of Severo's teacher Otto Meyerhof, discovered in 1929 what we now know as the central compound of bioenergetics—adenosine triphosphate.[1]

A great achievement of the flourishing European schools of biochemistry in the 1930s was the complete elucidation of the mechanism of the splitting of glucose into lactic acid in muscle and into ethanol and carbon dioxide in yeast. These discoveries established the role of adenine nucleotides as carriers of energy from energy-yielding degradation reactions to energy-requiring reactions, such as muscle contraction. With the discovery of acyl phosphate (diphosphoglyceric acid) as the intermediate in the oxidative phosphorylation reaction of glycolysis Warburg closed in 1939 a chapter in the history of biochemistry,[2] that had begun in 1906 with Harden and Young's discovery of the stoichiometric requirement for phosphate for yeast fermentation.[3] It was W. J. Young, while Professor of Biochemistry at the University of Melbourne, who guided my first steps in biochemistry.

The mechanism of action of glyceraldehydephosphate dehydrogenase, the enzyme that catalyses the synthesis of the acyl phosphate, is not, however, a closed book. It continues to fascinate a select band of biochemists, who simply refer to it as "the enzyme", and this research has been given a new dimension by the recently published three-dimensional structure.

The significance of phosphate esters and of the difference between "energy-rich" and "energy-poor" phosphate compounds was spelled out in a beautiful classical review by Fritz Lipmann in the first volume of *Advances in Enzymology* in 1941.[4] The squiggle ($\sim$) was born in this article.

Almost simultaneously with the discovery of the way in which the energy made available from the splitting of glucose to lactic acid is coupled with the synthesis of ATP came the recognition that quantitatively this reaction is relatively unimportant in aerobic organisms. Already in 1920, Meyerhof[5] had shown that respiration is efficiently used in the recovery period following muscular contraction. He showed that the oxygen consumed during recovery is sufficient to account for only about one-fifth of the lactic acid produced during contraction, but that the energy made available by respiration is utilized to convert the remaining lactic acid to glycogen.

With these experiments, Meyerhof foreshadowed the discovery of oxidative phosphorylation, but who was the real discoverer? Was it the Eggletons and, Fiske and SubbaRow, who discovered in 1927 and 1929, respectively, the rapid synthesis of creatine phosphate during the recovery of muscle in oxygen? Was it Engelhardt who in 1930[6] found that pigeon erythrocytes rapidly form inorganic phosphate, at the expense of labile "pyrophosphate" (7-min P), when cyanide is added and that 7-min P is restored after washing away the cyanide? Certainly Engelhardt drew the correct conclusion, namely that splitting of ATP occurs both in the presence and absence of respiratory inhibitors, but that in their absence the splitting is compensated by the resynthesis of ATP. Why did Engelhardt's work seem to have little impact on contemporary biochemists? Was it because they required the demonstration of a phosphorylation reaction in a cell-free system where it is often easier to separate the components of coupled systems? Or did they have first to become more receptive to the idea of oxidative phosphorylation by Warburg's work on the glycolytic system? Whatever the answer to these questions, it is clear that "oxidative phosphorylation" was firmly established by Herman Kalckar who, in 1937–1939, showed the phosphorylation of glucose, glycerol or AMP during the aerobic oxidation of a variety of compounds by homogenates of kidney and other tissues.[7]

It took a long time to recognize that oxidative phosphorylation, or respiratory-chain phosphorylation as it later became known, is fundamentally a different process from glycolytic phosphorylation and the final recognition came quite recently with the demise of the postulated non-phosphorylated high-energy intermediates of oxidative phosphorylation. The first step in this direction was taken by Severo while working in Rudolph Peter's laboratory in Oxford[8] and by Belitser and Tsybakova[9] in Engelhardt's laboratory. In a courageous and, despite contemporary criticism, now seen to be a successful attempt to correct for the hydrolysis of ATP occurring simultaneously with its synthesis by oxidative phosphorylation, Severo found that the P:O ratio for the oxidation of pyruvate is 3. Both he and the Russian workers drew the correct conclusion, namely that phosphorylation must occur not only when the substrate is dehydrogenated—as in the oxidation reaction of glycolysis—but also during the further passage of the hydrogen atoms (or electrons) along the respiratory chain to oxygen. However, a realization of the full significance of this conclusion had to await further elucidation of the chain between NADH and oxygen, particularly the rejection of Albert Szent-Gyorgyi's proposal that dicarboxylic acids act as hydrogen carriers, and the direct demonstration by Albert Lehninger of phosphorylation linked with the aerobic oxidation of NADH.[10] Thereby, Lehninger demonstrated oxidative phosphorylation in the absence of oxidizable substrates of the type known as intermediary metabolites. While working in Severo's laboratory in New York in 1950, I was able to show that oxygen is also not necessary, since I was able to replace it with ferricytochrome *c*.

After many mistakes and missteps it became generally established that Severo's P:O ratio of 3 for the oxidation of pyruvate is due to the presence of 3 phosphorylation sites between NADH and oxygen, usually located (but not by orthodox disciples of Peter Mitchell) between NADH and ubiquinone, ubiquinol and ferricytochrome *c*, ferrocytochrome *c* and oxygen. But the 3.0 is an average of three times 3, one 2 and one 4. The lower ratio with succinate and the higher with 2-oxoglutarate is in line with the higher and lower redox potentials, respectively, of these substrates. The demonstration of the extra phosphorylation reaction with 2-oxoglutarate, and its exceptional nature (in that energy conservation does occur when the substrate is dehydrogenated) gave the necessary headaches. The relatively easy clarification of its mechanism by Seymour Kaufman in

Severo's laboratory was also possibly for some of us unfortunate in that it focused thinking on the mechanism of "substrate-linked phosphorylation".

The recognition of the universal role of the pyrophosphate bond in the transfer of energy from energy-yielding to energy-requiring reactions made it urgent to establish the value in energy units of the so-called energy-rich bond, the energy currency of the cell. Indeed Lipmann gave much attention to this in his 1941 article. In the intervening years the exchange value of this currency, like some others in the English-speaking world, has gradually declined (see Fig. 1). A minimum was reached in 1959, but towards the end

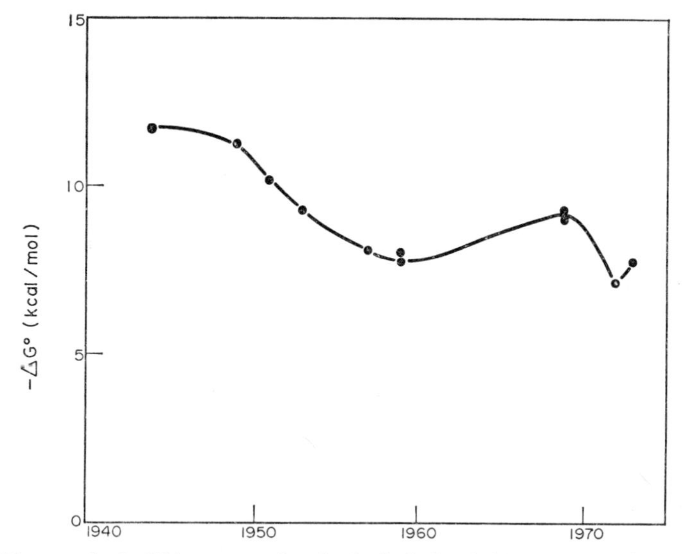

FIG. 1. The standard Gibbs energy for the hydrolysis of ATP, measured at pH 7.5, I 0.2 M, 5 mm Mg^{2+} and 25°, plotted according to the year in which it was reported. Recent data on ionization and complex-forming constants were used to adjust the original data to these standard conditions.

of the 1960s recalculations of old equilibria data and new determinations of acid dissociation constants started what on the stock exchanges is called a calf market. Since this gave us some problems, we decided to reinvestigate the matter and came out with a much lower value, even a little lower than the 1959 minimum. We do not believe that a more recently published slightly higher value presages the beginning of a new bull market.

Of course, the amount of Gibb's energy that is made available by the splitting of ATP (or is required for its formation from ADP and phosphate) depends upon temperature, ionic strength and the concentration of complexing ions including protons and Mg^{2+}. The conditions selected for the presentation of Fig. 1 are arbitrary, and were chosen because many determinations and calculations appropriate for these conditions have been described in the literature. In the absence of Mg^{2+}, the ΔG_0 for the hydrolysis is more negative since magnesium binds more firmly to ATP than to ADP and phosphate. More important, the actual change in Gibbs energy, ΔG, depends upon the concentrations of ATP, ADP and phosphate according to the relation (at 25°C)

$$\Delta G = \Delta G° + 1.36 \log \frac{[ATP]}{[ADP][P_i]} \text{ kcal/mol}$$

where ΔG° is the standard Gibbs energy change, when the concentrations of ATP, ADP and P_i are all 1 M. Respiring mitochondria or bacterial membranes, or illuminated chloroplasts, are able to make ATP at high concentrations of ATP and low concentrations of ADP and phosphate. For example, mitochondria respiring succinate at pH 7.7 and 25°C, in the absence of added magnesium, can reach a quasi-equilibrium or steady state (called State 4 by mitochondriacs) when the concentration of ATP in the suspending medium is 2×10^{-2} M, that of ADP 1.5×10^{-4} M and of P_i 5×10^{-4} M. Since ΔG_0 under these conditions equals 8.7 kcal/mol,

$$\Delta G = 8.7 + 1.36 \log \frac{2 \times 10^{-2}}{1.5 \times 10^{-4} \times 5 \times 10^{-4}} = 16.1 \text{ kcal/mol.}$$

Under these conditions, $[NADH]/[NAD^+] = 1.6$ and $[O_2] = 0.1$ atm, from which it can be calculated that ΔG for the reaction—

$$NADH + \tfrac{1}{2}O_2 + H^+ \rightleftharpoons NAD^+ + H_2O$$

is 50.8 kcal/mol. Since this reaction may be coupled with the synthesis of three molecules of ATP, the ΔG for the total reaction

$$NADH + \tfrac{1}{2}O_2 + H^+ + 3ADP + 3P_i \rightleftharpoons NAD^+ + H_2O + 3ATP + 3H_2O$$

is $-50.8 + 48.3 = -2.5$ kcal/mol, which is not far from equilibrium. It is indeed quite remarkable that thermodynamic equilibrium is approached so closely, when one considers that ATP is a competitive inhibitor of ADP entry into the mitochondria, where phosphorylation takes place, so that the reaction approaches equilibrium quite slowly, and ATP-splitting enzymes present in microsomal contamination or in damaged mitochondria are quite active.

Calculations of the so-called efficiency of oxidative phosphorylation, calculated from the quotient between the ΔG_0 values of the phosphorylation and redox reactions, which have dotted the literature in the last decades and which are still to be found in widely used textbooks are, then, in my opinion, quite meaningless. In the absence of side reactions all chemical reactions will finally reach an equilibrium, so that $\Delta G = 0$. Thus, all coupled reactions will reach an equilibrium in which the ΔG of the energy-liberating reaction will be exactly balanced by that of the energy-consuming reaction. The "efficiency" of the coupling, insofar as this is a meaningful term, must be 100%.

The true nature of oxidative phosphorylation, i.e. the synthesis of ATP linked to the transfer of electrons along a chain of electron (or hydrogen) carriers, was emphasized by the discovery of photosynthetic phosphorylation in chloroplasts by Arnon in 1954.[11] Here neither oxidizable substrates nor oxygen play a role. Electrons derived from the photolytic cleavage of water are fed into an electron-transfer chain and withdrawn at the other end of the chain by the oxidizing equivalents formed by the photolytic reaction.

As already mentioned, the mechanism of substrate-linked phosphorylation strongly influenced the thinking of those interested in the mechanism of phosphorylation linked with electron transfer, including those gathered for coffee or lunch in the canteen of the Medical School of New York University in 1949–1950. Although not belonging to Severo's department (Pharmacology), Ef Racker from Microbiology usually joined these gatherings that were remarkable more for the vigour and from time to time perhaps the quality of the intellectual exchanges, than for the food. The idea that high-energy intermediates are formed before phosphate came into the picture was emerging from Ef's work on the active thiol group in glyceraldehyde-phosphate dehydrogenase. Lynen's identification shortly afterwards of the "active C-2 fragment" of pyruvate oxidation as acetyl-coenzyme A firmly established this concept.

Since I knew from the work in Keilin's laboratory that electron transfer along the respiratory chain can take place in the absence of phosphate,[12] it was attractive to assume that energy is also conserved in the respiratory chain before the intervention of phosphate. This was where the hypothesis that I proposed in 1953[13] differed from that put forward earlier by Lipmann. I was also much attracted to the idea that high-energy intermediates accumulated in the absence of phosphate could be used directly for energy-consuming reactions in the cell, without having first to be converted into ATP. This is indeed the case, as became firmly established in the early 1960s mainly as a result of work by Chance, Ernster and ourselves. Indeed for a time the pendulum swung too far away from ATP but it is now realized that, although the energy from electron transfer in the mitochondrial membrane can be used to drive energy-requiring reactions in the mitochondrion in the absence of phosphate, the physiologically important reaction is the synthesis of ATP for export to extramitochondrial energy-requiring compartments. Isaac Harary, Severo's only graduate student when I was with him in 1949–1950, and I working together in Amsterdam in 1965 showed, for example, that this is the case in isolated and beating heart cells.[14]

The confirmation of one of the predictions of the so-called chemical hypothesis of oxidative phosphorylation inspired the active search for the postulated high-energy (~) intermediates. Indeed, it was surprising how easy it appeared to be to prove their existence, but they seemed to run out through your fingers when you tried to grab them and isolate them. One by one claims were made, but were either withdrawn or shown by others to rest on artifacts. Not without justification, new claims were greeted with suspicion and you would now probably have to submit crystals of such an intermediate with your paper if you hoped to have it accepted for publication.

It is, indeed, now generally accepted that high-energy intermediates of the type envisaged in my 1953 hypothesis, which postulated the formation of a covalent bond been a component of the electron-transfer chain and an external ligand, do not exist. In 1964 Boyer suggested that energy is conserved in a conformation change of the protein of the electron carrier without forming a covalent bond.[15] This variation of the chemical hypothesis, called the conformation hypothesis, is still very much alive.

The demise of the high-energy intermediates, or rather their non-existence, was already predicted in 1961 by Mitchell,[16] who put forward this as the simplest explanation for our failure to isolate them. At the same time he proposed an alternative hypothesis, namely that the electron and hydrogen carriers of the respiratory chain are so located in the membrane that electron transfer from substrate to oxygen is coupled with the translocation of protons from inside to outside the mitochondrion. The ATPase present in the membrane also couples the hydrolysis of ATP with the translocation of protons in the same direction, and the proton gradient generated by electron transfer is used to drive this ATPase in reverse, leading to synthesis of ATP. This ingenious and stimulating hypothesis has dominated developments in the field in the last decade. It has been discouraging to the chemists who have nothing to isolate except a proton gradient or a membrane potential which is as elusive as the non-existent high-energy intermediates.

All this takes place in the mitochondrial inner membrane, which really does exist and contains a richness and variety of proteins to keep the protein chemists busy for some time to come. The strongly hydrophobic nature of many of the polypeptides and their strong association with the phospholipids that make up about 30% of the weight of the membrane has made it difficult to apply many of the techniques successful in other fields. As Ef Racker showed, however, early in the fifties, the ATP-synthesizing enzyme (Factor 1, F-1, not Ef-1) and certain proteins necessary for its binding to the membrane (the so-called OSCP and Fc_2) can be more easily isolated. David Green concentrated on the electron-transferring components. It was his great success to show that the many polypeptides present in the membrane are associated together in four macromolecules or complexes, catalysing the transfer of electrons from NADH to Q, from succinate to Q, from QH_2 to ferricytochrome *c* and from ferrocytochrome *c* to oxygen, respectively. Ef Racker has been able to reconstitute oxidative phosphorylation by bringing together an electron-transferring macromolecule, the ATPase, hydrophobic proteins, lipids and proteins required for binding the ATPase to the membrane in an orientation appropriate for coupling.

PRESENT

At present considerable progress is being made in identifying and, in many cases, isolating the individual polypeptides that make up the energy-transducing membranes. Progress has been made possible by the introduction of new separation techniques, particularly affinity chromatography and new methods of detection, especially electrophoresis in polyacrylamide gels of the protein mixtures after dispersal in dodecyl sulphate. Genetics and the selective inhibition of the biosynthesis of polypeptides on the cytosol ribosome, leaving the synthesis of polypeptides on the mitochondrial ribosome intact, have also been of great help is assigning certain properties to specific polypeptides, and in isolating certain proteins. Weiss, for example, has isolated cytochrome *b* from *Neurospora* mitochondria by selectively labeling with ^{14}C the relatively few polypeptides that are synthesized on the mitochondrial ribosomes. Mutants of *Escherichia coli* have been isolated either lacking the ATPase proper or with defects in the segment of the membrane to which it is bound.

Although opinions differ on the precise nature of the coupling between the electron-transfer proteins and the phosphorylating machinery, all agree that the synthesis of ATP takes place on the ATPase proper or Ef's F-1. The ATPases isolated from mitochondria, chloroplasts and bacteria are strikingly alike. We have been able to show that this enzyme contains firmly bound ATP and ADP and that the residual adenine nucleotides found in washed mitochondrial, chloroplast or bacterial membranes are, in fact, bound to this ATPase.[17] Both Paul Boyer[18] and ourselves have proposed that the energy delivered by the electron-transfer chain is largely used to bring about a conformation change in the ATPase protein in such a way that the firmly bound ATP is dissociated. This hypothesis is supported by two lines of evidence, namely (i) that according to several criteria the ATPase *does* change its conformation when electrons pass along the electron-transfer chain, and (ii) that the ATP firmly bound to chloroplast ATPase becomes exchangeable with added nucleotide when electrons are transferred along the chain. Peter Mitchell believes, however, that protons transmitted through a hole to the active site of the enzyme play a direct rôle in the phosphorylation reaction.

Whichever hypothesis is correct, it is clear that more needs to be learned about the way in which the ATPase binds ATP and ADP, and catalyses the hydrolysis of ATP to ADP and P_i. The two largest subunits of the ATPase are sufficient for this reaction. At least five polypeptides (subunits C, D and F of the ATPase proper, OSCP and Fc_2) are necessary for binding of the ATPase to the membrane. An additional polypeptide (subunit E of the ATPase) inhibits the enzymic activity of the ATPase and has to be removed before the isolated enzyme hydrolyses ATP. Presumably, it would also inhibit the reverse reaction—the synthesis of ATP, but it never gets a chance to do so, because, as van de Stadt and van Dam showed, the transfer of electrons along the electron carriers induces the dissociation of this subunit from the ATPase bound to the same membrane.[19]

The armory of protein chemistry is now being turned on these subunits.

Four additional polypeptides have been identified in the membrane sector. One of these, a strongly hydrophobic protein, seems to contain the binding site of dicyclohexylcarbodiimide, an inhibitor of oxidative phosphorylation, and probably also of oligomycin. These four polypeptides are coded for by the mitochondrial genome, the others by the nuclear genome.

Membranes lacking the ATPase, either because it has been experimentally detached or because of a mutation, are "leaky" towards protons. The leak is plugged by dicyclohexylcarbodiimide or oligomycin. It is possible, then, that the polypeptide containing the binding site for these inhibitors or an interacting polypeptide is the proton hole leading to the ATPase, as envisaged by Mitchell.

Ef Racker is continuing successfully to reconstitute parts of the energy-transducing system. Of particular interest is the light-driven ATP synthesis brought about in vesicles made by bringing together bacterial rhodopsin, the ATPase, membrane proteins sufficient for the formation of a vesicle, and the polypeptides specifically involved in binding the ATPase to the membrane and lipids.[20] Since light induces proton translocation across membranes containing bacterial rhodopsin, this experiment gives strong support to Mitchell's hypothesis.

THE FUTURE

The main line of future work on the mechanism of the phosphorylation linked with electron transfer is now laid down. Within 5 years at the outside we should know if the conformational or the chemiosmotic hypothesis, or both, or (dare we think it) neither, gives a correct description. Taking apart the membrane, separating the polypeptides, sequencing them, putting them together again with the lipids in the right place will probably keep the next generation busy. Surely we shall have by the end of the century a complete description of the path taken by electrons within and between the polypeptides. Before then we should have an answer to the question: why are so many different electron-accepting groups present in energy-transducing membranes?

The mechanism by which man obtains and utilizes energy—bioenergetics—is clearly one of the most fundamental problems in molecular biology. Its intrinsic interest is sufficient to justify the use of scientific and technical resources to unravel this mechanism. But the so-called energy crisis that the world is facing may make it urgent that some problems of bioenergetics are solved more rapidly than would be predicted by the present tempo of research.

All the energy that man consumes is derived from the Sun. To stay alive, the average human (including babies and children in the average) needs about 2000 kcal per day in the form of food. This is equivalent to about 100 joules per second, that is to a continuous input of about 100 W or the same as a fairly large electric light bulb. This amount of energy is sufficient to make about 1.4 nmoles ATP per second.

Considering what man can do, 100 W energy expenditure does not seem much. He is, in fact, a very efficient machine. However, since there are 3.8×10^9 of us, the world population already needs the continuous production of 3.8×10^{11} W in the form of food. This energy is derived from the Sun which has been absorbed by the chlorophyll in the chloroplasts and used to synthesize carbohydrates and other foodstuffs from carbon dioxide, water and nitrogen. We eat these plants or animals that have eaten plants.

The agricultural production in the world is just about sufficient to supply this 3.8×10^{11} W continuous energy, and it could be geared to produce a little more, but not very much. It is painfully obvious that it will not be able to support the exponentionally growing population much longer.

This amount of 3.8×10^{11} W is, however, only sufficient to keep a naked man alive and reproducing. He needs a lot more energy to be clothed, stay warm in winter or cool in summer, live in houses, enjoy himself and to travel other than by foot. The total energy consumption over the whole world is about 20 times as much, in prosperous countries like the United States 100 times.

It may seem paradoxical that man is reaching the limits of the possibilities of growing sufficient food, but at the same time is consuming 20 times (100 times in advanced countries) as much energy for non-vital purposes. This is especially so when we consider that the energy that he is consuming was also laid down as a result of the ability of chlorophyll-containing energy-transducing membranes to convert solar energy into combustible fuel. The coal, oil and natural gas that we now burn is derived from the carbohydrate of forests growing millions of years ago. This fossil Sun energy is our energy capital and we are using it up now at a rate 20 times faster than our solar-energy income derived from our present agriculture.

Our total reserves of fossil Sun energy are about 1000 times our present yearly use. This seems, at first sight, a reassuringly large amount. However, the simple graph in Fig. 2 shows that it is alarmingly small. In the first place, only 10% of these reserves are in the form of oil and natural gas, another 5% is in the form of shale oil and the remaining 85% in the form of coal, much of it difficult to mine. Secondly, the world consumption of energy is (or was until very recently) increasing at the rate of about 5% per year, and it is difficult to see how this can be reduced much in the next decades, given the increase in the world population and the desire for increasing living standards among the poorer populations. Even a relatively small annual accress of 5% has frightening effects over quite a short period, measured by the lifetime of man. Already by 2010, reserves equivalent to all our oil and gas will be gone, the reserves of shale oil will follow a few years later and in 80 years, within lifetime of some of our grandchildren, everything will be gone.

Of course, we must do something about it. The first requirement, if an enormous catastrophe is to be prevented, is to stop the population increase. If we cannot do this, we are finished anyway. Even if we are successful, however, we have to find new forms of energy and, if we would like man to continue to enjoy this planet for a few hundred years more, new sources of energy income will have to be sought.

In the short term, we might use up some other energy capital to give us time to work out ways of increasing our energy income. Theoretically, geothermic energy, revealed to us in hot springs and volcanos, might be used. It has been estimated that these stores

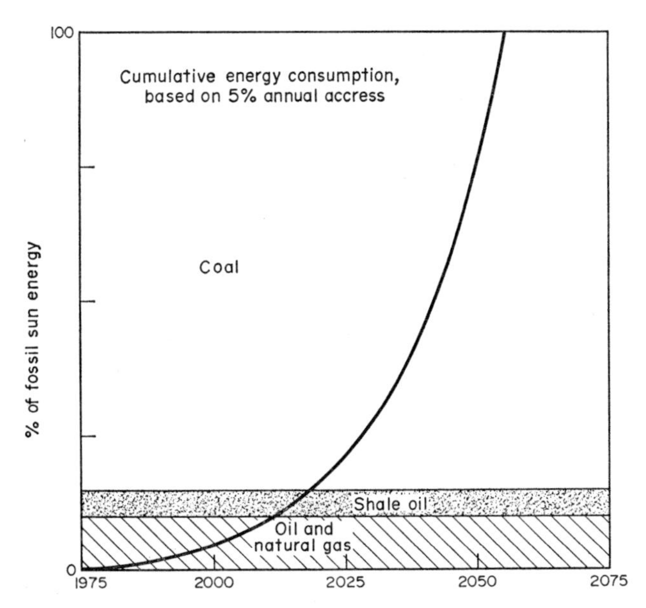

FIG. 2. The cumulative consumption of the world's reserves of fossil Sun energy, calculated on the basis of the consumption in 1975 and a 5% annual accress.

amount to about 80% of the fossil Sun energy (see Fig. 3). However, we do not know yet how to tap this source. We do know how to use the energy made available by splitting heavy atoms, and, despite the political dangers, it seems inevitable that we must introduce nuclear reactors on a large scale. However, the amount of thorium and uranium is limited and, with the reactors now in use, only sufficient to supply the world with energy at the 1975 rate for about 3 years, unless uranium can be obtained from the sea. This becomes financially feasible only if the energy yield of the reactors is increased, by the use of breeding reactors. In that case, the total reserves of uranium and thorium are equivalent to about 7 times our fossil Sun energy. If it is not possible to extract uranium from the sea, it is only one-fifth the amount of fossil Sun energy.

Theoretically, it is possible to obtain energy by the fusion of light atoms, e.g. deuterium with tritium. However, tritium has to be made from lithium, and this is also present in restricted amounts, sufficient to yield energy equivalent to about 60% of our fossil Sun reserves. The fusion of deuterium and deuterium would give virtually unlimited reserves of energy and if the physicists ever succeed in this, they have made a sun on earth, and the energy crisis will be over.

In case they do not, however, and until they do, we have to make use of the Sun we have. This is our energy income. (It is true that we can harness the wind and the tides, but these can at best make a very small contribution to our total needs.) The Sun radiates 1.3×10^{17} W energy, 20,000 times our present total consumption. And it sends out this energy whether we use it or not, so we are not robbing future generations by using it now.

We can harness directly the heat of the Sun, not only for heating houses, but also to drive electrical turbines or to split water into hydrogen and oxygen. But this is rather costly and the bioenergeticist would like to copy the way certain algae do it. It is indeed perfectly feasible to allow a chloroplast suspension to reduce a low-potential dyestuff with

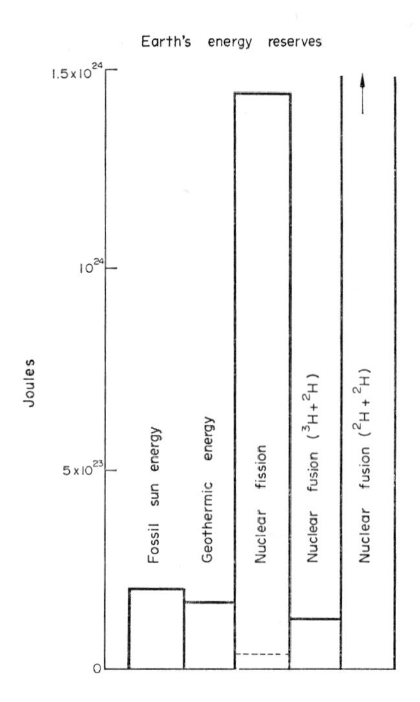

FIG. 3. The Earth's energy reserves. The dotted line in the block representing nuclear fusion refers to reserves of uranium and thorium in the land masses, calculated for breeding reactors. The solid block takes into account the amounts in sea water. The reserves of deuterium are virtually unlimited.

reducing equivalents derived from the photolysis of water (the oxidizing equivalents are evolved as oxygen) and to feed these reducing equivalents to hydrogenase which converts them to molecular hydrogen. The system chloroplasts + dyestuff + hydrogenase can bring about the photolysis of water to hydrogen and oxygen. Hydrogen is, in many ways, the ideal fuel since the product of its combustion is water.

In a workshop held in Gatlinburg in the U.S.A. in 1972, it was calculated that, if the solar energy could be utilized by chlorophyll with 10% efficiency, 20 g hydrogen could be made per m^2 per day in south-west U.S.A.[21] Since this is equivalent to 30 W, the present world energy use would require 2.2×10^{11} m^2 or 2.2×10^5 km^2 or 9×10^4 square miles, i.e. a square of about 300 miles. Since this is only 1.5% of the area at present under cultivation and deserts not suitable for cultivation could be used, there should be enough room on the Earth for biophotolysis factories of this nature. Although a factory 300 mile square is unimaginably large, it works out to only about 58 km^2 for a city of 1 million people, not much more than what the city fathers usually think necessary for an airfield.

The bioenergeticist does not yet have enough information on the feasibility of stabilizing the catalysts and of obtaining a sufficiently high energy conversion (only 1% is reached in agriculture, compared with 10% used in the above calculation) to be able to talk to the engineer. However, he should be able to obtain this information quite quickly if he is interested enough or can be made interested enough to get it.

We do not have much choice. Either we trust the physicists to make us a sun without blowing us up, or we let the bioenergeticists use our present one. Even if we do have complete confidence in our physicist friends, it might not be a bad idea to take out an insurance policy. Otherwise, we won't last more than a hundred years or so.

This is an exciting challenge for bioenergetics.

REFERENCES

1. LOHMANN, K. (1929) Über die Pyrophosphatfraktion im Muskel. *Naturwissenschaften*, **17**, 624–625.
2. WARBURG, O. and CHRISTIAN, W. (1939) Isolierung und Kristallisation des Proteins des Oxydierenden Gärungsferments. *Biochem. Z.* **303**, 40–68.
3. HARDEN, A. and YOUNG, W. J. (1906) The alcoholic ferment of yeast-juice. *Proc. R. Soc.* (*Lond.*) B, **77**, 405–420.
4. LIPMANN, F. (1941) Metabolic generation and utilization of phosphate bond energy. *Adv. Enzymol.* **1**, 99–162.
5. MEYERHOF, O. (1920) Die Energieumwaltlungen im Muskel. III. Kohlenhydrat—um Milchsäureumsatz in Froschmuskel. *Pflügers Arch. Ges. Physiol.* **185**, 11–32.
6. Engelhardt, W. A. (1930) Ortho- und Pyrophosphat im aeroben und anaeroben Stoffwechsel der Blutzellen. *Biochem. Z.* **227**, 16–38.
7. KALCKAR, H. (1937) Phosphorylation in kidney tissue. *Enzymologia*, **2**, 47–52.
8. OCHOA, S. (1940) Nature of oxidative phosphorylation in brain tissue. *Nature*, **146**, 267.
9. BELITSER, V. A. and TSYBAKOVA, E. T. (1939) The mechanism of phosphorylation as related to respiration. *Biokhimiya*, **4**, 516–535.
10. FRIEDKIN, M. and LEHNINGER, A. L. (1948) Phosphorylation coupled to electron transport between dihydrodiphosphopyridine nucleotide and oxygen. *J. Biol. Chem.* **174**, 757–758.
11. ARNON, D. I., WHATLEY, F. R. and ALLEN, M. B. (1954) Photosynthesis by isolated chloroplasts. II. Photosynthetic phosphorylation, the conversion of light into phosphate bond energy. *J. Am. Chem. Soc.* **76**, 6324–6328.
12. BONNER, W. D., JR. (1951) Activation of the heart-muscle succinic oxidase system. *Biochem. J.* **49**, viii–ix.
13. SLATER, E. C. (1953) Mechanism of phosphorylation in the respiratory chain. *Nature*, **172**, 975–978.
14. HARARY, I. and SLATER, E. C. (1965) Studies *in vitro* on single beating heart cells. VIII. The effect of oligomycin, dinitrophenol, and ouabain on the beating rat heart. *Biochim. Biophys. Acta*, **99**, 227–233.
15. BOYER, P. D. (1965) Carboxyl activation as a possible common reaction in substrate-level and oxidative phosphorylation and in muscle contraction. In *Oxidases and Related Redox Systems* (T. E. KING, H. S. MASON and M. MORRISON, eds.), vol. 2, pp. 994–1008, Wiley, New York.
16. MITCHELL, P. (1961) Coupling of phosphorylation to electron and hydrogen transfer by a chemi-osmotic type of mechanism. *Nature*, **191**, 144–148.
17. SLATER, E. C. (1974) Electron transfer and energy conservation. In *Dynamics of Energy-transducing Membranes* (L. ERNSTER, R. W. ESTABROOK and E. C. SLATER, eds.), pp. 1–20, Elsevier, Amsterdam.
18. BOYER, P. D. (1974) Conformational coupling in biological energy transductions. In *Dynamics of Energy-transducing Membranes*, (L. ERNSTER, R. W. ESTABROOK and E. C. SLATER, eds.), pp. 289–301. Elsevier, Amsterdam.
19. VAN DE STADT, R. J., DE BOER, B. L. and VAN DAM, K. (1973) The interaction between the mitochondrial ATPase (F_1) and the ATPase inhibitor. *Biochim. Biophys. Acta*, **292**, 338–349.
20. RACKER, E. and STOECKENIUS, W. (1974) Reconstitution of purple membrane vesicles catalyzing light-driven proton uptake and adenosine triphosphate formation. *J. Biol. Chem.* **249**, 662–663.
21. An enquiry into Biological Energy Conversion (1972), The University of Tennessee.

ENZYMES REVISITED*

Hugo Theorell

Nobel Institute of Medicine, Solnavagen 1, Stockholm 60, Sweden

FROM BERZELIUS TO SUMNER

Practically all chemical reactions in living nature are started and directed in their course by enzymes. This being the case, man has of course since time immemorial seen examples of what we now call enzymatic reactions, e.g. fermentation and decay. It would thus be possible to trace the history of enzymes back to the ancient Greeks, or still further for that matter. But it would be rather pointless, since to observe a phenomenon is not the same thing as to explain it. It is more correct to say that our knowledge of enzymes is essentially a product of twentieth century research.

Enzymes are a sort of *catalyzers*, and in this connection a reminder of the origin of the concept catalysis may be in place. It was put forward by a school-fellow of mine from Linköping High School. I never had the honour of meeting him, as he was 124 years older than I. It was one of the founders of Karolinska Institutet and the Swedish Medical Society, Jöns Jacob Berzelius, who in 1835 wrote in his year-book:

> This is a new force producing chemical activity and belonging as well to inorganic as organic nature, a force which is undoubtedly more widespread than we have hitherto imagined, and whose nature is still concealed from us. When I call it a new force I do not thereby mean to say that it is a capacity independent of the electrochemical relations of matter, on the contrary, I cannot but presume it to be a particular manifestation of these, but as long as we cannot understand their reciprocal connections it will facilitate our researches to regard it for the time being as an independent force, just as it will facilitate our discussion thereof if it be given a name of its own. I shall therefore, to use a derivation well-known in chemistry, call it the *catalytic force* of bodies, decomposition through this force *catalysis*, just as with the term *analysis* we describe the separation of the constituent parts of bodies by means of ordinary chemical affinity. The catalytic force appears actually to consist therein that through their mere presence and not through their affinity bodies are able to arouse affinities which at this temperature are slumbering. . . .

Enzymes are the catalyzers of the biological world, and Berzelius' description of catalytic force is surprisingly far-sighted—one is tempted to say prophetic. Especially is one struck by his expressly refusing to believe that other than chemical forces are here in play; no, if one could once understand the mechanism it would doubtless prove that the forces of ordinary chemistry would suffice to explain also these as yet mysterious reactions.

Almost 100 years were to pass before it became clear that Berzelius had been right. The year 1926 was a memorable one. The German chemist Richard Willstätter gave a lecture then in Deutsche Chemische Gesellschaft, in which he summarized the experiences gained in his attempts over many years to produce pure enzymes. Through various adsorption methods he had removed more and more of the impurities in some enzymes; especially he had worked with a so-called peroxidase, an enzyme of general occurrence in the vegetable kingdom. Finally, there was so little substance left that on ordinary analysis for protein,

* Part of this essay is excerpted from the Nobel Lecture by Hugo Theorell (*The Nature and Mode of Action of Oxidation Enzymes*, The Nobel Foundation, Stockholm, 1955).

sugar or iron the solutions gave negative results. But the "catalytic" enzyme effect was still there. Willstätter drew the conclusion that the enzymes could contain neither protein, carbohydrate nor iron and that they did not belong to any known class of chemical substance at all, and he was even inclined to believe that the effects of the enzymes derived from a new natural force, thus the view that 90 years earlier Berzelius had dismissed as improbable.

That same year, through an irony of fate, the American researcher J. B. Sumner published work in which he claimed to have crystallized in pure form an enzyme, urease, from "jack-beans". It splits urea into carbon dioxide and ammonia. Sumner had got his crystals in rather considerable quantities with the help of much simpler methods than those applied by Willstätter in purification experiments on other enzymes. Sumner's crystals consisted of colourless protein. In the ensuing years J. H. Northrop and his collaborators crystallized out a further three enzyme preparations, pepsin, trypsin and chymotrypsin, like urease, hydrolytic enzymes that split linkages by introducing water.

If these discoveries had been undisputed from the outset it would probably not have been 20 years before Sumner, together with Northrop and Stanley, received a Nobel prize. But it was not so easy to show that the beautiful protein crystals really were the enzymes themselves and not merely an inactive vehicle for the actual enzymes. Both Sumner and Northrop adduced many probable proofs that what they had produced really were pure enzymes, but no absolutely conclusive experiment could be brought forward, and as a matter of fact this was at that time probably not possible, for the simple reason that their preparation appeared to consist of only colourless protein. At that time, and even today, for the rest, the methods of separation and analysis were scarcely sufficiently refined definitely to exclude the occurrence of small quantities of impurities in a protein preparation. From many quarters, accordingly, objections were raised to Sumner's and Northrop's results, and for obvious reasons especially the Willstätter school made itself heard in this connection.

ALCOHOL DEHYDROGENASE IN 1955

Alcohol dehydrogenases occur in both the animal and the vegetable kingdoms, e.g. in liver, in yeast and in, for example, peas. They are colourless proteins which together with DPN may either oxidize alcohol to aldehyde, as occurs chiefly in the liver, or conversely reduce aldehyde to alcohol, as occurs in yeast.

The yeast enzyme was crystallized by Negelein and Wulff (1936) in Warburg's institute, the liver enzyme (from horse liver) by Bonnichsen and Wassén at our institute in Stockholm in 1948.

These two enzymes have come to play a certain general rôle in biochemistry on account of the fact that it has been possible to investigate their kinetics more accurately than is the case with other enzyme systems. The liver enzyme especially we have on repeated occasions studied with particular thoroughness, since especially favourable experimental conditions here presented themselves. For all reactions with the DPN-system, it is possible to follow the reaction $DPN^+ + 2H \leftrightharpoons DPNH + H^+$ spectrophotometrically, since DPNH has an absorption-band in the more long-wave ultra-violet region, at 340 mμ, and thousands of such experiments have been performed all over the world. A couple of years ago, moreover, we began to apply our fluorescence method, which is based on the fact that DPNH but not DPN fluoresces, even if considerably more weakly than the flavins. As regards the liver enzyme there is a further effect, which proved extremely useful for certain spectrophotometrical determinations of reaction speeds; together with Bonnichsen I found in 1950 that the 340 mμ band of the reduced coenzyme was displaced, on combination with liver alcohol dehydrogenase, to 325 mμ, and together with Britton Chance we were thus able with the help of his extremely refined rapid spectrophotometric methods to determine the velocity constant for this very rapid reaction.

I shall not go into further details, but simply point out that extremely complicated reactions result from the fact that we are here dealing with a three-body problem containing the enzyme protein, the coenzyme and the substrate, where, furthermore, both the coenzyme and the substrate occur in both oxidized and reduced forms.

A more or less complete system may be written thus that only after nine steps does the enzyme become free to begin a new cycle; the net result is that the alcohol has given two of its hydrogen atoms to the coenzyme. Even this simplified scheme means that one must determine eighteen speed constants, two for each part-reaction, which is of course a formidable task. We have succeeded, however, in determining some of them as regards the yeast enzyme.

The kinetics of the liver enzyme is quite other than that of the yeast enzyme. Here we have to do with a simpler reaction process which can be expressed with only three equations and six velocity constants.

We have here been able to determine all the six constants at different degrees of acidity and with different salt concentrations, so that the reaction velocities of this enzyme system are probably at present the best known of all (Theorell, Bonnichsen, Chance, Nygaard).

The differences between the yeast and liver enzymes indicated here explain why the yeast enzyme produces alcohol from aldehyde, while the liver enzyme does the contrary.

It is a curious whim of nature that the same coenzyme which in the yeast makes alcohol by attaching hydrogen to aldehyde also occurs in the liver to remove from alcohol the same hydrogen, so that the alcohol becomes aldehyde again, which is then oxidized further.

When we had studied the kinetics of the alcohol dehydrogenases it was a simple matter to use these to determine alcohol quantitatively, e.g. in blood samples. This so-called "ADH"-method is about as accurate as Widmark's method, only more sensitive and above all practically specific for ethyl alcohol. It is now legally introduced in forensic chemical practice in Sweden and in West Germany.

ALCOHOL DEHYDROGENASES REVISITED TWENTY YEARS LATER

The *horse liver alcohol dehydrogenase*, crystallized by my collaborators Bonnichsen and Wassén in 1948, has been subject to much work the last 20 years, both in my lab and others. Its amino acid sequence (374 residues per subunit = $\frac{1}{2}$ molecule) was cleared up by my young collaborator Hans Jörnvall working with I. J. Harris in Cambridge from 1967 and then with us in Stockholm. Alcohol dehydrogenases from other sources were also studied in our and many other peoples laboratories. It is at present one of the most intensively studied enzymes in the world.

Of special interest it seems to be that its three-dimensional structure has been very intensely studied by X-ray crystallography in Uppsala by C. Brändén and his group, working on enzyme material purified by us in Stockholm. By now the whole structure is cleared up in great detail.

On Fig. 1 you can see the two identical peptide chains of the dimeric molecule, each with 374 amino acid residues. One of the monomers is coloured, the other not. We can see different regions of helices, pleated sheets and irregular folding. Each monomer contains two Zn atoms. One, the bigger, lower to the left is the "catalytic zinc", which binds the substrate and keeps it in optimal position to react with the nicotinamide in the coenzyme. This zinc has three bonds to the enzyme, cysteine 46-S, cysteine 174 and histidine 67-N. The fourth bond of the zinc couples the oxygen atom of the substrate, alcohol.

The other zinc helps to maintain the three-dimensional structure of the enzyme molecule by being bound to four sulphur atoms in cysteine residues Nos. 97, 100, 103 and 111.

Figure 2 shows the whole (upper right) subunit of a molecule and, lower to the left darker coloured, a bit of the other subunit.

The dark-coloured part in the middle indicates in the upper part the coenzyme binding site: the adenine moiety at "Au" up to the left, the phosphate at "Pt", the nicotinamide part at 1 in the middle. The dark sack in the lower part is the substrate pocket. The walls around this pocket contain many amino acid residues with lipophilic side chains, resulting in the enzyme having high affinity for both substrates and competitive inhibitors rich in alkyl groups.

Figure 3 illustrates how the substrate is held in its pocket by the "catalytic Zn" in an optimal position for transferring a H-atom to C_4 in the coenzymes nicotinamide ring. You may ask what is the practical advantage of finding out all these facts about an enzymes three-dimensional structure? Well, let us take one example. From the shape and dimensions of the substrate–coenzyme pocket you can predict which substrate-competitive inhibitors and which substrates can find space in it. Such work is now going on by us in collaboration with Brändén's crystallographic group in Uppsala.

CYTOCHROME *c* IN 1955

One of my early interests was the purification of a hemin proteid, viz. cytochrome *c*, one of the "histohematins" or "myohematins" observed by the Irishman MacMunn in his home-made spectroscope at the end of the 1880s. MacMunn's "hematins", after a period of obscure existence in petit type in larger textbooks, had been brought out into the light again by David Keilin in Cambridge in 1925. In 1936 we had obtained the cytochrome approximately 80% pure, and in 1939 close to 100%.

It is a beautiful red, iron-porphyrin-containing protein which functions as a link in the chain of the cell-respiration enzymes, the iron atom now taking up and now giving off an electron, and the iron thus alternating valency between the 3-valent ferri and the 2-valent ferro stages. It is a very pleasant substance to work with, not merely because it is lovely to look at, but also because it is uncommonly stable and durable. From 100 kg heart-meat of horse one can produce 3–4 g of pure cytochrome *c*. The molecule weighs about 12,000 and contains one mol iron porphyrin per mol.

Experiment—Two cuvettes each contain a solution of ferricytochrome *c*. The colour is blood-red. To the one are added some grains of sodium hydrosulphite: the colour is changed to violet-red (ferrocytochrome). Oxygen is now bubbled through the ferrocytochrome-solution: no visible change occurs. The ferrocytochrome can thus not be oxidized by oxygen. A small amount of cytochrome oxidase is now added: the ferricytochrome colour returns.

From this experiment we can draw the conclusion that reduced cytochrome *c* cannot react with molecular oxygen. In a chain of oxidation enzymes it will thus not be able to be

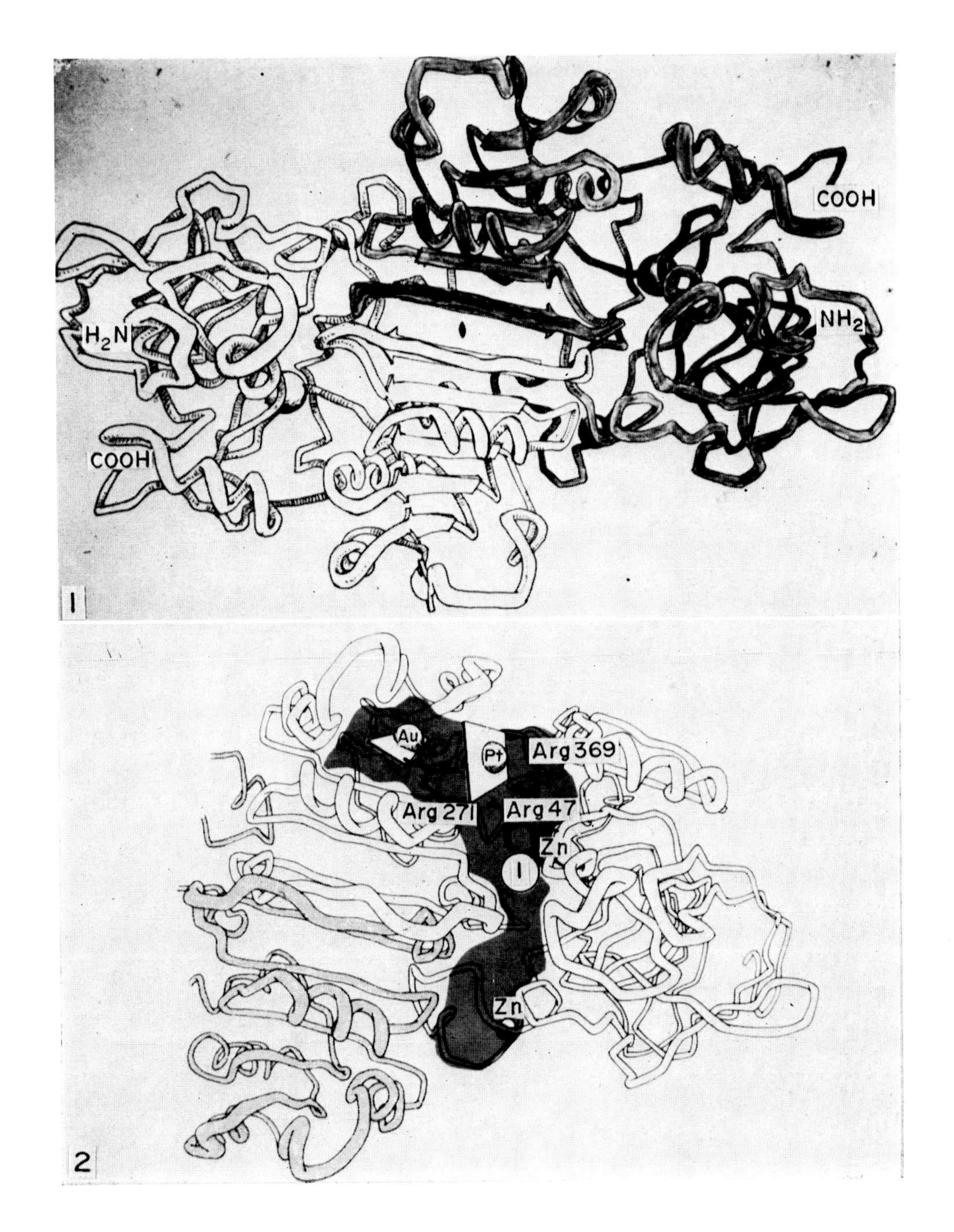

FIG. 1. Horse liver alcohol dehydrogenase. Peptide chain folding in the two subunits.
FIG. 2. Coenzyme and substrate pocket in the ADH molecule (explanation see text).

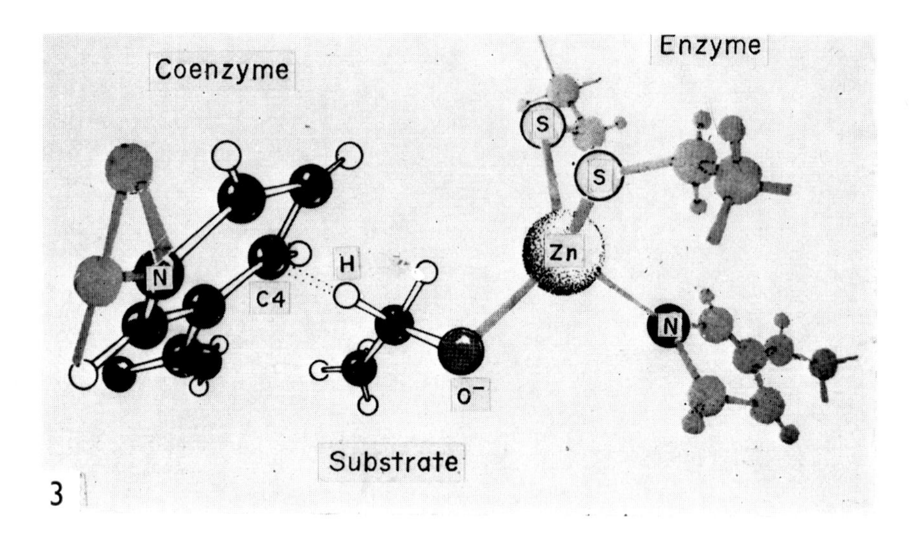

FIG. 3. Coupling of substrate between the catalytic Zn and the nicotinic acid amide of coenzyme.

next to the oxygen. The incapacity of cytochrome to react with oxygen was a striking fact that required an explanation. Another peculiarity was the extremely firm linkage between the red hemin pigment and the protein part; in contradistinction to the majority of other hemin proteids, the pigment cannot be split off by the addition of acetone acidified with hydrochloric acid. Further, there was a displacement of the light-absorption bands which indicated that the two unsaturated vinyl groups occurring in ordinary protohemin were saturated in the hematin of the cytochrome. In 1938 we succeeded in showing that the porphyrin part of the cytochrome was linked to the protein by means of two sulphur bridges from cysteine residues in the protein of the porphyrin in such a way that the vinyl groups were saturated and were converted to α-thioethyl groups. The firmness of the linkage and the displacement of the spectral bands were herewith explained. This was the first time that it had been possible to show the nature of chemical linkages between a "prosthetic" group (in this case iron porphyrin) and the protein part in an enzyme. Karl-Gustav Paul has since found an elegant method whereby to split the sulphur bridges with silver salts, and he has with organic chemical methods conclusively confirmed the constitution in this respect.

The light-absorption bands of the cytochrome showed that it is a so-called hemochromogen, which means that two as a rule nitrogen-containing groups are linked to the iron, in addition to the four pyrrol–nitrogen atoms in the porphyrin. From magnetic measurements that I made at Linus Pauling's institute in Pasadena and from amino acid analyses, titration curves and spectrophotometry together with Å. Åkeson it emerged (1941) that the nitrogen-containing, hemochromogen-forming groups in cytochrome *c* were histidine residues, or to be more specific, their imidazole groups.

Recently we have got a bit farther. Tuppy and Bodo in Vienna began last year with Sanger's method to elucidate the amino acid sequence in the hemin-containing peptide fragment that one obtains with the proteolytic breaking down of cytochrome *c*, and succeeded in determining the sequence of the amino acids nearest the hemin. The experiments were continued and supplemented by Tuppy, Paléus and Ehrenberg at our institute in Stockholm with the following result (Fig. 4):

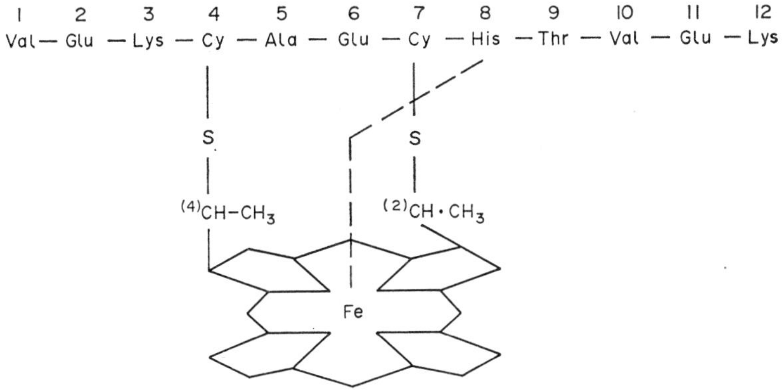

1 2 3 4 5 6 7 8 9 10 11 12
Val— Glu —Lys — Cy — Ala — Glu — Cy — His —Thr — Val — Glu — Lys

S S
(4)CH–CH3 (2)CH·CH3
Fe

FIG. 4.

The peptide chain 1–12 ("Val" = the amino acid valine, "Glu" = glutamine, "Lys" = lysine and so forth) is by means of two cysteine-S-bridges and a linkage histidine-Fe linked to the hemin.

When in 1954 Linus Pauling delivered his Nobel lecture in Stockholm he showed a new kind of models for the study of the steric configuration of peptide chains, which as we know may form helices or "pleated sheets" of various kinds. It struck me then that it would be extremely interesting to study the question as to which of these possibilities might be compatible with the sulphur bridges to the hemin part and with the linkage of nitrogen-containing groups to the iron. Pauling was kind enough to make me a present of his peptide-model pieces, which I shall show presently. This is thus the second time they figure in a Nobel lecture.

Anders Ehrenberg and I now made a hemin model on the same scale as the peptide pieces and constructed models of hemin peptides with every conceivable variant of hydrogen bonding. It proved that many variants could be definitely excluded on steric grounds, and others were improbable for other reasons. Of the original at least twenty alternatives finally only one remained—a left-twisting α-helix with the cysteine residue no. 4 linked to the porphyrin side chain in 4-position, and cysteine no. 7 to the side chain in 2-position. The imidazole residue fitted exactly to linkage with the iron atom. The peptide spiral becomes parallel with the plane of the hemin disc (Fig. 5).

I think it may be said that it was of considerable interest to have Pauling's and Corey's most important spiral confirmed with purely chemical methods, which in our case of course was possible thanks to the unique circumstance that we had a short peptide linked at no fewer than three places to a rigid structure, the hemin. After we had sent this work to the printer's there arrived from Arndt and Riley in England an X-ray-crystallographic confirmation that cytochrome *c* contains left-twisting α-helices (Fig. 6).

Through calculations on the basis of the known partial specific volume of the cytochrome we now consider it extremely probable that the hemin plate in cytochrome *c* is surrounded by peptide spirals on all sides in such a way that the hemin iron is entirely screened off from contact with oxygen; here is the explanation of our experiment in which we were unable to oxidize reduced cytochrome *c* with oxygen-gas. The oxygen simply cannot get at the iron atom. There is, on the other hand, a possibility for electrons to pass in and out in the iron atom via the imidazole groups.

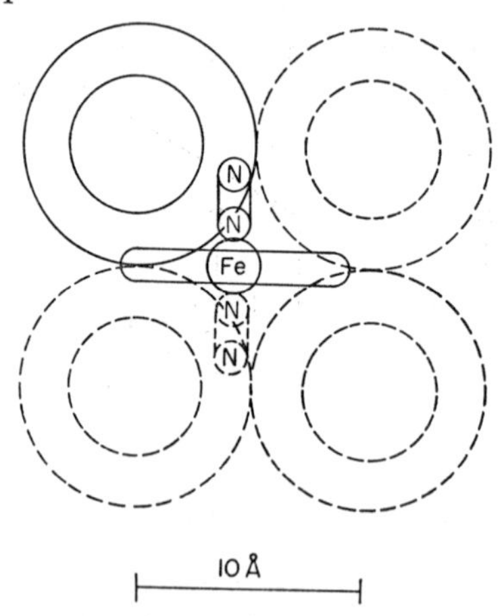

FIG. 7. Hypothetical section through a cytochrome-*c* molecule. Whole-drawn lines: the hemopeptide. Dashed lines: parts of the natural molecule split off with pepsin. The region between the outer and inner circles is taken up by the lateral chains of the amino acids. The four peptide chains surround the iron atom (Fe), making it inaccessible to oxygen. (*Acta Chem. Scand.* (1955) **9**, 1193.)

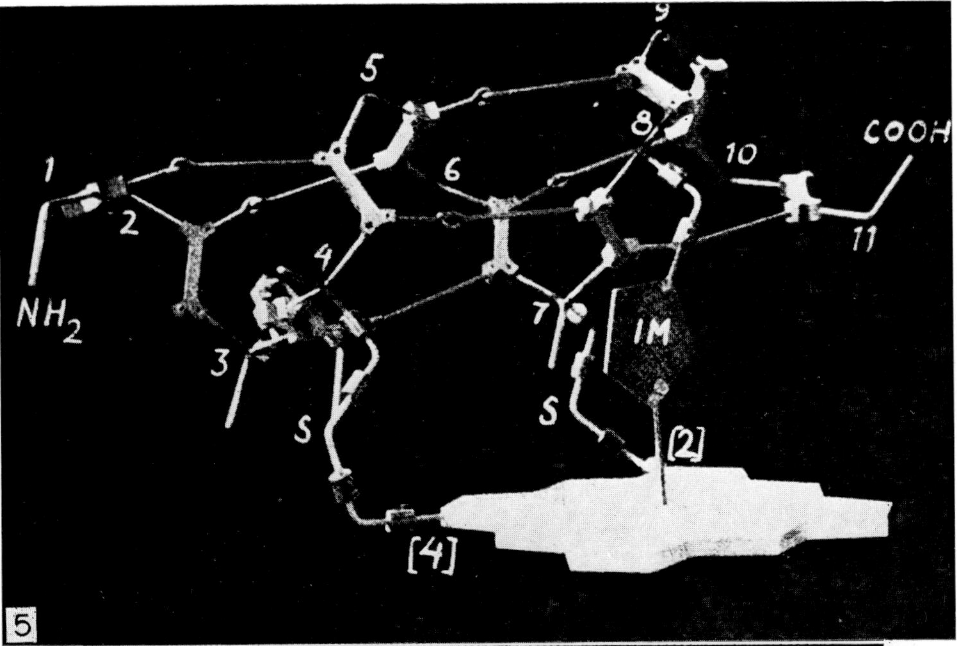

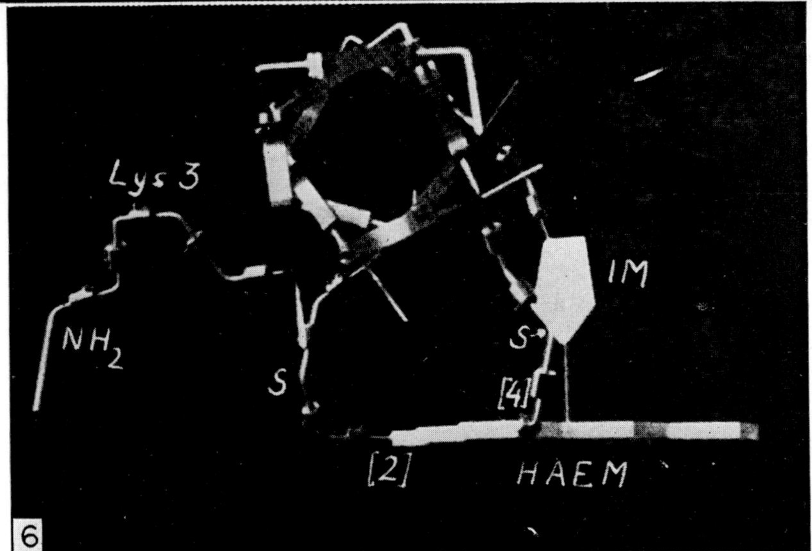

FIG. 5. Steric model of the hemopeptide remaining after pepsin-digestion of cytochrome *c*. constructed of metal parts representing the atomic distances and valency directions. In the figure the peptide chain is disposed by means of hydrogen bonds as a left-twisting α-helix, seen at right angles to the longitudinal axis. Both the suplhur bridges (S) and the imidazole group (IM) of the histidine then fit their correct linkage-positions, and the peptide chain becomes parallel with the hemin plate (the light, polygonal metal plate at the bottom of the picture). (*Acta Chem. Scand.* (1955) **9**, 1193.)

FIG. 6. The model seen in the longitudinal direction of the α-spiral. For the sake of survey-ability the side chains have been fitted only to the amino acids Lys (3), Cy (4), Cy (7) and His (8). (*Acta Chem. Scand.* (1955) **95**, 1193.)

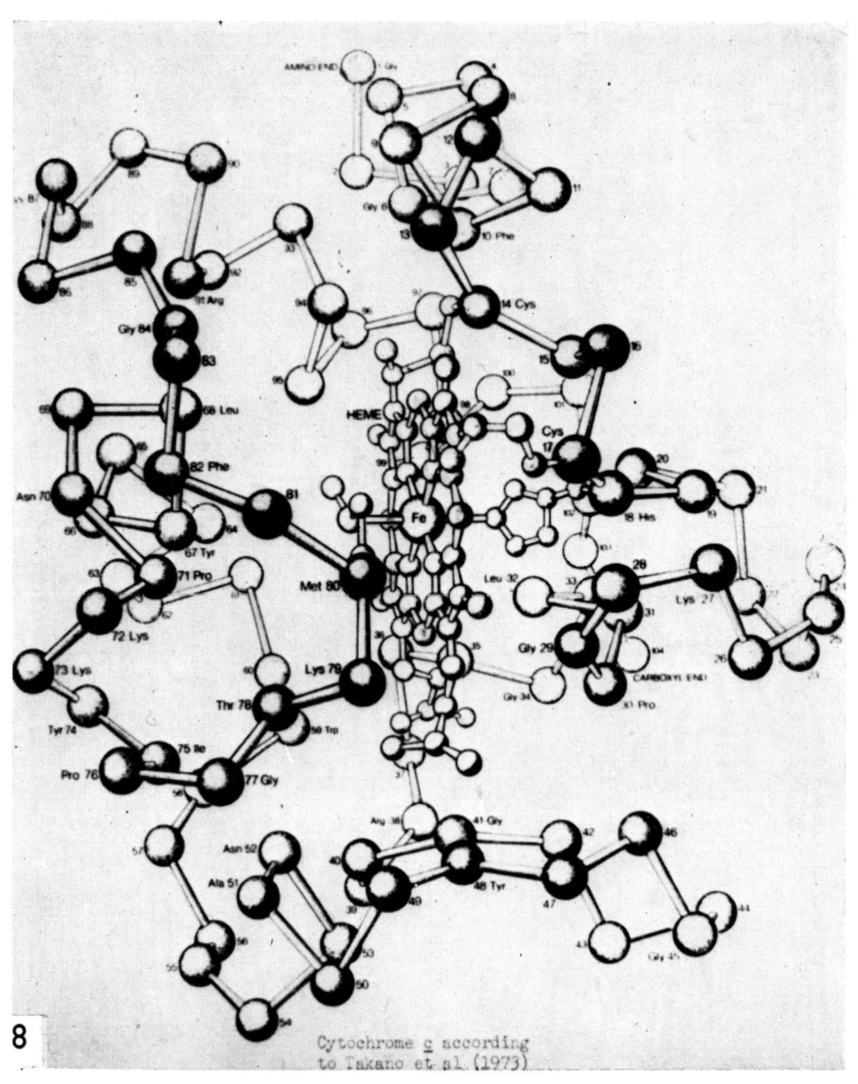

FIG. 8. Model of cytochrome *c* molecule (Dickerson and Jimkovich (1975) *The Enzymes*, **11**, 397). In the middle hemin disc seen from the edge. Upwards to the right: animo acid residues 1–22.

One-sixth of the entire steric structure of the cytochrome molecule is herewith elucidated, and we glimpse further possibilities of gradually elucidating the rest. It strikes us as interesting that even at this stage the special mode of reacting of the cytochrome is beginning to be understood from what we know of its chemical constitution.

CYTOCHROME *c* REVISITED TWENTY YEARS LATER

In 1955 we found that the tryptic hemopeptide had the cysteine residues, now numbered 14 and 17, coupled to the α-carbon atoms of the vinyl side chains 4- and 2- of the hemin, and imidazol of His 18 bound to the iron.

These restrictions were found to be compatible with all these residues forming according to Linus Pauling, a left-handed α-helix. Later it was found that most, if not all proteins, are right-handed. Ehrenberg in our laboratory built a right-handed cytochrome *c* 11-22 hemopeptide and found it just as possible as the left-handed.

The next question was: is total, or even partial hydrogen bonding in the residue region 14 to 18 necessary for Cys 14 and 17 binding to the hemin side chains, 4 and 2, His 18 to Fe?

The X-ray crystallographic work of Dickerson and his collaborators has contributed greatly to elucidate this problem, though one essential point still remains obscure: is the oxidoreduction accompanied by any structure change? But another problem they have solved: the heme-coupling residues His 18, Cys 17 and Cys 14 are situated in a some what extended right-handed helix, which bends at 18 and goes as right-handed α-helix down to 1 (Fig. 8).

Our suggestion 20 years ago that the whole cytochrome *c* molecule could consist of four symmetrically arranged α-helices was obviously only partly correct; as Warburg once said to me: "Bei der Nachkontrolle gehen die schönsten Entdeckungen verloren"—"the most beautiful discoveries get lost in the further control". Of course we have to be glad that the foundation for our theory—the bonds of Cys 14 and 17, and His 18 to the rigid hemin disc remain undisputed (Fig. 9), but this is obviously no reason to squeeze all the over 100 amino acid residues into α-helix conformation.

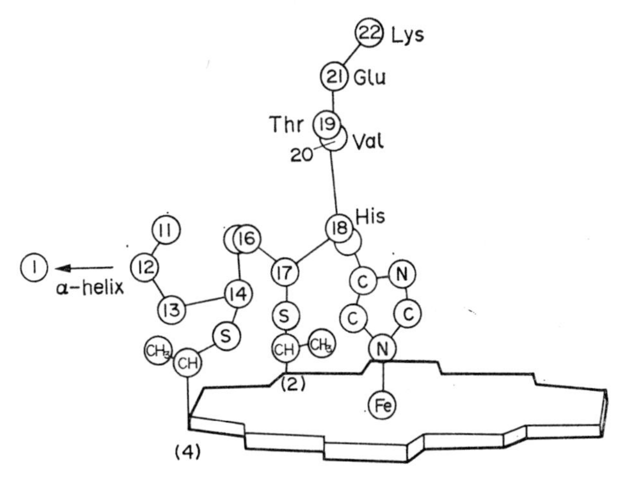

FIG. 9. Peptide 11–22 with cysteines 14 and 17 bound to hemins vinyl α-C_4 and C_2, histidine 18 to Fe. (Note: 14, 17 and 18 are numbered 4, 7 and 8 in our old model, Fig. 4.)

UNRAVELLING THE PENTOSE PHOSPHATE PATHWAY

B. L. HORECKER

Roche Institute of Molecular Biology, Nutley, New Jersey 07110

In 1931 Otto Warburg discovered that the oxidation of glucose 6-phosphate in red blood cells required both an enzyme and a heat stable coenzyme, and he later referred to the latter as the "Wasserstoffubertragendes Co-Ferment",[1] to distinguish it from cozymase, the coenzyme of yeast fermentation, which had been discovered many years earlier by Sir Arthur Harden. He found that the functional group in both coenzymes was nicotinamide, which acted as acceptor for the protons removed from the substrate. In addition to nicotinamide, each coenzyme also contained one adenine and two ribosyl residues, but they differed in their phosphorus content. These observations led him to name the cozymase "diphosphopyridine nucleotide" (DPN), and the hydrogen-transferring coenzyme "triphosphopyridine nucleotide" (TPN).*

Thus from the time of their discovery the two coenzymes were considered to be concerned with fundamentally different processes: DPN was regarded as the coenzyme of fermentation (in 1918 Otto Meyerhof[2] showed that the same coenzyme was required for lactic acid production in muscle extracts), whereas TPN was considered to be the coenzyme of respiration. It was expected, therefore, that a link would eventually be found between TPNH and the cytochrome system (Fig. 1). David Keilin and E. F. Hartree[3] had shown

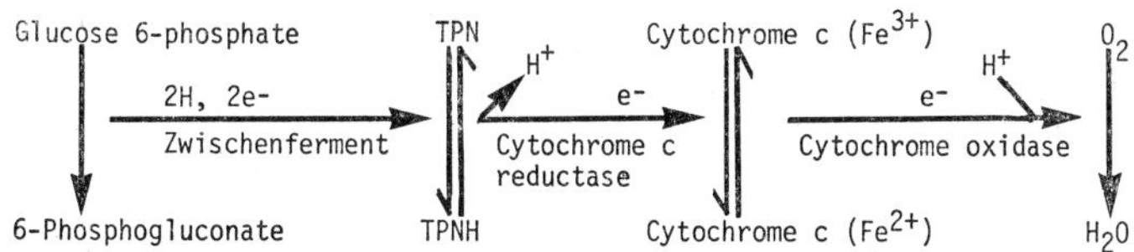

FIG. 1. The proposed "respiratory" pathway for the oxidation of glucose 6-phosphate.

that cytochrome *c* was oxidized very rapidly by O_2 in the presence of cytochrome oxidase, but no one had described an enzyme that would catalyze the reduction of cytochrome *c* by TPNH or DPNH. In 1939, while I was a graduate student in T. R. Hogness' laboratory at the University of Chicago, Erwin Haas arrived from Warburg's laboratory in Dahlem, and took up the problem of isolating this missing link that we thought would complete the hydrogen and electron transport system from substrate to oxygen. "Taking up the problem" meant first isolating TPN and DPN from sheep liver, and hexosemonophosphate from yeast, and purifying Zwischenferment (see below) from yeast and cytochrome *c* from horse heart. I was fortunate to be invited to work with Haas, who was a meticulous scientist and a great teacher, and during the following 18 months we succeeded in isolating

* In this historical account, I will keep the original names for these pyridine nucleotides. In 1964 the Enzyme Nomenclature Commission proposed the new names "nicotinamide adenine dinucleotide" (NAD) and "nicotinamide adenine dinucleotide phosphate" (NADP) and the new names, although less convenient and evocative, have unfortunately tended to displace the names proposed by the original workers.

the enzyme from brewer's yeast, showed it to be a flavoprotein, and named it TPNH-cytochrome *c* reductase.[4] These results were reported at the first Enzyme Symposium in Wisconsin in 1941, and it was there that I first met Severo Ochoa, who had recently come from Rudolph Peters' laboratory in Oxford to work with Carl Cori in St. Louis.

In our assay for TPNH-cytochrome *c* reductase, we used glucose 6-phosphate and Zwischenferment (yeast glucose 6-phosphate dehydrogenase) to generate reduced TPN, and thus began my interest in the oxidative pathway. This also introduced me to the powerful tool of absorption spectrophotometry, 3 years before the Beckman spectrophotometer. Warburg, who was probably the first to build an instrument for enzyme studies, had discovered the absorption bands of the reduced coenzymes, but he was devoted to his manometric techniques, and it remained for Haas and myself to apply the new technique to the routine assay of an enzyme-catalyzed reaction, which we measured by following the rate of appearance of the reduced band of cytochrome *c*.

During the Second World War I was assigned to work on a variety of war-related programs, first at the U.S. Civil Service Commission, and later at the National Institutes of Health, and it was not until 1945 that I was able to return to enzymology. One of my first goals was to isolate the mammalian TPNH-cytochrome *c* reductase, and the rat liver enzyme was isolated in 1950. By then, however, the earlier view of the physiological roles of DPN and TPN had begun to change, largely because of the discovery of the respiratory function of mitochondria, and especially following the demonstration by Albert Lehninger[5] that the coenzyme of oxidative phosphorylation was DPN. TPN has now been established as the coenzyme for reductive biosynthesis and for drug metabolism, and the physiological acceptor for TPNH-cytochrome *c* reductase has been shown to be cytochrome P450, rather than cytochrome *c*.

Before Lehninger's demonstration of the role of DPN in oxidative phosphorylation, it was generally thought that the oxidation of glucose 6-phosphate was the first step in a pathway for the aerobic metabolism of glucose, as opposed to its fermentation by the Embden–Meyerhof pathway. Metabolism via these two pathways was thought to be controlled by oxygen, based on Pasteur's observation that oxygen inhibited the formation of lactic acid,[6] a phenomenon that came to be known as the "Pasteur Effect". This notion of independent pathways for glycolysis and respiration was reinforced by Einar Lundsgaard's discovery[7] that iodoacetate, which was an effective inhibitor of glycolysis, had very little effect on respiration. In addition, phosphofructokinase was found by Engelhardt and Barkach to be highly sensitive to oxidizing agents, including oxygen, and this, together with the earlier observation of Lundsgaard, led them to propose the "hexosemonophosphate shunt" for the aerobic utilization of glucose.[8]

It was natural, therefore, that many investigators were attracted to the study of glucose 6-phosphate oxidation, and by the time I returned to this problem in 1949 considerable progress had been made. Based on the pioneering studies of Warburg, Frank Dickens and Fritz Lipmann, we knew (1) that the first product of the oxidation of glucose 6-phosphate was 6-phosphogluconate and (2) that with crude fractions from yeast this product was further oxidized, with TPN again as the coenzyme, to yield CO_2 and a mixture of phosphorylated products that appeared to include pentose phosphate and triose phosphate.[9] Lipmann, then at the Carlsburg laboratories in Copenhagen, suggested that a series of oxidations and decarboxylations might occur in which hexose monophosphate would be converted to C_5, C_4 and finally C_3 sugar. Indeed, evidence for such a process was provided by Dickens, who showed that ribose 5-phosphate was rapidly metabolized by yeast extracts. However, arabinose 5-phosphate, the pentose that would be expected to arise from the oxidative decarboxylation of 6-phosphogluconate, was not formed, and this posed an interesting problem as to the true nature of the "pentose" ester formed from 6-phosphogluconate.

The final proof that ribose 5-phosphate was formed from 6-phosphogluconate came from my laboratory in 1951. With a partially purified preparation of 6-phosphogluconate dehydrogenase from yeast, and with the oxidation of this substrate coupled to the reduction of pyruvate by lactic dehydrogenase,† Polly Smyrniotis, Hans Klenow and I were able to obtain stoichiometric yields of pentose phosphate from 6-phosphogluconate, and from the reaction mixture we crystallized and identified the benzylphenylhydrazone of ribose. During the early phase of the reaction, a precursor of ribose phosphate was detected and separated from ribose 5-phosphate by chromatography on Dowex 1-formate. This precursor was identified as the ketopentose ester, D-ribulose 5-phosphate, by hydrolysis and crystallization of the *o*-nitrophenylhydrazone. Thus the enigma of the inversion of C-3 of glucose to yield ribose 5-phosphate as the ultimate product was solved; oxidation of 6-phosphogluconate at the 3-position and decarboxylation of a 3-keto intermediate would yield D-ribulose 5-phosphate, which was then converted to ribose 5-phosphate by a pentose phosphate isomerase, traces of which were found to be present in our purified yeast phosphogluconate dehydrogenase preparations.

A few years earlier Ochoa and his coworkers had described two similar oxidative decarboxylations, of isocitrate to α-ketoglutarate and of malate to pyruvate, both specific for TPN, and demonstrated the reversibility of these reactions.[11] Inspired by Ochoa's observations, we carried out a similar experiment with 6-phosphogluconate dehydrogenase and found that in the presence of ribulose 5-phosphate and TPNH, CO_2 was fixed into 6-phosphogluconate in significant quantities. This led us to wonder whether this might not be the long-sought mechanism for CO_2 fixation in photosynthesis. When we found that phosphogluconolactone was also readily reduced to glucose 6-phosphate, the theory became even more attractive, particularly since Wolf Vishniac and Ochoa[12] had demonstrated TPNH formation and coupled CO_2 fixation with illuminated chloroplast fragments. However, the theory was short-lived, because at about that time Calvin's group identified phosphoglyceric acid as the primary product of CO_2 fixation in photosynthesis.‡

Having defined the oxidative steps for hexosemonophosphate metabolism in yeast, we asked whether these reactions could also be demonstrated in animal tissues. E. J. Seegmiller, who joined my group in 1950, used our coupled system to study the oxidation of 6-phosphogluconate with rat liver extracts. Not only did he identify ribulose and ribose phosphates as products of the reaction, but he also observed that with continued incubation the pentose phosphates disappeared, and in their place hexosemonophosphate accumulated.[13] This was the first indication that we were dealing with a cyclic process, in which pentose phosphate formed by the oxidation of hexosemonophosphate could be converted

† At the time there was no convenient way of regenerating TPN from TPNH, so that the coenzyme could be used catalytically, except with highly reactive substances such as phenazine methosulfate. The suggestion to use lactic dehydrogenase came from Arthur Kornberg; while he was in Ochoa's laboratory they had discovered that this enzyme would utilize not only DPN, but also TPN, although the rate was slower.

‡ Years earlier, Rubin, Kamen and Hassid[12] had shown that an early product of photosynthesis in *Chlorella* contained a carboxyl group and at least one hydroxy group; either phosphoglyceric acid or phosphogluconic acid would have satisfied the properties of their compound.

back to hexosemonophosphate in a non-reductive process. Many years before, Zacharias Dische[13] had made the important observation that chick red cell hemolysates could convert adenosine to triose and hexose phosphates, and Fritz Schlenk had shown that rabbit liver homogenates would form glucose 6-phosphate from ribose 5-phosphate. The appearance of triose phosphate suggested an obvious mechanism for the formation of hexose phosphate: pentose phosphate was cleaved to yield triose phosphate and a C_2-compound with two molecules of the former condensing to form fructose diphosphate. However, Dische later showed that fructose diphosphate was not an intermediate, since it was not converted to hexosemonophosphate in his red cell homogenates, and in addition Gertrude Glock, in Dickens' laboratory, reported that the yield of hexose monophosphate was far in excess of the amount that could be formed from triose phosphate alone.[15] We were beginning to realize that the mechanism of pentose phosphate breakdown was not a simple C_2–C_3 cleavage.

From my laboratory and the laboratories of Ef Racker, Frank Dickens and Melvin Calvin, the years that followed witnessed a series of parallel and often highly synergistic discoveries on the nature of the pentose phosphate pathway and the path of carbon in photosynthesis. Andrew Benson and others in Calvin's laboratory, had shown that phosphate esters of ribulose and sedoheptulose were early products of CO_2 fixation in photosynthesis,[16] and the immediate precursor of phosphoglyceric acid, and therefore the primary CO_2 acceptor, appeared to be ribulose diphosphate. The major problems became: (1) to find the enzyme or enzymes that catalyzed the formation of phosphoglyceric acid from ribulose diphosphate; and (2) to define the reactions leading to the synthesis of ribulose diphosphate from triose and hexose phosphates.

A most important clue to the nature of the steps between pentose phosphate and hexosemonophosphate, and thus to the role of the pentose phosphate pathway in photosynthesis, came from our discovery in 1953 of sedoheptulose 7-phosphate as the first product formed from pentose phosphate. The enzyme transketolase had been purified from rat liver and spinach in my laboratory and crystallized from yeast by Racker and his coworkers and the two laboratories simultaneously discovered that this enzyme contained thiamine pyrophosphate as its functional group.[9] Isotope studies in my laboratory showed that sedoheptulose was formed by the transfer of a C_2 group ("active glycolaldehyde") from one molecule of pentose phosphate to another, and that the reaction was fully reversible; thus sedoheptulose 7-phosphate was also a C_2-donor. In addition, Racker's laboratory made the important finding that fructose 6-phosphate would also yield active glycolaldehyde, and Arturo Bonsignore and his coworkers discovered that rat liver extracts catalyzed the rapid non-oxidative conversion of hexose phosphate to sedoheptulose phosphate.[18]

In the studies described above, we had employed ribose 5-phosphate as the substrate, but the requirement for pentose phosphate isomerase in addition to transketolase suggested that the true substrate was ribulose 5-phosphate. However, this hypothesis was incompatible with the observations that sedoheptulose and fructose phosphates were substrates, since ribulose 5-phosphate possessed the opposite configuration at the critical C-3 position. The answer to this puzzle was provided by the discovery by Gilbert Ashwell and Jean Hickman[19] at the National Institutes of Health of a new pentose ester, D-xylulose 5-phosphate, which Paul Stumpf in my laboratory also obtained as an intermediate in pentose fermentation in *Lactobacillus plantarum*.[20] Jerard Hurwitz, then a post-doctoral fellow in my laboratory at the National Institutes of Health, showed almost simultaneously with workers in Racker's and Dickens' laboratories that the substrate for transketolase was xylulose 5-phosphate, not ribulose 5-phosphate, and that an additional enzyme, ketopentose 3-epimerase, was required for the conversion of ribulose 5-phosphate to sedoheptulose 7-phosphate.[21]

We had already shown that sedoheptulose 7-phosphate was an intermediate in the conversion of pentose phosphate to hexosemonophosphate in crude liver extracts, since it accumulated early in the reaction and then disappeared, coincident with the formation of hexose phosphate.[22] The enzyme that catalyzed the interconversion of sedoheptulose 7-phosphate and fructose 6-phosphate was discovered in my laboratory in 1953, and with Paul Marks and Howard Hiatt, we showed that it catalyzed a C_3-transfer from the substrate to a suitable aldehyde donor.[23] This led to the discovery of another new phosphate ester, erythrose 4-phosphate, formed in the C_3-transfer from sedoheptulose 7-phosphate to glyceraldehyde 3-phosphate. Finally, erythrose 4-phosphate could be shown to be the acceptor for a second reaction catalyzed by transketolase, with xylulose 5-phosphate as the donor. All of the steps in the pentose phosphate pathway had now been elucidated (Fig. 2). Erythrose 4-phosphate was also shown to condense with glyceraldehyde 3-phosphate to

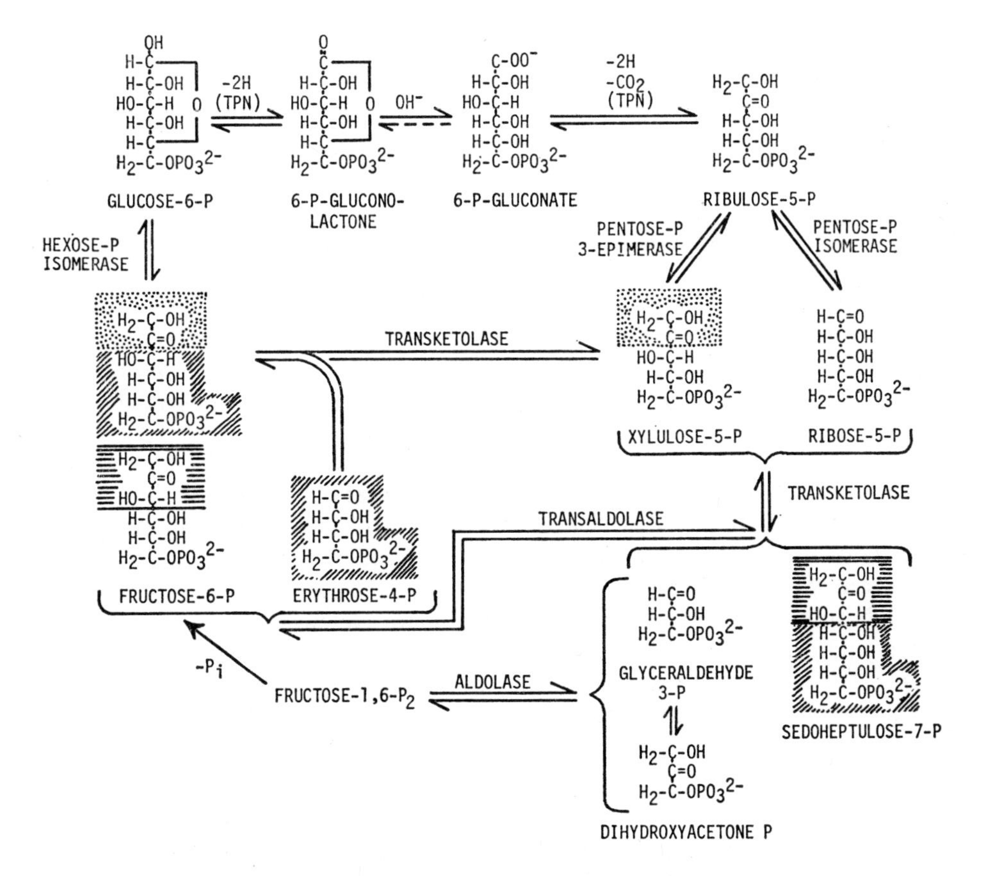

FIG. 2. Reactions of the pentose phosphate pathway.

form sedoheptulose 1,7-diphosphate, which can be hydrolyzed to the 7-phosphate by a specific phosphatase.

In photosynthesis, phosphoglyceric acid is reduced to glyceraldehyde 3-phosphate and the latter converted to fructose 1,6-diphosphate by the action of triose phosphate isomerase and aldolase (Fig. 3). An additional step, catalyzed by fructose diphosphatase, was

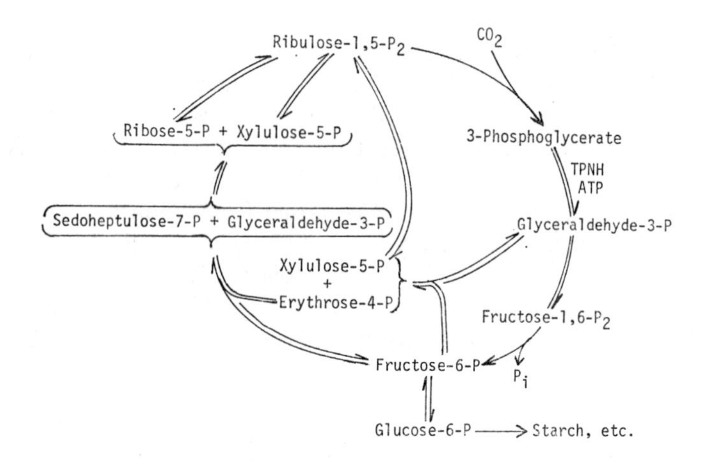

FIG. 3. The role of the pentose phosphate pathway in photosynthesis. One-sixth of the fructose 1,6-diphosphate formed is assimilated and five-sixths is converted back to ribulose 1,5-diphosphate. An alternative reaction to the hydrolysis of fructose 1,6-diphosphate is the condensation of glyceraldehyde 3-phosphate and erythrose 4-phosphate to form sedoheptulose 1,7-diphosphate, which is hydrolyzed by a specific phosphatase. However, this would not account for the formation of starch and other assimilation products.

required for the synthesis of sucrose or starch, and also for regenerating ribulose phosphate by the reversible pentose phosphate pathway. The synthesis of ribulose 1,5-diphosphate by a specific kinase was first described in my laboratory by Arthur Weissbach and Polly Smyrniotis,[24] and with Jerard Hurwitz they also discovered ribulose diphosphate carboxylase, the enzyme that catalyzes the fixation of CO_2 into 3-phosphoglyceric acid,[25] simultaneously with Rodney Quayle, Clint Fuller and Benson in Calvin's laboratory, and William Jakoby and Dewey Brummond in Ochoa's laboratory.

Calvin's laboratory had shown that hexose phosphate is first labeled in carbon atoms 3 and 4, and that this label would later be distributed into other carbon atoms.[26] These data were readily explained in terms of the operation of the pentose phosphate pathway, as described above, to regenerate ribulose 1,5-diphosphate. Noteworthy was their observation that the α- and β-carbon atoms of phosphoglyceric acid always contained equal amounts of isotope; this fact is also readily explained by the pentose phosphate pathway, since it predicts that both of these carbon atoms arise uniquely from C-3 of hexose.

CONCLUSION

The work of many laboratories was required to develop our present knowledge of the hexosemonophosphate oxidation and the pentose phosphate pathways and their role in the metabolism of higher plants and animals. Space does not permit a discussion of the role of these or related reactions in microbial systems (see ref. 9), which provide interesting examples of the versatility of living forms and their ability to adapt to specific ecological niches. Indeed, George Wald[27] has proposed that the hexosemonophosphate oxidation pathway may represent a primitive metabolic system that antedates the appearance of the glycolytic pathway.

Finally, it is a pleasure to recognize the great debt that biochemists of my generation owe to Severo Ochoa. Along with Erwin Haas and Otto Warburg, he taught me how the purification of enzymes and the careful characterization of the products formed with purified enzymes can lead to clarification of even the most complex pathways and mechanisms. During the 1950s, his service on NIH panels brought him frequently to Washington, and his visits to our laboratory on those occasions were a time of active and fruitful discussion. Later, when I joined the faculty at N.Y.U., our association with Severo and his charming wife Carmen became even closer, and remained so after I moved to the Albert Einstein College of Medicine. Presently, at the Roche Institute, Severo and I once again find ourselves on adjacent floors in the same Institution, in a pleasant and stimulating environment that is reminiscent of the "good old days".

REFERENCES

1. Warburg, O., Christian, W. and Griese, A. (1935) Wassersoffübertragendes Co-Ferment, seine Zusammensetzung und Wirkungsweise. *Biochem. Z.* **282**, 157–198.
2. Meyerhof, O. (1918) Presence of the coenzyme of the alcoholic yeast fermentation in muscular tissue and its significance in the respiratory metabolism. *Z. Physiol. Chem.* **101**, 165–175.
3. Keilin, D. and Hartree, E. F. (1938) Cytochrome oxidase. *Proc. R. Soc. (Lond.)* **B105**, 171–186.
4. Haas, E., Horecker, B. L. and Hogness, T. R. (1940) The enzymatic reduction of cytochrome *c*: cytochrome *c* reductase. *J. Biol. Chem.* **136**, 747–774.
5. Lehninger, A. L. (1951) Phosphorylation coupled to oxidation of diphosphopyridine nucleotide. *J. Biol. Chem.* **190**, 345–359.
6. Pasteur, L. (1861) Expériences et vues nouvelles sur la nature des fermentations. *Comptes Rendu*, **52**, 1260–1264.
7. Lundsgaard, E. (1930) Über die Einwirkung der Monojodessigsäure auf den Spaltungs-und Oxydationsstoffwechsel. *Biochem. Z.* **220**, 8–18.
8. Engelhardt, W. A. and Barkash, A. P. (1938) Oxidative breakdown of phosphogluconic acid. *Biokhimiya* **3**, 500–521.
9. References to this earlier work can be found in the review by Gunsalus, I. C., Horecker, B. L. and Wood, W. A. (1955) Pathways of carbohydrate metabolism in microorganisms. *Bacteriol. Rev.* **19**, 79–128.
10. Dickens, F. (1938) Yeast fermentation of pentose phosphoric acids. *Biochem. J.* **32**, 1645–1653.
11. Ochoa, S. (1951) Biological mechanisms of carboxylation and decarboxylation. *Physiol. Rev.* **31**, 56–106.
12. Vishniac, W. and Ochoa, S. (1953) Fixation of carbon dioxide coupled to photochemical reduction of pyridine nucleotides by chloroplast preparations. *J. Biol. Chem.* **195**, 75–93.
13. Seegmiller, J. E. and Horecker, B. L. (1952) Metabolism of 6-phosphogluconic acid in liver and bone marrow. *J. Biol. Chem.* **194**, 261–268.
14. Dische, Z. (1938) Phosphorylierung der im Adenosin enthaltenden *d*-Ribose und Nachfolger Zerfall des Esters unter Triocephosphatbildung im Blute. *Naturwiss.* **26**, 252.
15. Glock, G. E. (1952) The formation and breakdown of pentosephosphates by liver fractions. *Biochem. J.* **52**, 575–583.
16. Benson, A. A., Bassham, J. A., Calvin, M., Hall, A. E., Hirsch, H. E., Kawaguchi, S., Lynch, V. and Tolbert, N. E. (1952) The path of carbon in photosynthesis. XV. Ribulose and sedoheptulose. *J. Biol. Chem.* **196**, 703–716.
17. Buchanan, J. G., Bassham, J. A., Benson, A. A., Bradley, D. F., Calvin, M., Daus, L. L., Goodman, M., Hayes, P. M., Lynch, V. H., Norris, L. T. and Wilson, A. T. (1952) The path of carbon in

photosynthesis. XVII. Phosphorus compounds as intermediates in photosynthesis. In *Phosphorus Metabolism*, Vol. II (W. D. McElroy and B. Glass, eds.), pp. 440–466, Johns Hopkins Press, Baltimore.
18. Bonsignore, A., Pontremoli, S., Fornaini, G. and Grazi, E. (1957) Non-oxidative heptose formation in enzyme preparations of rat liver. *Italian J. Biochem.* **VI**, 227–238.
19. Ashwell, G. and Hickman, J. (1954) Formation of xylulose phosphate from ribose phosphate in spleen. *J. Am. Chem. Soc.* **76**, 5589.
20. Stumpf, P. K. and Horecker, B. L. (1956) Role of xylulose 5-phosphate in xylose metabolism of *Lactobacillus pentosus. J. Biol. Chem.* **218**, 753–768.
21. Horecker, B. L., Smyrniotis, P. Z. and Hurwitz, J. (1956) Role of xylulose 5-phosphate in the transketolase reaction. *J. Biol. Chem.* **223**, 1009–1019.
22. Horecker, B. L., Gibbs, M., Klenow, H. and Smyrniotis, P. Z. (1954) The mechanism of pentose phosphate conversion to hexose monophosphate. *J. Biol. Chem.* **207**, 393–403.
23. For references to these publications see ref. 9.
24. Weissbach, A., Smyrniotis, P. Z. and Horecker, B. L. (1954) Enzymatic formation of ribulose diphosphate. *J. Am. Chem. Soc.* **76**, 5572.
25. Weissbach, A., Horecker, B. L. and Hurwitz, J. (1956) Enzymatic formation of phosphoglyceric acid from ribulose diphosphate and carbon dioxide. *J. Biol. Chem.* **218**, 795–810.
26. Bassham, J. A., Benson, A. A., Kay, L. D., Harris, A. Z., Wilson, A. T. and Calvin, M. J. (1954) The path of carbon in photosynthesis. XXI. The cyclic regeneration of carbon dioxide acceptor. *J. Am. Chem. Soc.* **76**, 1760–1770.
27. Wald, G. (1964) The origins of life. *Proc. Nat. Acad. Sci.* **52**, 595–611.

REDUCING POWER AND THE REGULATION OF PHOTOSYNTHESIS

MANUEL LOSADA
Departamento de Bioquimica, Facultad de Ciencias y C.S.I.C.,
Universidad de Sevilla, Sevilla, Spain

INTRODUCTION

From an energetic point of view, the basic difference among living organisms resides in the ultimate source of energy—either chemical or electromagnetic—they use for their metabolism. For a nomenclature based on energy sources we proposed some years ago [1,2] the terms chemoergonic and photoergonic to describe the different classes of organisms. Photoergonic organisms (green plants) are able to convert light into physiological chemical energy and can thus grow on substrates with no useful chemical potential. On the contrary, chemoergonic organisms (bacteria in general, fungi, animals) cannot use light as energy source and depend entirely on the chemical energy of the substrates—whether inorganic or organic—they transform.

Life is beautiful, uncertain, fragile and fleeting as the beams of light that wonderfully sustain and keep it going. As Arnon has pointed out,[3] photosynthesis is first and foremost a process for converting the radiant energy of sunlight into physiological chemical energy, namely, reducing power $\{H_2\}$ and high-energy pyrophosphate bonds $\sim$P. $\{H_2\}$ represents reducing power about the level of the hydrogen electrode rather than molecular hydrogen proper. Green plants are in consequence the only organisms which can simultaneously use as nutrients the biogenic elements in their most oxidized form, i.e. completely devoid altogether of useful chemical energy: carbon as carbon dioxide (CO_2), nitrogen as nitrate (NO_3^-) or molecular nitrogen (N_2), sulfur as sulfate (SO_4^{2-}), hydrogen as water (H_2O) and phosphorus as phosphate (PO_4^{3-}). Water supplies reducing power $\{H_2\}$ by photodecomposition, with the result of oxygen (O_2) being evolved as a by-product. Phosphorus rather than change its valence is photochemically incorporated into the energy-rich compound adenosine triphosphate (ATP) during a process coupled to a light-induced electron flow known as photophosphorylation. Carbon, nitrogen and sulfur are then reduced in the dark, at the expense of the assimilatory power, $\{H_2\}$ and $\sim$P, synthesized in the light, to the level of sugar (CH_2O), ammonia (NH_3) and hydrogen sulfide (H_2S), and are further assimilated into cell material (carbohydrates, amino acids, proteins, fats, nucleic acids) through reactions which take place also in the dark.[1–3]

Since only green plants among living organisms can split water into hydrogen and oxygen, lysis of water by radiation at an energy level of visible light can be considered as the most simple and primordial endergonic reaction of bioenergetics, a reaction exclusive of photosynthesis on which life depends on our planet. Equally simple and essential for life is the opposite exergonic reaction that chemoergonic organisms carry out in the dark during respiration in order to transform the redox chemical energy stored in hydrogen and oxygen into pyrophosphate bonds.

The basic reactions of photosynthesis and respiration involve in fact a series of uphill and downhill hydrogen transfers via sophisticated electron transport chains eventually resulting in the splitting of water into hydrogen and oxygen or in its synthesis from them. For our purpose, however, these reactions can be summarized in a simplified manner in accord with the two overall equations:

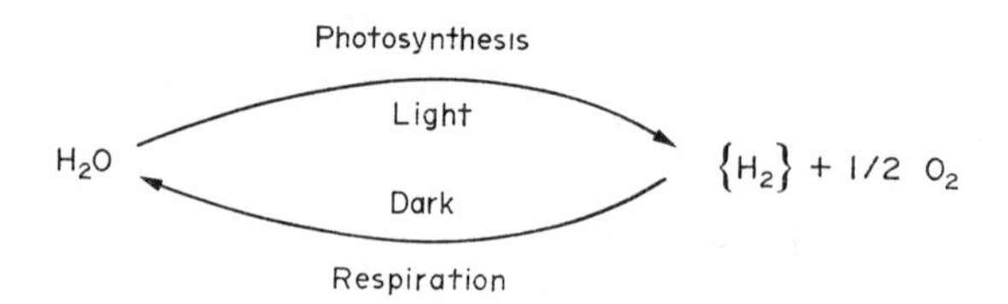

Water is thus the primary substrate of green plant photosynthesis for the generation of reducing power and the final product of the aerobic respiration of hydrogen.

The fact that the most evolved photoergonic and chemoergonic organisms have differentiated highly specialized cytoplasmic organelles—named, respectively, chloroplasts and mitochondria—in order to convert with remarkable efficiency the potentially useful electromagnetic or chemical energy stored in light or in chemical nutrients into physiological chemical energy is only a proof of the fundamental significance cells pay to their own energetic processes.

On the other hand, and limiting our discussion to the enzymatic machinery of green cells, the complexity of the photosynthetic pathways for the generation of reducing power and ATP and for the further assimilation of carbon, nitrogen and sulfur, as well as the intricate interrelations among the various metabolic processes makes it necessary the existence of multiple regulatory sites to coordinate the rates of the key reactions. As will be summarized in the present article, a generalized type of regulation in photosynthetic cells seems to be induced primarily by light and mediated first of all by a change in the redox status of some key coenzyme(s), that becomes more reduced, and also by fluctuations in the energy charge, in the pH and in the concentration of key metabolites, ions, substrates and products.

PATHWAYS OF CARBON, NITROGEN AND SULFUR PHOTOSYNTHETIC REDUCTION

Historical background

In 1920 Warburg made the fundamental discovery[4] that living *Chlorella* cells suspended in a solution of nitrate to which no CO_2 was added develop oxygen in the light with the concomitant reduction of nitrate to ammonia, according to the equation:

$$HNO_3 + H_2O \xrightarrow{\text{light}} NH_3 + 2O_2$$

Warburg, however, pointed out[4] that although the balance of the reaction is a photolysis of water, nitrate reduction can be resolved into a reduction of nitrate in the dark by oxidation of cell carbohydrate to CO_2, and a splitting of CO_2 in the light to carbohydrate and oxygen, according to the equations:

$$HNO_3 + 2C + H_2O \xrightarrow{\text{dark}} NH_3 + 2CO_2$$

$$2CO_2 \xrightarrow{\text{light}} 2C + 2O_2$$

Warburg's conclusion was therefore[4] that the balanced Hill reaction[5] written above is deceptive in the sense that what appears to be a photolysis of H_2O is in its mechanism a photolysis of CO_2. Warburg[4] extended his reasoning to all Hill reactions,[5], even to the reduction in the light of triphosphopyridine nucleotide, discovered in 1951 by Vishniac and Ochoa.[6]

Van Niel, however, based on his own hypothesis of the photochemical dehydrogenation of water, offered an alternate explanation to the one advanced by Warburg and interpreted the reduction of nitrate in the light as a reaction essentially analogous to the photochemical reduction of CO_2, with nitrate replacing CO_2 as the final hydrogen acceptor. Thus, according to Van Niel, light energy is used in photosynthesis to photodecompose water, and the released hydrogen is used to reduce either CO_2 or nitrate by dark enzymic reactions.[7]

In 1965, Warburg[8] repeated his classical experiment with more refined techniques and reached identical results, but in discussing the light-dependent production of O_2 by living *Chlorella* in the presence of nitrate he became even more committed to his early interpretation. Notwithstanding, the hypothesis of CO_2 photolysis is by now generally rejected in favor of H_2O photolysis, as discussed in the introduction.

Arnon and his group[3,9] have definitively established that ferredoxins (iron–sulfur proteins noted for their strongly electronegative redox potentials) are the primary electron acceptors in photosynthesis, and that they are essential electron carriers for the light-induced generation of reducing power and ATP formed in the processes of cyclic and non-cyclic photophosphorylation. Reducing power—either reduced ferredoxin or reduced nicotinamide adenine dinucleotides, NAD(P)H—and ATP constitute the assimilatory power required for the further assimilation in the dark of carbon dioxide, nitrate and sulfate.[1–3]

Carbon dioxide reduction

The pathway of carbon dioxide fixation and assimilation to the level of sugar consists of a cyclic series of reactions sometimes referred to as the reductive pentose cycle (in contrast to the dissimilatory oxidative pentose cycle), the carbon reduction cycle or the Calvin cycle.[10]

The overall reaction by which glucose is catabolized by the oxidative pentose cycle is:

$$C_6H_{12}O_6 + 6H_2O \xrightarrow{\sim P} 6CO_2 + 12\,\{H_2\}$$

The process is catalyzed in a series of consecutive dark reactions and its primary purpose is to generate reducing power in the form of NADPH. The most characteristic enzymes of the cycle are the first and second ones, which catalyze, respectively, the dehydrogenation of glucose-6-phosphate (G6P) to 6-phosphogluconate and the oxidative decarboxylation of 6-phosphogluconate to ribulose-5-phosphate (Ru5P). Since the overall equilibrium of these reactions lies far in the direction of dehydrogenation and decarboxylation, the reversal of the phosphogluconate pathway must avoid these two steps. Therefore, the reductive pentose cycle uses instead a combination of phosphorylating, reducing and dephosphorylating enzymes in order to reduce CO_2 to sugar, according to the equation:

$$6CO_2 + 12\{H_2\} \xrightarrow{18 \sim P} C_6H_{12}O_6 + 6H_2O$$

Chloroplasts are the organelles which carry on the total synthesis of carbohydrates from CO_2 through the reductive pentose cycle driven by the assimilatory power formed in the light phase of photosynthesis and made up of NADPH and ATP, but the oxidative pentose cycle can also operate inside the chloroplasts and regenerate NADPH in the dark.[3,11]

The cyclic sequence of dark reactions by which carbon is reduced to carbohydrate through the reductive pentose cycle [10,11]consists of three phases.[1-3] In the carboxylative phase two unique enzymes of carbon assimilation, ribulose-5-phosphate kinase (Ru5P kinase) and ribulose-1,5-diphosphate carboxylase (RuDP carboxylase), participate consecutively in two irreversible reactions: the phosphorylation of Ru5P to RuDP with ATP and the carboxylation of RuDP with CO_2 and H_2O to two molecules of 3-phosphoglyceric acid (PGA). In the reductive phase PGA is phosphorylated by ATP and reduced by glyceraldehyde-3-phosphate dehydrogenase to glyceraldehyde-3-phosphate (GA1d3P) with NADPH by a reversal of the well-known glycolytic reactions. Finally, in the regenerative phase triose phosphate is converted into storage carbohydrate and the Ru5P needed to start another turn of the cycle. In this last phase two hydrolytic enzymes participate which catalyze also irreversible reactions: fructose-1,6-diphosphatase (FDPase) that catalyzes the cleavage of fructose-1,6-diphosphate (FDP) to fructose-6-phosphate (F6P) and sedoheptulose-1,7-diphosphatase (SDPase) that liberates phosphate from sedoheptulose-1,7-diphosphate (SDP) in a similar manner.

Some plants, such as corn and sugar cane, have evolved an auxiliary C_4-dicarboxylic acid cycle[12,13] that cooperates with the reductive pentose cycle in the photosynthetic assimilation of CO_2. In plants with this cycle (sometimes referred to as the Hatch and Slack cycle), chloroplasts in the mesophyll cells near the surface on the leaf contain three C_4-pathway specific enzymes: pyruvate, phosphate-dikinase that directly converts pyruvate into phosphoenolpyruvate (PEP) with ATP, PEP carboxylase that catalyzes the carboxylation of PEP to oxaloacetate, and malate dehydrogenase that finally reduces oxaloacetate to malate with NADPH. The purpose of these steps is apparently to incorporate CO_2 and NADPH into malate in order to translocate them to the vascular bundle sheath cells, where they are again released by the action of a NADP-dependent malic enzyme. The malic enzyme is located in the bundle sheath chloroplasts together with the enzymes of the Calvin cycle. CO_2 is then reduced to carbohydrates while pyruvate is presumably transported back to the mesophyll cells. Besides the malate-type C_4-plants, there is a second and larger group of species (aspartate type) that contains little malic enzyme and utilizes aspartate as the CO_2 carrier.

Arnon and coworkers[9] have postulated a new cyclic pathway, named reductive carboxylic acid cycle, that provides another mechanism independent of the reductive pentose cycle for the assimilation of CO_2 in bacterial photosynthesis. This cycle generates acetyl-CoA from two molecules of CO_2 and is driven by reduced ferredoxin, reduced nucleotides, and ATP, according to the overall simplified equation:

$$2CO_2 + 4\{H_2\} \xrightarrow{2 \sim P} CH_3\text{—}COOH + 2H_2O$$

In its overall effect the new cycle is a reversal of the oxidative tricarboxylic acid cycle, but endergonic in nature and hence includes two reactions (α-ketoglutarate synthase and citrate lyase) that can by-pass the irreversible steps of the Krebs cycle. Photosynthetic bacteria also contain pyruvate synthase that brings about the direct synthesis of pyruvate from acetyl-CoA, CO_2 and reduced ferredoxin, as well as PEP synthase that catalyzes the direct synthesis of PEP from pyruvate and ATP.

Nitrate and molecular nitrogen reduction

There has been a long controversy about the mechanism, pathway, regulation and localization of nitrate reduction in green cells. Much of the literature pertaining to this subject has recently been reviewed [2,14,15] and no comprehensive survey of it is undertaken here. By now it is well established that the assimilatory reduction of nitrate to ammonia proceeds in two separate and well-defined steps: (1) The reduction of nitrate to nitrite, catalyzed by the molybdoprotein nitrate reductase, and (2) the reduction of nitrite to ammonia, catalyzed by the heme-protein nitrite reductase:

$$NO_3^- \xrightarrow{2e} NO_2^- \xrightarrow{6e} NH_3$$

Nitrate reductase from green algae and higher plants is an enzyme complex of high molecular weight (m. w. about 500,000) that specifically requires NADH as electron donor. The complex contains FAD, cytochrome b_{557} and Mo as prosthetic groups. In blue-green algae, however, the initial FAD-dependent NADH-diaphorase moiety of the nitrate reductase complex seems to be lacking, reduced ferredoxin acting as electron donor.

Nitrite reductase is a much smaller protein (m. w. about 60,000) than nitrate reductase, and in all cases studied utilizes reduced ferredoxin as electron donor. It contains two atoms of iron per molecule and exhibits a characteristic spectrum determined by its prosthetic group, recently identified as siroheme.

It seems to be in general agreement with the intracellular localization of nitrite reductase in green algae and higher plants as a soluble chloroplast enzyme, but there is contradictory evidence regarding nitrate reductase. In the blue-green alga *Anacystis nidulans* both enzymes are bound to the chlorophyll-containing particles, which can catalyze the ferredoxin-dependent gradual photoreduction of nitrate to nitrite and ammonia, in what appears to be one of the most simple and relevant examples of photosynthesis.

Some photosynthetic bacteria[1,9] and blue-green algae [16]can also catalyze the photoreduction of molecular nitrogen to ammonia, a strongly endergonic reaction known to depend on ATP and reduced ferredoxin:

$$N_2 \xrightarrow[6e]{n\sim P} 2NH_3$$

Sulfate reduction

Trebst and Schmidt have reported that sulfate reduction by leaves is localized in the chloroplast and that complete photoreduction of sulfate to sulfide and S-containing amino acids can be accomplished *in vitro* in isolated chloroplasts.[17] Assimilatory sulfate reduction has been assumed to occur by sulfate activation with ATP to 3′-phosphoadenosine-5′-phosphosulfate (PAPS), reduction of PAPS to sulfite presumably by NADPH and further reduction of sulfite to free sulfide in a ferredoxin-dependent reaction.[17,18] However, recent investigations on the mechanism of sulfate reduction in cell-free systems of *Chlorella* and spinach chloroplasts seem to indicate that free sulfite and free sulfide are not intermediates in the sequence of sulfate reduction, which is cyclic rather than linear.[19] The new pathway involves activation of sulfate to adenosine-phosphosulfate (APS), transfer of sulfate from APS to a thiosulfonate reductase and reduction by reduced ferredoxin of the sulfate bound to the enzyme (i.e. bound sulfite) to bound sulfide in a six-electron reaction. Bound sulfide then accepts a serine residue from *o*-acetylserine and is transformed into bound cysteine that is finally split in a redox reaction involving two electrons and the cycle starts again.

REGULATION OF THE PHOTOSYNTHETIC PATHWAYS OF CARBON AND NITROGEN ASSIMILATION

The investigations carried out in different laboratories during the last few years at the cellular, subcellular and enzymatic levels have accumulated increasing evidence on the metabolic regulation and interrelations of photosynthetic carbon dioxide and nitrate assimilation. It appears that the key enzymes of the reductive pentose cycle and the C_4-dicarboxylic acid cycle are reversibly regulated *in vivo* by light intensity, being rapidly activated upon illumination—mainly as the result of an increase in the reduced form of some photosynthetic redox coenzyme(s) and in the charged form of adenine nucleotide—and inactivated in the dark. This is logical, since the operation of the cycles is beneficial in the light when there is an abundance of assimilatory power and detrimental in the dark, when there is a shortage of chemical energy. On the other hand, the key enzymes of the nitrate-reducing system are reversibly regulated in the opposite way, i.e. they are inactivated by reducing power and ADP in what appears to be a very sophisticated mechanism dependent on light, nitrate, ammonia and carbon dioxide. In summary, the intensity of incident light would primarily control, via photoreduction and photophosphorylation, the redox status, energy charge and ions concentration in the photosynthetic organelles. As a consequence, the proportion of active to inactive enzymes present in the chloroplasts would change in response to the redox status of electron carriers or coenzymes (ferredoxin, nicotinamide-nucleotides) and to the prevailing cellular concentrations of adenosine phosphates and inorganic phosphate, metabolites, ions, substrates and products.

The reductive pentose cycle and related pathways

Despite the strong experimental support for the operation of the reductive pentose cycle in photosynthetic organisms, it has been challenged among other reasons because the activities of several of its key enzymes (RuDP carboxylase, FDPase, SDPase) are not high enough to satisfy the rate of overall CO_2 fixation normally observed in intact cells or in isolated chloroplasts. The discovery of several activation mechanisms which operate in the light by either increasing the V_{max} or the affinity of the critical enzymes has solved the puzzle.

Although many of the intermediate compounds of the Calvin cycle, e.g. PGA, FDP, F6P, are identical to metabolites of the glycolytic or gluconeogenic pathway used by nonphotosynthetic organisms to break down or synthesize sugars, regulation follows quite different mechanisms in each case. Among other important differences, to be discussed below, the following one can serve to illustrate the point: chloroplast FDPase differs from its counterpart in non-photosynthetic cells in being activated by reduced ferredoxin rather than inhibited by adenosine monophosphate (AMP).

Bassham and collaborators[11] have studied *in vivo* the regulation of photosynthetic carbon observing the responses of metabolite concentrations to external stimuli, both with unicellular algae and with isolated chloroplasts. Since it is the nature of *in vivo* metabolic studies that single experiments are seldom conclusive, many indications of kinetic data must be pieced together in order to draw some conclusion. The techniques of suddenly interrupting the steady-state photosynthesis and following the transient effects on metabolite concentrations have, however, been very valuable in locating the sites of metabolic regulation in the photosynthetic pathways of carbon metabolism.

From many kinetic tracer studies of metabolite concentrations during light–dark and dark–light transitions and upon addition of inhibitors, Bassham inferred[11] that carboxylation of RuDP, hydrolysis of FDP and SDP, and phosphorylation of Ru5P are reactions subject to activation by light. He also suggested the following specific sites of metabolic regulation in the light: the synthesis of sucrose, probably at the reaction between uridine diphosphoglucose and F6P, and the conversion of PEP to pyruvate. The last step, catalyzed by the enzyme pyruvate kinase, controls the flow of carbon from the photosynthetic carbon cycle to the synthesis of most amino acids and fatty acids. This point of control is also activated in *Chlorella* by the addition of ammonium ion. Added ammonium completely stops, besides, sucrose synthesis and switches therefore metabolism from sucrose to protein production.

Buchanan *et al.*[20] have found that activity of FDPase during photosynthesis by isolated chloroplasts is controlled photochemically through reduced ferredoxin. In the presence of a protein factor and Mg^{2+}, reduced ferredoxin activates alkaline FDPase. Reduced ferredoxin can be replaced by reduced methyl viologen or dithiothreitol. Spinach FDPase has a molecular weight of 145,000 and contains 2% carbohydrate. The amino acid composition of chloroplast FDPase resembles that of mammalian FDPases, except for a remarkably high half-cystine content. The protein factor has been partially purified and has a molecular weight of 40,000. They have suggested that the mechanism of activation of FDPase may involve a reduction by reduced ferredoxin of specific S–S groups.

SDPase, like its FDPase counterpart, is a regulatory enzyme whose activity in chloroplasts is controlled by light via ferredoxin.[21] Activation by photoreduced ferredoxin depends on a protein factor—which appears to be the same as that required for FDPase—and is optimal at pH 7.8 and at a Mg^{2+} concentration of 5 mM. SDPase can also be activated by the nonphysiological sulfhydryl reagent dithiothreitol.

The finding by Buchanan *et al.*[22] that RuDP carboxylase activity is stimulated by F6P and inhibited by FDP suggested that these two intermediates of the carbon cycle could regulate carboxylase activity during photosynthesis. Activation of RuDP carboxylase by F6P would solve a serious problem, namely the low affinity of RuDP carboxylase for CO_2. This difficulty is particularly noteworthy because photosynthesis in nature operates at low partial pressures of CO_2. To overcome this limitation the enzyme would require a concentration of CO_2 about 100-fold greater than that normally present in air.

RuDP carboxylase is present at very high concentration in leaves (about half of the soluble protein in chloroplasts) and catalyzes the carboxylation of RuDP to 2 PGA with CO_2 rather than CO_3H^- as the active carbon species. As shown by work in the laboratories of Calvin, Horecker, Ochoa and Racker, this enzyme accounts for the entry of CO_2 into the metabolism of photosynthetic cells. Notwithstanding, at high O_2 tension and no CO_2 the enzyme behaves as an oxidase and catalyzes the oxidative cleavage of RuDP to phosphoglycolate and PGA, a reaction that may be the key to the puzzling phenomenon of photorespiration.

RuDP carboxylase has been purified to homogeneity from algae and higher plants. The spinach enzyme has a molecular weight of 560,000 and dissociates into nonidentical subunits of molecular weight about 60,000 and 16,000. At saturating levels of Mg^{2+} (K_M = 1 mM), the carboxylase shows the following affinities for the substrates: RuDP, K_M = 0·1 to 0·25 mM; CO_2, K_M = 0·45 mM. The concentration of CO_2 in a solution equilibrated with air is, however, about 100-fold smaller than this latter value. Such a low concentration would therefore be incapable of supporting appreciable rate of CO_2 fixation by the carboxylase in the absence of a mechanism of activation.

F6P induces a shift from sigmoidal to Michaelian kinetics and decreases the K_M for CO_2 (up to a factor of 6) and also for Mg^{2+}. Increase in the affinity for CO_2 is most striking at a limiting level of Mg^{2+} or at neutral or acidic pH. FDP deactivates the carboxylase and effects, in general, a restoration of the original kinetic characteristics. Like F6P, 6-phosphogluconate activates RuDP carboxylase at low CO_2 concentrations. At present it is not known whether this activation, that is also reversed by FDP, is of physiological significance in chloroplasts.

Buchanan's results[(22)] provide evidence for a mechanism that increases the affinity of RuDP carboxylase for CO_2 and renders the enzyme active at a CO_2 concentration that approximates that of a solution in equilibrium with air. If the carboxylase is indeed regulated *in vivo* by the relative concentrations of the fructose derivatives, its activity would depend directly on FDPase. It is therefore possible that RuDP carboxylase is activated in the light by F6P formed via the ferredoxin-dependent FDPase and is deactivated in the dark by FDP. PGA formed by the activated carboxylase would in turn activate another regulatory enzyme, adenosine diphosphoglucose pyrophosphorylase, which, as discovered by Preiss and collaborators,[(23)] is stimulated by PGA and inhibited by inorganic phosphate. This enzyme forms adenosine diphosphoglucose (ADPG), the precursor of starch, from glucose-1-phosphate and ATP. In such a mechanism, the ferredoxin-activated FDPase reaction emerges as the initial regulatory reaction for the total synthesis of starch from CO_2 by chloroplasts. Such a role seems well suited to the FDPase because it renders the regulation of photosynthetic CO_2 assimilation light-dependent through its requirement for reduced ferredoxin. RuDP carboxylase and ADPG pyrophosphorylase would represent additional control points that are governed, respectively, by the levels of F6P and FDP and of PGA and phosphate. Starch synthesis may stop in the dark not only because the concentration of PGA decreases but because that of phosphate increases when photophosphorylation is not functioning.

Phosphoribulokinase, the other unique enzyme of the reductive pentose cycle together with RuDP carboxylase, is also located exclusively in the chloroplasts and is likewise subject to activation by light, but apparently by a different and more direct mechanism. Kinetic studies by Gibbs and collaborators [(24)]have shown that Ru5P kinase is activated 2- to 4-fold by illumination of intact chloroplast preparations, with a half-time of less than 15 seconds. The photoactivated state of the enzyme decays in the dark with a half-time of about 8 minutes. Since dark incubation of broken chloroplasts with dithiothreitol causes the same activation that light exposure of intact preparations and, moreover, activation by illumination is selectively inhibited by inhibitors of photosynthetic electron transport, it has been concluded that activation is due to the effect on the enzyme of a photoproduced reductant in a site preceding ferredoxin.

The work done in Ziegler's laboratory[25] has shown that irradiation of green leaves and isolated chloroplasts causes a rapid and reversible increase (up to 4,5-fold) in the activity of the NADP-dependent GA1d3P dehydrogenase. The *in vivo* light-induced activation of the enzyme is closely connected with the operation of the non-cyclic electron flow of photosynthesis, as shown by experiments with inhibitors.[25] NADPH causes a specific activation (up to about 10-fold) of the inactive "dark-enzyme"[25], and ATP is also a similarly effective activating agent *in vitro.*[26] Anderson[27] has achieved the conversion of one electrophoretically distinct form of NADP–GA1d3P dehydrogenase to a second form by irradiation of intact dark-treated pea plants with light or by treatment of extracts with dithiothreitol. She has concluded that light activation involves reduction of the enzyme resulting in a change in charge and hence in conformation rather than in aggregation,[27] but other authors claim conversion between high and low molecular weight forms.[28]

By contrast with the reactions subject to metabolic regulation in the photosynthetic pathways of carbon, the conversion of G6P to 6-phosphogluconate by G6P dehydrogenase—the first enzyme of the oxidative pentose phosphate cycle—is inactivated in the light and activated in the dark. Light inactivation may be a reductive process since dithiothreitol treatment of crude extracts produces similar inactivation.[29]

The purpose of the activation of FDPase, SDPase, RuDP carboxylase, Ru5P kinase and GA1d3P dehydrogenase in the light and of their inactivation in the dark appears to be to permit the operation of the reactions unique to the reductive pentose cycle in the light and their blocking in the dark. On the other hand, the purpose of the activation of G6P dehydrogenase in the dark and its inactivation in the light seems to be to allow the functioning of the oxidative pentose cycle in the dark and its prevention in the light.

The C_4-dicarboxylic acid cycle

Hatch and Slack[12] have found that the activity of pyruvate phosphate-dikinase rapidly falls on transferring illuminated plants to darkness, with a half-time of 15 minutes. Illumination of dark-treated plants results in an immediate rapid increase in activity up to 25-fold in 25 minutes, the maximum rate of increase occurring during the first $1\frac{1}{2}$ minutes. Studies *in vitro* suggest that activation and inactivation of the enzyme *in vivo* involve the reduction and oxidation of thiol groups. In fact, activation of the inactive enzyme isolated from dark-treated leaves requires a thiol, inorganic phosphate and a heat-labile large-molecular weight component and is inhibited by AMP. On the other hand, the active enzyme isolated from illuminated leaves is inactivated by ADP in an enzyme-catalyzed oxygen-dependent reaction. The enzyme so inactivated can be again reactivated with a thiol, phosphate and the protein component. Interconversion of the enzyme between an active (reduced) form and an inactive (oxidized) form may regulate the rate of formation of the substrate for CO_2 fixation via the photosynthetic C_4-pathway.

NADP-malate dehydrogenase in green leaves is regulated *in vivo* in a manner similar to the other key mesophyll chloroplast enzyme, pyruvate phosphate-dikinase, i.e. it is rapidly inactivated in the dark and reactivated upon illumination.[13] Maize plants kept in darkness for 30 minutes exhibited only 2% of the activity of illuminated plants. Upon transfer from darkness to illumination, activity increased almost instantaneously about 50-fold. The mechanism of the interconversion is apparently also similar for the two enzymes, involving the reversible oxidation of vicinal thiol groups but the requirements for activation and inactivation of NADP-malate dehydrogenase *in vitro* are relatively simple compared with pyruvate phosphate-dikinase, since activation of the inactive enzyme extracted from darkened leaves is achieved simply by adding dithiothreitol. When leaf extracts from illuminated plants were depleted of the thiol compound, activity declined rapidly to zero, but could be restored by adding it back.

It can be assumed that pyruvate phosphate-dikinase and NADP-malate dehydrogenase have no useful function in the dark and may even have detrimental effects on the metabolic status quo of chloroplasts if they use the ATP or reducing power available.

The nitrate-reducing system

The literature pertinent to the metabolic interconversion of nitrate and nitrite reductase of the assimilatory type has been recently reviewed.[14,30,31] The phenomenon was first discovered in green algae and has since then been intensively studied in these organisms, with special reference to the nitrate reductase complex. According to present evidence, interconversion of nitrate reductase appears to be primarily induced by a fluctuation in the redox status of the cell in response to a change in some of a variety of environmental conditions. In order to understand this exciting interconversion phenomenon a brief description of the main properties of the enzyme complex involved may be helpful.

It is well established that in the transfer of electrons from NADH to nitrate, catalyzed by the enzyme complex NADH-nitrate reductase from green cells, two moieties participate sequentially that can be independently assayed: the first is a FAD-dependent NADH-diaphorase, which is not affected by interconversion, and the second is the molybdoprotein nitrate reductase proper or terminal nitrate reductase, which is subject to this peculiar kind of control. Both activities respond very differently to selective treatments and inhibitors and can be specifically protected against inactivation or reactivated once inactive by their respective substrates or analogs. In this regard, the diaphorase moiety is very sensitive to heating and —SH binding reagents in the absence of NADH or FAD, whereas the terminal reductase can be totally inhibited by metal binding compounds in the absence of nitrate, particularly when the enzyme is kept in its reduced state by natural or artificial reducing agents. The inactivation of the second moiety of nitrate reductase is especially effective with cyanide as the metal chelating compound, and the inactive enzyme can be rapidly reactivated with ferricyanide as the oxidizing agent.

In algae both enzymes of the nitrate-reducing system, i.e. nitrate reductase and nitrite reductase, are nutritionally repressed by ammonia, the end product of the assimilatory pathway of nitrate reduction. The addition of ammonia to alga cells growing in the light on nitrate promotes besides the rapid conversion of the active form of the nitrate reductase complex into its inactive form by indirectly causing the reduction of its second moiety. Inactivation by ammonia requires light and carbon dioxide. Apparently ammonia acts *in vivo* as an uncoupler of photosynthetic phosphorylation, thus inducing an increase in the cellular levels of reducing power and ADP. No inactivation is observed upon the addition of ammonia in the presence of catalytic amounts of vitamin K_3—which is readily photo-oxidized by air and uses up the photosynthetic reducing power available—or of traces of an inhibitor of the non-cyclic electron flow. Arsenate and methylamine behave in the same manner as ammonia. The inactivation process is reversible, and upon ammonia (or the uncoupler) removal from the cell culture the enzyme becomes again oxidized and active. The metabolic inteıconversion of the enzyme can also be achieved *in vivo* by changing from aerobic to anaerobic conditions, but then the process is light and carbon dioxide independent.

Vennesland's group has concluded[31] that the inactivation *in vivo* after the cells have been treated with ammonia involves the formation of a firmly bound complex of reduced enzyme and cyanide. They have speculated that the product of ammonium assimilation which inhibits nitrate reductase is cyanide. In this respect the CO_2 requirement for inactivation by ammonia in the light could be relevant. On the other hand, CO_2 could potentiate the stimulating effect of ammonia on the photosynthetic non-cyclic electron flow.[4,5] If this interpretation would be correct, the activation of the ADP-dependent pyruvate kinase reaction by ammonia discovered by Bassham in *Chlorella*[11] might be better explained by an increase in the cellular ADP level induced by the uncoupling effect of ammonia on photophosphorylation than by a direct activation of the kinase by the ammonium cation, as it has been postulated.

The conversion *in vitro* of the active form of *Chlorella* nitrate reductase into the inactive form depends on its reduction by NAD(P)H, and is reversible by reoxidation with ferricyanide. Inactivation by NADH requires the first moiety of the complex to be active and proceeds much faster at high pH or when ADP at low concentration (0·3 mM) is simultaneously present. This synergistic effect is quite specific for NADH and ADP. Nitıate, as well as several of its competitive inhibitors, completely prevents and even reverses inactivation by NADH and ADP. In fact, Vennesland and co-workers have demonstrated the presence of cyanide in the *in vivo* inactivated enzyme by overnight incubation with nitrate and phosphate.[31]

The experimental work from the author's laboratory reported herein was supported in part by grants from the Phillips Research Laboratories (The Netherlands) and the National Science Foundation (U.S.A.).

REFERENCES

1. Whatley, F. R. and Losada, M. (1964) The photochemical reactions of photosynthesis. In *Photophysiology* (Giese, A. C. ed.) vol. 1. pp. 11–154, Academic Press, New York.
2. Losada, M. (1972) *La Fotosíntesis del Nitrógeno Nítrico*, Real Academia de Ciencias, Madrid.
3. Arnon, D. I. (1971) The light reactions of photosynthesis. *Proc. Nat. Acad. Sci. USA*, **68,** 2883–2892.
4. Warburg, O. (1964) Prefactory chapter. *Ann. Rev. Biochem.* **33,** 1–14.
5. Hill, R. (1965) The biochemists' green mansions: the photosynthetic electron-transport chain in plants. In *Essays in Biochemistry* (Campbell P. N. and Greville, G. D., eds.), vol 1. pp. 121–151, Academic Press, New York.
6. Vishniac, W., Horecker, B. L. and Ochoa, S. (1957) Enzymic aspects of photosynthesis. *Adv. Enzymol.* **29,** 1–77.
7. Van Niel, C. B. (1941) The bacterial photosyntheses and their importance for the general problem of photosynthesis. *Adv. Enzymol.* **1,** 263–328.
8. Warburg, O., Krippahl, G. and Jetschmann, C. (1965) Widerlegung der Photolyse des wassers und Beweis der Photolyse der kohlensäure nach versuchen mit lebender chlorella und den hill-reagentien Nitrat und $K_3Fe(CN)_6$. *Z. Naturforschg.* **20b,** 993–996.
9. Buchanan, B. B. and Arnon, D. I. (1970) Ferredoxins: chemistry and function in photosynthesis, nitrogen fixation, and fermentative metabolism. *Adv. Enzymol.* **33,** 120–176.

10. CALVIN, M. and BASSHAM, J. A. (1962) *The Photosynthesis of Carbon Compounds*, W. A. Benjamin, New York.
11. BASSHAM, J. A. (1971) Photosynthetic carbon metabolism. *Proc. Nat. Acad. Sci. USA*, **68,** 2877–2882.
12. HATCH, M. D. and SLACK, C. R. (1969) Studies on the mechanism of activation and inactivation of pyruvate, phosphate dikinase. *Biochem. J.* **112,** 549–558.
13. JOHNSON, H. S. and HATCH, M. D. (1970) Properties and regulation of leaf nicotinamide-adenine dinucleotide phosphate-malate dehydrogenase and "malic" enzyme in plants with the C_4-dicarboxylic acid pathway of photosynthesis. *Biochem. J.* **119,** 273–280.
14. LOSADA, M. (1975) La reducción fotosintética del nitrato y su regulación. In *Libro Homenaje al Profesor Lora Tamayo*, pp. 589–614, Real Academica de Ciencias, Madrid.
15. MANZANO, C., CANDAU, P., GOMEZ-MORENO, C., RELIMPIO, A. M. and LOSADA, M. (1976) Ferredoxin-dependent photosynthetic reduction of nitrite and nitrate by particles of *Anacyctis nidulans*. *Molec. Cell. Biochem. 10,* 161-169.
16. FOGG, G. E., STEWART, W. D. P., FAY, P. and WALSBY, A. E. (1973) *The Blue-Breen Algae*. Academic Press, New York.
17. TREBST, A. and SCHMIDT, A. (1969) Photosynthetic sulfate and sulfite reduction by chloroplasts. In *Progress in Photosynthesis Research* (METZNER, H. ed.), vol. 3. pp. 1510–1516, Liechtenstein, Tübingen.
18. BANDURSKI, R. S. (1965) Biological reduction of sulfate and nitrate. In *Plant Biochemistry* (BONNER, J. and VARNER, J. E., eds.), pp. 467–490, Academic Press, New York.
19. SCHMIDT, A. (1973) Sulfate reduction in a cell-free system of chlorella. *Arch. Mikrobiol.* **93,** 29–52.
20. BUCHANAN, B. B., SCHÜRMANN, P. and KALBERER, P. P. (1971) Ferredoxin-activated fructose diphosphatase of spinach chloroplasts. *J. Biol. Chem.* **246,** 5952–5959.
21. SCHÜRMANN, P. and BUCHANAN, B. B. (1975) Role of ferredoxin in the activation of sedoheptulose diphosphatase in isolated chloroplasts. *Biochim. Biophys. Acta*, **376,** 189–192.
22. BUCHANAN, B. B. and SCHÜRMANN, P. (1973) Ribulose 1,5-diphosphate carboxylase: a regulatory enzyme in the photosynthetic assimilation of carbon dioxide. In *Current Topics in Cellular Regulation* (HORECKER, B. L. and STADTMAN, E. R., eds.), pp. 1–20. Academic Press, New York.
23. PREISS, J. and KOSUGE, T. (1970) Regulation of enzyme activity in photosynthetic systems. *Ann. Rev. Plant. Physiol.* **21,** 433–466.
24. AVRON, M. and GIBBS, M. (1974) Properties of phosphoribulokinase of whole chloroplasts. *Plant. Physiol.* **53,** 136–139.
25. MÜLLER, B., ZIEGLER, I. and ZIEGLER, H. (1969) Lichtinduzierte, Reversible Aktivitätssteigerung der NADP-abhängigen Glycerinaldehyd-3-Phosphat-Dehydrogenase in Chloroplasten. *Europ. J. Biochem.* **9,** 101–106.
26. MÜLLER, B. (1970) On the mechanism of the light-induced activation of the NADP-dependent glyceraldehyde phosphate dehydrogenase. *Biochim. Biophys. Acta*, **205,** 102–109.
27. ANDERSON, L. E. and LIM, T. (1972) Chloroplast glyceraldehyde-3-phosphate dehydrogenase: light-dependent change in the enzyme. *FEBS Letters*, **27,** 189–191.
28. PAWLIZKI, K. and LATZKO, E. (1974) Partial separation and interconversion of NADH- and NADPH-linked activities of purified glyceraldehyde 3-phosphate dehydrogenase from spinach chloroplasts. *FEBS Letters* **42,** 285–288.
29. ANDERSON, L. E. LIM, T. and PARK, K. (1974) Inactivation of pea leaf chloroplastic and cytoplasmic glucose 6-phosphate dehydrogenase by light and dithiothreitol. *Plant Physiol.* **53,** 835–839.
30. LOSADA, M. (1974) Interconversion of nitrate and nitrite reductase of the assimilatory type. In *Metabolic Interconversion of Enzymes* 1973 (E. H. FISCHER, E. G. KREBS, H. NEURATH and E. R. STADTMAN, eds.), pp. 257–270, Springer-Verlag, Berlin.
31. LORIMER, G. H., GEWITZ, H., VÖLKER, W., SOLOMONSON, L. P. and VENNESLAND, B. (1974) The presence of bound cyanide in the naturally inactivated form of nitrate reductase of *Chlorella vulgaris*. *J. Biol. Chem.* **249,** 6074–6079.

STUDIES ON HIGH-ENERGY PHOSPHATE BONDS: FROM BIOSYNTHETIC TO MECHANOCHEMICAL REACTIONS

YOSHITO KAZIRO
Institute of Medical Science, University of Tokyo, Takanawa, Minatoku, Tokyo

It was a rainy day of March 1959, in Tokyo, when I met Severo Ochoa for the first time. Since then, he has been my teacher and he will be for ever. I met him at the Imperial Hotel in Tokyo right after his arrival from the United States. In that year, he came over to give a lecture at the 15th General Assembly of the Japan Medical Congress. I was introduced to him by Norio Shimazono, then Professor of Biochemistry at the Faculty of Medicine, University of Tokyo. After graduating from the Medical School of the University of Tokyo, I was finishing my Ph.D work in his laboratory. I brought the application form of the Foreign Postdoctoral Fellowship of the U.S. Public Health Service and asked Severo Ochoa whether I could come to his laboratory as a postdoctoral fellow, and he kindly agreed. I believe that, in that moment, my career as enzymologist was destined.

N.Y.U. AND PROPIONYL-CoA CARBOXYLASE

I left Tokyo on 25 August 1959, full of hope, to join Severo Ochoa at the New York University Medical Center. Several weeks before my departure, he wrote to me that he would like to have me work on propionyl-CoA carboxylase. He wanted to show that the enzyme contained biotin as prosthetic group. At that time, Lynen and his coworkers[1] had just demonstrated the presence of biotin in their enzyme, β-methylcrotonyl-CoA carboxylase. I replied I would be pleased to work on the carboxylase and biotin, although I think that any other problem he might have proposed would have also interested me. For my Ph.D. thesis, I was working on the pyrophosphorylation of thiamine, the reaction in which the pyrophosphate group is transferred from ATP to thiamine.[2] Through this work, of course, I became very much interested in the mechanism of utilization of the energy of ATP for biosynthetic reactions, and very often had a thought on the possible mechanism of CO_2 activation by ATP, as well as the presumed involvement of biotin as had already been suggested from early nutritional experiments.

There is not enough time and space to describe my early work on thiamine pyrophosphokinase, but I would like to point out that the pyrophosphorylation is a reaction less common than either phosphorylation or adenylylation. So far, only three examples are known. The one first reported by A. Kornberg[3] was the synthesis of phosphoribosyl pyrophosphate and the most recent one was the formation of ppGpp from ATP and GDP as demonstrated by Sy and Lipmann.[4] I feel it no mere coincidence that I am now working a little on the role of this unusual nucleotide in the stringent control.

To come back to propionyl-CoA carboxylase, I was more than happy when I was working in New York University with this enzyme. I had the pleasure of collaborating with an Italian biochemist, Enzo Leone, who arrived at Ochoa's laboratory about the same time as myself to work on the same subject. He had a lot of experience in enzymology and a good sense of humor. Since he spoke good English, I used to communicate with other people in the lab through him. Soon after we started, we could purify the enzyme to a considerable extent and demonstrated the presence of biotin using the bioassay technique. One day in October 1959 the whole biochemistry department was very much excited on hearing that Severo Ochoa had received a telegram from Stockholm that he and Kornberg were to share the Nobel prize. We immediately stopped working and celebrated with champagne. I was glad that I could witness the happiest moment in the life of this most distinguished scientist.

If I continue to describe our life in N.Y.U. around the 1960s I would never finish this paper. I must concentrate the rest of my essay on the pure scientific subject. But I would like to add, at that time all of us were very happy and the whole department was surrounded with a warm, very pleasant, and friendly atmosphere. Whenever we encountered some difficulties in research, we would go to Severo Ochoa and he would give us, always, adequate and very precise advice. We often said that he has a very good nose, which means that he could exactly smell out what had and what had not to be done.

Enzo Leone returned to Italy after a few months, and I continued the work on propionyl-CoA carboxylase with the help of Morton Schneider, who had lived almost all of his laboratory life with Severo. We were fortunate in succeeding in the purification of the enzyme and obtaining it in a crystalline form.[5] Actually this was the first biotin-enzyme that had been crystallized. The enzyme had a molecular weight of 700,000 and contained 4 moles of biotin per mole of protein. However, we could not analyze its subunit, since the advanced technique of sodium dodecyl sulfate polyacrylamide gel electrophoresis had not yet been developed at that time.

In studying the mechanism of the reaction catalyzed by propionyl-CoA carboxylase, we were again fortunate in isolating the $CO_2{\sim}$biotin-enzyme complex as an intermediate of the reaction.[6] This was an enzyme complex which had a carboxyl group at an N-1 position of biotin covalently linked to the ϵ-amino group of lysine of the carboxylase protein. The $[^{14}C]CO_2 \sim$ biotin-enzyme complex was formed by incubating the labeled bicarbonate, ATP, and the enzyme, in the presence of Mg^{2+}, and was isolated by gel filtration. The isolated complex contained 4 moles of ^{14}C-radioactivity per mole of protein, or 1 mole per mole of biotin. Using this complex, we could then demonstrate that the carboxyl group can either be transferred from the complex to propionyl-CoA to form methylmalonyl-CoA, or be released from the enzyme, in the presence of ADP and ^{32}Pi, with the formation of a stoichiometric amount of $[\gamma - {}^{32}P]$ATP.[6]

$$\text{ATP} + \text{HCO}_3^- + \text{biotin-enzyme} \rightleftharpoons \text{CO}_2 \sim \text{biotin-enzyme} + \text{ADP} + \text{Pi} \quad (1)$$

$$\text{CO}_2 \sim \text{biotin-enzyme} + \text{propionyl-CoA} \rightleftharpoons \text{methylmalonyl-CoA} + \text{biotin-enzyme} \quad (2)$$

$$\text{Sum:} \quad \text{Propionyl-CoA} + \text{HCO}_3^- + \text{ATP} \leftrightharpoons \text{methylmalonyl-CoA} + \text{ADP} + \text{Pi} \quad (3)$$

Throughout this work, I realized the importance of using a substrate amount of enzyme for the study of the reaction mechanism. Since then, I have always relied on this methodology. That is to isolate the pure enzyme in a large quantity and to study the formation and degradation of the reaction intermediate involving enzyme. Through this approach, it is possible to write the reaction sequences, bringing the enzyme into the equations as one of the substrates. You will see later the extension of this methodology to my recent work on protein synthesis.

In 1963, Severo Ochoa and I wrote two review articles, one on the metabolism of propionic acid in *Advances in Enzymology*,[7] and one on the carboxylases and the role of biotin in *Comprehensive Biochemistry*.[8] These two articles are the summary of my 4 years' work in N.Y.U. with Severo Ochoa and I went back to Japan in the fall of 1963, almost with the feeling as if my "Heimat" were left behind.

STUDIES ON THE POLYPEPTIDE CHAIN ELONGATION AND A CHALLENGE TO MECHANOCHEMICAL REACTIONS

It has been known for some time that GTP, besides ATP, is required for protein synthesis. Later, it became more apparent that each process of protein synthesis, i.e. initiation, elongation, and termination (for termination at least in eukaryotic system) of the polypeptide chain has a specific requirement for GTP, and that in each reaction GTP is split into GDP and Pi (see refs. 9 and 10).

The strict dependency of protein synthesis on GTP held my attention ever since its discovery by Keller and Zamecnik[11] in the middle of 1950s. At first I was attracted rather instinctively, without knowing its true significance. However, after working for many years on the ATP-dependent biosynthetic reactions, I became more and more interested in this problem. Since the energy for making a peptide bond is already supplied by ATP in the process of amino acid activation, the energy derived from GTP must be utilized for a purpose other than the synthesis of a covalent bond. The investigation on the enzymatic mechanism of this novel type of utilization of the high-energy phosphate bonds seemed to me highly attractive. I imagined that the mode of utilization of GTP in protein synthesis might share a common characteristic with other mechanochemical reactions involving ATP, i.e. active transport, muscle contraction, and even oxidative phosphorylation.

Although I had long been thinking of working on this fascinating problem, I could not start this until the fall of 1967. One year after that I moved to the Institute of Medical Science where for the first time I had my own laboratory. At that time, several groups of investigators had already been working on the enzymology of protein synthesis, and partially purified preparations of EF-G and EF-T were obtained (see refs. 9 and 10). However, the role of GTP split in protein synthesis had not often been elaborated from the viewpoint of energy transformation. Soon after we started, we became aware that the system has a certain advantage for studying the molecular mechanism of mechanochemical reactions. For example, the protein factors participating in the polypeptide chain elongation in prokaryotes are present in abundance in the cell, and are purified without much difficulty. The purified factors are of relatively simple structure, soluble in an aqueous medium, and easily characterized. Furthermore, the various intermediates of the reaction are prepared and isolated in stable forms, and it was possible to study the partial reactions using these intermediates and substrate amounts of highly purified protein factors. Below, I would like to describe the results of our recent studies on the role of GTP in the polypeptide chain elongation, which revealed that the conformational transition of the protein molecules in association with the change of their nucleotide ligand is of primary importance for this type of reactions.

Figure 1 summarizes the current knowledge of polypeptide chain elongation reactions in *E. coli.* The three protein factors, EF-Tu, EF-Ts and EF-G, are involved in these reactions. The complex of EF-Tu and EF-Ts, the EF-Tu·EF-Ts complex, reacts with GTP

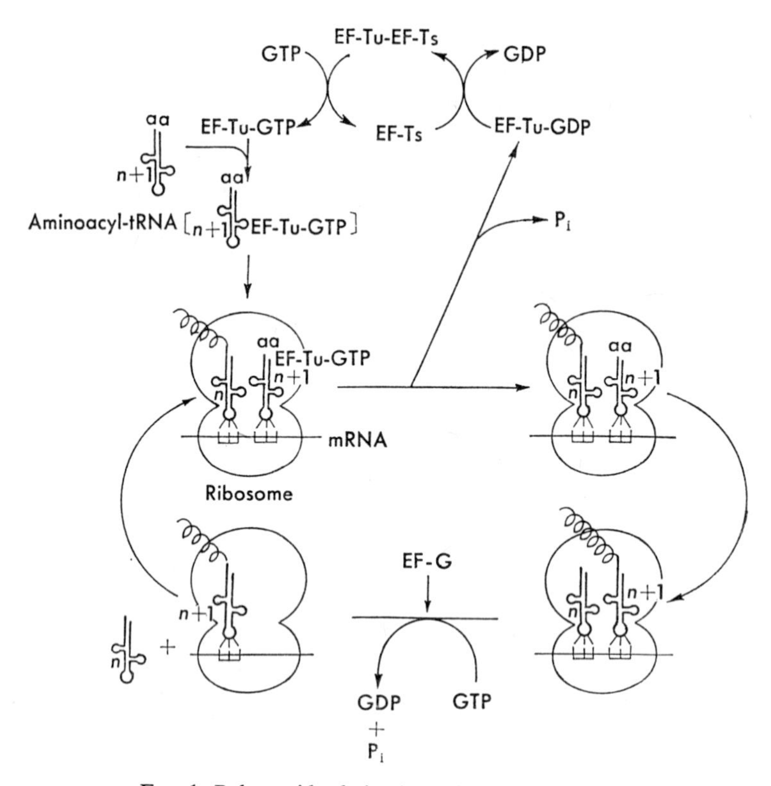

FIG. 1. Polypeptide chain elongation in *E. coli.*

to yield EF-Tu·GTP which immediately interacts with aminoacyl-tRNA to form a ternary complex, aminoacyl-tRNA·EF-Tu·GTP. The complex is then transferred to the A site of a ribosome having a peptidyl-tRNA prebound to its P site, GTP is hydrolyzed and EF-Tu·GDP is released. EF-Tu·GDP subsequently reacts with EF-Ts to regenerate EF-Tu·EF-Ts complex with the displacement of GDP.

Then a new peptide bond is formed by the transfer of the carboxyl group of peptidyl-tRNA to the adjacent amino group of newly bound aminoacyl-tRNA. The resulting ribosome possesses a deacylated tRNA in its P site, and a peptidyl-tRNA, having a chain one amino acid longer, in its A site. EF-G catalyzes the translocation of peptidyl-tRNA from the A site to the P site with concomitant release of deacylated tRNA from the P site and the movement of mRNA by a three-nucleotide distance on the 30S ribosomal subunit. One mole of GTP is hydrolyzed in the translocation reaction catalyzed by EF-G. The net result of the above sequences of reactions is the elongation of one peptide bond at the expense of two molecules of GTP.

The EF-Tu-promoted, GTP-dependent binding of aminoacyl-tRNA may be considered as a process analogous to active transport. In this reaction, EF-Tu functions as a carrier protein which promotes the energy-dependent, unidirectional transport of aminoacyl-tRNA to ribosomes. On the other hand, the translocation reaction may be compared to muscle contraction since mRNA moves or slides on the 30S ribosomal subunit. What I wish to show here is that both reactions are mediated through essentially similar mechanisms and that the role of GTP in these two reactions is quite comparable.

In an earlier attempt to study the mechanism of translocation reaction, it was found that no high-energy intermediate like X ~ P, or X ~ GDP, is formed during the reaction. In order to obtain more insight into the role of GTP in the above two reactions and to understand the molecular mechanism by which the energy derived from GTP is utilized, it seemed essential to obtain the elongation factors in a highly purified form and in substrate quantity. We have therefore attempted a large-scale purification of the elongation factors from *E. coli*, and have succeeded in obtaining these factors in crystalline form. The details of the purification have already been reported.[12,13] Some physicochemical and catalytic properties of the purified factors are summarized in Table 1. I do not have space

TABLE 1. PROPERTIES OF THE POLYPEPTIDE CHAIN ELONGATION FACTORS

Elongation factors	State	Molecular weight	SH required	Stabilizer	K_d for GTP	K_d for GDP
Prokaryotic:					M	M
EF-Tu	Crystalline	47,000	Yes	GDP	3.6×10^{-7}	4.9×10^{-9}
EF-Ts	Homogeneous	36,000	Yes		—	—
EF-G	Crystalline	83,000	Yes	Sucrose	1.4×10^{-5}	1.1×10^{-5}
Eukaryotic:					M	M
EF-1α	Homogeneous	53,000	Yes	Glycerol	2×10^{-7}	5×10^{-7}
EF-1β	Nearly homogeneous	~90,000	Yes	Phosphate	—	—
EF-2	Homogeneous	100,000	Yes	Sucrose	1.4×10^{-5}	4.1×10^{-7}

to describe the mammalian system in detail, but I would like to note that Iwasaki *et al.*[14] have succeeded in resolving EF-1 activity into two complementary fractions, EF-1α and EF-1β. EF-1α was purified to a homogeneous state as judged by several criteria,[15] and EF-1β to near homogeneity (Motoyoshi *et al.*, unpublished). I will discuss below the function of the new eukaryotic elongation factor, EF-1β.

Role of GTP in the EF-Tu-promoted binding of aminoacyl-tRNA to ribosomes

The most remarkable feature of the EF-Tu-promoted aminoacyl-tRNA binding was that only EF-Tu·GTP and not EF-Tu·GDP could interact with aminoacyl-tRNA to form a ternary complex, aminoacyl-tRNA·EF-Tu·GTP (see refs. 9 and 10). EF-Tu·GTP binds selectively with an aminoacylated form of tRNA in preference to its deacylated form, whereas EF-Tu·GDP interacts with neither aminoacyl-tRNA nor deacylated tRNA. Recently, it was further shown that in the absence of aminoacyl-tRNA, EF-Tu·GTP and not EF-Tu·GDP interacts with ribosomes.[16] Therefore, it was suggested that the binding of GTP to EF-Tu alters its conformation in a manner that facilitates its interaction with aminoacyl-tRNA and ribosomes.

The proposed conformational change has been verified by Arai and Kawakita, using hydrophobic and spin-label probes.[17,18] In the latter case, a spin-label analogue of N-ethylmaleimide was specifically introduced to the sulfhydryl group of EF-Tu essential for interaction with aminoacyl-tRNA.* It was observed that ESR spectra of the spin-labeled EF-Tu·GTP were markedly different from those of the spin-labeled EF-Tu·GDP. Furthermore, the reactivity of SH_2 toward several SH-reagents was found to be markedly increased in EF-Tu·GTP compared to EF-Tu·GDP. The results which were obtained with hydrophobic probes also indicated that a conformational change occurred in EF-Tu when its ligand is altered from GDP to GTP, so as to create one additional binding site for a hydrophobic dye, 1-anilino-8-naphthalenesulfonate, near the SH_2 group.

Now the question arises as to when and for what the fission of GTP is required. To solve this, we have studied the aminoacyl-tRNA binding reaction in the presence of a non-hydrolyzable analogue of GTP, Gpp(CH_2)p or Gpp(NH)p. When a ribosomal complex having N-acetyl-[^{14}C]Phe-tRNA at the P site (Complex I, or post-translocational complex) was incubated with [^{14}C]Phe-tRNA in the presence of EF-Tu·EF-Ts and GTP, a new complex (Complex II, or pretranslocational complex) was formed which possessed N-acetyl-di[^{14}C]Phe-tRNA and uncharged tRNA at the A and P sites, respectively. On the other hand, in experiments in which either Gpp(CH_2)p or Gpp(NH)p was substituted for GTP, only negligible dipeptide formation was observed between the newly bound Phe-tRNA and the prebound N-acetyl-Phe-tRNA, and EF-Tu·Gpp(CH_2)p or EF-Tu·Gpp-(NH)p remained bound on ribosomes.[21] This suggested that the hydrolysis of GTP is required in some step prior to the formation of the new peptide bond.

We have demonstrated that GTP hydrolysis is required for the removal of EF-Tu, the presence of which could be inhibitory to the subsequent peptidyl transfer. When the complex formed in the presence of Gpp(CH_2)p was centrifuged through sucrose solution, Tu·Gpp(CH_2)p was released from ribosomes and the peptidyl transfer reaction took place.[21] Apparently, the masking of the amino group of aminoacyl-tRNA by EF-Tu·Gpp-(CH_2)p on ribosomes is the reason for the inability of the bound aminoacyl-tRNA to engage in the peptidyl transfer reaction.

The role of GTP in EF-Tu-promoted binding of aminoacyl-tRNA could be formulated as follows. First, a unique conformation of EF-Tu induced by GTP can select exclusively the aminoacylated form of tRNA in preference to its deacylated form. Second, the ternary complex, aminoacyl-tRNA·EF-Tu·GTP, is transferred to a precise location on the 50S ribosomal subunit through the conformation of EF-Tu·GTP favorable for interaction with ribosomes. The conformational change of aminoacyl-tRNA induced by complexing with EF-Tu·GTP may also serve for this interaction. Third, after the transfer of aminoacyl-tRNA to the A site of ribosomes, EF-Tu is to be released from ribosomes to reinitiate a new cycle of reactions. This could be accomplished by the hydrolysis of bound GTP to GDP. An additional advantage of the split of GTP is to shift the equilibrium irreversibly

* As has been reported by Miller and Weissbach[19] and by Arai *et al.*,[20] EF-Tu has two reactive and one unreactive sulfhydryl groups; of the former two, one (SH_1) is essential for interaction with guanine nucleotides and EF-Ts, and the other one (SH_2) is required for the binding with aminoacyl-tRNA and ribosomes.

toward the binding of aminoacyl-tRNA to ribosomes. It was shown that Phe-tRNA bound to Complex I in the presence of EF-Tu and Gpp(CH_2)p is exchangeable with external Phe-tRNA while one bound with GTP is not.[22]

Role of GTP in EF-G-promoted translocation reaction

A remarkable similarity has been found in the mechanism of utilization of GTP energy for the translocation promoted by EF-G. Although bound less tightly than EF-Tu, EF-G can bind with guanine nucleotides to form a binary complex, EF-G·GTP or EF-G·GDP.[23] The affinity of EF-G toward ribosomes is again regulated by the species of guanine nucleotide; in the presence of GTP, EF-G can interact with ribosomes to form a stable ternary complex, whereas in the presence of GDP, it has practically no affinity to ribosomes (see refs. 9 and 10).

The conformational change of EF-G was again demonstrated by the use of hydrophobic and spin-label probes (N. Arai *et al.*, unpublished). EF-G possesses three sulfhydryl groups per mole of protein; one sulfhydryl group which is reactive and essential for interaction with ribosomes, and two sulfhydryls which are unreactive under native conditions. When EF-G was modified with spin-labeled analogues of N-ethylmaleimide, a considerable difference in ESR spectra of EF-G·GTP and EF-G·GDP was observed, indicating the presence of a conformational change near the active site interacting with ribosomes.

The EF-G·GTP complex promoted translocation when bound to Complex II (or pretranslocational complex). The process does not require the splitting of GTP, since a single turnover translocation could be demonstrated by the use of a nonhydrolyzable analogue of GTP, Gpp(CH_2)p or Gpp(NH)p.[24] In the presence of a stoichiometric amount of EF-G, Gpp(CH_2)p or Gpp(NH)p was shown to replace GTP in promoting the release of uncharged tRNA from ribosomes concomitant with the shift of N-acetyl-diPhe-tRNA from the A to the P site. Therefore, it is evident that not the hydrolysis of GTP, but the binding of EF-G·GTP to ribosomes is the trigger for translocation. The hydrolysis of GTP is again required for the release of EF-G from ribosomes, in order to proceed to the subsequent step of polypeptide chain elongation, and to reutilize EF-G for a new cycle of reactions. It must be noted that these experiments were carried out with substrate amounts of EF-G. When a catalytic amount of EF-G was employed, neither Gpp(CH_2)p nor Gpp(NH)p could replace GTP, because the catalytic reutilization of EF-G is inhibited in the absence of GTP hydrolysis.

Speculations and Extrapolations

As has been described above and also elsewhere,[25] the role of GTP in EF-Tu- and EF-G-promoted reactions appears to be quite analogous. In both cases, the conformation as well as reactivity of the protein molecules are reversibly and qualitatively altered by the change of its nucleotide ligands. A single turnover reaction is accomplished utilizing the specific conformation induced by GTP, and the splitting of GTP is required to shift the protein to an alternate conformation.

An analogous mechanism of GTP cleavage was also found in IF-2-promoted binding of fMet-tRNA to ribosomes.[26] Formyl-Met-tRNA bound to ribosomes with IF-2 and GTP was able to react with puromycin whereas the same complex formed with Gpp(CH_2)p was not reactive with puromycin. However, after removing IF-2 and Gpp(CH_2)p from ribosomes, the fMet-tRNA became reactive to puromycin. Therefore, in this case also, the hydrolysis of GTP seems to be required for the removal of IF-2. A similar mechanism might also be found in the release factor-promoted termination reaction. It is conceivable that the four GTP-utilizing reactions in protein synthesis, catalyzed by IF-2, EF-Tu, EF-G, and RF, respectively, are probably mediated through similar and common reaction mechanisms.

I should like to further point out that a similar reaction mechanism is also applicable to eukaryotic systems, naturally with several modifications. As can be seen in Table 1, the dissociation constant of EF-2·GDP was smaller than that of the prokaryotic EF-G·GDP by two orders of magnitude. Thus, the EF-2·GDP binary complex could be detected on the nitrocellulose membrane. Furthermore, EF-2·GDP had a higher affinity toward ribosomes than EF-G·GDP and could form a ternary complex in the absence of fusidic acid. However, the stability of the complex was lower than that of the corresponding Gpp(CH_2)p complex.[27] On the other hand, the eukaryotic EF-1α·GDP has a dissociation constant two orders of magnitude greater than that of EF-Tu·GDP, and hence, it is not likely that it requires EF-1β for the displacement of bound GDP. A preliminary experiment indicated that, in the absence of EF-1β, EF-1α was rather slowly released from ribosomes after the hydrolysis of GTP (Nagata *et al.*, unpublished). If this is the case, the function of the eukaryotic EF-1β is more similar to the prokaryotic IF-1 which promotes the release of IF-2 from ribosomes.

You may recall that I have pointed out that EF-Tu-promoted aminoacyl-tRNA binding and EF-G-promoted translocation are somewhat comparable to the processes of active transport and contraction, respectively. Although this might be imprudent speculation, I would like to try to extrapolate the above arguments to other systems.

In various models dealing with the reaction mechanism of mechanochemical reactions, it was often arbitrarily assumed that the hydrolysis of ATP precedes the mechanochemical work. This is mostly based on the intuitive view that the chemical energy is first derived from ATP and then is transformed to the mechanical energy. However, this is a concept obtained from the macroscopic view of the overall reaction, and hence, it may not be very useful for resolving the sequences of events occurring as the partial reactions. Furthermore, it was rather difficult to visualize the mechanism in which the energy derived from the hydrolysis of a chemical bond at the active site of an enzyme is conserved and then subsequently transformed.

When I envisaged the function of elongation factors, I was impressed by the remarkable difference between the properties of EF-G·GTP and EF-G·GDP, and of EF-Tu·GTP and EF-Tu·GDP, in terms of reactivity toward ribosomes and also in the latter case toward aminoacyl-tRNA. It was remarkable that the reactivity of the protein is converted qualitatively and reversibly by "phosphate-energy level" of its ligands. Then it became more plausible for me to assume that the transformation of the energy is achieved by the conformational transition induced by "high-energy phosphates".

I have deliberately emphasized in this article that EF-Tu-promoted aminoacyl-tRNA binding and EF-G-promoted translocation precede the hydrolysis of GTP. In order to avoid any misinterpretation of the above statement, I would like to call again the readers' attention to the fact that we are using a substrate amount of factors for these experiments, and that factors as well as GTP analogue are still retained on the ribosomal complex. Therefore, the reaction is not perfectly completed but is still in an "intermediate state". It must be noted that such an "intermediate state" could only be detected by the use of a "substrate amount" of enzyme and an "unsplittable" analogue of GTP. I do not think that this kind of approach has seriously been attempted in any other energy-transducing systems. It may not be so easy, because in active transport, for example, the "intermediate state" is analogous to the state of carrier proteins facing inside of the membrane, but still associated with the substrate to be transported. In this state, the reaction could still be *reversible.*

In protein synthesis, it is the conformation of the functional proteins associated with GTP (E·GTP) which is utilized to drive the reaction. The factor is then released from ribosomes as E·GDP, and the exogenous energy is fed into this system to regenerate E·GDP to E·GTP. In other energy-transducing systems, similar but variant types of conformational transition are naturally observed. For example, in active transport, two kinds of conformational transitions could be found, namely E·ATP and $E \sim P$, and $E \sim P$ and E. It might be assumed that the former transition is utilized for the transport of a cation from outside to inside of the membrane and the latter transition is utilized to the counter-transport of another cation to the opposite direction.

Presumably, the reversible and energy-dependent conformational transitions of the functional protein are of prime importance for the energy transduction in biological systems. The conformational transitions in a protein molecule are achieved again by the reversible and energy-dependent change in ligands covalently or noncovalently associated with the protein. These could be the general characteristics of the mechanism by which various energy-transducing reactions are catalyzed.

REFERENCES

1. Lynen, F., Knappe, J., Lorch, E., Jüting, G. and Ringlemann, E. (1959) Die biochemisches Funktion des Biotins. *Angew. Chem.* **71**, 481–486.
2. Kaziro, Y., Tanaka, R., Mano, Y. and Shimazon, N. (1961) On the mechanism of transpyrophosphorylation in the biosynthesis of thiamine diphosphate. *J. Biochem. (Tokyo)*, **49**, 472–476.
3. Kornberg, A. (1957) Pyrophosphorylases and phosphorylases in biosynthetic reactions. *Advances in Enzymology*, vol. 18, pp. 191–240. Interscience Publishers, John Wiley & Sons, New York.
4. Sy, J. and Lipmann, F. (1973) Identification of the synthesis of guanosine tetraphosphate (MS I) as insertion of a pyrophosphoryl group into the 3′-position in guanosine 5′-diphosphate. *Proc. Nat. Acad. U.S.A.* **70**, 306–309.
5. Kaziro, Y., Ochoa, S., Warner, R. C. and Chen, J. Y. (1961) Metabolism of propionic acid in animal tissues. VIII. Crystalline propionyl carboxylase. *J. Biol. Chem.* **236**, 1917–1923.
6. Kaziro, Y. and Ochoa, S. (1961) Mechanism of the propionyl carboxylase reaction, I. Carboxylation and decarboxylation of the enzyme. *J. Biol. Chem.* **236**, 3131–3136.
7. Kaziro, Y. and Ochoa, S. (1964) The metabolism of propionic acid. *Advances in Enzymology*, vol. 26, pp. 283–378. Interscience Publishers, John Wiley & Sons, New York.
8. Ochoa, S. and Kaziro, Y. (1965) Carboxylases and the role of biotin. *Comprehensive Biochemistry*, vol. 16, pp. 210–249, Elsevier, Amsterdam–London–New York.
9. Lucas-Lenard, J. and Lipmann, F. (1971) Protein biosynthesis. *Ann. Rev. Biochem.* **40**, 409–448.
10. Haselkorn, R. and Rothman-Denes, L. B. (1973) Protein synthesis. *Ann. Rev. Biochem.* **42**, 397–438.
11. Keller, E. B. and Zamecnik, P. C. (1956) The effect of guanosine diphosphate and triphosphate on the incorporation of labeled amino acids into proteins. *J. Biol. Chem.* **221**, 45–59.
12. Kaziro, Y., Inoue-Yokosawa, N. and Kawakita, M. (1972) Studies on polypeptide elongation factor from *E. coli*, I. Crystalline factor G. *J. Biochem.* **72**, 853–863.
13. Arai, K., Kawakita, M. and Kaziro, Y. (1972) Studies on polypeptide elongation factors from *E. coli*, II. Purification of factors Tu-GDP, Ts, and crystallization of Tu-GDP and Tu-Ts. *J. Biol. Chem.* **247**, 7029–7037.

14. IWASAKI, K., MIZUMOTO, K., TANAKA, M. and KAZIRO, Y. (1973) A new protein factor required for polypeptide elongation in mammalian tissues. *J. Biochem.* (*Tokyo*), **74**, 849–852.
15. IWASAKI, K., NAGATA, S., MIZUMOTO, K. and KAZIRO, Y. (1974) The purification of low molecular weight form of polypeptide elongation factor 1 from pig liver. *J. Biol. Chem.* **249**, 5008–5010.
16. KAWAKITA, M., ARAI, K. and KAZIRO, Y. (1974) Interactions between elongation factor Tu-guanosine triphosphate and ribosomes and role of ribosome-bound transfer RNA in guanosine triphosphatase reaction. *J. Biochem.* (*Tokyo*), **76**, 801–809.
17. ARAI, K., ARAI, T., KAWAKITA, M. and KAZIRO, Y. (1975) Conformational transitions of polypeptide chain elongation factor Tu. I. Studies with hydrophobic probes. *J. Biochem.* (*Tokyo*), **77**, 1095–1106.
18. ARAI, K., KAWAKITA, M., KAZIRO, Y., MAEDA, T. and OHNISHI, S. (1974) Conformational transition in polypeptide elongation factor Tu as revealed by electron spin resonance. *J. Biol. Chem.* **249**, 3311–3313.
19. MILLER, D. L., HACHMANN, J. and WEISSBACH, H. (1971) The reaction of the sulfhydryl groups on the elongation factors Tu and Ts. *Arch. Biochem. Biophys.* **144**, 115–121.
20. ARAI, K., KAWAKITA, M., NAKAMURA, S., ISHIKAWA, I. and KAZIRO, Y. (1974) Studies on the polypeptide elongation factors from *E. coli.* VI. Characterization of sulfhydryl groups in EF-Tu and EF-Ts. *J. Biochem.* (*Tokyo*), **76**, 523–534.
21. YOKOSAWA, H., INOUE-YOKOSAWA, N., ARAI, K., KAWAKITA, M. and KAZIRO, Y. (1973) The role of GTP hydrolysis in elongation factor Tu promoted binding of aminoacyl-tRNA to ribosomes. *J. Biol. Chem.* **248**, 375–377.
22. YOKOSAWA, H., KAWAKITA, M., ARAI, K., INOUE-YOKOSAWA, N. and KAZIRO, Y. (1975) Binding of aminoacyl-tRNA to ribosomes promoted by elongation factor Tu: further studies on the role of GTP hydrolysis. *J. Biochem.* (*Tokyo*), **77**, 719–728.
23. ARAI, N., ARAI, K. and KAZIRO, Y.(1975) Formation of a binary complex between elongation factor G and guanine nucleotides. *J. Biochem.* (*Tokyo*), **78**, 243-246.
24. INOUE-YOKOSAWA, N., ISHIKAWA, C. and KAZIRO, Y. (1974) The role of guanosine triphosphate in translocation reaction catalyzed by elongation factor G. *J. Biol. Chem.* **249**, 4321–4323.
25. KAZIRO, Y. (1973) The role of GTP in the polypeptide elongation reaction in *E. coli.* In *Organization of Energy Transducing Membranes* (M. NAKAO and L. PACKER, eds.), pp. 187–200, University of Tokyo Press, Tokyo.
26. BENNE, R. and VOORMA, H. O. (1972) Entry site of formylmethionyl-tRNA. *Fed. Eur. Biochem. Soc. Lett.* **20**, 347–351.
27. MIZUMOTO, K., IWASAKI, K. and KAZIRO, Y. (1974) Studies on polypeptide elongation factor 2 from pig liver. III. Interaction with guanine nucleotide in the presence and absence of ribosomes. *J. Biochem.* (*Tokyo*), **76**, 1269–1280.

GLUTAMATE FERMENTATION AND THE DISCOVERY OF B_{12} COENZYMES

H. A. BARKER
University of California, Berkeley, California 94720

The discovery of the coenzyme forms of vitamin B_{12} and related corrinoid compounds resulted from studies on the fermentative degradation of glutamate by *Clostridium tetanomorphum* that I and my associates had pursued off and on for many years. My interest in glutamate fermentation, and indeed in anaerobic amino acid degradation in general, was aroused by Professor C. B. van Niel who introduced me to the world of bacteria and later sponsored my work as a postdoctoral fellow at the Hopkins Marine Station in 1934–35. During that year van Niel started enrichment cultures for soil anaerobes able to ferment various amino acids and one of his students isolated a pure culture of a clostridium from a glutamate enrichment, but did not demonstrate that it was able to ferment glutamate.

The following year I took up this problem while working as a postdoctoral fellow in the laboratory of Prof. A. J. Kluyver in Delft, Holland, and isolated another organism, later identified as *Cl. tetanomorphum*, that uses glutamate as its main energy source and converts it to acetate, butyrate, carbon dioxide, ammonia and hydrogen.[(1)] Quantitative analysis showed that only 1 mole of carbon dioxide was formed per mole of glutamate fermented. This and other considerations seemed to exclude a degradation of glutamate to acetate via α-ketoglutarate, succinate, fumarate, malate, oxaloacetate and pyruvate since this sequence would yield at least 3 moles of carbon dioxide per mole of glutamate. On the basis of the stoichiometric relations among the fermentation products an alternative pathway was postulated that involved a C_1–C_4 cleavage of the glutamate molecule to give carbon dioxide and a four-carbon monocarboxylic acid intermediate that could be either oxidized to 2 moles of acetate or reduced to butyrate. Such a pathway is used by several glutamate-fermenting bacteria, but not by *Cl. tetanomorphum*.[(2)]

Further investigation of the pathway of glutamate fermentation was put off for many years while I became involved in what appeared to be more interesting problems, carbon dioxide utilization by heterotrophic bacteria, biological methane formation, sucrose synthesis, and the formation of caproic acid from ethanol and acetate by *Cl. kluyveri*. I did not return to the problem of glutamate fermentation until 1953 when several kinds of specifically ^{14}C-labeled glutamate had become available. Then I encouraged one of my students, Joseph T. Wachsman, to determine the fate of individual carbon atoms during the fermentation of glutamate by *Cl. tetanomorphum*. He showed that glutamate carbon atoms 1 and 2 are converted mainly to the carboxyl and methyl carbon atoms of acetate, respectively; glutamate carbon 4 is mainly incorporated into carbon atoms 1 and 4 of butyrate; and glutamate carbon 5 goes exclusively to carbon dioxide[(3)] (Fig. 1). This

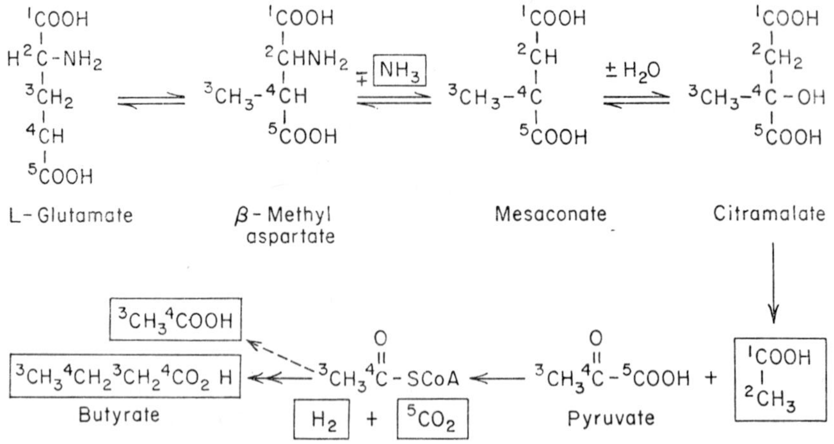

FIG. 1. Path of glutamate fermentation by *Cl. tetanomorphum*. The two reactions involved in the interconversion of glutamate to mesaconate are readily isolated from the subsequent reactions by the addition of dipyridyl which inhibits the Fe^{2+}-dependent conversion of mesaconate to citramalate. The latter reaction is also strongly inhibited by O_2.

labeling pattern was inconsistent with any previously described pathway of glutamate degradation. It suggested that glutamate fermentation involves a cleavage between carbon atoms 2 and 3 to form acetate (from carbon atoms 1 and 2) and pyruvate (from carbon atoms 5, 4 and 3), followed by an oxidation of the latter to carbon dioxide (carbon 5) and acetyl-CoA (carbon atoms 4 and 3) which is then converted largely to butyrate. Wachsman confirmed this suggestion in part by investigating the action of cell-free extracts of *Cl. tetanomorphum* on glutamate under anaeorbic conditions and showing that the products were mainly pyruvate, acetate, carbon dioxide and hydrogen. When a similar experiment was done in the presence of oxygen and the products were examined by paper chromatography, he detected the formation of an additional non-volatile acidic product that absorbed ultraviolet light.[(4)] This compound had several properties in common with fumaric acid, but the R_f values were consistently higher in several solvents. This suggested that it was an analog of fumaric acid containing an additional methyl or methylene group. Three unsaturated C_5 dicarboxylic acids (citraconic, itaconic, glutaconic) were tested and found to differ from the enzymatic product. Finally, I obtained a small sample of a fourth compound, mesaconic acid (2-methyl fumaric acid) from the Emil Fischer collection of chemicals through the assistance of Professor H. O. L. Fischer. Wachsman compared this compound with his enzymatic product and found it to be identical. Authentic mesaconate was also readily converted to pyruvate, acetate and carbon dioxide by the bacterial extract. The identification of mesaconate as an intermediate in glutamate dehradation was an unexpected development, since it evidently involved the conversion of the linear carbon skeleton of glutamate into a branched-chain compound. The only other somewhat similar reaction known at that time was the conversion of methylmalonyl-CoA to succinyl-CoA which Flavin, Ortiz and Ochoa[(5)] and Katz and Chaikoff[(6)] had recently found to participate in the metabolism of propionate by animal tissues.

The nature of the carbon skeleton rearrangement in the glutamate–mesaconate conversion was more clearly defined by using [4-^{14}C]glutamate as substrate and determining the position of the isotope in the resulting [^{14}C]mesaconate.[7] Chemical degradation of the mesaconate showed that it was labeled exclusively in the carbon atom adjacent to the methyl group. This result, in combination with other tracer experiments, proved that the methyl group of mesaconate is derived from carbon atom 3 of glutamate and that the formation of the branched chain structure of mesaconate results from cleavage of the bond between glutamate carbon atoms 2 and 3, and the formation of a new bond between carbon atoms 2 and 4. In other words carbon atoms 1 and 2 are transferred from carbon 3 to carbon 4, leaving carbon 3 in a methyl group (Fig. 1).

Since the conversion of glutamate to mesaconate appeared to involve at least two distinct reactions, a rearrangement of the carbon skeleton and a deamination, I made several attempts to detect the presumed intermediate. This was done first by allowing unfractionated extracts to partially convert [^{14}C]glutamate to mesaconate in the presence of dipyridyl (which blocks the further metabolism of mesaconate) and then looking for the accumulation of a new radioactive product by a combination of paper chromatography and radio-autography. These experiments gave negative results; only glutamate, mesaconate and some small radioactive impurities in the ^{14}C-glutamate preparation were detected. I then tried to see whether a similar experiment could be done with the reverse reaction starting with ammonium mesaconate. The equilibrium proved to be favorable for the reverse reaction and a considerable amount of a dicarboxylic amino acid accumulated that was identified as glutamate by paper and column chromatography. Only about a third of the amino acid was L-glutamate, judged by its reactivity with an L-glutamate decarboxylase preparation. The rest was thought to be D-glutamate which might be the primary product of the reverse reaction. The presence of both D- and L-glutamate was not unexpected since the *C. tetanomorphum* extract was known to contain a glutamate racemase.

At about this time (May 1956), Agnete Munch-Petersen completed experiments on the carbon skeleton rearrangement and I persuaded her to start an investigation of the cofactor requirements for the conversion of glutamate to mesaconate and ammonia during the few remaining months of her fellowship. To assay the reaction, she measured the rate of ammonia formation. She soon found that treatment of cell-free extract with charcoal rendered the extract almost inactive and the activity could be fully restored by adding a suitable amount of a heated, protein-free extract. By adding a rate-limiting amount of heated extract, the effects of adding various known cofactors could be tested. In this way she found that a number of coenzymes, including ATP, UTP, DPN, TPN and coenzyme A, did not stimulate glutamate decomposition. Of the inorganic cofactors tested, only Mg^{+2}, Co^{2+}, or Mn^{2+} accelerated the reaction slightly. Since none of the known cofactors tested could replace heated extract of *Cl. tetanomorphum*, Munch-Petersen began to investigate some ionic properties of the cofactor by passing heated extract through small columns of cation- or anion-exchange resins. She found that the activity was not retained by either type of resin at neutral pH, indicating that the cofactor activity was associated with an uncharged molecule. This was a valuable result since passage of heated extract through cation- and anion-exchange resins removed many charged molecules, including most of the common coenzymes.

These early experiments on the properties of the coenzyme were greatly complicated by the instability of both the enzyme system and the cofactor. The activity of the latter often declined in a most erratic manner as a result no doubt of a variable exposure to light which, as we later discovered, causes relatively rapid destruction of the cofactor. Consequently most experiments had to be done several times to obtain sufficiently reliable data to justify firm conclusions.

Although Munch-Petersen's results indicated that further study of the cofactor might be rewarding, I did not immediately continue this problem, partly because the isolation of a relatively unstable cofactor seemed like a formidable task, and partly because the chemical reactions involved in the conversion of glutamate to mesaconate needed to be further clarified. So I returned to the study of the latter conversion with the expectation that knowledge of the cofactor requirement might assist in the elucidation of the reaction sequence. I began this phase of the investigation by looking for a more convenient assay for the reversible conversion of glutamate to mesaconate than the formation or utilization of ammonia. I first tried a radiochemical assay based upon the conversion of ^{14}C-glutamate, a cation in acid solution, to the anionic ^{14}C-mesaconate which could be readily estimated by radioactivity measurements after passage through a cation-exchange resin. This type of assay was convenient for following glutamate decomposition but not mesaconate utilization because a commercial source of ^{14}C-mesaconate was not available. I next tried a spectrophotometric assay that proved to be rapid and convenient for following either the formation or the utilization of mesaconate. This assay was based on the measurement of the end absorbance of mesaconate at 240 nm. An initial difficulty with the assay was the high background absorbance caused by the nucleic acids and other materials in the bacterial extract, but this was overcome by cleaning up the enzyme by treatment with protamine and an ammonium sulfate precipitation.

With the spectrophotometric assay I quickly confirmed Munch-Petersen's observation that the conversion of glutamate to mesaconate is prevented by charcoal treatment of the extract and is restored by addition of heated extract, but found to my surprise that the reverse reaction was not affected by charcoal treatment and was not stimulated by heated extract. The amino acid formed from ammonium mesaconate by charcoal-treated extract had almost the same R_f as glutamate in paper chromatography in two solvents, but was not decarboxylated by L-glutamate decarboxylase. The earlier conclusion that the amino acid was D-glutamate was now excluded by the observation that this compound, unlike L-glutamate, was not decomposed by a partially purified enzyme preparation in the presence of the cofactor. Furthermore, the isolated amino acid formed from ammonium mesaconate by charcoal-treated extract, unlike L- or D-glutamate, was rapidly deaminated to mesaconate by the same extract. Therefore, I concluded that the amino acid formed from ammonium mesaconate by charcoal-treated extract must be a direct amination product, either 2-methyl- or 3-methylaspartic acid. Subsequent work identified the product as *threo*-3-methyl-L-aspartic acid.[8,9] This result, in combination with previous work, led to the conclusion that the cofactor-dependent reaction is the reversible conversion of glutamate to 3-methylaspartate (Fig. 1). When the enzyme responsible for this reaction was later purified the L-isomer of glutamate was found to be the actual substrate.

Following the detection of 3-methylaspartic acid, almost a year was devoted to the isolation and characterization of the amino acid,[10] the purification of the enzyme β-methylaspartase catalyzing its reversible deamination,[11] and study of the kinetics, specificity and equilibrium of the reaction. In January 1958 I returned to the study of the cofactor with the assistance of Robert Smyth. For the first time we used the spectrophotometric assay extensively for measuring cofactor activity. This was done by providing glutamate as substrate and determining the rate of mesaconate formation in the presence of excess glutamate mutase and β-methylaspartase and a rate-limiting level of cofactor. This assay had the great advantage over those used earlier that it permitted several experiments to be completed in a single day and therefore minimized complications caused by cofactor and enzyme instability. In a few weeks we were able to determine a number of properties of the cofactor. We found that it can be extracted from cells with hot 80% ethanol and can be concentrated at low pressure without much loss of activity. We also found that although the cofactor from the alcohol extracts is not retained by anion or cation-exchange resins at neutral pH, it is firmly held by Dowex-50-Na^{+} at pH 3 or below and can be eluted from the resin by sodium acetate buffers of higher pH. By means of differential adsorption and elution from a small column of this type we were able to separate the cofactor activity in a relatively narrow peak from most impurities. The active fractions at this stage of purity were colorless and showed no significant absorption in the ultraviolet region, but the activity was correlated with a small amount of a ninhydrin-reactive material. When the active fractions from the column were pooled and concentrated a faint pink color was observed, but since we had previously seen pink fractions in an elution pattern that did not correlate with cofactor activity, we concluded that the color was probably present in a contaminating compound. During concentration of the peak fractions, all the cofactor activity was lost.

We next undertook to purify the cofactor on a larger scale, using 23 g of lyophilized cells. The original extract was active, but during differential elution from the Dowex-50 column, the activity was again lost. The next day we started another larger preparation and this time obtained an orange elution peak that contained about 40% of the starting activity. In attempting to account for the great difference in cofactor yield from the two columns, I noted that the first column was eluted in the daytime when the cold room was in constant use and the lights were always on, whereas the second column was eluted mainly at night while the lights were off. This suggested that the cofactor might be decomposed by light. The next day I tested this hypothesis by comparing the activities of three samples of a cofactor solution, one exposed to a tungsten lamp, another to an ultraviolet lamp and the third kept in the dark. The solutions exposed to light lost all of their cofactor activity in about 2 hours, whereas the dark solution remained fully active. When a cofactor solution was exposed to direct sunlight the activity was completely destroyed within a few minutes. After making these observations we darkened our laboratories and shielded all cofactor solutions with black cloth and aluminum foil. Thereafter we had no difficulty with cofactor instability; the purification progressed rapidly and within 2 weeks we obtained solutions of highly purified cofactor.

The active compound had a bright orange color and a complex absorption spectrum (Fig. 2) with a high absorbance in both visible and ultraviolet regions.[12,13] For some time we had no idea what kind of compound we had isolated. The orange color was reminiscent of vitamin B_{12} (cyanocobalamin) which is deep red. However, I found that the absorption spectrum of cyanocobalamin differs markedly from that of the cofactor (Fig. 2). Cyano-cobalamin has major absorption peaks at about 360 and 500–570 nm, whereas the cofactor has peaks at about 260 and 460 nm. Although the spectra of the two compounds have some common features, a low and rather broad absorbance peak in the visible region and a high, narrow peak in the ultraviolet, the differences impressed me more than the similarities, and I did not recognize the relation of the compounds. This was delayed for

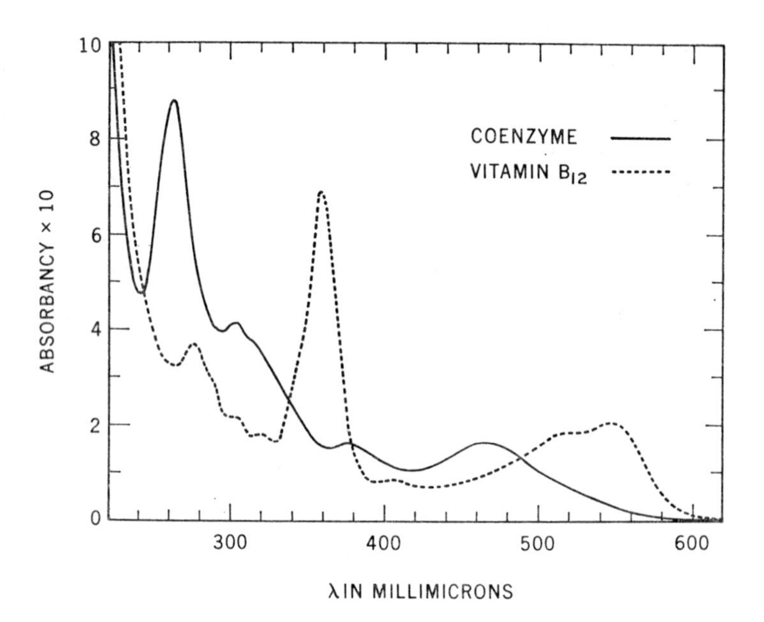

FIG. 2. Absorption spectra of the coenzyme form of pseudovitamin B_{12} (coenzyme) and of cyanocobalamin (vitamin B_{12}) in neutral aqueous solution.

about 7 weeks, until more was learned about the chemical and physical properties of the cofactor and its degradation products.

The first properties to be examined were the spectral and chemical changes associated with inactivation of the cofactor by various reagents. When the cofactor was exposed to light, the absorption spectrum underwent extensive changes which were directly proportional to the loss of activity. The most conspicuous change was a shift of the broad 460-nm peak to about 525 nm and the appearance of a conspicuous new peak at 351 nm. Examination of the products of the photolysis reaction by paper ionophoresis showed the formation of a new pink compound and a colorless UV quenching compound. Acid hydrolysis of the cofactor also gave red and colorless UV quenching products which, however, differed from those formed by photolysis. These experiments were done by Herbert Weissbach who was spending a year in my laboratory as a postdoctoral fellow. At about the same time, I tried adding KCN to the cofactor to see whether it could cause inactivation and produce a spectral change such as that which occurs with oxidized DPN. The activity was indeed destroyed; the solution changed from orange to purple and the spectrum underwent a dramatic change with the appearance of new peaks at 367, 545 and 579 nm. Had we been familiar with the chemistry of vitamin B_{12}, this spectrum would have immediately established the identity of the cofactor as a corrinoid compound. As it was, we did not recognize this relation until a month later when Weissbach went over all the spectra we had collected and noted the similarity between the spectra of cyanocobalamin in water and the coenzyme in alkaline cyanide solution. He then added cyanide to solutions of vitamin B_{12} and coenzyme and found that the spectra were identical from about 300 to 600 nm and differed only at shorter wavelengths (Fig. 3). The higher absorption of the coenzyme in the 260-nm region indicated that it contains a light-absorbing component not present in vitamin B_{12}. This component, which is split off from the coenzyme by acid hydrolysis or by treatment

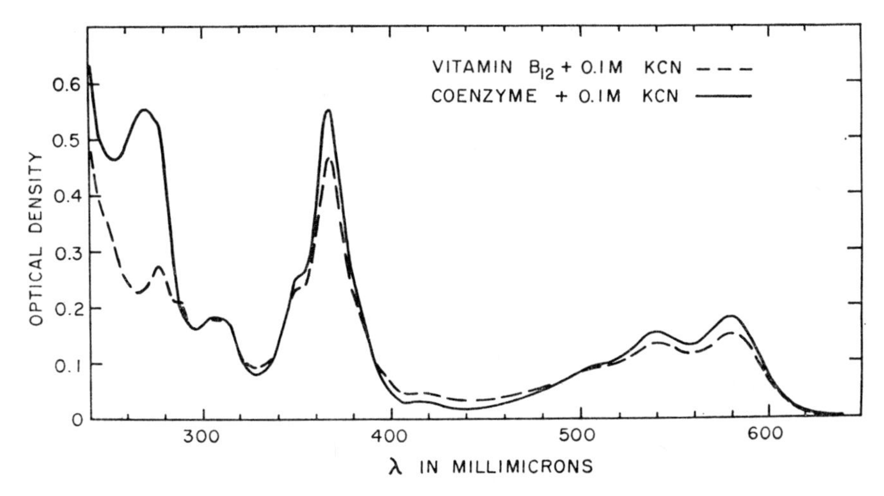

FIG. 3. Absorption spectra of the coenzyme form of pseudovitamin B_{12} (coenzyme) and cyanocobalamin (vitamin B_{12}) in 0.1 M KCN.

with alkaline cyanide, was soon identified as adenine. Indeed we found that the coenzyme contained 2 moles of adenine, only one of which was removed by cyanide treatment. The other adenine moiety remained attached to the resulting corrinoid compound that was identified as pseudovitamin-B_{12}; this differs from vitamin B_{12} by having adenine in place of dimethylbenzimidazole in the nucleotide side chain.

After the isolation and partial characterization of the coenzyme form of pseudovitamin-B_{12} (adenylcobamide (AC) coenzyme), Weissbach and John Toohey undertook to prepare the coenzyme form of vitamin B_{12} by growing *Cl. tetanomorphum* in the presence of 3,5-dimethylbenzimidazole. Earlier investigations in other laboratories had shown that several bacteria that normally form pseudovitamin B_{12} can be induced to synthesize a benzimidazole-containing corrinoid compound by adding a benzimidazole to the growth medium in suitable concentration. Since 3,5-dimethylbenzimidazole was not immediately available to us, unsubstituted benzimidazole was used in the first experiment. The result was that the bacteria grown with benzimidazole contained a six-fold higher level of coenzyme activity than those grown without benzimidazole, although the actual yield of coenzyme was not increased. We then grew a larger quantity of bacteria with benzimidazole and isolated the benzimidazolylcobamide (BC) coenzyme by the same methods previously used for the adenylcobamide coenzyme. An aqueous solution of the new coenzyme which was allowed to concentrate during several days in a desiccator in the cold, deposited small, highly pleochroic crystals that were yellow when viewed in one direction and red in the other. This was the first crystalline coenzyme preparation we obtained. A little later we purified a small amount of the coenzyme form of cobalamin (coenzyme B_{12}; DMBC coenzyme) from cells of *Cl. tetanomorphum* grown in the presence of dimethylbenzimidazole.[14] This coenzyme proved to be much less active than the other two analogs in the glutamate mutase assay, but later studies in other laboratories showed that it is more active in a number of other enzymatic systems, particularly those derived from animal tissues. Larger amounts of coenzyme B_{12} were soon obtained from cells of *Propionibacterium shermanii* generously supplied by David Perlman, Squibb Institute of Medical Research.[15]

This coenzyme readily crystallized from aqueous acetone. The crystals were dark red and lacked the conspicuous pleochroism of the benzimidazole analog.

Following the isolation of the B_{12} coenzymes the major chemical problem that remained was the elucidation of the mode of attachment of the adenine moiety to the corrinoid structure. Examination of the products of photolysis of the adenylcobamide coenzyme by Weissbach *et al.*[16] established that two adenosine-like nucleosides are formed. The molar quantity of adenine in the two nucleosides formed by photolysis was equal to the quantity of adenine liberated by treatment of coenzyme with alkaline cyanide. This evidence indicated that the extra adenine, characteristic of the coenzyme, is joined to a sugar moiety, probably a modified pentose. The adenine nucleoside could be linked to the corrinoid either through the adenine or the sugar. Our initial speculation and that of other investigators[17] favored a linkage between N7 of adenine and the cobalt atom since this type of bond was well known in corrinoid compounds. The assumption of the participation of cobalt in the bond was supported by the occurrence of the photolytic reaction. Previous investigations had shown that the bond between cobalt and cyanide in cyanocobalamin is labilized by light and it seemed probable that a similar type of reaction would be involved in the photolytic inactivation of the coenzyme.

The identification of the linkage between the adenine nucleoside and the corrinoid as a cobalt–carbon bond involving the 5′-carbon of a 5′-deoxyadenosyl moiety was made by Lenhert and Hodgkin[18,19] by means of the x-ray diffraction method (Fig. 4). At almost

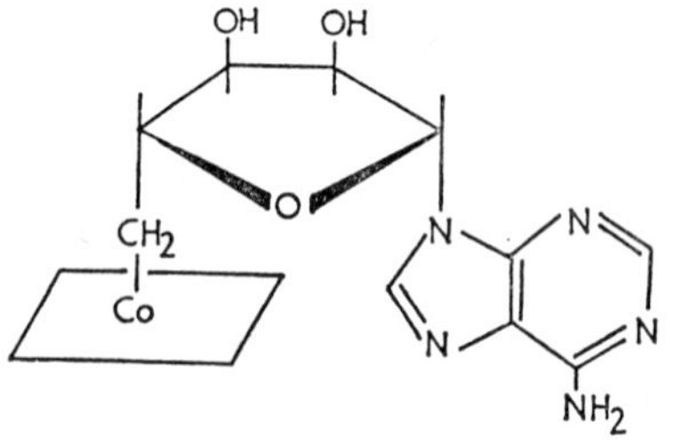

FIG. 4. Attachment of the deoxyadenosyl group to the cobalt of the corrinoid structure.

the same time, Johnson and Shaw[20] reported the identification of one of the products of photolysis as adenosine-5′-carboxylic acid which was presumably formed by oxidation of the 5′-deoxyadenosyl radical produced in the photolysis of coenzyme B_{12}. Adenosine-5-carboxylic acid has not subsequently been observed as a photolysis product, but the detection of this compound led Hogenkamp *et al.*[21] to the identification of adenosine-5′-aldehyde as one of the two photolysis products normally formed in the presence of oxygen. The second aerobic photolysis product was shown to be 8,5′-cyclic-adenosine by Hogenkamp.[22] This is the only adenine nucleoside formed by photolysis in the absence of oxygen.

The clarification of the structure of coenzyme B_{12} stimulated attempts to synthesize the compound from vitamin B_{12} and an activated adenosine derivative. Smith *et al.*[23] described a chemical synthesis of the coenzyme from fully reduced vitamin B_{12} (B_{12s}) and the 5′-tosyl-2′,3′-isopropylidene derivative of adenosine. At the same time they reported the synthesis of Co-methyl-cobalamin from B_{12s} and methyl iodide. Soon thereafter a biological function of methylcobalamin in the methylation of homocysteine was demonstrated by Guest *et al.*[24] Only one other Co-alkyl-corrinoid compound has so far been found in biological material, namely the carboxymethyl derivative isolated from *Cl. thermoaceticum* by Ljungdahl *et al.* (1965)

As a sidelight on the history of science, I should mention that we were not the first group to isolate a B_{12} coenzyme. During correspondence with E. Lester Smith in 1959, he mentioned that he and his associates at Glaxo Laboratories had isolated a similar compound in November 1953. Examination of the absorption spectrum of this compound provided convincing evidence that it was the deoxyadenosyl derivative of a benzimidazolylcobamide, possibly cobalamin. The biological significance of this compound was not recognized at that time because of the lack of a B_{12} coenzyme-specific assay system. The compound was regarded merely as one of the numerous minor corrinoids that occur in biological materials. A description of the compound was never published.

Finally, it should be noted that coenzyme B_{12} would probably have been discovered by Beck *et al.*[26] during their studies of the conversion of methylmalonyl-CoA to succinyl-CoA by the sheep kidney mutase (isomerase) were it not for the fact that this enzyme binds the coenzyme very tightly in such a way that it is not susceptible to inactivation by light. Because of these properties they were unable to obtain a coenzyme-free mutase and so concluded that the reaction probably has no cofactor requirement.

REFERENCES

1. BARKER, H. A. (1937) On the fermentation of glutamic acid, *Enzymologia* **2,** 175–182.
2. BUCKEL, W., BARKER, H. A. (1974) Two pathways of glutamate fermentation by anaerobic bacteria. *J. Bacteriol.* **117,** 1248–1260.
3. WACHSMAN, J. T. and BARKER, H. A. (1955) Tracer experiments on glutamate fermentation by *Clostridium tetanomorphum. J. Biol. Chem.* **217,** 695–702.
4. WACHSMAN, J. T. (1956) The Role of α-ketoglutarate and mesaconate in glutamate fermentation by *Clostridium tetanomorphum. J. Biol. Chem.* **223,** 19–27.
5. FLAVIN, M., ORTIZ, P. J. and OCHOA, S. (1955) Metabolism of propionic acid in animal tissues. *Nature,* **176,** 823–826.
6. KATZ, J. and CHAIKOFF, I. L. (1955) The metabolism of propionate by rat liver slices and the formation of isosuccinic acid. *J. Am. Chem. Soc.* **77,** 2659–2660.
7. MUNCH-PETERSEN, A. and BARKER, H. A. (1958) The origin of the methyl group in mesaconate formed from glutamate by extracts of *Clostridium tetanomorphum. J. Biol. Chem.* **230,** 649–653.
8. BARKER, H. A., SMYTH, R. D. and BRIGHT, H. J. (1961) β-methyl-L-aspartic acid. *Biochem. Prep.* **8,** 89–92.
9. WINITZ, M., BIRNBAUM, S. M. and GREENSTEIN, J. P. (1961) L- and D-*threo*-β-methylaspartic acid. *Biochem. Prep.* **8,** 96–99.
10. BARKER, H. A., SMYTH, R. D., WAWSZKIEWICZ, E. J., LEE, M. N. and WILSON, R. M. (1958) Enzymic preparation and characterization of an α-L-β-methylaspartic acid. *Arch. Biochem. Biophys.* **78,** 468–476.
11. BARKER, H. A., SMYTH, R. D., WILSON, R. M. and WEISSBACH, H. (1959) Purification and properties of β-methylaspartase. *J. Biol. Chem.* **234,** 320–328.
12. BARKER, H. A., WEISSBACH, H. and SMYTH, R. D. (1958) A coenzyme containing pseudovitamin B_{12}. *Proc. Nat. Acad. Sci. U.S.A.* **44,** 1093–1097.
13. BARKER, H. A., SMYTH, R. D., WEISSBACH, H., MUNCH-PETERSEN, A., TOOHEY, J. I., LADD, J. N., VOLCANI, B. E. and WILSON, R. M. (1960) Assay, purification and properties of the adenylcobamide coenzyme. *J. Biol. Chem.* **235,** 181–190.
14. WEISSBACH, H., TOOHEY, J. I. and BARKER, H. A. (1959) Isolation and properties of B_{12} coenzymes containing benzimidazole or dimethylbenzimidazole. *Proc. Nat. Acad. Sci. U.S.A.* **45,** 521–525.
15. BARKER, H. A., SMYTH, R. D., WEISSBACH, H., TOOHEY, J. I., LADD, J. N. and VOLCANI, B. E. (1960) Isolation and properties of crystalline cobamide coenzymes containing benzimidazole or 5,6-dimethylbenzimidazole. *J. Biol. Chem.* **235,** 480–488.
16. WEISSBACH, H., LADD, J. N., VOLCANI, B. E., SMYTH, R. D. and BARKER, H. A. (1960) Structure of the adenylcobamide coenzyme: degradation by cyanide, acid and light. *J. Biol. Chem.* **235,** 1462–1473.
17. JOHNSON, A. W. and SHAW, N. (1960) Some structural observations on the vitamin B_{12} coenzyme. *Proc. Chem. Soc.*, pp. 420–421.

18. LENHERT, P. G. and HODGKIN, D. C. (1961) Structure of the 5,6-dimethylbenzimidazolylcobamide coenzyme. *Nature* **192,** 937–938.
19. LENHERT, P. G. and HODGKIN, D. C. (1962) The X-ray crystallographic investigation of the 5,6-dimethylbenzimidazole-cobamide coenzyme. *In*: *Vitamin B_{12} and Intrinsic Factor* (H. C. HEINRICH, ed.) pp. 105–110, Enke Verlag, Stuttgart.
20. JOHNSON, A. W. and SHAW, N. (1961) Isolation of a crystalline nucleoside from the vitamin B_{12} coenzyme. *Proc. Chem. Soc.*, pp. 447–448.
21. HOGENKAMP, H. P. C., LADD, J. N. and BARKER, H. A. (1962) The identification of a nucleoside derived from coenzyme B_{12}. *J. Biol. Chem.* **237,** 1950–1952.
22. HOGENKAMP, H. P. C. (1963) A cyclic nucleoside derived from Coenzyme B_{12}. *J. Biol. Chem.* **238,** 477–480.
23. SMITH, E. L., MERVYN, L., JOHNSON, A. W. and SHAW, N. (1962) Partial synthesis of vitamin B_{12} coenzyme and analogues. *Nature* **194,** 1175.
24. GUEST, J. R., FRIEDMAN, S., WOODS, D. D. and SMITH, E. L. (1962) A methyl analogue of cobamide coenzyme in relation to methionine synthesis by bacteria. *Nature* **195,** 340–342.
25. LJUNGDAHL, L., IRION, E. and WOOD, H. G. (1965) Total synthesis of acetate from CO_2. **I.** Co-methyl-cobyric acid and Co-(methyl)-5-methoxybenzimidazolyl-cobamide as intermediates with *Clostridium tetanomorphum*. *Biochem.* **4,** 2771–2780.
26. BECK, W. S., FLAVIN, M. and OCHOA, S. (1957) Metabolism of propionic acid in animal tissues. II. Formation of succinate. *J. Biol. Chem.* **249,** 997–1010.

TRAILING THE PROPIONIC ACID BACTERIA

Harland G. Wood
Department of Biochemistry, Case Western Reserve University,
Cleveland, Ohio 44106

I decided I better look up the meaning of trailing in the dictionary after I sent in this title and found the following explanation: (1) (a) To drag or let drag behind one, especially on the ground, etc.; (b) to bring along behind; as he trailed dirt in the house. (2) (a) To make or mark (a path, track, etc.); (b) to make a path in (grass, etc.). (3) To follow the tracks of; track. (4) To hunt by tracking. (5) To follow behind, especially in a lagging manner. I was raised in Minnesota, where the Sioux Indians hunted and trailed and I intended to give the impression by the title that I had been stalking the propionic acid bacteria like a hound follows the scent of game. Not that I had been dragging dirt in the house or even that I was lagging behind the propionic acid bacteria, although, come to think of it, maybe I have been all these years.

I have trailed the propionic acid bacteria, admittedly with a few diversions such as investigating: the effect of poliomyelitis on the anaerobic glycolysis of brain from cotton rats, the conversion of CO_2 to both carbons of acetate by *Clostridium thermoaceticum*, determining the distribution of ^{14}C in the carbon chain of glucose of liver glycogen as an indicator of the intermediary metabolism of labeled compounds given *in vivo* to rats, the use of labeled glucose to estimate the role of the pentose cycle *in vivo* and for a while I milked the left half of an udder into one bucket and the right half into another bucket to study lactose synthesis following injection of ^{14}C tracers in the left pudic artery of a cow. Aside from a few such diversions, I have been trailing the propionic acid bacteria for over 40 years. It has been an exciting trail but I imagine if one studied any subject for 40 years, he would be bound to find something that is exciting. I suppose the same would have been true with *E. coli*, for example, but with the propionic acid bacteria, one had the field more or less to himself and could investigate them serenely without fear of being overcome by a host of other investigators.

The propionic acid bacteria have many interesting properties.

1. They metabolize glycerol and some other compounds with a net uptake of CO_2. Very few species aside from the autotrophs have this property.

2. They ferment 3, 4, 5 or 6 carbon compounds yielding essentially the same products from each, i.e. propionate, acetate, succinate and CO_2.

3. They ferment glucose labeled in any one of its six carbons to products which contain the tracer in every carbon of the product. This is in marked contrast to many fermentations which are so faithful that the location of the carbon in a product may be used as an exact indicator of the origin of that carbon in the glucose.

4. They contain a host of interesting enzymes. One contains a corrinoid, another biotin. One provided the first clearly demonstrated example of a pyrophosphoryl form of

an enzyme as well as a phosphoryl form and with a novel kinetic mechanism which is tri uni uni ping pong. Another of its enzymes provided the first example of a hybrid or non-classical ping pong mechanism in which substrates occupy distinctly separate sites. This same enzyme, in contrast to most enzymes, which are held in more or less rigid structure by multiple weak bonds, has a flexible structure presenting a variety of profiles in the electron microscope, one of which has a profile like the head of Mickey Mouse; thus, the enzyme has become known as the Mickey Mouse enzyme. This same enzyme is a metallo-enzyme containing cobalt, which is rare among enzymes.

5. To cap it off, the propionic acid bacteria may utilize inorganic pyrophosphate as a source of energy. Recently, it has been found that fructose-6-P is converted to fructose-1,6-diP using PPi and that the usual ATP-dependent phosphofructokinase is of very low activity in these bacteria. In addition, PPi is involved in two other reactions, one in CO_2 fixation and another in which pyruvate is converted to P-enolpyruvate. These reactions may be vestiges surviving from primitive forms which utilized inorganic pyrophosphate rather than ATP as a form of energy. This metabolism also may in part account for the fact that the propionic acid bacteria yield an exceptionally large amount of cells per mole of fermented glucose.

My trailing of the propionic acid bacteria began at Iowa State College (now Iowa State University) at Ames, Iowa, in the early 1930s during the depression. The only time I have ever lost a few days, work due to contamination of a fermentation was during one of the dust storms which we had in those days. This occurred even though the inoculation was done in a closed room in which the air had been washed down with a sprinkling system. Dust simply sifted into every nook and corner of the home or laboratory during those storms.

I have written previously in some detail about how we discovered fixation of CO_2 using the propionic acid bacteria[1] and will only describe it briefly here. At that time it was customary to study fermentations by determining carbon and oxidation-reduction balances. A substrate was fermented and the products were determined quantitatively. The purpose was to determine if all the carbon of the fermented substrate was accounted for in the products and whether the oxidized products equalled the reduced products as they should in an anaerobic fermentation when oxygen is not present as an oxidant. We used $CaCO_3$ to neutralize the acids produced by the fermentation and to maintain a neutral pH. At the conclusion of the fermentation, the medium was acidified and boiled to liberate the CO_2 from the excess $CaCO_3$. The CO_2 produced by the fermentation is the total CO_2 minus the CO_2 added as $CaCO_3$. This procedure had worked very well in fermentations of glucose and we obtained good carbon and oxidation-reduction balances. However, in a set of glycerol fermentations, the results were very disappointing because negative values for the CO_2 were obtained, i.e. the total CO_2 was less than that added initially as $CaCO_3$. I considered that I had made some sort of foolish error in weighing out the $CaCO_3$ but there was still another nagging fact. Although the carbon balances were quite good (if anything, a bit more carbon was recovered in the products than could be accounted for by the fermented glycerol), the oxidation reduction balances were very poor, there being a large excess of oxidized products.

About a year later, while writing my Ph.D. thesis, I decided that even though I thought the CO_2 values were wrong I should try and write something about the glycerol fermentations. Suddenly the idea hit me, if the missing CO_2 which I had considered to be in error was actually correct and the CO_2 had been reduced and utilized, it should be taken into account in calculating the balances. I hurriedly calculated the O/R balances and they were almost perfect. Then and there I knew that CO_2 is utilized by the propionic acid bacteria. But, the old dogma that CO_2 is used only by autotrophs was strong at that time and it was not until radioactive carbon became available as a tracer in the 1940s that it was shown that most, if not all, forms of life utilize CO_2. In fact I am told on good authority that when I presented the first results at a North Central Branch Meeting of the Society of American Microbiologists in Minneapolis in 1935 that someone sitting next to Professor Werkman, my mentor, said, "I don't believe a word of it" and Professor Werkman, is said to have replied, "I don't either".

When tracer carbon became available, our interest in CO_2 and the mechanism by which it is utilized naturally lead us to the use of this tool. We found the tracer in the carboxyls of succinate and from that and other results proposed that CO_2 is fixed as follows:

$$\text{Pyruvate} + CO_2 \rightarrow \text{oxalacetate} \quad (1)$$

This reaction became known as the Wood and Werkman reaction. Actually, the Wood and Werkman reaction does not occur in the propionic acid bacteria, although, pyruvate carboxylase which does catalyze this reaction but with utilization of ATP, has been found in numerous other tissues and organisms,

$$\text{Pyruvate} + CO_2 + \text{ATP} \leftrightharpoons \text{oxalacetate} + \text{ADP} + \text{Pi} \quad (2)$$

Strangely, we had never turned our attention to determining how CO_2 is fixed by the propionic acid bacteria and in 1961 when we did get to this problem we discovered, to our surprise, a new enzyme for fixation of CO_2. There were three enzymes which were known to fix CO_2 into oxalacetate at that time, P-enolpyruvate carboxylase discovered by R. S. Bandurski and G. M. Griener in 1953,

$$\text{P-enolpyruvate} + CO_2 \leftrightharpoons \text{oxalacetate} + \text{Pi} \quad (3)$$

P-enolpyruvate carboxykinase described by M. F. Utter and K. Kurahashi in 1953,

$$\text{P-enolpyruvate} + CO_2 + \text{GDP} \leftrightharpoons \text{oxalacetate} + \text{GTP} \quad (4)$$

and pyruvate carboxylase discovered by M. F. Utter and D. B. Keech in 1960 (reaction 2 above). We thought the enzyme of propionibacteria would be one of these three and probably pyruvate carboxylase. In addition, fixation of CO_2 into malate was known to occur by the malate enzyme which had been discovered by S. Ochoa, A. H. Mehler and A. Kornberg in 1948,

$$\text{Pyruvate} + CO_2 + \text{TPNH} \leftrightharpoons \text{malate} + \text{TPN} \quad (5)$$

and propionyl CoA carboxylase described by M. Flavin, P. J. Ortiz and S. Ochoa in 1955,

$$\text{Propionyl CoA} + \text{ATP} + CO_2 \leftrightharpoons \text{methylmalonyl CoA} + \text{ADP} + \text{Pi} \quad (6)$$

So we had these latter enzymes under consideration as well. Rune Stjernholm and I made crude extracts of the propionic acid bacteria and using the proper substrates we looked for $^{14}CO_2$ conversion to the expected compounds. As a test of $^{14}CO_2$ fixation we simply bubbled $^{12}CO_2$ through the incubation mixture after making it acid and then determined if there was radioactivity remaining in the mixture. We soon found with P-enolpyruvate and $^{14}CO_2$ that there was fixation of CO_2 and the product was oxalacetate. But then an amusing incident occurred. All progress by Stjernholm and me came to a halt. As soon as we tried to purify the enzyme by ammonium sulfate fractionation it lost all activity. However, we found activity was restored if we added some boiled crude extract to the ammonium sulfate fraction. We then thought there must be a co-factor and we tested all the co-factors we could find, but to no avail. Finally, we ashed the boiled extract and to our surprise even this restored the activity. Then we tested a long series of metals, even concentrated tap water but all of these failed to give activity. We could consistently repeat the results and there certainly was a clear-cut problem for someone to solve.

At that time we became involved full time working on transcarboxylase so I asked Pat Siu, one of my graduate students, to take a hand with this problem. About 3 days later he came to me and said, "I don't know what is wrong with you and Stjernholm, I can fractionate the crude extract with ammonium sulfate and the fraction is active with no further additions". I said, "Ah, come on, what are you talking about, Rune and I have done those experiments a dozen times and never got activity". But Pat refused to be swayed, so we went over his protocol step by step. Then we discovered that Pat Siu had dissolved the ammonium sulfate fraction in phosphate buffer whereas Stjernholm and I had used Tris buffer. Thus, Siu had shown that phosphate is required for this reaction and then and there we wrote down the reaction:

$$\text{P-enolpyruvate} + CO_2 + \text{Pi} \leftrightharpoons \text{oxalacetate} + \text{PPi} \qquad (7)$$

This was based on the thought that our enzyme was catalyzing a reaction similar to that of P-enolpyruvate carboxykinase (reaction 4) which was being studied by Utter's group in the department, only in our case Pi was the phosphate acceptor rather than GDP and PPi was the product rather than GTP. Since there is a transphosphorylation from P-enolpyruvate to Pi and a CO_2 fixation, the enzyme is called carboxytransphosphorylase. Siu soon verified the reaction and went on to complete the study for his Ph.D. thesis. He, at the same time, taught me that one should not keep making the same mistake over and over!

Some time later, one of my graduate students, Herbert Evans, began a search for pyruvate carboxylase in the propionic acid bacteria. He did not find pyruvate carboxylase but he did find a second enzyme which utilized inorganic pyrophosphate. The rationale back of this study was that the propionic bacteria might convert pyruvate to P-enolpyruvate through the combined action of pyruvate carboxylase [reaction (2)] and carboxytransphosphorylation [reaction (7)]. This idea was based on the studies of M. F. Utter and his coworkers who had shown that P-enolpyruvate is formed in liver by a mechanism involving pyruvate carboxylase [reaction (2)] and carboxykinase [reaction (4) using GTP].

$$\text{Pyruvate} + \text{ATP} + CO_2 \leftrightharpoons \text{oxalacetate} + \text{ADP} + \text{Pi} \qquad (2)$$

$$\text{Oxalacetate} + \text{PPi(GTP)} \leftrightharpoons \text{P-enolpyruvate} + \text{Pi(GDP)} + CO_2 \qquad (4 \text{ or } 7)$$

$$\text{Sum:}\quad \text{Pyruvate} + \text{ATP} + \text{PPi(GTP)} \leftrightharpoons \text{P-enolpyruvate} + \text{ADP} + 2\text{Pi(Pi} + \text{GDP)} \qquad (8)$$

We were intrigued by the idea that the propionic acid bacteria might use carboxytransphosphorylase [reaction (7) with PPi] and thus substitute PPi for GTP and conserve the energy of PPi instead of wasting it in hydrolysis as is usually proposed. We considered this conservation of energy might in part account for the high growth yields of the propionic acid bacteria. Evans, therefore, looked for pyruvate carboxylase. For a time, he thought he had it, he used ATP, pyruvate and $^{14}CO_2$ and found oxalacetate was formed from these substrates. But, then he found Pi was required for the reaction (shades of Pat Siu!). Evans then found that oxalacetate is not formed from pyruvate by pyruvate carboxylase in propionic acid bacteria but is formed by the following sequence of reactions:

$$\text{Pyruvate} + \text{ATP} + \text{Pi} \leftrightharpoons \text{P-enolpyruvate} + \text{AMP} + \text{PPi} \qquad (9)$$

$$\text{P-enolpyruvate} + CO_2 + \text{Pi} \leftrightharpoons \text{oxalacetate} + \text{PPi} \qquad (7)$$

$$\text{Sum:}\quad \text{Pyruvate} + \text{ATP} + CO_2 + 2\text{Pi} \leftrightharpoons \text{oxalacetate} + \text{AMP} + 2\text{PPi} \qquad (10)$$

No wonder there was a need for Pi, it is required for both reactions. The enzyme catalyzing reaction (9) is called pyruvate, phosphate dikinase, since both pyruvate and Pi are phosphorylated by ATP. Thus, the propionibacteria instead of forming oxalacetate directly from pyruvate as does liver, form it by the combined action of two enzymes while liver forms oxalacetate directly from pyruvate but forms P-enolpyruvate indirectly from oxalacetate. Comparative biochemistry does have its limitations! Independently, in the same year, R. E. Reeves found pyruvate, phosphate dikinase in *Entamoeba histolytica* and M. D. Hatch and C. R. Slack found it in tropical grasses such as sugar cane.

Pyruvate, phosphate dikinase has proven to be an exceedingly interesting enzyme. The reaction sequence involves three partial reactions and three forms of the enzyme as shown below:

$$\text{Enzyme} + \text{ATP} \leftrightharpoons \text{enzyme-PP} + \text{AMP} \qquad (11)$$

$$\text{Enzyme-PP} + \text{Pi} \leftrightharpoons \text{enzyme-P} + \text{PPi} \qquad (12)$$

$$\text{Enzyme-P} + \text{pyruvate} \leftrightharpoons \text{P-enolpyruvate} + \text{enzyme} \qquad (13)$$

$$\text{Sum:}\quad \text{Pyruvate} + \text{ATP} + \text{Pi} \leftrightharpoons \text{P-enolpyruvate} + \text{AMP} + \text{PPi} \qquad (9)$$

Yoram Milner came to my laboratory shortly thereafter and he demonstrated that the mechanism involves tri uni uni ping pong kinetics as predicted for a reaction involving three different forms of the enzyme. He also showed there are three independent substrate sites on this dimeric enzyme of molecular weight ~160,000, one for each substrate pair (ATP, AMP; Pi, PPi; and pyruvate, P-enolpyruvate). The mechanism may be pictured as follows where — is a covalent bond and · is a Michaelis–Menten type complex:

$$\begin{array}{cccccccc}
 & \text{ATP}\downarrow & \text{AMP}\uparrow & \text{Pi}\downarrow & \text{PPi}\uparrow & \text{Pyruvate}\downarrow & \text{P-enolpyruvate}\uparrow & \\
\hline
E & E^{-PP}_{\cdot AMP} & E^{-PP} & E^{-PP}_{\cdot Pi} \rightarrow E^{-P}_{\cdot PPi} & E^{-P} & E^{-P}_{\cdot \text{pyruvate}} \rightarrow E_{\cdot \text{P-enolpyruvate}} & & E
\end{array}$$

Furthermore, the pyrophosphoryl and phosphoryl forms of the enzyme were isolated and demonstrated to be fully active. This was the first unequivocal proof of a pyrophosphoryl enzyme as intermediate in a reaction.[2]

Although PPi is involved in reactions (7) and (9), there is no indication that the enzymes function physiologically in the direction of utilization of PPi in propionibacteria. However, recently a third enzyme has been found by William E. O'Brien in propionibacteria which apparently does use PPi for phosphorylation. It catalyzes the following reaction:

$$\text{PPi} + \text{fructose-6-P} \leftrightharpoons \text{Pi} + \text{fructose-1,6-diP} \qquad (14)$$

and its activity is much greater in propionibacteria than the ATP-dependent phosphofructokinase.[11] It appears that there is insufficient ATP-dependent phosphofructokinase in the bacteria to fulfill the requirements for glucose metabolism and that the PPi dependent phosphofructokinase serves this function. Richard Reeves had reported the presence of this enzyme in *Entamoeba histolytica*[3] and it was this finding which led us to look for the enzyme in propionibacteria. It is indeed interesting that the eukaryotic amoeba and the prokaryotic propionibacteria both contain three enzymes in which inorganic pyrophosphate has a role [reactions (7), (9) and (14)]. Reeves has presented a strong case to support the view that *Entamoeba histolytica* produces and utilizes PPi as an energy source. With propionibacteria, the case is not as strong since we have not found a reaction on the main catabolic pathway of glucose which provides the PPi for phosphorylation of fructose-6-P. We think there must be such a reaction. Although PPi is generated during the anabolic syntheses of proteins, fats and nucleic acids and by the anaplerotic reactions (7) and (9), one wonders if these reactions are sufficient to serve as a source of PPi for a main catabolic pathway. Perhaps PPi generation is linked to electron transfer during the reduction of fumarate to succinate which is on the main pathway of propionate formation.

Let us now turn back in time again to about 1950 to review how the discovery of the enzyme transcarboxylase in propionibacteria came about. Of all enzymes that we found in trailing the propionic acid bacteria, transcarboxylase is perhaps the most intriguing. No one could have predicted that it would have a multitude of interesting features which would lead us to a study of the role of biotin in the reaction, to a study of ping pong kinetics, to the use of electron microscopy, to amino acid sequencing, to studies of protein–protein interactions and even to a bit of immunology. It had generally been considered that the propionic acid fermentation occurs as diagrammed in abridged form in Fig. 1.

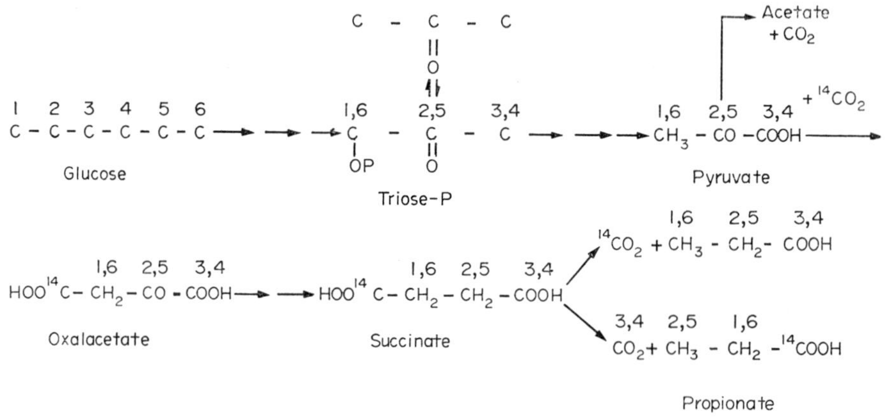

Glucose (1 2 3 4 5 6: C – C – C – C – C – C) → → → Triose-P (1,6 C(OP) – 2,5 C=O – 3,4 C ⇌ C – C(=O) – C) → → → Pyruvate (1,6 CH_3 – 2,5 CO – 3,4 COOH) → Acetate + CO_2 ; + $^{14}CO_2$ →

Oxalacetate (1,6 2,5 3,4: $HOO^{14}C–CH_2–CO–COOH$) → → Succinate (1,6 2,5 3,4: $HOO^{14}C–CH_2–CH_2–COOH$) → $^{14}CO_2 + CH_3–CH_2–COOH$ (1,6 2,5 3,4) or $CO_2 + CH_3–CH_2–^{14}COOH$ (3,4 2,5 1,6)

Propionate

FIG. 1. Mechanism which had been postulated for the formation of propionate from glucose. The numbers above the compounds indicate the source of the carbons from the glucose.

According to this scheme, CO_2 is fixed in oxalacetate which is then converted to succinate. The succinate in turn is decarboxylated to propionate. Since succinate is a symmetrical molecule there is an equal chance of either carboxyl being removed, the one which originated from carbon of the glucose or the one from the $^{14}CO_2$ that had been fixed in oxalacetate (if the fermentation is done in the presence of a pool of $^{14}CO_2$) (see

Fig. 1). Therefore, for every mole of propionate formed there should be half a mole of $^{12}CO_2$ produced from carbon of the glucose and in addition for every mole of acetate formed there should be one mole of $^{12}CO_2$ from the glucose (see Fig. 1). Fred Leaver and I decided to see if this is true and we determined the amount of $^{12}CO_2$ produced by measuring the dilution of the ^{14}C in the $^{14}CO_2$ pool by $^{12}CO_2$. To our surprise the $^{12}CO_2$ produced was far below that predicted for the proposed fixation of CO_2 in oxalacetate, its reduction to succinate and the decarboxylation to CO_2 and propionate.

We also undertook a series of studies on the distribution of ^{14}C in the products formed from a variety of labeled substrates. The results showed that, although succinate was not decarboxylated to CO_2, the tracer patterns in the products were in accord with the assumption that succinate is a precursor of propionate. For example, if propionate were formed from succinate then glucose-1-^{14}C, glucose-2-^{14}C or glucose-6-^{14}C should give equal labeling in the 2 and 3 positions of propionate (see Fig. 1). Such results were uniformly observed. In fact, the isotope was randomized in the products even beyond that expected from the occurrence of just a symmetrical C_4 dicarboxylic acid. It also appears that a symmetrical C_3 compound is formed during the fermentation, perhaps by dephosphorylation of dihydroxyacetone phosphate (Fig. 1). For evidence, Stjernholm and I used trehalose as indicator of metabolism.[4] Between these two symmetrical compounds the ^{14}C of any position of glucose becomes partially randomized to every position of the products. I shudder to think what would have transpired if the propionic acid rather than the lactic acid bacteria had been selected for the first determination of isotope patterns in glucose by fermentation.

From these tracer studies, we were quite convinced that propionate is formed from succinate by decarboxylation but the CO_2 turnover studies showed that this decarboxylation did not involve free CO_2. This forced us to consider that the decarboxylation might occur by a transcarboxylation from succinate to pyruvate to form oxalacetate, which would account for the observed isotope patterns and for the low turnover of CO_2.

R. W. Swick undertook the job of looking for this enzyme. At that time, Flavin and Ochoa had shown that methylmalonyl CoA is formed from propionyl CoA by CO_2 fixation and Swick took a clue from this work and found that methylmalonyl CoA rather than succinyl CoA is the carboxyl donor in the transcarboxylation (see ref. 5 for a more complete review of transcarboxylase). The reaction is as follows:

$$\underset{\displaystyle CH_3{\cdot}CH{\cdot}COSCoA}{\overset{COO^-}{|}} + CH_3{\cdot}CO{\cdot}COO^- \rightleftharpoons CH_3{\cdot}CH_2{\cdot}COSCoA + {}^-OOC{\cdot}CH_2{\cdot}CO{\cdot}COO^- \qquad (15)$$

In the fermentation the sequence occurs as shown in Fig. 2. The carboxyl is transferred from methylmalonyl CoA to pyruvate yielding oxalacetate and propionyl CoA. The oxalacetate is reduced to succinate. CoA then is transferred from the propionyl CoA to succinate and the methylmalonyl CoA is regenerated from the resulting succinyl CoA. The net result of this cycle is the conversion of pyruvate to propionate. In studying the cycle we investigated the conversion of succinyl CoA to methylmalonyl CoA by the corrinoid enzyme, methylmalonyl CoA mutase, but space limitation prevents discussion of this interesting enzyme.

Shortly after we discovered transcarboxylase I went to Feodor Lynen's laboratory in Munich, Germany, in 1962 for a sabbatical leave. While there, S. Wakil and M. Waite caused quite a stir. They reported that in biotin enzymes the biotin acts as a carboxyl carrier in CO_2 fixation through turnover of the uriedo carbon rather than as Lynen's group had reported, by CO_2 becoming a carboxamide at the 1′-nitrogen of the biotin. There followed a flurry of activity which resulted in convincing evidence from several groups that biotin enzymes function as carboxyl or CO_2 carriers via the 1′-nitrogen. Later, after I returned to Cleveland, together with George Allen and Rune Stjernholm, by use of biotin labeled with ^{14}C in the uriedo carbon, it was shown that there is no turnover of the uriedo carbon of the biotin of transcarboxylase during transcarboxylation.

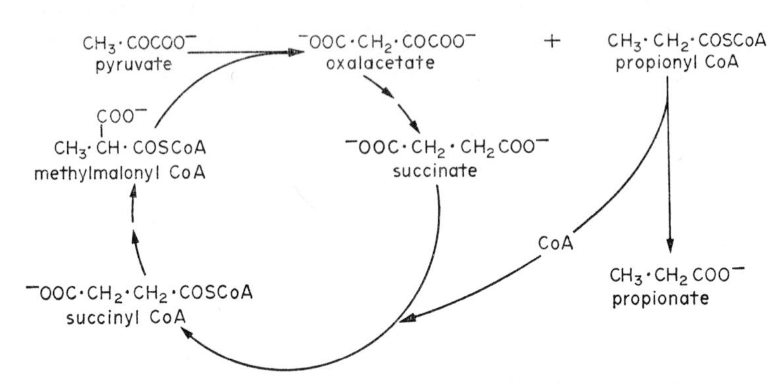

FIG. 2. Role of transcarboxylase in the propionic acid fermentation.

It was then that I was introduced to ping pong kinetics. Michael Scrutton, a graduate student with Dr. M. F. Utter, came to me full of enthusiasm and said that transcarboxylase should exhibit ping pong kinetics if the reaction involves a free and carboxylated form of the enzyme. His enthusiasm prevailed and I assigned one of my graduate students, Dexter Northrop, to undertake the study. Northrop showed in a rather short time that all the required parallel lines could be demonstrated for ping pong kinetics. However, there was one troublesome problem. Previous enzymes exhibiting ping pong kinetics, such as transaminase, had a single substrate site and the kinetic theory had been developed on the basis of compulsory binary complexes. For example, in transamination, aspartate combines with the enzyme and the amino group is transferred to the pyridoxal of the enzyme and the resulting oxalacetate is released. Pyruvate then occupies the site and accepts the amino group yielding alanine. When it was suggested that transcarboxylase might likewise have only one substrate site, I was very skeptical because methylmalonyl CoA and propionyl CoA are structurally very different than oxalacetate and pyruvate. Consequently, I urged Northrop to try to prove that there are two substrate sites on transcarboxylase even though it did display ping pong kinetics. Northrop then found by use of inhibitors that transcarboxylase does, in fact, have two independent binding sites, one for the CoA esters and one for the keto acids. He thus demonstrated for the first time the existence of the "two site" hybrid, or non-classical ping pong mechanism and showed that with two sites, parallel lines are theoretically possible.[6]

The quaternary structure of transcarboxylase is pictured in Fig. 3. This structure has been deduced from electron microscope studies and other information.[7] The protein, avidin, complexes very tightly with biotin and it proved very useful in locating the biotins in transcarboxylase.[7] In the electron microscope the avidin of the complexes is seen to lie between each set of peripheral subunits thus showing that the biotins are located at each end of these subunits. The hole in the central $12S_H$ subunit likewise becomes evident. Recently, trypsin has been used to advantage in these studies. On treatment of the enzyme

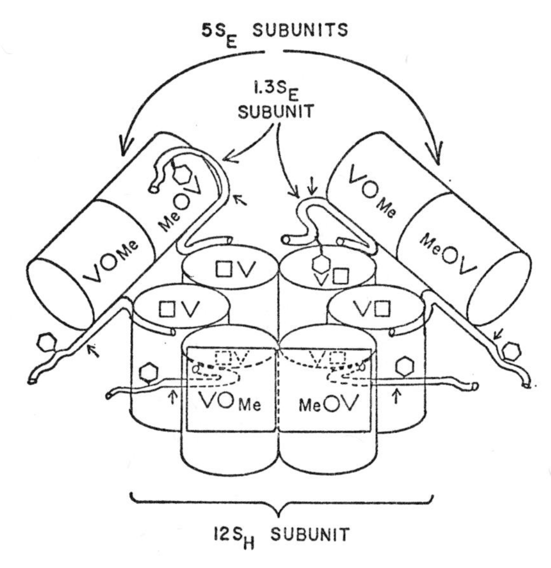

FIG. 3. Diagrammatic representation of the structure of transcarboxylase and the mechanism of the transcarboxylation reaction.[9] The peripheral dimeric $5S_E$ subunits are linked to the central hexameric $12S_H$ subunit by the $1.3S_E$ biotinyl carboxyl carrier subunits (⬡ = biotin ring). This 18S form of the enzyme is thus made up of eighteen peptides, six of each type. The keto acid and divalent metal sites (indicated by O and Me) occur on the peripheral subunits and the CoA ester sites (□) are on the central subunit. The binding sites for the biotinyl ring (V) are near the substrate sites so that the biotinyl groups can serve as carboxyl carriers between the different substrate sites. The arrows indicate cleavage by trypsin to yield the biotinyl peptides.

with trypsin 65 and 46 residue peptides are released which contain biotin but the main protein structure remains intact.[8] We have shown that the remaining non-biotinyl portion of the $1.3S_E$ subunits provides the amino acid sequence which serves to bind the $12S_H$ and $5S_E$ subunits together. The non-biotinyl peptide which contains 42 residues has been isolated and it has been shown that it or the $1.3S_E$ subunit (see Fig. 3) is required for assembly of the $5S_E$ and $12S_H$ subunits.[8]

Gene Zwolinski and Botho Bowien of our department, in collaboration with Hans Neurath's group of the University of Washington and particularly with Lowell Ericsson and Kenneth Walsh, have determined most of the sequence of the 123 residue $1.3S_E$ subunit.[12] Thus, we now are in a position to explore the subunit–subunit interaction using known structures to form the complexes.

We recently have shown[9] that the CoA ester site is on the $12S_H$ subunit and the keto acid site is on the $5S_E$ subunit (Fig. 3). Like other biotin enzymes, the transcarboxylase mechanism involves two partial reactions. The $12S_H$ subunit in combination with the 1.3_E biotinyl subunit catalyzes the first partial reaction and the $5S_E$ subunit the second partial reaction as follows:

$$\underset{\substack{|\\ \text{COO}^-}}{\text{CH}_3\cdot\text{CH}}\cdot\text{COSCoA} + \text{biotinyl subunit} \xrightleftharpoons{12\text{S}_\text{H}\text{ subunit}} \text{CH}_3\cdot\text{CH}_2\cdot\text{COSCoA} + {}^-\text{OOC-biotinyl subunit} \quad (13)$$

$${}^-\text{OOC-biotinyl subunit} + \text{CH}_3\cdot\text{CO}\cdot\text{COO}^- \xrightleftharpoons{5\text{S}_\text{E}\text{ subunit}} \text{biotinyl subunit} + {}^-\text{OOC}\cdot\text{CH}_2\cdot\text{CO}\cdot\text{COO}^- \quad (14)$$

$$\text{Sum: } \underset{\substack{|\\ \text{COO}^-}}{\text{CH}_3\cdot\text{CH}}\cdot\text{COSCoA} + \text{CH}_3\cdot\text{CO}\cdot\text{COO}^- \rightleftharpoons \text{CH}_3\cdot\text{CH}_2\cdot\text{COSCoA} + {}^-\text{OOC}\cdot\text{CH}_2\cdot\text{CO}\cdot\text{COO}^- \quad (12)$$

In the assembled form of the enzyme all the subunits are in juxtaposition (Fig. 3) and therefore transcarboxylation is not diffusion dependent. For that reason the intact enzyme is a very effective catalyst when compared to the catalysis by the unassembled subunits via reactions (13) and (14) in which diffusion of the $1.3S_E$ subunit between the $12S_H$ subunit and the $5S_E$ subunit is required for the overall reaction (12). Recently, A. S. Mildvan, C. H. Fung and R. J. Feldman using EPR and NMR have shown that the pyruvate on the $5S_E$ subunit is about 8 Å from the propionyl CoA on the $12S_H$ subunit.[10] Since the biotinyl ring probably is positioned between the substrate sites it is clear that a very small oscillation would permit it to serve as a very effective carboxyl carrier between the sites.

The most surprising recent development has been the discovery that the form of the enzyme we have been studying is apparently an incomplete structure.[12] We now have evidence that the native enzyme may have two sets of peripheral subunits, one set of three each on both faces of the $12S_H$ subunit. Thus, this 26S form of the enzyme is made up of twelve $1.3S_E$ subunits, six $5S_E$ subunits and one $12S_H$ subunit or a total of thirty peptides. The one set of peripheral subunits is very labile and dissociates from the enzyme readily unless very special precautions are used during the isolation.

We now are left with the problem of trying to determine what causes the one set of subunits to have such different properties than the other set. We know that all twelve biotinyl groups are carboxylated with {3-^{14}C}methylmalonyl CoA as a substrate, so all the biotinyl groups have access to CoA ester sites on the $12S_H$ subunit. This raises a host of questions about the structure of the $12S_H$ subunit. Are there twelve substrate sites on the $12S_H$ subunit, two on each peptide? Has there been gene duplication and fusion with formation of repeating sequences in the peptides of the $12S_H$ subunit so that both faces are similar and thus can provide six sites on each face for combination with the $1.3S_E$ subunits? Tryptic peptide mapping has provided no evidence of repeating sequences since the number of peptides obtained are equivalent to the total number of lysines and arginines. Thus, if there is a repeating sequence, the homology has been sufficiently modified by mutation to make it impossible to detect the homology by peptide mapping. Why is such a big molecule required to catalyze such a relatively simple reaction? Does the easily dissociable set of peripheral subunits have something to do with control mechanisms? From these and other questions it is obvious that there is quite a long trail ahead before we can catch up with the propionic acid bacteria.

REFERENCES

1. Wood, H. G. (1972) My life and carbon dioxide fixation. In *The Molecular Basis of Biological Transport* (J. F. Woessner, Jr. and F. Huijing, eds.), pp. 1–54, Academic Press Inc., N.Y.
2. Milner, Y. and Wood, H. G. (1972) Isolation of a pyrophosphoryl form of pyruvate, phosphate dikinase from propionibacteria. *Proc. Nat. Acad. Sci. U S.A.* **69**, 2463–2468.
3. Reeves, R. E., South, D. J., Blytt, H. J. and Warren, L. G. (1974) Pyrophosphate: D-fructose 6 phosphate 1-phosphotransferase. *J. Biol. Chem.* **249**, 7737–7741.
4. Stjernholm, R. and Wood, H. G. (1960) Glycerol dissimilation and the occurrence of a symmetrical three-carbon intermediate in the propionic acid fermentation. *J. Biol. Chem.* **235**, 2757–2761.
5. Wood, H. G. (1972) Transcarboxylase. In *The Enzymes* (P. D. Boyer, ed.), **6**, 83–115, Academic Press, N.Y.
6. Northrop, D. B. (1969) Transcarboxylase. VI. Kinetic analysis of the reaction mechanism. *J. Biol. Chem.* **244**, 5808–5819.
7. Green, N. M., Valentine, R. C., Wrigley, N. T., Ahmad, F., Jacobson, B. and Wood, H. G. (1972) Transcarboxylase. XI. Electron microscopy and subunit structure. *J. Biol. Chem.* **247**, 6284–6298.
8. Ahmad, F., Jacobson, B., Chuang, M., Brattin, W. and Wood, H. G. (1975) Isolation of peptides from the carboxyl carrier subunit of transcarboxylase. Role of the non-biotinyl peptide in assembly. *Biochemistry*, **14**, 1606–1611.
9. Chuang, M., Ahmad, F., Jacobson, B. and Wood, H. G. (1975) Evidence that the two partial reactions of transcarboxylase are catalyzed by two dissimilar subunits of transcarboxylase. *Biochemistry*, **14**, 1611–1619.
10. Mildvan, A. S., Fung, C. H. and Feldman, R. J. (1975) Conformation of enzyme-bound propionyl coenzyme A determined by proton and phosphorous relaxation rates. *Fed. Proc.* **34**, 690.
11. O'Brien, W. E., Bowien, S. and Wood, H. G. (1975) Isolation and characterization of a pyrophosphate-dependent phosphofructokinase from *Propionibacterium shermanii*. *J. Biol. Chem.* **250**, 8690.
12. Wood, H. G. (1976) Subunit-subunit interactions of transcarboxylase. *Fed. Proc.* in press (June issue).

BIOSYNTHESIS OF CELL COMPONENTS FROM ACETATE

H. L. Kornberg*

Department of Biochemistry
University of Leicester
Leicester, LE1 7RH, England

The view that one forms of a scientific discovery in retrospect is aptly illustrated by the well-known story of Columbus' egg: in order to achieve a solution, a novel approach must be tried and even firmly held preconceptions have to be discarded; after the solution has been reached, one is amazed at one's blindness at not seeing it before. The history of the events that led to the formulation of the glyoxylate cycle manifests all these and other characteristics. Since contributors to this volume have been urged to be autobiographical, a rehearsal of these events and my role in them may prove instructive, if not edifying, to the reader.

Two of the three strands of the thread that led to my involvement in the recognition of this novel pathway were woven on two social occasions, separate in time and space. The first occasion, in 1953, was one of the delightful Friday afternoon tea parties where the good ladies of New Haven (Connecticut) sustained, with tea, cookies and kindly solicitude, the members of the various Departments of Yale Medical School, sorely in need of sustenance after a week of intellectual toil. It was over such a convivial cup that Ef Racker, in whose lab. I was a Research Fellow, demanded to know from me (who had recently arrived from H. A. Krebs' Department in Sheffield) how it was that bacteria could grow on ethanol or acetate as sole carbon source? At that time, it was not clear whether the Krebs cycle operated in micro-organisms—indeed, the evidence then available was so confusing that it was even suggested that *Aerobacter* had a Krebs cycle when grown on citrate but not when grown on acetate.[1] However, H. A. Krebs and colleagues had recently demonstrated the cycle to be a major means for the provision of cell components in yeast[2] even if its role in energy supply was still being questioned. If the Krebs cycle did operate in a micro-organism, Ef argued, then acetate should be totally combusted to carbon dioxide and water and no growth should occur unless the organism were able either to fix carbon dioxide or to have some other reaction of acetate that led to the formation of a precursor of cell components. At the time, I could not make any sensible answer to Ef's question, and it was not until much later that I even remembered that it had been put to me.

The second occasion was an even more convivial and considerably more alcoholic gathering in New York one or two years later, at a meeting of the Enzyme Club. I had had the good fortune to meet Severo Ochoa whilst I was still an undergraduate at Sheffield, when Carmen and he attended a Symposium on photosynthesis there. Presuming on this

* Present address: Department of Biochemistry, University of Cambridge, Tennis Court Road, Cambridge CB2 1QW, U.K.

slender introduction, I approached Severo and asked him whether his recently discovered "malic" enzyme could effect the net formation of malate from pyruvate and carbon dioxide or whether it acted predominantly to decarboxylate malate. Severo thought the latter more probable. What was the particular reason for my question? I wondered whether Severo's enzyme might play a part in microbial growth on acetate: all that would be needed would be a way of converting acetate to pyruvate. Severo, with that gentle firmness which characterises his conversations with the ignorant, pointed out some of the more elementary thermodynamic objections to the path that I so glibly suggested and bade me to reflect again. There was, in any case, one reaction, that might fulfil the purpose I sought, already respectably enshrined in the textbooks: this was the so-called "Thunberg condensation":

$$\begin{matrix} CH_3\cdot COOH & & CH_2\cdot COOH & \\ + & \rightarrow & | & +[2H] \\ CH_3\cdot COOH & & CH_2\cdot COOH & \\ 2\times \textit{acetic acid} & & \textit{succinic acid} & \end{matrix} \qquad (1)$$

It was not until much later that I realised this reaction to be depressingly similar to the dragon's cave in the fairy story: although the footprints of many workers could be seen to go in, none appeared to come out.

After 2 years in the USA, and when both conversations had been long since pigeon-holed in some corner of my subconscious, I returned to England as a (temporary) member of staff of the MRC Unit for Research in Cell Metabolism, that H. A. Krebs had moved from Sheffield to Oxford. Professor Krebs suggested that I might like to begin my research by studying the fate of proline in the liver: by what enzymic steps is it converted to glutamate? I certainly found this interesting, established myself in a quiet corner of the laboratory and began work; through no causal connection, Professor Krebs left shortly after on his summer holidays. Imagine my consternation, therefore, when, left all alone with this problem, I discovered, on picking up the latest issue of *Biochimica et Biophysica Acta* to arrive, that in a paper of less than two pages, H. J. Strecker and P. Mela[3] had solved the question and that there was little left for me to do except to utter a despairing cheer. What does any research worker do if his problem disappears from under his nose? If he has any sense, he finds another problem.

I immediately wrote to Professor Krebs to report the situation but expressed the hope that, whatever fresh problem would be chosen for study, it should be a problem in which Professor Krebs had a direct interest. He replied at once and reminded me of the confusion still surrounding the status of the tricarboxylic acid cycle in micro-organisms, and of the difficulties in envisaging how that cycle could fulfil its dual function to supply carbon skeletons for biosyntheses[2] and also the energy required for such biosyntheses—if C_2-units were all that entered it. He also reminded me that I had spent the summer of 1954 working in Melvin Calvin's group at Berkeley and had thus acquired some knowledge of the techniques used by them to elucidate the path of carbon in photosynthesis. Similar procedures—brief incubations of whole cells with labelled substrates and analysis by two-dimensional chromatography of labelled products—might be used in order to unravel the path of acetate in cells growing on that substrate as sole source of carbon. It was thus that the third (and major) impetus was given, to propel my future work in the direction it still takes today. It was thus also that the two conversations of long ago were stimulated to rise into my conscious: of course, I enthusiastically accepted Professor Krebs' suggestions.

Fortunately, my cousin, Margot Kogut, working in Sidney Elsden's Department in Sheffield, had recently published some fascinating work done with a pseudomonad (designated KB1) capable of growing on all manner of organic materials, including acetate. I wrote to her to ask for some of this organism and also for some elementary instruction in the techniques of microbiology. These she readily supplied. Equally fortunate, another member of Professor Krebs' Unit was David Hughes (now Professor of Microbiology at the University of Wales, Cardiff) who took pity on me and showed me how to grow the organism I wanted with reasonable certainty that it was free from contaminants and other unwanted beasties. By the time Professor Krebs returned, I was feeling optimistic that there would be no major experimental snags to this endeavour: by that time, I had committed nearly all the mistakes that it was possible to commit . . . . Or nearly all: it was still open to me to forget Severo's lesson in elementary thermodynamics and to postulate that "CO2-fixation reactions must play a major role in the synthesis of protein [from acetate and ammonia] and that most of the amino acids synthesised pass through the stage of CO_2 or compounds in ready equilibrium with CO_2." It was not until a year later that the significance of the observations, wrongly interpreted in this brief Note,[4] became clear.

Measurements of the fate of labelled acetate in *Pseudomonas* KB1, exposed to this substance for periods from 3 seconds onwards, showed that isotopic label appeared solely in intermediates of the Krebs cycle and in amino acids derived from them, even at the earliest times of incubation. Yet the initial premise of our work made it obvious that the Krebs cycle, which from these data appeared to be the sole pathway for acetate metabolism, could not account for growth on acetate. The semantic nature of this paradox (a logical absurdity which G. K. Chesterton described as "Truth standing on her head in order to attract attention") was hidden from me until a naïve question from my wife (not a biochemist) brought it to my attention. Feeling particularly glum at having thus apparently come up against a brick wall, I outlined the reason for my despondency to her. At the end of my recital, she asked "How does one see a Krebs cycle?" This brought me up short: I was, of course, assuming that any intermediate of the Krebs cycle that I saw was an intermediate in only that cycle and was not participating in other and possibly novel reactions.

I therefore re-examined my chromatograms and indeed found that the distribution of label, amongst the products formed from acetate at early times, was entirely inconsistent with their formation via the Krebs cycle. In particular, the proportion of isotope present in malate was initially higher than that in citrate and both were much higher than that in succinate, although progressively more labelled succinate was formed. It was evident that, at these early times, labelled malate could not have been formed from succinate; moreover, it also appeared that labelled acetate had entered the cycle at two points, to form citrate at one and malate at the other.

At this time, I was fortunate to have as collaborator Neil Madsen, who had recently come to spend a postdoctoral year in the Oxford Biochemistry Department after graduating from Carl Cori's Department in St. Louis, and who had chosen to risk possible disappointment by sharing with me the struggle with an apparently intractable problem. Neil had already succeeded in showing that acetate *per se* was not metabolized by extracts of our organisms but that such extracts were rich in acetate thiokinase activity and were thus able rapidly to form acetyl co-enzyme A from acetate, ATP and CoASH. However, despite all manner of tricks, we were unable to persuade such extracts either to oxidise acetate to CO_2

and water, or to catalyse the reductive condensation of 2 mols of acetate to succinate, as required by reaction (1).

Neil was also seized at once with the significance of the anomalous isotope distributions on our chromatograms and, having by chance some fresh extract of acetate-grown cells handy, we set up a simple experiment to look for a second point of entry of acetyl coenzyme A into the Krebs cycle. This experiment was indeed so simple and seemed so much of a "blunderbuss" approach that we did not report it until a year later,[5] when the interpretation to which it pointed so clearly had been adequately substantiated. The experiment consisted of taking some extract of acetate-grown *Pseudomonas* and placing it, together with labelled acetate, ATP, glutathione, a magnesium salt and phosphate buffer, into each of eleven test tubes. To these were added, respectively, DPN, or TPN, or one of each of the intermediates of the tricarboxylic cycle. It would be expected that the occurrence of the "Thunberg" reaction[1] might lead to the formation of labelled succinate in the presence of either of the electron acceptors: however, such was not found. It would also be expected that the initial reaction of the Krebs cycle, the formation of citrate from oxaloacetate and acetyl co enzyme A, should lead to a major incorporation into acid-stable products of label from acetate and that this label should be found largely if not exclusively as citrate: such was indeed observed. However, there is no reaction in the Krebs cycle in which a second molecule of acetyl co-enzyme A enters to form an acid-stable product. It was thus with a mixture of disbelief and awe that Neil and I watched our Geiger counter burst into a veritable paroxysm of activity when the product formed in the tube containing isocitrate as reactant was submitted to it. Chromatographic analysis showed this labelled product to be overwhelmingly one compound: malate.

It may be recalled that the strain of *Pseudomonas* that we were using had been obtained from Dr. Margot Kogut, then working in the laboratory of Sidney Elsden. A previous visitor to Professor Elsden's group was Dr. Howard Saz who, 2 years previously,[6] had studied an enzyme then termed "isocitritase" which catalyses the splitting of isocitrate to glyoxylate and succinate. He found this enzyme, which is an aldolase, in *Pseudomonas* KB1—the very strain with which we were working. (The enzyme had also been found to be present in fungi and in several bacteria.) The reaction it catalyses is readily reversible but its physiological role was obscure. It had been suggested that it might provide glyoxylate for the biosynthesis of glycine or, acting in reverse, to provide a cyclic mechanism for the oxidation of compounds more highly oxidised than acetate, such as glycine or glycollate.[7] Another suggestion was that the glyoxylate formed through its action was oxidised via formate to carbon dioxide and water and that this pathway represented a mechanism alternative to the Krebs cycle for the total oxidation of acetate. We had known of the existence of this enzyme but it was one of several that catalysed peculiar reactions not explicable at the time and we had therefore not paid particular attention to it.

However, by an extraordinary coincidence, a paper (to which our attention was drawn by Professor Krebs) appeared at just this time in the *Journal of the American Chemical Society* which brought the possible metabolism of glyoxylate very sharply to our notice. D. T. O. Wong and S. J. Ajl reported[8] that extracts of *E. coli* could promote the condensation of acetyl coenzyme A and glyoxylate to form malate: this reaction was, of course, formally analogous to that whereby citrate was formed from acetyl coenzyme A and oxaloacetate. The authors termed the novel enzyme "malate synthetase". The "blunderbuss" experiment we had done suggested that this enzyme was present also in the *Pseudomonas* extract. We hastily repeated the experiment with a tube containing sodium glyoxylate as well as the system for generation of labelled acetyl coenzyme A: as much labelled malate was now formed as had been formed in the tube incubated with isocitrate. Thereafter, it was a comparatively simple matter for Neil and myself to show that there was not only an incorporation of isotope from acetate into malate but a *net* formation of this C_4-acid from acetyl coenzyme A and isocitrate, and that this was accompanied by the stoicheiometric formation of succinate. The results thus obtained form a sequence, the operation of which enables decarboxylating steps of the Krebs cycle to be by-passed, which enables the net content of C_4-dicarboxylic acids to be increased and which thus forms a means of maintaining the dual function of the Krebs cycle—to catalyse the total oxidation of acetate and also to supply precursors for cellular biosyntheses (Fig. 1). Neil and I reported our work at a meeting of the Biochemical Society in April 1957 and also in a paper that referred to this sequence as a "glyoxylate bypass"; this paper(9) was published in *Biochimica et Biophysica Acta* also in 1957.

The biological significance of these observations was greatly enlarged by another happy coincidence. A young British plant physiologist, Harry Beevers, had emigrated to the United States but had returned to Oxford for a period of study leave. Professor Krebs knew that Harry was working with castor beans, and recalled that the fatty seeds of *Ricinus*, during germination, converted their content of castor oil virtually quantitatively to sugars: no mechanism for this net conversion of fat to carbohydrate was then known. Professor Krebs went to see Dr. Beevers and told him of our work with bacteria, which might also explain the conversion of acetyl coenzyme A to precursors of sugars in the castor beans. Would Harry be interested in collaborating with me? Of course, he was. And so it was that someone who had a problem, and someone who had a solution but did not know of the problem, came to meet each other and to work together at just the right time. Harry and I immediately found that castor-bean extracts were rich in isocitritase and that they also contained malate synthetase: by the end of 1957, our two papers on the conversion of fat to carbohydrate in castor beans had already appeared.(10,11) Harry Beevers' subsequent distinguished work on the cellular location of these enzymes and their physiological control is well known and needs no rehearsal from me.

A further circumstance brought our work to public attention much more effectively than might otherwise have been possible. At about this time, the editors of *Nature* invited H. A. Krebs to write a general article on some aspect of carbohydrate metabolism. Since the outcome of our work on acetate utilization represented a biosynthetic counterpart to the degradative tricarboxylic cycle and was thus of general interest, Professor Krebs proposed to discuss the nature and broader significance of these two cyclic pathways; he invited me (and I was delighted to accept) to collaborate on such a survey. The article(12) was published in *Nature* in May 1957. It drew a distinction between the cycle that oxidized acetyl coenzyme A totally to CO_2 and water (the Krebs cycle) and the novel route, incorporating the by-pass reactions: it was thus that this novel route received the name "Glyoxylate cycle" that is still used.

We were fortunate that our work soon became widely known by means other than that *Nature* article. Prior to submitting our paper to *Nature*, we had sent a typescript of it to Dr. Ajl, who (with D. T. O. Wong) had not only discovered the malate synthetase reaction in *E. coli*(8) but had also published on isocitritase in this organism.(7) He was kind enough to thank us for sending him our work and to state that he was "certainly glad to learn of the

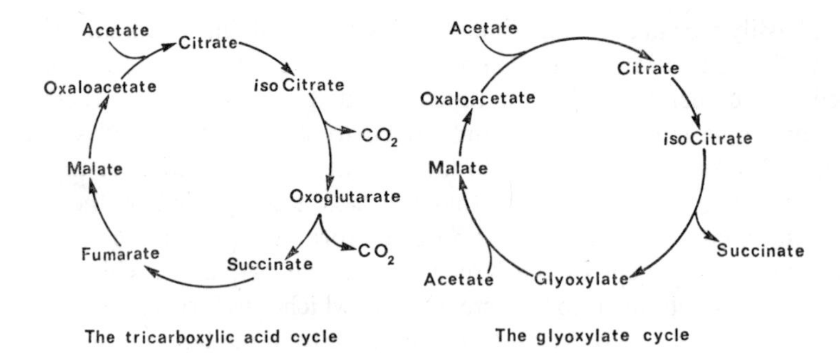

The tricarboxylic acid and glyoxylate cycles

apparent overall significance of the malate synthetase reaction"; in a major review published in *Physiological Reviews*(13) exactly a year after his letter, he described our work in an Addendum. Moreover, the article on the metabolism of carbohydrates in *Annual Reviews of Microbiology* for 1958 happened to be written by S. R. Elsden and J. L. Peel of Sheffield: they most generously devoted a considerable proportion of their space to a discussion of our work. We were thus gratified to achieve a virtually instantaneous acceptance of the validity of the "Glyoxylate cycle" and were not beset by the opposition that so often seems to be aroused by the formulation of new concepts.

Although we had an easy passage from problem to solution to acceptance of that solution, the description of the intermediate steps in that sequence that I have here set down will not, I hope, conceal the extraordinary proportion of sheer good luck that formed part of it. I am acutely conscious of the fact that I would probably not have chosen to work on this topic if several events had not conspired to make me stumble into it; I am equally conscious of my good fortune in having had colleagues, and (in Hans Krebs) a mentor, who made me see things that others had also seen, but made me notice what others had missed. And that, after all, is the essence of research.

REFERENCES

1. Ajl, S. J. and Wong, D. T. O. (1951) Studies on the mechanism of acetate oxidation by bacteria. IV. Acetate oxidation by citrate-grown *Aerobacter aerogenes* studied with radioactive carbon. *J. Bact.* **61**, 379–387.
2. Krebs, H. A., Gurin, S. and Eggleston, L. V. (1952) The pathway of oxidation of acetate in baker's yeast. *Biochem. J.* **51**, 614–627.
3. Strecker, H. J. and Mela, P. (1955) The interconversion of glutamic acid and proline. *Biochim. Biophys. Acta* **17**, 580–581.
4. Kornberg, H. L. (1956) CO_2-fixation during protein synthesis from ammonium acetate. *Biochim. Biophys. Acta* **22**, 208–210.
5. Kornberg, H. L. (1958) Synthesis of cell constituents from acetate via the glyoxylate cycle. *Proc. 4th Int. Cong. Biochem.* **13**, 251–266.
6. Saz, H. J. (1954) The enzymic formation of glyoxylate and succinate from tricarboxylic acids. *Biochem. J.* **58**, xx–xxi.
7. Wong, D. T. O. and Ajl, S. J. (1955) Isocitritase in *Escherichia coli*. *Nature* (*Lond.*) **176**, 970–971.
8. Wong, D. T. O. and Ajl, S. J. (1956) Conversion of acetate and glyoxylate to malate. *J. Am. Chem. Soc.* **78**, 3230.
9. Kornberg, H. L. and Madsen, N. B. (1957) Synthesis of C_4-dicarboxylic acids from acetate by a "glyoxylate by-pass" of the tricarboxylic acid cycle. *Biochem. Biophys. Acta* **24**, 651–653.

10. KORNBERG, H. L. and BEEVERS, H. (1957) A mechanism of conversion of fat to carbohydrate in castor beans. *Nature* (*Lond.*) **180**, 35–36.
11. KORNBERG, H. L. and BEEVERS, H. (1957) The glyoxylate cycle as a stage in the conversion of fat to carbohydrate in castor beans. *Biochim. Biophys. Acta* **26**, 531–537.
12. KORNBERG, H. L. and KREBS, H. A. (1957) Synthesis of cell constituents from C_2-units by a modified tricarboxylic acid cycle. *Nature* (*Lond.*) **179**, 988–991.
13. AJL, S. J. (1958) Evolution of a pattern of terminal respiration in bacteria. *Physiol. Rev.* **38**, 196–214.

LESSONS IN LEARNING*

I. C. GUNSALUS

Department of Biochemistry
University of Illinois at Urbana–Champaign
Urbana, Illinois 61801

Historical documentation, ever fragile, is especially so in biological systems, including memory. Living matter being so diverse in organized form, with her exploration initiated simultaneously on numerous phenomena, couched in multiple terms, the convergence required for communication may be long delayed. History, thus being that perceived or remembered, rather than what happened, and often written only later, I offer apology for any dissonance of recollection or oversight of these observations with the historical record of others.

Severo Ochoa first entered my remembrance in September 1941, via his translation to Otto Meyerhof of the words and meaning of a five-tined fork award at a presocial evening in the now University of Wisconsin-Madison biochemistry lecture hall on Henry Mall. On that occasion, "A Symposium on Respiratory Enzymes," many young Americans were exposed to the Meyerhof School, among others, and to exciting news on the processes of glycolysis and respiration.

A vital sojourn I term "Severo, Seymour and Alice" occurred a decade later. Severo's 1950 hospitality in the old Twenty-sixth Street Pharmacology Laboratories of the New York University School of Medicine provided exposure to experiments and to understanding generated by the flow of ideas during stimulating daily exchanges, luncheon and tea conference-debates and the Saturday noons. The pivotal role of pyruvate furnished a most appropriate experimental vehicle; *E. coli* provided separable enzymes for chain cleavage, oxidation and acetyl transfer; the known cofactors were purchased, or given by friends, and phosphate or oxalacetate served as acyl acceptor to form acetyl phosphate[1] or citrate.[2] Earl Stadtman with Fritz Lipmann was pursuing coenzyme A-acyl transfer,[3] and O'Kane and I had already discovered, but not identified, the "pyruvate oxidation factor,"[4] later isolated, synthesized and labeled lipoic acid.[5]

Several occasions in the decade between these two events, and many since, stand out for reasons far beyond the ever-important experiments. Particularly memorable meetings occurred at Pacific Grove in 1948 and Berkeley in 1949. The first was during an unusually open summer in an environment devoted to learning, humanity and an intellectual view of the microbes as living creatures selected to nature's environmental niches with the "Sage of the Western Rocks", Kees van Niel.[6] Though Severo's visit was short, the

*Or "What have the Microbes Taught Man Recently?"

interaction of these two superb people was especially stimulating and memorable. The second more extended occasion was a summer in Berkeley as Severo lectured on biochemistry and I on microbial metabolism and there followed often luncheon discussions with Barker, Doudoroff, Hassid, Roger Stanier. As part of Roger's laboratory family I was afforded the first of three intensive encounters with the pseudomonads and what Otto Rahn would call their "amazing appetities".[7]

The Berkeley and 1950 New York sojourns generated an ease and regard that permeate meetings and conversations with Severo over topics as deep as the meaning of the human endeavor and contribute in major importance to my learning and maturation. They flow too much together to be enumerated in this short record, nor can the true flavor be told by a list of the names of other important and memorable participants.

Recall of the 1941 events half-way through my first decade of professional science, that coincides with half of the 70 years we celebrate here in tribute to Severo, refocuses recollection on my trail of learning. This process has accommodated for me at least one turning per decade, with the emphasis somehow planted and nurtured to seek new concepts, new meaning and the tools to carry each succeeding inquiry to deeper levels of definition. As the facts and interpretation emerge to reveal their general importance and enlist the skills of others better suited for the next phase, to see the problem redefined and carried far beyond the initial vision, as my attention to new problems dilutes the effort given to ongoing inquiries as the excitement of new leads encroaches. The prime significance continues to center on people, ideas, observations, and the motivation to retain a personal and professional freedom and responsibility that accepts each new experience or opportunity.

A partition by decades of the important events in my learning coincides first with the era at Cornell, learning with J. M. Sherman from his streptococci of a potential far broader than the term "homolactic" as then used. Second, the restrictive nutrition technique, the amino acid roles of pyridoxal, the 5-phosphate coenzyme and the move to enzyme experiments. Later, when applied to dissimilatory processes, the "pyruvate oxidation factor," the cycles of the co-factors in energy coupling during keto acid oxidation. Third, a chemistry role in biology and, as a bridge to the microbes, the five volumes of *The Bacteria*.[8] Meanwhile exploring for new processes, enzymes, systems and addition of a genetic wing. Fourth, since 1965, work permeated with physics and genetics probes on dioxygen reduction, iron proteins, and the organization and regulation of these systems in cells. Fifth, from 1975—what now?

I. GO AWAY, LEARN

Leo Szilard left his Ten Commandments,[9] the ninth: "Do your work for six years; but in the seventh, go into solitude or among strangers, so that the memory of your friends does not hinder you from being what you have become." Lawrence Carruth, on this principle, provoked me on the base of two college years of chemistry, mathematics, biology, art and dramatics, with a year of physics and a wish to study with Ross Gortner at Minnesota, to a change of course—not a new tack; a new course. Larry's probing—"Why leave a small Norwegian college for a large one? If you move, why not also a measured change?"—and his subsequent actions tumbled me into the life of the microbes.

A period for saturation in variety, growth and behavior, battered by constantly more exciting ideas, catapulting forward term on term in an increasing variety and number of hours and courses. At a point in the continuum, Cornell assigned, by her rules a bachelor's degree as I raced to include with biology more chemistry, organic, physical and bio, physics, including a laboratory with Bill Cady that taught me a great deal on setting up laboratory teaching and how to ignore people when they are learning; a course in qualitative organic with Bill Miller, unfailingly teetering on his toes at my shoulder when clumsiness or error encroached. R. C. Gibbs I found a dry lecturer in physics, but filled with a kindly wisdom and guidance, and J. G. (Jack) Kirkwood an exciting one in chemistry, far out ahead, but patient. J. R. (Jack) Johnson's impeccable organic lectures were spiced by eyes dancing from a delight and excitement for the next question, the new problem, unforgettable until today, on my Ph.D. examination.

James B. (Jimmy) Sumner, full of biochemical lore, helpful to all, disdaining the slothful, full of a dry and pungent wit, "always willing to accept criticism from qualified students." The stakes for "qualified" came very high. The laboratory tools—Duboscq colorimeter, microscope to view crystals, especially of proteins, a Thunberg tube. Unstressed in the background, hydrogen and quinhydrone electrodes on a Leeds and Northrop potentiometer and, barely mentioned, Barcroft manometers and bath. Somewhere along came a first smell of genetics and in bacteriology, Georges Knaysi's lectures on structure, a laboratory on fungi, patterned after A. T. Henrici, Pauline Stark teaching how to handle a laboratory class—cultures ready, cells in precisely the proper state—and J. M. (Jimmy) Sherman's lectures, conversations and streptococci.

Otto Rahn, a physicist from Danzig via Kiel, made an impact delayed in total effect—until my first attempt to teach metabolism to students not first prepared by his bacterial physiology. He introduced mathematics to the analysis of growth rates and endpoints, energy and death at Michigan State before 1910, at Illinois in 1913 and 1914, and after a long interruption initiated by the First World War, at Cornell in 1926. *Physiology of Bacteria.* 1927,[10] that won the respect of many thinking biologists and furnished the base for our growing knowledge of metabolism. Otto Rahn's clarity, his punchy sentences, command attention! From the preface[10] . . .

> *I believe that some of the principles of biology can be found and studied only with the simplest forms of life, and that general physiology has much to learn from the physiology of bacteria.*
>
> (Emphasis added)

And, a typical sentence closing a paragraph:

> All that is necessary is just a habit of thinking, which is easily acquired.

Reading Fritz Lipmann's *Wanderings . . .*[11] recalled two landmarks: the primers, near bibles, that prepared a neophyte interested in the chemistry of microbial processes to be excited by Severo and his colleagues at Madison; and in transition from the physiology of Rahn, probably both influenced by Kluyver[12,13]—Fritz' confession from his pre-1940 pyruvate oxidation days jumped at me from page 28: "One of the lessons I had learned in Meyerhof's laboratory was that if mammalian systems show difficulties, one should turn to microorganisms." The early primers two Cambridge monographs—1923, Harden, *Alcoholic Fermentation*;[14] 1930, Marjory Stephenson, *Bacterial Metabolism*[15]—and from Kiel in 1924, Otto Meyerhof, *Chemical Dynamics of Life Phaenomena.*[16] By 1937 Szent-Gyorgyi, *Studies on Biological Oxidation*,[17] as 1940 dawned Cold Spring Harbor Symposia on Quantitative Biology VII, with a Dean Burk summary on page 450, the

Embden–Meyerhof scheme, and page 250, plus or minus a couple, Lipmann's previously mentioned analysis of pyruvate oxidation in a lactic acid bacterium,[18] and only a year later "Wiggle P," ~P on page 99, *Advances in Enzymology, I.*[19]

In the fall of 1940, not sensing the imminence of Pearl Harbor, we were quickly at work. This period is difficult to condense in recount for the richness one attributes to early events. I discovered my tendency to think in groups rather than items or lines and fully realized that logic, a powerful process for analysis, explanation and teaching, could not be a preferred route to detecting the weak signals of nature. To generate the hunches, called ideas, the leads to experimentation, if facilities and time coincide, each individual must learn his own means to enhance the willingness of nature to yield her secrets—from solitude, reading, working with one's hands, a conversation with the informed, or merely interested. Years later, Hans Frauenfelder, a collaborator in current studies and a Professor of Physics at Illinois, taught me that I use two computers, one that responds immediately and the other with a long search time. When my recollection hesitates for a name or fact, Hans says, "Go on, it will surface." And half-way through the conversation or 10 minutes later the missing item appears—a useful lesson.

Sherman's streptococci, in fermentation balances, had taught us much. Early probings for dehydrogenases with various dyes preceded a break by Betty Greisen (née Chase) into Dave Hand's seven-manometer Warburg to find that group B streptococci not only oxidize ethanol with O_2 as acceptor, but also with H_2O_2. Kay Miller almost simultaneously probed the group D, *Streptococcus faecalis* for growth and reaction conditions favorable to pyruvate oxidation. These early learnings are traceable from my review with B. L. Horecker and W. A. Wood, or by the latter in *Bacteria II* (Chapter 2).[20]

A stage-setting landmark to one versed in the lactic bacteria a Snell and Strong quantitative assay for riboflavin with *Lactobacillus casei*[21] encouraged our pursuit of the nutritive requirements among streptococci which led to our use of restricted nutrition to detect function. A detour and two examples of success. The detour, later to prove scientifically useful, followed the December 1941 Sunday morning when Uncle Jimmy arrived at the laboratory with the shock of Pearl Harbor that brought a quietness rare to the group at work as usual. Soon we were working days on problems supplied by others, nights on those we designed and, in the hours between, taught the students who remained. We were held by the University Manpower Czar from accepting "off-campus" assignments. Dex Bellamy, Jack Campbell and I had begun to look at organic acid metabolism in the lactics. Jack, a Canadian, was urgently requested by his government to return. He finished quickly the citrate studies; Bellamy and I turned to other work.

One "supplied" problem concerned mild intestinal upsets and the ballooning of young (green) American cheese shipped as concentrated food to England. Knowing that *Streptococcus faecalis* present during the initial lactic fermentation can decarboxylate amino acids freed during cheese ripening, the resulting amines being implicated in mild food poisoning, we pursued the conditions for tyrosine decarboxylation. When grown on a vitamin supplemented dilute medium, decarboxylase induction at low pH by tyrosine was found along with nicotinic acid and pyridoxine demand at increased levels. To concentrate on pyridoxine was a simple hunch—a nicotinic acid function was known in the pyridine nucleotides. Soon we achieved a superb Warburg assay with cell suspensions, later, with dried cells. A "pseudopyridoxine" interest of Esmond Snell, now in Texas, with wartime travel restricted, prompted us to mail him the pencil drawings from which our final graphs were prepared. Return mail brought two vials to be identified in a Letter to the Editor in the next issue of the *Journal of Biological Chemistry*—the two following issues carried our Letters identifying pyridoxal, then with dried cells pyridoxal-5-phosphate as tyrosine apodecarboxylase activator and coenzyme.[22] I continued coenzyme synthesis guided by Wayne Umbreit's activity measurements; he explored other potential B_6 requiring enzymes. Many tales lie buried in this telescoped recounting of the exciting days with our students Bellamy, Foust, Wood, O'Kane, Jeffs (now Doreen O'Kane) and Lichstein, among the principals.

An application of the same procedure to pyruvate, building on data from Kay Miller's thesis, led Dan O'Kane to an additional requirement for pyruvate oxidation we termed "pof," pyruvate oxidation factor. Recognized as an unidentified substance, the next decade led us to the isolation, synthesis and understanding of energy coupling in keto acid oxidation.

Acknowledgement of the critical value to my education of these 10 years without orders, with assignments of responsibility where my questions brought the reply, both spoken and implied, "If you were not the best equipped for the job you would not have been asked; your performance will certainly be appropriate and exceed our expectations," a method designed, and often successful, in bringing forth supreme effort, excellence, innovation and the building of investigators. One could hope this attitude were more widespread, and be encouraged to find it not entirely unique, as attested by van Niel's quotation from Eric Hoffer, *The Ordeal of Change*, opening the Prefatory Chapter of *The Education of a Microbiologist: Some Reflections*, written in his 70th year.[23]

II. SEARCH AND MIGRATE

Invitations and opportunities emerged with the excitement following our discovery of vitamin B_6 functions, of pyridoxal, the synthesis and coenzyme role of pyridoxal phosphate and the variety of α-amino acid forming and degrading enzymes that require this coenzyme. The recognition by colleagues, new to me, provided additional education. A debate with Paul Karrer, resolved in our favor, concerning the position of the phosphate on the pyridine ring of pyridoxal, increased Wayne's and my confidence. Karrer's deduction from bulk properties that 3 (phenol) phosphate possesses coenzyme activity was discarded when a comparison showed less than 0.1% the activity of the 5-phosphate we had prepared.[25] Pyridoxal experiments carried to Indiana included spectral analysis of the coenzyme, an inducible alanine racemase and transaminases for most amino acids in bacteria, the latter by methods given to me by Snell, who was again in Madison. Summaries of this era half-way through the decade appear in *Annual Reviews* 1948, and symposia of the Society of Biological Chemists 1949, and at McCullum Pratt in 1951.[24]

At Indiana, the opportunity to absorb genetics from the pioneers in seminars, lectures and conversations, the friendship with Luria initiated by his defining a relationship that has carried great meaning for nearly 30 years, the excitement of adequate housing for my family and a new pair of twins kept life from being in any sense dull, although the stress of moving after 15 pleasant years in upstate New York was considerable. A group of excellent students soon developed, a stimulating interaction with the talented genetic-virology people around Luria, and access to others in the Delbruck–Luria–Hershey "Phage

Group" promoted another go-away-and-learn period. A joint course with Luria on the general biology of microbes was initiated, and greatly extended after our simultaneous move to Illinois in the fall of 1950.

The wisdom of Fernandus Payne, retiring dean of the graduate school and the builder of biological quality and strength, his interaction with Herman Wells and Herman Briscoe, had a tremendous impact on the atmosphere of Indiana University. His counsel, guidance and permissively wielded strength smoothed the way and helped solve problems in a short time without the necessity of formal procedures.

The blocked-out purification of "pyruvate oxidation factor" was taken with us to Indiana and the early steps were scaled up for us at the Eli Lilly Research Laboratories in Indianapolis. Later, at the time of my transfer to Illinois, a realignment of the Lilly research management revealed to them the identity of the "pyruvate oxidation" and the "acetate replacing" factors, the latter supplied to Lester Reed at Texas. A pointed encouragement came from Lilly for us to join forces, which we did, continuing with the best features from each effort.[26]

At Illinois, an initiation to the chemistry department research attitude was soon revealed through Dr. Marvel, who invited me to join the organic division around their twice-daily consultant and staff coffee table, and Dr. Fuson, who introduced two of his graduate students into my laboratory. The acceptance, advice, facilities and helpfulness encouraged me, by the beginning of the next decade, to attempt the chemistry–biology mixing requested of me. The lipoic acid story was summarized,[5] new advances recorded,[27,28] and the superb story of the keto acid enzyme complexes of microbial and mammalian origin provided by Lester Reed.[29]

The Severo, Seymour and Alice period at NYU that contributed so much to my scientific and personal growth occurred between the Indiana and Illinois posts. On my early February arrival, Severo suggested that I join Seymour Korkes, assisted by Alice del Campillo, on pyruvate oxidation studies in *E. coli* contemplated to yield the necessary intermediate for citrate generation. The laboratory Severo had intended for me was now occupied and a suitable office desk was not available. But, since Severo was quiet and did not occupy his office for long periods, I might find it convenient to use a table there. His kindness proved extremely compatible. Even when Irma arrived to take dictation and I rose to leave, Severo would caution that he spoke quietly, would take only a moment and, he felt sure, I would not be disturbed if I continued to work. In the laboratory I took up residence with Seymour and Alice and the work went well. Severo provided a stress-free, nontaxing environment that quickly reoriented my personal emotions and thinking. I became enmeshed in the interchange with the working family of diverse backgrounds, temperaments and aims.

Discussions with Seymour were frequent, quick, informative and to the point. My prejudices from earlier studies with pyruvate and the lactic bacteria were tolerated and modified; we soon formed a lasting bond. We would argue each morning about which of us was in charge. Alice, who had arrived early and prepared the needed solutions, would smile until the tirade subsided. Then the one in charge invariably would inquire of the other his preferences for the day's experiments. In mid-summer, Severo and Carmen departed for Spain and Seymour for a windjammer holiday in the Caribbean. Alice and I remained to complete the experiments and put notes in proper shape. Each morning, while I drummed with the eraser of a new pencil on Severo's table and watched the variety of craft moving on the East River or the ivy rippling on the wall across the court, Alice would ask about the day's experiments, "Have you decided yet?" Frequently, I had not and she would reappear in half an hour with the same gentle question. With protocols prepared, I was no longer surprised by her cheerful, "I have the solutions ready." We worked until the last possible day. I drove to Ithaca for a few days sailing and put the data together. Carol typed tables and a suitable letter, which were in New York to greet Severo and Seymour on their return. Such experiences are as indelible as they are infrequent.

III. CHEMISTRY–BIOLOGY BRIDGE

September 1955 and transfer to the chemistry department introduced the problem of pointing biochemistry toward 1975 and beyond, from the mountain of the accomplishments of W. C. Rose over 30 years—from 1922. In this decade of service, supplying the division with appropriate staff and facilities and myself for new teaching and student assignments, and coping with the internal forces of a successful and solidly based department, I wrote fewer than twenty full papers. With excellent colleagues, however, we completed experiments en route and learned from exploration the firm bases in genetics, organisms, growth and soluble enzyme systems that flowered in the following decade.

With reduced access to her two top stars, Will Rose by retirement and H. E. Carter to become head of the Department of Chemistry, biochemistry was most fortunate in Herb Carter's continued lectures to the general course, providing orientation to an increased biological component and vice versa. His infectious warmth, good humor, regard for individuals of all ages, students and staff alike, in the halls, seminars, classrooms, was a constant stabilizing and supportive force.

Carl Vestling's "Come in, my friend, have a chair—what can I do for you?" brightened my spirit and lightened all burdens. The resulting consultation that followed, Carl's knowledge of pertinent history, appropriate comments, good will and infectious good humor, and his ability to handle touchy situations smoothed many a craggy snare to leave all parties in improved spirits.

Fortunately, external funds were increasing with added federal attention to science and the availability of biological problems to physical and chemical probes and interpretation. In a decade the staff doubled, research support increased more than ten fold and by 1965 all of the biochemistry staff were new save for Herb Carter who soon continued his support from the vice chancellor's office. In addition to biochemistry's broader functional and structural expertise, the educational quality of other departments was increased by attracting staff and students who serve as professors and department heads in universities from Los Angeles to San Francisco and Seattle, and from Minneapolis to Boston and Houston. Our post-doctorates and graduate students were growing in many posts. From my own laboratory, for example, in addition to those employed in the States, colleagues were working in staff posts in Paris, Marseilles, Aberystwyth, Hull, Copenhagen, Jerusalem, Bangalore, Poona and Kanazawa, to name a few. Their temporary residence had educated our students, ourselves, contributed to the interaction with biology and enriched our endeavors as we, in turn, gained in appreciation for their problems and the kinds of information they needed.

The superb quality of biologists in many of our adjoining departments, their need for biochemical techniques, expertise and advice, and in return, their willingness to acquaint our students with problems and appropriate organisms as biological tools opened unusually strong and effective cooperative interaction.

The bridge for chemists seeking biological agents to solve problems was strengthened by the five source volumes on *The Bacteria*, coedited with Roger Stanier, in the areas of Structure, Metabolism, Biosynthesis, Growth and Heredity.[8]

A few brief remarks on my research interest and progress with problems completed and those developed are documented, especially the new areas to become increasingly visible and rewarding, as reported, beyond 1965. The lipoic studies were concluded as Lowell Hager equated one *E. coli* pyruvate enzyme fraction with lipoic dehydrogenase[27] and Will Gruber demonstrated the enzymatically active acetyl and succinyl to be thioesters of the secondary, 6, mercapto group.[28] Succinic thiokinase as an energy couple was followed by Roberts Smith, Pete Knight and Jane Gibson, a visiting staff member from Sheffield who provided a purification still used[30]; the details of this acyl coupling mechanism escaped our leads and continues to be pursued by Paul Boyer.

Terpene dissimilation studies were initiated in a search for manageable hydrocarbon active enzymes with coupled dioxygen reduction-oxygenation. Bradshaw,[31] by enrichment, isolated two classes of organisms able to grow on D-camphor as a sole carbon source. The other mono- and bicyclic terpenes employed in enrichments also yielded organisms to be left in limbo while the intermediates from camphor dissimilation were pursued, with awareness of 30-year-old data on the products isolated from the urine of dogs fed camphor as a clue to systems with possible mammalian counterparts. The two Bradshaw organisms, a fluorescent pseudomonad, later assigned to the *Pseudomonad putida* species by Stanier, and a *Mycobacterium* related to the soil diptheroids identified at a National Collection of Industrial Bacteria in the species *rhodochrous* were soon found by vapor phase chromatography to produce different intermediates. After identification by E. J. Corey and students,[31] ordering the compounds in a pathway revealed that the pseudomonad initiates oxidation by *exo*hydroxylation of methylene carbon 5 and *Mycobacterium* by *endo*hydroxylation on methylene carbon 6,[31–33] trivially termed the 1,4- and the 1,3-diketone pathways. By 1960, E. J. Corey from the organic division, and H. E. Conrad who had joined me, had identified all intermediates to cleave both carbocyclic rings. Clear evidence of mixed function hydroxylases, analogous to the steroid 11 β and 20, 22 systems was at hand.

The prime credit for advancing the enzymology of mixed function oxygenation rests with H. E. Conrad,[34] who identified and characterized the components and the process of ketolactonase I, and Jens Hedegaard,[35] who in 1961 accomplished the solubilization of the first oxygen reducing hydroxylase. These data and the relevant biological experiments[36] are summarized in the symposia in Japan[33] and Israel[32] in 1965. The biology and the development of the genetic system, resting on summers of active work at Cold Spring Harbor, 1963–1966, were on firm ground for expansion.

IV. GENETICS AND PHYSICS PROBE

As 1965 opened, Crawford and I[37] completed the identification of the enzymatic defects of a series of tryptophan auxotrophic mutants on the fluorescent pseudomonads used in terpene oxidation. All clues needed for their genetic analysis were at hand. Dioxygen reduction of the participation of the 2 electron monoxygenase stoichiometry were available in soluble systems with evidence of similarities to the mammalian oxygenases of *Pseudomonas putida* strain with terpene oxidation competence.

Approaches to *P. putida* grew from a single phage isolated on our degrading strain in 1963 by Bertland.[36] Chakrabarty was borrowed from India in 1965 and contributed much insight and development in Pseudomonas genetics. Collaboration on tryptophan auxotrophs with Crawford during 1965[37] yielded an effective transducing system[38] and conjugation in 1968.[39] By 1973 Rheinwald had demonstrated that the structural genes for camphor metabolism resided on an extra chromosomal plasmid, and J. Johnston, joining the effort in 1974, developed a transformation system for plasmid identification.[40–45]

Two collaborative surprises were ahead that were to enhance our precision, stimulate oxygenases studies, and educate a new generation of investigators with physics knowledge, awareness of the biochemical concepts, tools, lore and phenomena—hopefully without succumbing to our pitfalls. The first, initiated by a visit to Helmut Beinert and to meet in his laboratory also Bill Orme-Johnson, identified the $g = 1.94$ EPR signal of the first clear sulfur "nonheme iron" protein to answer the question he had been pursuing since the mitochondrial studies with David Green. The path was open to quantify iron chalcogenide interaction in the iron sulfur proteins. Second, a meeting with a physics graduate student after a seminar on Mössbauer spectroscopy on cytochrome *c* was to build a bridge to physics on our campus with a growing collaboration between our research group and those of Professors Frauenfelder and Debrunner which, with their post-doctoral and graduate students, was to enormously enrich the tools and the understanding of sulfide and heme iron active centers, later their energetics and dynamics over time, temperature and dynamic range far greater than previously available.[46–50] Although the summary of these areas would be premature, the hunch—perhaps even the opinion—may be advanced, pending their refinement and tests by qualified colleagues. It appears not overbold to feel that languages and the theoretical and factual frameworks of the chemistry of enzyme structure and function can accommodate to bridge with the impact of physics.

Thus, the link, from the early 1940s, through learning, migration and search. Nature is found to be forever subtle, but never malicious. Her secrets on the molecular details of species propagation, development and fundamental interactions are being revealed by concerted efforts of those biologists, chemists and physicists who shun central dogma and follow the spark of an idea to ask and to explain the exciting and mystifying questions of life.

REFERENCES

From 1940—to Sojourn with Severo

1. KORKES, S., DEL CAMPILLO, A., GUNSALUS, I. C. and OCHOA, S. (1951) Enzymatic synthesis of citric acid. IV. Pyruvate as acetyl donor. *J. Biol. Chem.* **193**, 721–735.
2. KORKES, S., STERN, J. R., GUNSALUS, I. C. and OCHOA, S. (1950) Enzymatic synthesis of citrate from pyruvate and oxalacetate. *Nature*, **166**, 439–440.
3. STADTMAN, E. R. and BARKER, H. A. (1949) Fatty acid synthesis by enzyme preparations of *Clostridium kluyveri*. II. The aerobic oxidation of ethanol and butyrate with the formation of acetyl phosphate. *J. Biol. Chem.* **180**, 1095–1116; CHANTRENNE, H. and LIPMANN, F. (1950) Coenzyme A dependence and acetyl donor function of the pyruvate-formed exchange system. *ibid.* **187**, 757.
4. O'KANE, D. J. and GUNSALUS, I. C. (1948) Pyruvic acid metabolism. A factor required for oxidation by *Streptococcus faecalis*. *J. Bacteriol.* **56**, 499–506.
5. GUNSALUS, I. C. (1954) Group transfer and acyl-generating functions of lipoic acid derivatives. In *The Mechanisms of Enzyme Action* (W. D. MCELROY and B. GLASS, eds.), pp. 545–580, Johns Hopkins, Baltimore.

6. STANIER, R. Y. and DOUDOROFF, M. (1967) Professor C. B. van Niel on his 70th birthday. *Arch. für Mikrobiol.* **59**, 1–3.
7. RAHN, O. (1945) *Microbes of Merit*, 277 pp., Jaques Cattell Press, Lancaster, Pa.
8. GUNSALUS, I. C. and STANIER, R. Y., eds. (1960–1964) *The Bacteria*, 5 vols., Academic Press, New York.
9. SZILARD, L. (1961) *The Voice of the Dolphins* (German edition).

Primers or Bibles?

10. RAHN, O. (1932) *Physiology of Bacteria*, 438 pp., P. Blakiston & Son, Philadelphia.
11. LIPMANN, F. (1971) *Wanderings of a Biochemist*, 229 pp., Wiley–Interscience, New York.
12. KLUYVER, A. J. (1959) *His Life and Work*, 567 pp., North-Holland, Amsterdam.
13. KLUYVER, A. J. (1931) *The Chemical Activities of Micro-organisms*, 109 pp., University of London Press, London.
14. HARDEN, A. (1923) *Alcoholic Fermentation*, 194 pp., Longmans, Green, London.
15. STEPHENSON, M. (1930) *Bacterial Metabolism*, 320 pp., Longmans, Green, London.
16. MEYERHOF, O. (1924) *Chemical Dynamics of Life Phaenomena*, 110 pp., J. B. Lippincott, Philadelphia.
17. SZENT-GYÖRGYI, A. (1937) *Studies on Biological Oxidation and Some of its Catalysts*, 98 pp., Eggenbergersche Buchhandlung, Karl Rényi, Budapest.
18. *Cold Spring Harbor Symposia on Quantitative Biology* (1939), vol. VII, 463 pp., Darwin Press, New Bedford, Mass.
19. LIPMANN, F. (1941) Metabolic generation and utilization of phosphate bond energy. *Adv. Enzymol.* **1**, 99–162.

Pyridoxal Phosphate and the Lactics

20. GUNSALUS, I. C., HORECKER, B. L. and WOOD, W. A. (1955) Pathways of carbohydrate metabolism in microorganisms. *Bacteriol. Rev.* **19**, 79–128; WOOD, W. A. (1961) Fermentation of carbohydrates and related compounds. In *The Bacteria* (I. C. GUNSALUS and R. Y. STANIER, eds.), vol. II, pp. 59–149.
21. SNELL, E. E. and STRONG, F. M. (1939) A microbiological assay for riboflavin. *Ind. Eng. Chem. Anal. Ed.* **11**, 346–350.
22. SNELL, E. E. (1944) The vitamin activities of "pyridoxal" and "pyridoxamine". *J. Biol. Chem.* **154**, 313; GUNSALUS, I. C. and BELLAMY, W. D. (1944) A function of pyridoxal. *J. Biol. Chem.* **155**, 357–358.
23. VAN NIEL, C. B. (1967) Prefatory chapter: The education of a microbiologist; some reflections. *Ann. Rev. Microbiol.* **21**, 1–30.
24. GUNSALUS, I. C. (1948) Bacterial metabolism. *Ann. Rev. Biochem.* **17**, 627–656; *Ann. Rev. Microbiol.* **2**, 71–100; GUNSALUS, I. C. (1950) Decarboxylation and transamination. *Fed. Proc.* **9**, 556–561; GUNSALUS, I. C. (1951) The structure of pyridoxal phosphate. In *A Symposium on Phosphorus Metabolism, I.* (W. D. MCELROY and B. GLASS eds.) pp. 417–420, Johns Hopkins, Baltimore.
25. UMBREIT, W. W. and GUNSALUS, I. C. (1949) Codecarboxylase not pyridoxal-3-phosphate. *J. Biol. Chem.* **179**, 279–281.

Lipoic Acid Acyl Generator

26. REED, L. J., DEBUSK, B. G., GUNSALUS, I. C. and HORNBERGER, C. S. JR., (1951) Crystalline alpha-lipoic acid: A catalytic agent associated with pyruvate dehydrogenase. *Science*, **114**, 93–94.
27. HAGER, L. P. and GUNSALUS, I. C. (1953) Lipoic acid dehydrogenase: The function of *Escherichia coli* fraction B. *J. Am. Chem. Soc.* **75**, 5767–5768.
28. GUNSALUS, I. C., BARTON, L. S. and GRUBER, W. (1956) Biosynthesis and structure of lipoic acid derivatives. *J. Am. Chem. Soc.* **78**, 1763–1766.
29. REED, L. J. and COX, D. J. (1970) Multienzyme complexes. *Enzymes*, **1**, 213–240.
30. GIBSON, J., UPPER, C. D. and GUNSALUS, I. C. (1967) Succinyl coenzyme A synthestase from *Escherichia coli:* I. Purification and properties. *J. Biol. Chem.* **242**, 2474–2477.

Dioxygen and Hydrocarbon Monoxygenases

31. BRADSHAW, W. H., CONRAD, H. E., COREY, E. J., GUNSALUS, I. C. and LEDNICER, D. (1959) Microbiological degradation of (+)-camphor. *J. Am. Chem. Soc.* **31**, 5507.
32. GUNSALUS, I. C., CHAPMAN, P. J. and KUO, J. F. (1965) Control of catabolic specificity and metabolism. *Biochem. Biophys. Res. Commun.* **18**, 924–931; GUNSALUS, I. C., CONRAD, H. E., TRUDGILL, P. W. and JACOBSON, L. A. (1965) Regulation of catabolic metabolism. *Israel J. Med. Sci.* **1**, 1099–1119.
33. GUNSALUS, I. C., TRUDGILL, P. W. and CUSHMAN, D. W. (1966) Components and reaction of ketolactonases and methylene hydroxylases. In *Biological and Chemical Aspects of Oxygenases* (K. BLOCH and O. HAYAISHI, eds.), pp. 339–342, Maruzen, Tokyo.
34. CONRAD, H. E., DUBUS, R. and GUSALUS, I. C. (1961) An enzyme system for cyclic ketone lactonization. *Biochem. Biophys. Res. Commun.* **6**, 293–297.
35. HEDEGAARD, J. and GUNSALUS, I. C. (1965) Mixed function oxidation: IV. An induced methylene hydroxylase in camphor oxidation. *J. Biol. Chem.* **240**, 4038–4043.

Plasmids and the Genetics of Diversity

36. BERTLAND, A. U. II, MINI, P., BROOKENS, E. and GUNSALUS, I. C. (1964) Isolation and properties of bacteriophages for fluorescent pseudomonads. *Bacteriol. Proc.* **1964,** 139.
37. CRAWFORD, I. P. and GUNSALUS, I. C. (1966) Inducibility of tryptophan synthetase in *Pseudomonas putida. Proc. Nat. Acad. Sci. U.S.A.* **56,** 717–724.
38. GUNSALUS, I. C., CHAKRABARTY, A. M., GUNSALUS, C. F. and CRAWFORD, I. P. (1967) Chromosomal organization in *Pseudomonas putida*: transduction with lytic bacteriophage. *Science* **156,** 538; CHAKRABARTY, A. M., GUNSALUS, C. F. and GUNSALUS, I. C. (1968) Transduction and the clustering of genes in fluorescent *pseudomonads. Proc. Nat. Acad. Sci. U.S.A.* **60,** 168–175; GUNSALUS, I. C., GUNSALUS, C. F., CHAKRABARTY, A. M., SIKES, S. and CRAWFORD, I. P. (1968) Fine structure mapping of the tryptophan genes in *Pseudomonas putida. Genetics* **60,** 419–435.
39. GUNSALUS, I. C., and CHAKRABARTY, A. M. (1967) Transductional heterogenote from interstrain transfer of "mandelate" genes in *Pseudomonas putida. Phage Information Service Abstracts,* August 1967, Cold Spring Harbor, New York, p. 8; CHAKRABARTY, A. M. and GUNSALUS, I. C. (1970) Trandsuction and genetic homology between *Pseudomonas* species *putida* and *aeruginosa. J. Bacteriol.* **103,** 830–832.
40. CHAKRABARTY, A. M. and GUNSALUS, I. C. (1969) Defective phage and chromosome mobilization in *Pseudomonas putida. Proc. Nat. Acad. Sci. U.S.A.* **64,** 1217–1223.
41. RHEINWALD, J. G., CHAKRABARTY, A. M. and GUNSALUS, I. C. (1973) A transmissible plasmid controlling camphor oxidation in *Pseudomonas putida. Proc. Nat. Acad. Sci. U.S.A.* **70,** 885–889.

Cytochrome P450 and the Physics of Iron Enzymes

42. CUSHMAN, D. W., TSAI, R. L. and GUNSALUS, I. C. (1967) The ferroprotein component of a methylene hydroxylase. *Biochem. Biophys. Res. Commun.* **26,** 577–583.
43. DERVARTANIAN, D. V., ORME-JOHNSON, W. J., HANSEN, R. E., BEINERT, H., TSAI, R. L., TSIBRIS, J. C. M., BARTHOLOMAUS, R. C. and GUNSALUS, I. C. (1967) Identification of sulfur as a component of the EPR signal at $g = 1.94$ by isotopic substitution. *Biochem. Biophys. Res. Commun.* **26,** 569–576.
44. COOKE, R., TSIBRIS, J. C. M., DEBRUNNER, P., TSAI, R. L., GUNSALUS, I. C. and FRAUENFELDER, H. (1968) Mössbauer studies on putidaredoxin. *Proc. Nat. Acad. Sci. U.S.A.* **59,** 1045–1052.
45. KATAGIRI, M., GANGULI, B. N. and GUNSALUS, I. C. (1968) A soluble cytochrome P-450 functional in methylene hydroxylation. *J. Biol. Chem.* **243,** 3543–3546.

Systems, Energetics and Flash Kinetics

46. LIPSCOMB, J. D. and GUNSALUS, I. C. (1973) Structure and reactions of a microbial monoxygenase: The role of putidaredoxin. In *Iron–Sulfur Proteins* (W. LOVENBERG, ed.), vol. 1, pp. 151–172, Academic Press, New York.
47. GUNSALUS, I. C., MEEKS, J. R., LIPSCOMB, J. D., DEBRUNNER, P. G. and MÜNCK, E. (1974) Bacterial monoxygenases—The P450 cytochrome system. In *Molecular Mechanisms of Oxygen Activation* (O. HAYAISHI, ed.), pp. 559–613, Academic Press, New York.
48. LIPSCOMB, J. D. (1974) Energy transfer and segregation: mixed function oxidation by cytochrome $P450_{cam}$ and putidaredoxin. Ph.D. Thesis, University of Illinois.
49. SLIGAR, S. G. (1975) A kinetic and equilibrium description of camphor hydroxylation by the $P450_{cam}$ monoxygenase system. Ph.D. Thesis, University of Illinois.
50. AUSTIN, R. H., BEESON, K. W., EISENSTEIN, L., FRAUENFELDER, H. and GUNSALUS, I. C. (1975) Dynamics of ligand binding to myoglobin. *Biochemistry 14,* 5355–5373.

THE IMPACT OF THE ATP-GENERATING SYSTEMS

SANTIAGO GRISOLIA

Department of Biochemistry, University of Kansas Medical Center
Kansas City, Kansas 66103

It was the firm belief of Otto Warburg that the advancement of biochemistry depends primarily on the development and proper use of new techniques for, in this way, it is possible to approach problems which otherwise might be insoluble. In this sense the development of modern biochemistry, and particularly the flourishing of intermediary metabolism, is indebted to a few but powerful techniques, e.g. the use of isotopes, the development of optical methods, chromatographic procedures, the development of the homogenizing technique by Potter and Elvehjem[1] and the ATP-generating system which was originated by Severo Ochoa as I have pointed out briefly before.[2]

It is the purpose of this brief essay to stress the importance of the ATP-generating technique first used by Severo and a few years later recommended by him to Sarah Ratner who developed, refined and popularized its use and to myself for our studies in biosynthesis of urea, and to indicate how with its help many reactions, which were at the time poorly understood, were clarified.

It should be made quite clear that the acceptance of ATP-generating systems suffered a time-delay element, as is generally the case for new findings in biochemistry or in any other science. On the other hand, the best indication of scientific success is when a finding loses its primary identification and is absorbed by and becomes a part of the overall body of knowledge; then it is truly a part of science. This is what has happened with the ATP-generating system.

It is important for the young reader to turn back the clock some 30 years and to realize that the simple homogenate technique developed by Potter and referred to above was revolutionary, for it was suddenly possible, with little effort, to study reactions under many experimental conditions, with fairly homogeneous preparations and with similar aliquots as contrasted with the tedious and painstaking technique of slices introduced by Otto Warburg. Then suddenly it was possible to have in a matter of minutes concentrated tissue suspensions with main components of the cell still intact. The reader must remember again that many cell components such as mitochondria and other particles were yet to be rediscovered!

However, it became evident that these homogenates often had to be fortified with a number of reagents, and even then the homogenates were not always active for one depended largely on oxidation to demonstrate a number of reactions. Indeed, there was some sort of vague feeling that ADP or AMP in some mysterious way were better donors than ATP. This, of course, was due to poor preparations; many of the ADP and AMP preparations at the time were made by the individual investigators who started from their own ATP so that in general the AMP and ADP were purer preparations. At any rate, given the fact, as we know it today, that many substances are not permeable through membranes, including mitochondrial membranes, one had often destroyed the potentiality for oxidative phosphorylation by the mitochondria with the often concomitant increase in ATPase liberated by mistreated mitochondria and homogenates. Moreover, it was frequently difficult to demonstrate a number of reactions by balance methods because of the endogenous values present in homogenates.

In the early 1940's, Severo was interested in oxidative phosphorylation.[3–6] It must be said that although his measurements (from 1940 to approximately 1944) were most accurate and his work very brilliant, his good sense led him to abandon this area in spite of his earlier strifes and successes. At the time of this writing, approximately 30 years later, the mechanism of oxidative phosphorylation is still to be solved.

At any rate, Severo was keenly aware of the need to correct for ATPase in order to obtain a good idea of the efficiency of oxidative phosphorylation as shown in one of his early papers.[4]

> *Transfer of Phosphate from Phosphocreatine to Glucose*—Here phosphocreatine is the donor of phosphate to adenylic acid forming adenosine triphosphate which passes its acid-labile phosphate over to the sugar. The adenylic acid system acts in a catalytic manner as a carrier of phosphate between phosphocreatine and glucose just as it acts catalytically in aerobic phosphorylation as a carrier of inorganic phosphate to the phosphate acceptor. If sufficiently active, adenosinetriphosphatase will interfere with the phosphorylation of the acceptor to a similar extent in both cases.

He showed in a number of papers[3–6] that on the addition of creatine phosphate and creatine phosphokinase, phosphoenol pyruvate or P-glycerates as donors of phosphate, he would obtain higher values for P/O ratios than hitherto estimated, for he was then able to correct for ATPase. In other words, he was able to demonstrate for the first time (a) that the uptake of oxygen produced more than one ATP per atom of oxygen, and indeed he was remarkably close to the accepted present value (i.e. he calculated 15 moles of ATP per mole of pyruvate oxidized *in the middle 1940's*) and (b) that the lower levels obtained by other people were probably due to interference by ATPase. In other words, the large amount of ATPase, which prevented the demonstration of reactions, or the impossibility of using very large amounts of ATP due to inhibitory effects, were by-passed by the use of an ATP-generating system. It should be noted that although these experiments were quite clear, in later years they were criticized by others, particularly by Ogston and Smithies.[7] Indeed, for a lively series of bad guesses including the suggestion that Ochoa's ATP was at best only 40% pure, the reader may find the Ogston and Smithies review of much interest.

It is of interest to recollect that while I talked about this with Severo, he calmly made a comment affirming his belief in his own data. He has always had a profound confidence in experimental data rather than in getting involved in interpretation. Naturally he was quite correct.

To my knowledge, the first demonstration of ATP coupling to a synthetic reaction was the study by the late John Speck of glutamine synthesis.[8] Again, Speck first used another Ochoa "trick", acetone powders, to decrease the amount of ATPase. Shortly thereafter, both Sarah Ratner and I demonstrated the direct coupling of ATP for the two main steps in the synthesis of urea, i.e. the arginino-succinic[9] and the citrulline synthesis step.[10] However, due to the now well-known inhibitory effects of high concentrations of ATP, the optimal conditions were difficult to set up; it was at this time that Severo first recommended the use of an ATP-generating system to Sarah and later on to me. It must be said that Sarah, with her quiet elegance, refined and simplified the technique.

Perhaps reflecting the more conservative attitude of the times (in those days research money was not yet abundant, which is again a turn of the wheel), she chose 3-phosphoglycerate as a relatively inexpensive, pure and easily made product, and a mixture of glycolytic enzymes which had been in use for quite some time, the so-called Racker Preparation, a simple ammonium sulfate fraction of a rabbit muscle extract which the ever-knowledgeable Efraim Racker knew contained many of the glycolytic enzymes (among others, P-glyceromutase, enolase and phospho(enol)pyruvatekinase). With these two preparations it was possible to study many of the ATP-requiring systems. Indeed, I myself, am indebted to Racker for his preparation, which led me to realize when trying to get better conditions for isolation of the intermediate Compound X,[10] later on identified as carbamyl phosphate, that phosphoglyceromutase was one of the enzymes that had never been purified. This led to the crystallization of this enzyme for the first time in my laboratory.[11] The delay in acceptance of findings which I mentioned at the beginning is best exemplified for the ATP-generating system by the following: I remember very clearly when the already outstanding Konrad Bloch who had been very nice to me in Chicago and was also a good friend of Severo's came to give a talk in Madison, Wisconsin, around 1950, when he first had evidence for the synthesis of glutathione as a model for peptide bond synthesis, and, incidentally, the ATP-generating system has been fundamental in the unravelling of protein synthesis, possibly influenced by the fact that Speck had been a very close friend and shared a laboratory with him (Bloch) in those years in Chicago. After the seminar, I suggested that he use an ATP-generating system. Without hesitation he replied, "That would not work, for you need oxidative phosphorylation to generate ATP". However, rather shortly, if one looks at Volume II of *Phosphorus Metabolism*, the series which was to metabolism as the big bands were to the forties, the ATP-generating systems came to be widely used, among others by Konrad Bloch himself[12] who by that time used P-glycerate as an ATP-generating system with increasing glutathione synthesis!

For brevity and as a further example, only the older outstanding work of Buchanan[13] in elucidating the biosynthesis of purines with the aid of ATP-generating systems will be mentioned. However, it is quite evident that the usefulness of the technique is still most prevalent as typified by work in protein synthesis. Indeed, Ochoa himself[14] is still using it for this as well as many others, e.g. even in marginally biochemical journals.[15] Another popular use is for the assays of ATPase, e.g. addition of PEP and pyruvate kinase with or without coupling to lactic dehydrogenase.[16] It is of interest then that the ATPase of the internal membrane of mitochondria is likely part of the oxidative phosphorylation mechanism.[17] As traced in this paper, the ATP-generating technique started with studies on oxidative phosphorylation; thus, it is still closely related to its inception, in a manner of speaking.

REFERENCES

1. Potter, V. R. (1949) The homogenate technique. In *Manometric Techniques*, (W. W. Umbreit, R. H. Burris and J. F. Stauffer, eds.), p. 170, Burgess Publishing Co. Minneapolis, Minn.
2. Grisolia, S., Carreras, J., Diederich, D. and Carache, S. (1970) Binding of 2,3-diphosphoglycerate (DPG) to oxyhemoglobin; levels and effect of DPG on oxygen affinity of normal and abnormal blood. In *Red Cell Metabolism and Function* (G. Brewer, ed.), p. 39, Plenum Press, New York.
3. Ochoa, S. (1941) 'Coupling' of phosphorylation with oxidation of pyruvic acid in brain. *J. Biol. Chem.* **138**, 751.
4. Ochoa, S. (1943) Efficiency of aerobic phosphorylation in cell-free heart extracts. *J. Biol. Chem.* **151**, 498.

5. OCHOA, S. (1941) Glycolysis and phosphorylation in brain extracts. *J. Biol. Chem.* **141,** 245.
6. OCHOA, S. (1944) α-Ketoglutaric dehydrogenase of animal tissues. *J. Biol. Chem.* **155,** 87.
7. OGSTON, A. G. and SMITHIES, O. (1948) Some thermodynamic and kinetic aspects of metabolic phosphorylation. *Physiol. Rev.* **28,** 283.
8. SPECK, J. F. (1947) The enzymic synthesis of glutamine. *J. Biol. Chem.* **168,** 403.
9. RATNER, S. and PETRACK, B. (1951) Biosynthesis of urea. III. Further studies on arginine synthesis from citrulline. *J. Biol. Chem* **191,** 693.
10. GRISOLIA, S. (1951) Mechanism of the biosynthesis of citrulline. In *Phosphorus Metabolism* (W. D. MCELROY and B. GLASS, eds.), vol. I, p. 619, Johns-Hopkins Press, Baltimore, Md.
11. RODWELL, V. W., TOWNE, J. C. and GRISOLIA, S. (1956) Crystalline phosphoglyceric acid mutase. *Biochim. Biophys. Acta*, **20,** 394.
12. BLOCH, K., SNOKE, J. E. and YARNARI, S. (1952) The role of phosphate in amino acid and protein metabolism. In '*Phosphorus Metabolism*' (W. D. MCELROY and B. GLASS, eds.) Vol. II, p. 82, Johns-Hopkins Press, Baltimore, Md.
13. BUCHANAN, J. M., LEVENBERG, B., FLAKS, J. G. and GLADNER, J. A. (1955) Interrelationships of amino acid metabolism with purine biosynthesis. In *A Symposium on Amino Acid Metabolism* (W. D. MCELROY and H. B. GLASS, eds.), p. 743, Johns-Hopkins Press, Baltimore, Md.
14. SIERRA, J. M., MEIER, D. and OCHOA, S. (1974) Effect of development on the translation of messenger RNA in artemia salina embryos. *Proc. Nat. Acad. Sci.* **71,** 2693.
15. SCHREIBER, S. S., ORATZ, M., ROTHSCHILD, M. A., REFF, F. and EVANS, C. (1974) Alcoholic cardiomyopathy. II. The inhibition of cardiac microsomal protein synthesis by acetaldehyde. *J. Molec. Cell. Cardiol.* **6,** 207.
16. PULLMAN, M. E., PENEFSKY, H. S., DATTA, A. and RACKER, E. (1960) Partial resolution of the enzymes catalyzing oxidative phosphorylation. *J. Biol. Chem.* **235,** 3322.
17. RACKER, E. (1975) Inner mitochondrial membranes: basic and applied aspects. In *Cell Membranes, Biochemistry, Cell Biology and Pathology* (G. Weissmann, ed.), p. 135, HP Publishing Co., Inc., New York, New York.

II. LIPIDS, SACCHARIDES, CELL WALLS

ON THE EVOLUTION OF A BIOSYNTHETIC PATHWAY

Konrad Bloch

James Bryant Conant Laboratory, Harvard University, Cambridge, Mass. 02138

Every generation of scientists tends to view its own era as the most important ever, a prejudice to which it is fully entitled. But even the objective future historian will probably look upon the past few decades as the first golden age of Biochemistry. We have come very close to understanding the chemical basis of the life process.

To contribute to "Reflections on Biochemistry" dedicated to my good friend Severo, a principal architect of modern Biochemistry, is indeed a privilege and a pleasure.

When I began my career in 1934 as a research assistant in the Schweizerisches Forschungsinstitut, Davos, Switzerland, my assignment was to resolve a controversy; an assignment prompted by contradictory reports in the literature whether or not sterols occurred in bacteria. According to investigators at Yale University headed by R. J. Anderson, sterols, compounds ubiquitous in animal and plant cells were not present in *Mycobacterium tuberculosis*.(1) Another report had claimed the contrary. My own findings, those of a neophyte in research, corroborated Anderson and his associates. In retrospect, I am struck by the coincidence that Rudolf Schoenheimer was one of the co-authors of the paper I confirmed. At the time I had, of course, no premonition that in years to come he was to be my principal mentor and that his influence would decisively shape my career. As for the absence of cholesterol in mycobacteria this did not seem to me especially significant. Biochemical unity and diversity were not prominent or even existing issues to someone trained in organic chemistry. Comparative biochemistry was to intrigue me only very much later and I doubt that my early encounter with one of its manifestations consciously contributed to this interest.

Working for my Ph.D. degree in the Department of Biochemistry at the College of Physicians and Surgeons of Columbia University, I shared one large laboratory with some twenty fellow graduate students. Ideas and experiences were freely exchanged and students working in various groups became involved in each others' problems as much as in their own. Here and during this period Schoenheimer and Rittenberg developed the methodology for biochemical studies with isotopic tracers, researches which to all of us seemed especially exciting. No wonder the promise of the tool as well as Schoenheimer's personality influenced the careers of many of my contemporaries as well as my own. When the opportunity came to join Schoenheimer's team I seized it only too eagerly. My apprenticeship in tracer research was spent synthesizing ^{15}N-labeled amino acids to be fed to rats for studies on the origin of creatine and on the metabolic relationship between creatine and creatinine. This work was satisfying in that it proved what was reasonable to expect on chemical grounds but the knowledge gained was relatively modest and unsurprising. Schoenheimer next asked me to take on an analytical problem which was to form part of a major study on the biosynthesis of cholesterol. This may have come about because of our earlier independent efforts to establish whether or not cholesterol was present in mycobacteria. Whatever the reason, Schoenheimer was evidently challenged by the magnitude of the problem of elucidating a biosynthetic pathway that was totally unknown and likely to be exceedingly complex. Specifically, I was to design a pyrolytic method for mass-spectrographic analysis of the hydroxyl oxygen of cholesterol. One may be sure that Schoenheimer wished to find out whether the source of oxygen in cholesterol was water or molecular oxygen. To raise such a question would be elementary today but in 1940, the possible existence of enzymatic reactions catalyzing the direct insertion of molecular oxygen into organic compounds—mono- and dioxygenases—had not yet entered biochemical consciousness. My efforts to develop the required methodology did not succeed but they made me aware of an intriguing problem. T. T. Tchen solved it many years later by proving that molecular oxygen is the source of the β-hydroxyl group in the sterol molecule.[2]

After Schoenheimer's untimely death in 1941, his associates fell heir to a wealth of problems and ideas which had been conceived and begun under his leadership. Continuing on our own, we may have divided the program according to personal preferences or by tossing coins. How these fateful decisions were made I do not recall. At any rate, D. Rittenberg and I began to collaborate on the mechanism of cholesterol biosynthesis.

Schoenheimer and Rittenberg had already recognized that the synthesis of cholesterol occurs totally and *de novo* in the animal body from low molecular weight precursors.[3] Independently, and in line with this conclusion, Sonderhoff and Thomas had found that "yeast grown in a medium supplemented with D_2-acetate had a deuterium content so high that a direct conversion of acetic acid to sterols had to be postulated."[4] Our experiments with rats and mice established the same role for acetic acid in cholesterol synthesis by animals and showed further that both the isooctyl side chain and the tetracyclic steroidal ring system were derived from this precursor.[5] To prove that acetic acid was the exclusive carbon source for the C_{27} structure in the animal body was a more difficult task. Acetic acid in the form of acetyl CoA arises in large amounts from the breakdown of carbohydrate, fatty acids and amino acids. On mixing with this large tissue pool, labeled dietary acetic acid will suffer isotope dilution to an extent which cannot be easily or accurately estimated. One way to solve the problem was to degrade biosynthetically labeled cholesterol carbon by carbon and to ascertain whether each of the twenty-seven positions had been tagged by one or the other acetate carbons. This feasible but formidable project was undertaken in the laboratories of Cornforth and Popjak and in our own but it took many years to complete it.[6,7] The mutant technique promised a more immediate and straightforward answer. Tatum and his associates had isolated an acetate-requiring mutant of *Neurospora crassa*, acetate-less because it lacked a key enzyme for metabolizing glucose to acetic acid. In the absence of metabolic acetate production, "isotope dilution" will not occur so that constituents of the mutant cell derived entirely from acetic acid will have the same isotope concentration as the external precursor. The outcome of the experiment was unambiguous; the sterol molecule—except for the extra carbon atom at C_{28} of ergosterol—was entirely acetate-derived.[8] No further attempts were made to use the powerful mutant technique for solving the problem at hand, at least not intentionally. The reasons were both practical and intrinsic. Practical because for mutant work, *E. coli* largely replaced *Neurospora crassa* as the microorganism of choice and intrinsic because *E. coli*, like *Mycobacterium tuberculosis* and in fact nearly all procaryotes do not manufacture sterols or require them. The work and race horse of modern biochemistry and molecular biology was a useless organism as far as the steroid biochemist was concerned. Still, one group of bacterial mutants has contributed crucially if inadvertently to the unravelling of sterol biosynthesis. At the Merck, Sharpe and Dohm Laboratories, a search for new bacterial growth factors unearthed a number of acetate-requiring strains of *Lactobacilus acidophilus*—reminiscent of the acetate-less *Neurospora* mutant—but apparently carrying different enzymatic lesions.[9] The acetic acid requirement of the mutant strains was so high as to suggest that a metabolite of acetic acid rather than acetate itself limited bacterial growth. Testing of various media known to be rich in vitamins (e.g. "distillers soluble") promptly led to the isolation of an "acetate replacing factor" and its chemical characterization as the lactone of β-methyl-β,δ-dihydroxyvaleric acid (mevalonic acid, MVA). Tavormina, Gibbs and Huff soon recognized the structural similarity between MVA and β-hydroxy-β-methylglutaric acid (HMG) a suspected but unproven intermediate on the route from acetate to sterol. Gratifyingly, liver homogenates converted MVA quantitatively to cholesterol[10] and in turn, MVA was shown to arise from acetic acid by way of the CoA derivatives of acetoacetate and HMG.[11] Following this important breakthrough the various phosphorylated and pyrophosphorylated intermediates that linked MVA to squalene were discovered in rapid succession. Almost simultaneously three research groups described several phosphorylated isoprenoid intermediates, including isopentenyl pyrophosphate, the "biological isoprene unit".[7,11,11a] Why MVA should be essential for the growth of lactobacillus species long remained a mystery. Anaerobic bacteria do not synthesize sterols or any of the quinone-linked polyisoprenoid chains found, e.g. in Coenzyme Q. Only in recent years was the puzzle solved when trace amounts of isoprene derivatives were discovered in all bacterial cells. These include N_6-isopentenyl adenine, one of the minor bases in some transfer-RNA molecules, and the undecaprenol phosphate derivatives which serve as lipophilic carriers in the synthesis of bacterial peptioglycans, of lipopolysaccharides and probably more generally in the assembly of hetero- and homopolymeric polysaccharide cell envelopes.

The chance discovery of MVA was of far-reaching significance. Its importance can be appreciated properly only by those who labored in the field of sterol biosynthesis and for years searched unsuccessfully for the "biological isoprene unit". The ways chosen by Nature for making organic compounds are easy enough to rationalize—by hindsight. Biochemistry remains, by and large, an empirical science and predicting the course of even the simplest biochemical reactions, on whatever grounds, remains a venture of risk.

STEROLS AND EVOLUTION

To speculate on the ultimate origin of biological molecules is a preoccupation that is rapidly growing in popularity. As Roger Stanier put it, "it has the same intellectual fascination for some biologists that metaphysical speculation possessed for some medieval scholastics". In that spirit a few thoughts on the evolution of the sterol structure will now be presented.

Some of the views on the evolution of biochemical capabilities and pathways are now widely shared and these will be summarized here briefly because they are necessary for the arguments that follow. According to the Oparin–Haldane hypothesis all major structural types that compose contemporary cells were created chemically before life began. From the rich prebiotic reservoir the ancestral cell took up organic compounds by a process of accretion, gradually depleting the supply of needed molecules. As living cells evolved, enzymes for making or replacing the components of the organic broth arose successively by single mutations, the first of these enzymes catalyzing the terminal event of a contemporary biosynthetic sequence (N. Horowitz). A corollary assumption is that the various precursors or intermediates were also present in the prebiotic broth. As they were successively depleted, new biosynthetic capabilities for replacing them were acquired in turn. In time a catalytic chain of whatever length would become established starting with the production of the last, most complex member of the series and resulting ultimately in complete auxotrophy (retroevolution).

Experimental successes in producing sugars, amino acids and nucleic acid bases under primitive earth conditions have made these hypotheses increasingly attractive. Unfortunately, all discussions dealing with the prebiotic environment and evolution tend to ignore the origin of sterols or for that matter of any lipid structures. Analyzing this problem one comes to realize that there are some fundamental reasons why chemical evolution could not have produced cholesterol or a related prototype and why synthesis of a sterol may be impossible to achieve under primitive earth conditions. It now appears axiomatic that the prebiotic and early terrestrial biosphere was anaerobic. In the presence of oxygen, organic compounds could not have survived. Oxygen is, however, obligatory for the biological synthesis of the sterol structure at least as we know it today. By no stretch of the imagination can one conceive of a chemical process that would generate cholesterol anaerobically. It is not difficult to visualize a prebiotic formation of isoprene units, isoprene polymers and perhaps even of squalene but there is no plausible way of cyclizing squalene to sterol (lanosterol) non-oxidatively (Fig. 1, I-III). (A proton-catalyzed

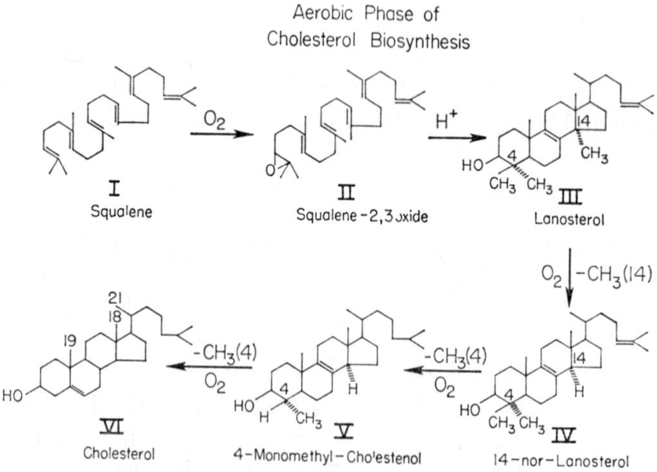

FIG. 1. The pathway from lanosterol to cholesterol.

cyclization of squalene is, of course, entirely feasible but the problem of introducing the oxygen function would remain.) On chemical grounds it is even less likely if not impossible to remove non-oxidatively the "extra" methyl groups of lanosterol for conversion to cholesterol (Fig. 1, III-VI). It therefore follows that any chemical evolution of the sterol pathway must have ended with squalene. Only this molecule, but no "later" ones could

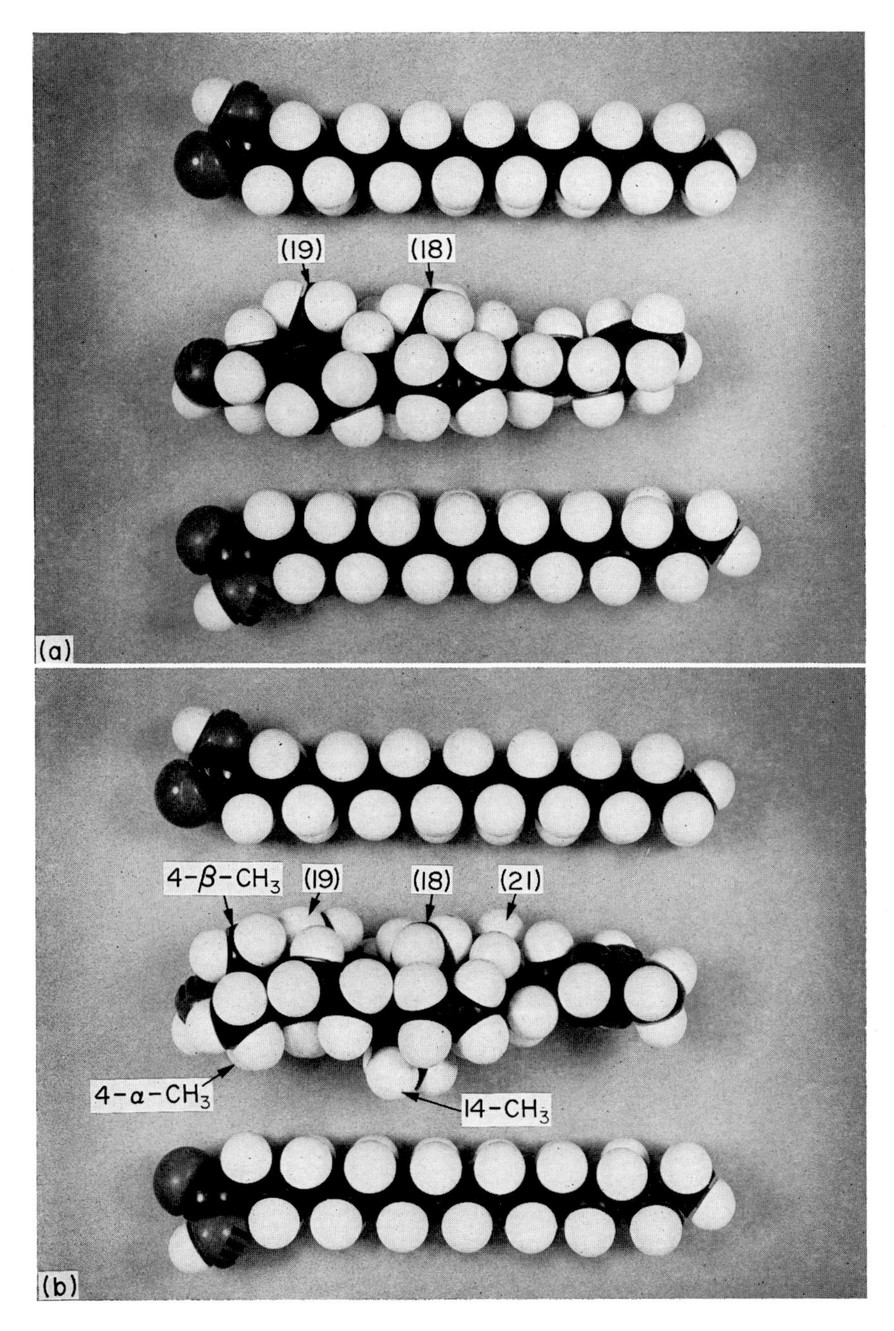

FIG. 2. Models for the spatial relationship between sterol and the fatty acid chains of phospholipid. Upper panel, cholesterol; lower panel, lanosterol.

have been available for accretion by a primitive anaerobic cell. The occurrence of squalene in anaerobic bacteria has not been reported but it is doubtful that it has been looked for systematically. However, at least two aerobic bacterial species, *Staphylococcus aureus*[12] and *Halobacterium cutirubrum*[13] contain squalene.

The procaryotic blue-green algae are generally thought to have been the first sources of molecular oxygen when the terrestrial atmosphere turned aerobic. Only after these organisms came into being could the respiratory process and oxygenase-catalyzed biosynthetic reactions begin to evolve. Arising in an anaerobic environment the most ancient blue-green algae like anaerobic bacteria must have lacked sterols but such restrictions disappeared once the oxygen-producing mode of photosynthesis became established.* Mutational events could now have given rise to oxygenase enzymes of the type required during the aerobic phase of sterol biosynthesis as we now know it. Recapitulating the contemporary pathway, squalene epoxidation (Fig. 1, I-II) must have been the first mutation that launched the sterol pathway. Per se, the acyclic squalene epoxide has no known function at least in extant cells. That it had survival value for an ancient organism that formed it as an end product seems doubtful. A further step, the cyclization of squalene epoxide to lanosterol (Fig. 1, II-III) had to evolve in order to generate a functionally useful molecule. However, cells that normally terminate sterol biosynthesis at the lanosterol stage are not known. The reason may be that the lanosterol structure is only marginally competent for membrane functions. Thus, while lanosterol supports the growth of anaerobic yeast, an artificial sterol auxotroph, it does so very poorly compared to cholesterol or ergosterol (R. Masters, Ph.D. thesis, Harvard University, 1963). The carbon skeleton of lanosterol differs from that of cholesterol by three additional methyl substituents, one attached at C_{14} and two at C_4 of the ring system. It is therefore tempting to argue that for improved function, the "extra" methyl groups of lanosterol had to be removed, cholesterol representing the end of the line, a molecule perfected by evolutionary pressure. The argument is structural, not one of esthetics. It takes its cues from models for cholesterol-containing membranes first proposed by Finean[17] and later variously refined.[18,19] These models emphasize the structural complementarity of phospholipids and cholesterol which permits highly specific 1:1 complexes to be formed between these molecules. The postulate is that cholesterol intercalated or sandwiched between fatty acyl chains of adjacent phospholipids modifies membrane mobility in several ways; by hydrogen bonding between the free hydroxyl group of cholesterol and the polar head group of phospholipid; close contacts and tight packing between the rigid planes of the steroidal ring system and the proximal half (C-atoms 1 to 8 or 9) of the fatty acid chains (solidifying effect); and less intimate hydrophobic contacts between the conformationally flexible and branched aliphatic sterol side chain and the distal or tail section of the fatty acid (liquefying effect). Inspection of molecular models (Fig. 2) makes it very clear why the structure of lanosterol is sterically much less favorable for these interactions than cholesterol. First and probably most importantly the axial α-methyl group of lanosterol at C_{14} bulges and protrudes from the planar underside of the steroidal ring system. This will have the effect of preventing or drastically reducing effective contacts with the fatty acyl chain. It should be noted that the methyl group at C_{14} is also the first to be removed in the biosynthetic sequence.

* While attempts in our laboratory to isolate sterols from a blue-green algae (*Anacystis nidulans*) were unsuccessful,[14] cholesterol and its 24-ethyl derivatives (in unstated amounts) have more recently been isolated from this organism.[15] *Porphyridium luridum* also contains traces of 24-ethyl cholesterol.[16]

Secondly, while the two "extra" methyls at C_4 of lanosterol do not visibly obstruct sterol packing with the fatty acid chain the model shows that they severely crowd the adjacent OH group. Conceivably, they interfere sterically in some manner with reactions that involve the 3-hydroxyl group of the sterol molecule. The processes affected adversely could be hydrogen bonding to the polar head group of phospholipid or to water at the bilayer interface; or, the neighboring methyl groups, by changing the polarity of the sterol hydroxyl, could upset the proper positioning of the sterol molecule relative to phospholipid. At any rate it appears a plausible and testable hypothesis that oxidative removal of the three methyl branches represents a streamlining process that adapted the sterol molecule structurally for optimal membrane function. No such evolutionary pressure seems to have existed for removing the two bridgehead methyl groups C_{18} and C_{19} (Fig. 1). They have remained a permanent structural feature of the sterol molecule. Naturally occurring sterols lacking either one or the other of these branches are not known. Reference to molecular models again provides a rationale. The two angular methyl groups together with the apposite side chain methyl (C_{21}) all lie in a plane rendering the upper sterol face (β-side) equally accessible to contacts with fatty acyl chains.

The route from lanosterol to cholesterol is long, requiring probably no less than fifteen steps. If fitness for membrane function was the evolutionary driving force then sequential mutations must have generated partially demethylated, progressively more useful lanosterol derivatives. However, organisms that accumulate such intermediates as *end* products of the sterol pathway (e.g. 14-des methyl- or 4-monomethyl sterols) are not known. In all cells capable of sterol synthesis the evolutionary process seems to have run its course. A systematic search for cells harboring more primitive microbial sterols is certainly needed before we dismiss the possibility that they exist. But it seems more likely that such hypothetical organisms have become extinct.

STEROLS AND THE PROCARYOTIC–EUCARYOTIC TRANSITION

In the beginning of this essay the point was made that sterols occur in procaryotic cells only rarely if at all. At one time it therefore seemed reasonable to argue that the sterol molecule was an invention of the eucaryotic cell made perhaps in parallel with the development of membrane-bounded organelles. This view is no longer tenable. In 1971 Bird *et al.* reported the isolation of squalene, of 4,4 dimethylcholestenols (C_{29}), 4-monomethyl-cholestenols (C_{28}) and zymosterol (C_{27}) from *Methylococcus capsulatus* grown on methane as the sole carbon source.[20] The amounts isolated were substantial, comparing in magnitude with the sterol content of yeast and fungi. The presence of the fully developed sterol pathway in bacteria is therefore clearly established at least in one bacterial species. Whether *Methylococcus* is phylogenetically primitive or advanced is irrelevant. It is more to the point that an organism adapted to metabolize hydrocarbons by oxygenase-type reactions has also the competence for oxidatively removing the methyl branches from the lanosterol nucleus. Still to be answered is the intriguing question whether these sterols are useful or essential for *Methylococcus*. Penicillin inhibits the growth of the organism[21] connoting the presence of the peptidoglycan cell wall typical of procaryotes. At the same time, polyene antibiotics such as rimicidin and candicin, which are sensitive probes for the presence of sterol in eucaryotic membranes, do not affect the growth of *Methylococcus*.[21]

This makes it unlikely that the sterols of this organism are needed for stabilizing a bacterial membrane bilayer.* (In certain strains of mycoplasma, bacteria that lack the typical procaryotic cell wall, sterols apparently do play such a role.) Nevertheless it is of interest that in some methylotrophic species, grown on methane as the sole carbon source, one can see an extensive system of peripherally arranged intracytoplasmic membranes,[22] and it is here where squalene and sterols seem to be localized (R. S. Hanson, private communication). The possibility that these structures are antecedents of eucaryotic organelles seems worth exploring. What remains baffling is why sterols are abundant in methanotrophic organisms and why otherwise their occurrence seems so exceedingly rare in the bacterial phylum. Even if more sensitive methods should detect sterols elsewhere and in procaryotes in which they have been missed so far, e.g. in *E. coli* or in *Mycobacteria*, one can confidently predict that in these organisms sterol concentrations will be very much lower than in eucaryotic cells, probably by at least 1-2 orders of magnitude.

Whether some procaryotes contain sterols and others do not has, of course, an important bearing on evolutionary relationships. One may ask, for example, whether *Methylococcus* and related methanotrophs were ancestral to a line of aerobic bacteria which then abandoned the sterol pathway or whether in another independently evolving branch of aerobes sterol biosynthesis was never invented. These issues are of interest not only by themselves but also because they bear on the matter of how the wide evolutionary gap between procaryotes and eucaryotic cells was bridged. As pointed out earlier, the possession of the sterol molecule is a universal feature of eucaryotic forms of life. For eucaryotes, phylogenetic relationships are much more certain than they are for procaryotic cells so that one can speak with some confidence of eucaryotes that are primitive and those that are more advanced. In the present context it is therefore significant that the red algae, supposedly the most primitive extant eucaryotes, not only contain sterols in substantial amounts but also that in these cells the biosynthetic pathway is modern, i.e. fully developed. Cholesterol or related cholestane derivatives are the dominant sterols in all members of the order Rhodophytae so far examined.[23]

This leads us to the question whether in eucaryotic cells sterol synthesis evolved independently or whether it was acquired from a procaryotic precursor; and to the more general problem of the origin of the eucaryotic cell.

It is now recognized that certain biochemical attributes of aerobic procaryotes strikingly resemble those of the O_2-consuming or producing organelles in eucaryotic cells. Attention has been focused especially on the similar properties of the respective machineries for protein synthesis. With the information that comparative biochemistry is now providing support is growing for evolutionary theories which postulate a symbiotic origin of the eucaryotic cell: the merging of a free-living aerobic bacterium (or blue-green alga) with another cell, the "protoeucaryote", by a process of endocytosis.[24] In time, so the theory goes, the ancestral aerobic bacterium became the mitochondrion and the blue-green algae the modern chloroplast, the procaryotic symbionts providing *inter alia* the genetic information and machinery for respiration and photosynthesis. The literature is less explicit about specific attributes of the "protoeucaryotic" host but logic would seem to dictate that its metabolic mode was fermentative, i.e. anaerobic. Had it been otherwise, no advantage would have been gained by capture of the aerobic procaryote which evolved

* This conclusion would be invalid if the bacterial cell wall should prove impermeable to polyene antibiotics.

into the respiratory organelle. By the same token any profitable O_2-dependent or utilizing biosynthetic processes—sterol biosynthesis among them—must necessarily have been the contribution of the aerobic, procaryotic partner. For this role one might propose *Methylococcus* or a related methylotroph as a plausible candidate. In this organism oxygen-dependent biosynthetic capacities are highly developed. The hypothesis, however, creates a new, if not necessarily insuperable predicament. In contemporary eucaryotes all enzymes of the sterol pathway and many oxygenase reactions are localized either in the cytosol or in the endoplasmic reticulum, not in the mitochondrion. This appears to be the prevailing pattern in animal tissues and in yeast but remains to be examined in other cells. However, one faces this difficulty with other pathways as well. The mitochondrion of extant eucaryotes is far from autonomous, unlike its presumed ancestral procaryote. Little of the genetic information and synthetic machinery for respiration has remained associated with contemporary mitochondria even though the respiratory enzymes themselves still reside in this organelle. Also, few of the reactions for lipid biosynthesis which bacterial cells are capable of have survived in the mitochondria of present-day cells. In the wake of the symbiotic relationship much of the informational and synthetic machinery may have been lost by gene transfer to extra-mitochondrial organelles.

BIBLIOGRAPHY

1. ANDERSON, R. J., SCHOENHEIMER, R., CROWDER, J. A. and STODOLA, F. H. (1935) *Z. Physiol. Chem.* **237**, 40.
2. TCHEN, T. T. and BLOCH, K. (1957) *J. Biol. Chem.* **226**, 921.
3. RITTENBERG, D. and SCHOENHEIMER, R. (1937) *J. Biol. Chem.* **121**, 235.
4. SONDERHOFF, R. and THOMAS, H. (1937) *Ann. Chem.* **530**, 195.
5. BLOCH, K. and RITTENBERG, D. (1942) *J. Biol. Chem.* **145**, 625.
6. CORNFORTH, J. W., GORE, I. Y. and POPJAK, G. (1957) *Biochem. J.* **65**, 94.
7. BLOCH, K. (1965) *Science*, **150**, 3692.
8. OTTKE, R. C., TATUM, E. L., ZABIN, I. and BLOCH, K. (1951) *J. Biol. Chem.* **189**, 421.
9. SKEGGS, M. R., WRIGHT, L. D., CRESSON, E. L., MACRAE, G. D., HOFFMAN, C. H., WOLF, D. E. and FOLKERS, K. (1956) *J. Bact.* **72**, 519.
10. TAVORMINA, P. A., GIBBS, M. H. and HUFF, J. W. (1956) *J. Am. Chem. Soc.* **78**, 4498.
11. LYNEN, F. (1964) *Les Prix Nobel*, p. 205.
11a. POPJAK, G. and CORNFORTH, J. W. (1960) *Adv. Enzymol.* **22**, 281.
12. SUZUE, G., TSUKADA, K., NAKAI, C. and TANAKA, S. (1968) *Arch. Biochem. Biophys.* **123**, 644.
13. TORNABENE, T. G., KATES, M., GELPI, E. and ORO, J. (1969) *J. Lipid Res.* **10**, 294.
14. LEVIN, E. Y. and BLOCH, K. (1964) *Nature*, **202**, 90.
15. REITZ, R. C. and HAMILTON, J. G. (1968) *Comp. Biochem. Physiol.* **25**, 401.
16. DESOUZA, N. J. and NES, W. R. (1968) *Science*, **162**, 363.
17. FINEAN, B. (1953) *Experientia*, **9**, 17.
18. RAND, R. P. and LUZZATI (1968) *Biophys. J.* **8**, 125.
19. ROTHMAN, J. E. and ENGLEMAN, D. M. (1972) *Nature*, **237**, 42.
20. BIRD, C. W., LYNCH, J. M., PIRT, F. J., REID, W. W., BROOKS, C. J. W. and MIDDLEDITCH, B. S. (1971) *Nature*, **230**, 473.
21. DAWSON, M. and BLOCH, unpublished.
22. PATT, T. E., COLE, G. C., BLAND, J. and HANSON, R. S. (1974) *J. Bact.* **120**, 955.
23. GOODWIN, T. W. (1973) in *Lipids and Biomembranes of Eukaryotic Microorganisms*, (J. A. ERWIN, ed.), p. 1, Academic Press, Inc.
24. MARGULIS, L. (1970) *Origin of Eukaryotic Cells*, Yale University Press, New Haven, Conn.

MY EXPEDITIONS INTO SULFUR BIOCHEMISTRY

FEODOR LYNEN
Max-Planck-Institut für Biochemie, D-8033 Martinsried, Germany

It is the purpose of my essay to please my good friend Severo at the occasion of his 70th birthday by reviewing a period in my scientific career which was very intimately interwoven with Severo's scientific activities and which even led to two important publications which bear both our names.[1,2] In addition I would like to demonstrate with my essay how it is possible to attack important problems in biochemistry by applying the thinking of an organic chemist in combination with simple experiments in which specific inhibitors and model substances are used.

My expeditions into sulfur biochemistry started by an observation made during the study of the oxidation of acetate by living yeast, which was initiated in 1932 by the work of Heinrich Wieland and Robert Sonderhoff in Munich. With the use of normal and isotopic acetate, the Munich laboratory had, quite early, accumulated rather strong evidence that acetate oxidation occurs through the citric acid cycle.[3] It was already recognized then that acetate first has to be "activated" in order to enter this cycle.[4] The reason for postulating the existence of "active acetate" was the observation that yeast cells "impoverished" in endogenous fuels by shaking under oxygen have some difficulty in using acetate as a substrate for oxidation. Only after a more or less prolonged "induction period", which was dependent on the duration of the starvation process, the oxidation of acetate begins. When it was found that this lag phase could be shortened and even eliminated by addition of small amounts of a readily oxidizable substrate such as ethanol, propanol or butanol, we came to the conclusion that the oxidation of acetate required an initial input of energy, this energy being furnished by the simultaneous oxidation of another substrate or by the oxidation of acetate itself, once this process has started.

As it turned out, Fritz Lipmann had come to a similar conclusion.[5] He had studied the oxidation of pyruvate by lactic acid bacteria and found that phosphate was required and acetyl phosphate was formed. This oxidation seemed to be analogous to the dehydrogenation of phosphoglyceraldehyde, as unravelled in Otto Warburg's laboratory, and the acetyl phosphate that was formed seemed to be the "active acetate" we were seeking. In order to prove this hypothesis, I synthesized the crystalline silver salt of acetyl phosphate by a new, rather complicated method.[6] But the experiment to achieve the synthesis of citrate, using acetyl phosphate and oxaloacetate with the aid of yeast cells, whose membranes had been made permeable to polybasic acids by freezing in liquid air, was without success. At this stage, which was reached in 1941, my work remained stagnant for many years due to the war conditions.

When scientific contact with the outside world was restored after a break of several years, I learned of the advances that had been made in the meantime in the investigation of the problem of "active acetate". The interest of biochemists in that metabolic intermediate increased enormously with the realization that the C_2-fragment was involved not only in citric acid synthesis but also in the acetylation of sulfonamide or choline and in the synthesis of acetoacetate, fatty acids and steroids. This insight was due in large part to the work of Rudolf Schoenheimer and his group. Coenzyme A was discovered and Fritz Lipmann, whose brilliant experiments contributed most to the clarification of this field, suggested that coenzyme A may act as an acetyl carrier.[7] This proposition was solidified through the work of Earl Stadtman[8] and of Severo Ochoa and his group.[9]

Experiments that I carried out with Ernestine Reichert and Luistraud Rueff on yeast cells agreed well with such a view. When in 1950 work on the activation problem was resumed in my laboratory in Munich, we found that during the "induction time" that precedes the oxidation of acetic acid by "impoverished" yeast, the respiration intensity and the coenzyme A content are proportional to each other. When the oxidation of acetic acid was in full progress, the coenzyme A content of the yeast had reached a high plateau. We therefore assumed that yeast cells in this phase must also be rich in acetylated coenzyme A, and would be a suitable starting material for its isolation.

We might not have attempted to isolate acetyl-CoA but for the fact that I was already possessed by the idea that it was a thiolester. I remember in every detail how this notion came about. My brother-in-law, Theodor Wieland, was on holiday in his parents' house next to ours. He had worked in Richard Kuhn's laboratory on pantothenic acid, the vitamin discovered by Lipmann to be a constituent of coenzyme A. We spent a whole night in shoptalk about the possible link between acetate and pantothenic acid but could not come to any conclusion. On my short way back to our house, it suddenly came to my mind that the acetyl residue might be bound, not to pantothenic acid at all, but to sulfur. I recalled that in Lipmann's last paper on the composition of the purified coenzyme A preparations he had mentioned the presence of sulfur, but he did not pay much attention to it because his preparations were not yet pure. In addition, it was known that all enzymatic reactions studied in which coenzyme A was involved required the addition of glutathione or cysteine, presumably as agent for the binding of inhibitory heavy metals. Nobody thought of the possibility that thiols were required to keep a sulfhydryl group in the reduced state. Third, as a chemist I knew that sulfhydryl groups are more acidic than hydroxyl groups, which means that acetic acid bound to sulfur must have the properties of an acid anhydride and this should have a capacity to acetylate amines or alcohols. The crucial step for me was that I combined the three facts. I became very excited, hurried into my study, and looked into Beilstein where I found that thioacetic acid was known to react with aniline to form acetanilide. I thus became completely convinced that "active acetate" must be a thiolester. Ernestine Reichert was very surprised when I walked into the laboratory next morning and told her that we would now embark on the isolation of acetyl CoA from yeast Kochsaft. In 2 months time we had reached this goal.[10,11]

Our assay for the detection of "active acetate" was based on the enzymic reaction with sulfanilamide. In the presence of the enzyme obtained from pigeon liver, the acetylation of sulfanilamide increased with increasing addition of "active acetate". In our purification work we used, however, another, much less time-consuming assay which was based on the delayed color development as given by thiolesters with sodium nitroprusside.[12] When ammonium hydroxide is added to the mixture of thiolesters and nitroprusside no color is seen in the first moment. It only gradually develops in the following minutes due to the hydrolysis of the thiolester in the alkaline medium. The color reaches a maximum in 2 to 3 minutes and then rapidly fades. Calculations of the amount of thiolester, taken from the point of maximum color development, were in excellent agreement with values obtained by sulfanilamide acetylation. To document this a comparison of different preparations of "active acetate", assayed by the two methods is presented in Fig. 1.

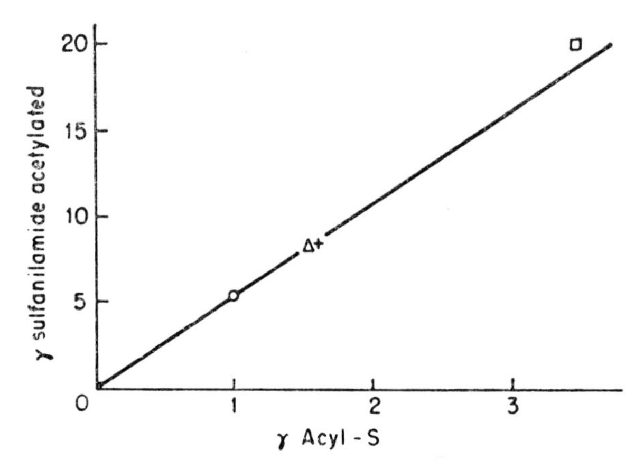

FIG. 1. Identity of acetyl donor with thiolester (= acyl-S).[11], Preparations of varying purity were assayed.
○ = decomposed barium salt; + = charcoal eluate; □ , △ = acetone precipitates.

It should be stressed that in order to use the delayed nitroprusside reaction for the detection of acetyl CoA it was necessary first to remove the large amounts of the free mercaptans, mostly glutathione, from the yeast Kochsaft which otherwise would have strongly interfered with the color reaction. We achieved this by titration of the Kochsaft with iodine, thereby oxidizing thiols to disulfides which do not any longer react with nitroprusside at alkaline pH. The identity of the sulfhydryl component of our active preparations with coenzyme A was finally proved in various ways which should not be discussed here in detail.

Following the isolation of "active acetate" and its identification with acetyl-CoA, we next showed that this was the long-sought acetyl donor, which provides the acetic acid residue not only for the acetylation of sulfanilamide but also for the formation of acetylcholine, citric acid, and acetoacetic acid.[13] In some of these experiments I cooperated with Severo and his group.[1] This way the function of coenzyme A as an acetyl transferring agent was chemically elucidated. The bonding of the acetyl residue to the thiol group of the coenzyme provides the acetyl group with the dual chemical reactivity required for its diverse metabolic reactions, i.e. reactivity in the carbonyl group with nucleophilic substituents and reactivity in the neighboring methyl group with electrophilic substituents.

Looking back at this period of my work on acetyl-CoA I realize that most of the biochemists working on "active acetate" were prejudiced by the idea that "active acetate" could only be some kind of a reactive but also labile phosphate derivative. The idea of the "energy-rich" phosphate bond was dominating the biochemical thinking. With the discovery of the thiolester bond in acetyl CoA the "energy-rich" phosphate bond had lost its unique role in metabolic reactions, and it was to be expected that still other "energy-rich" bonds would be discovered. Indeed, this has happened in the meantime.

Once the structure of acetyl CoA was known, a detailed chemical scheme of fatty-acid oxidation could be formulated, based on several experimental observations published in the literature. I presented it in my first publication on acetyl CoA[10,11] and called it later the "fatty acid cycle".[13] The substrate is regenerated in this repeated reaction sequence, which thus resembles the citric acid cycle or the cyclic process in the synthesis of urea. In these other cyclic processes, however, identical substrates are formed after each cycle, whereas a shorter homologous chain is formed in each repetition of the "fatty acid cycle". It would therefore be more appropriate to describe the oxidation of fatty acids by a spiral process rather than by a cycle.

The next step then was to isolate and identify the four enzymatic components participating in this process. Here again, my training as a chemist was a big help. The intermediates in the oxidative reaction are α,β-unsaturated-, β-hydroxy-, and β-keto-carboxylic acids bound to coenzyme A. In order to learn something about their chemical properties, Luise Wessely and Werner Seubert synthesized simple models in which these carboxylic acids were bound, not to coenzyme A, but to *N*-acetyl cysteamine (Fig. 2).

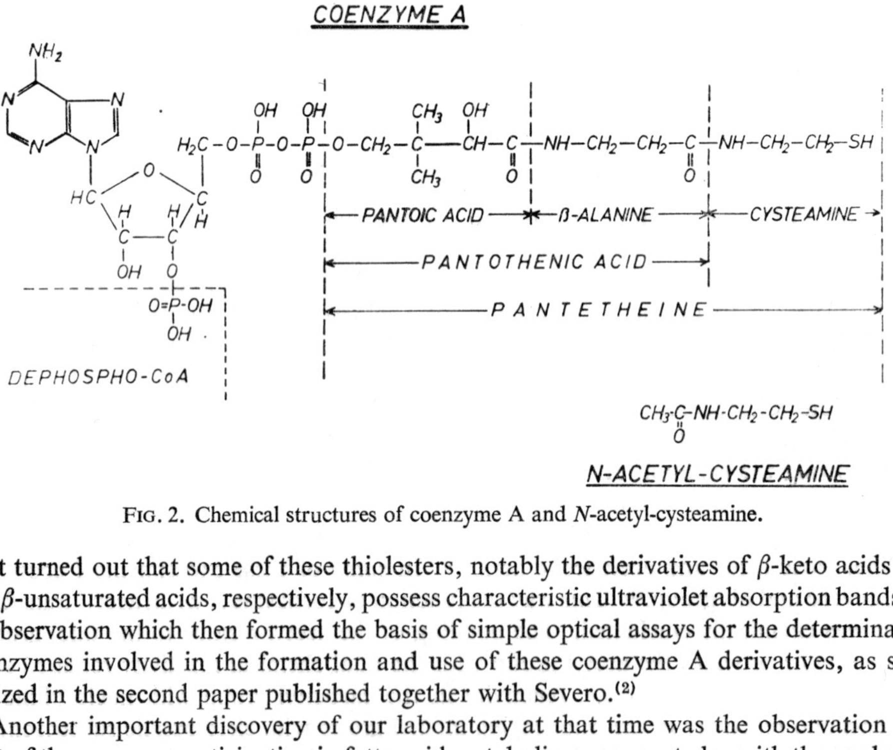

FIG. 2. Chemical structures of coenzyme A and *N*-acetyl-cysteamine.

It turned out that some of these thiolesters, notably the derivatives of β-keto acids and of α,β-unsaturated acids, respectively, possess characteristic ultraviolet absorption bands,[14] an observation which then formed the basis of simple optical assays for the determination of enzymes involved in the formation and use of these coenzyme A derivatives, as summarized in the second paper published together with Severo.[2]

Another important discovery of our laboratory at that time was the observation that most of the enzymes participating in fatty acid metabolism can react also with the analogues of the natural substrates, where coenzyme A is replaced by *N*-acetyl cysteamine. On this basis we developed optical assays with substrate models, which became very valuable in our work on the isolation and purification of the enzymes of fatty acid metabolism.

The enzyme which attracted most of our interest became the one which catalyzes the reversible thiolytic cleavage of β-ketoacyl derivatives of coenzyme A to acetyl-CoA and an acyl-CoA containing two fewer carbon atoms than the original β-keto acid. We found that the purified enzyme, which we named thiolase, is strongly inhibited by sulfhydryl blocking agents and therefore suggested that the active site of the enzyme is represented by a sulfhydryl group which participates in the chemical process in the following manner:[14]

$$\begin{array}{l} R\text{—}CH_2\text{—}CO\text{—}CH_2\text{—}CO\text{—}SCoA + HS\text{—}enzyme \\ \rightleftharpoons R\text{—}CH_2\text{—}CO\text{—}S\text{—}enzyme + CH_3\text{—}CO\text{—}SCoA \end{array} \quad (1)$$

$$R\text{—}CH_2\text{—}CO\text{—}S\text{—}enzyme + HSCoA \rightleftharpoons R\text{—}CH_2\text{—}CO\text{—}SCoA + HS\text{—}enzyme \quad (2)$$

The thiolytic cleavage is achieved by the HS-enzyme itself and leads to the formation of an acyl-enzyme intermediate which then can transfer its acyl group to coenzyme A. In collaboration with Ulrich Gehring and Christl Riepertinger[15] the thiolytic activity [equation (1)] and the acyl transferase activity [equation (2)] of the enzyme could be demonstrated separately. In one of the crucial experiments, as shown in Fig. 3, it could be

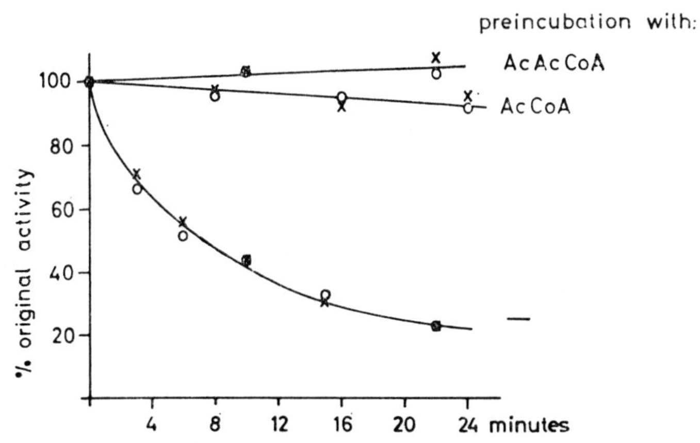

FIG. 3. Protection of thiolase against iodoacetamide inhibition by preincubation with acetoacetyl-CoA or acetyl-CoA.[15], Thiolase was incubated with 9×10^{-4} M acetoacetyl-CoA (AcAcCoA) or acetyl-CoA (AcCoA) for 10 minutes at 0°C and pH 8.2. 5×10^{-5} M iodoacetamide was then added and samples taken from the reaction mixture at the times indicated in the figure. The reaction of iodoacetamide with the enzyme was stopped by the addition of an excess of cysteine and the remaining thiolase (×——×) and acetyl transferase (○——○) activities were determined. In the control experiment (——) the enzyme was used without preincubation with the CoA derivatives.

demonstrated that the inhibition of thiolase and acyl transferase activities by iodoacetamide could be prevented by preincubation of the enzyme with acetoacetyl-CoA or acetyl-CoA. This could be explained by assuming that the functional sulfhydryl group in the active centre of the enzyme becomes acetylated either by thiolysis of acetoacetyl-CoA or by acyl transfer from acetyl-CoA and thus cannot any longer interact with the inhibitor iodoacetamide. This assumption could finally be proven later by the work of Ulrich Gehring and Ieuan Harris,[16] in which the amino acid sequences around the active cysteine residues labeled by preincubation with either ^{14}C-iodoacetamide or ^{14}C-acetyl-CoA were determined and found to be identical. The ^{14}C-acetyl-peptide isolated after trypsin digestion of the ^{14}C-acetyl-enzyme has the following structure:

$$\begin{array}{l} \quad\;\; S\text{—}Co\text{—}CH_3 \\ \quad\;\; | \\ Val\text{-}Cys\text{-}Ala\text{-}Ser\text{-}Gly\text{-}Mel\text{-}Lys \end{array}$$

The experience of our studies on thiolase, realizing that acyl groups are also bound to sulfhydryl groups of proteins, became of special interest in our work on fatty acid biosynthesis. We first assumed that fatty acid synthesis from acetyl-CoA is achieved through the reversal of the chemical reactions of fatty acid degradation. However, the pioneering studies of Salih Wakil led to the discovery of another enzyme system. He and his collaborators[17] isolated two protein fractions from pigeon liver extracts, which when combined can synthesize long-chain fatty acids, starting from acetyl-CoA, ATP and TPNH. The peculiarity of this system was the absolute requirement of bicarbonate as a cofactor of fatty acid synthesis. This lead to my suggestion, in 1958, at the Gordon Conference on Lipids, that CO_2 was required for the carboxylation of acetyl CoA to form malonyl CoA.[18] This hypothesis proved to be correct. Salih Wakil[19] could demonstrate that one of his purified fractions from pigeon liver indeed formed malonyl CoA which could be transformed by the second fraction in the presence of TPNH into long-chain fatty acids. The carboxylation reaction can be described as follows:

$$CH_3\text{—}CO\text{—}SCoA + HCO_3^- + ATP^{4-} = \underset{\displaystyle COOH}{\underset{|}{CH_2}}\text{—}CO\text{—}SCoA + ADP^{3-} + HPO_4^= + H^+ \qquad (3)$$

The use of malonyl-CoA for carbon chain synthesis appeared to be especially attractive from the thermodynamic point of view.[18] The condensation of acyl-CoA with a thiolester of malonic acid leads to the same β-ketoacyl derivative as with the thiolester of acetic acid, but in this case the simultaneous liberation of CO_2 [equation (4)] shifts the equilibrium toward synthesis.

$$R\text{—}CH_2\text{—}CO\text{—}SCoA + \overset{\displaystyle COO^-}{\overset{|}{CH_2}}\text{—}CO\text{—}SX + H^+$$
$$\rightleftharpoons R\text{—}CH_2\text{—}CO\text{—}CH_2\text{—}CO\text{—}SX + HSCoA + CO_2 \qquad (4)$$

The enzyme system which converts malonyl-CoA to higher fatty acids, referred to as "fatty acid synthetase", behaved as a single entity during the fractionation of tissue extracts and was isolated in a number of laboratories from various animal tissues. We ourselves obtained highly active enzyme fractions from yeast.[20,21]

Investigations of the catalytic nature of the purified enzyme preparation from yeast supported the following equation for fatty acid synthesis from malonyl-CoA:

$$CH_3\text{—}CO\text{—}SCoA + n\overset{\displaystyle COOH}{\overset{|}{CH_2}}\text{—}CO\text{—}SCoA + 2n\,TPNH + 2n\,H^+$$
$$\rightarrow CH_3\text{—}(CH_2\text{—}CH_2)_n\text{—}CO\text{—}SCoA + nCO_2 + nHSCoA + 2n\,TPN^+ + H_2O \qquad (5)$$

(for palmityl CoA: $n = 7$)

It was found by all workers in this field that acetyl-CoA acted as a "primer" of the synthetic process and became incorporated into the "tail" end of the fatty acid, which thus is formed by the successive addition of C_2-units derived from malonyl-CoA to the "primer" acetyl-CoA. In this function acetyl-CoA could be replaced by a great variety of saturated straight or branched chain acyl-CoA. In contrast, the oxidized intermediate compounds which occur in the course of fatty acid degradation, such as α,β-unsaturated-,

β-hydroxy- and β-keto-acyl-CoA derivatives of coenzyme A, were shown to be ineffective as "primers". Consequently, these compounds could not be intermediates in the course of fatty acid synthesis from malonyl-CoA.

To explain this observation, Wakil and Ganguly[22] suggested a mechanism of fatty acid synthesis in which dicarboxylic acid intermediates were postulated [equation (6)], which, after reduction to alkylmalonic acid derivatives, are decarboxylated [equation (7)]. The saturated acyl-CoA thus formed condenses, in turn, with the next molecule of malonyl CoA.

$$CH_3\text{—CO—SCoA} + \underset{}{\overset{\displaystyle COOH}{\overset{|}{C}}}H_2\text{—CO—SCoA} \rightarrow CH_3\text{—CO—}\overset{\displaystyle COOH}{\overset{|}{C}}H\text{—CO—SCoA} + \text{HSCoA} \qquad (6)$$

$$CH_3\text{—CO—}\overset{\displaystyle COOH}{\overset{|}{C}}H\text{—CO—SCoA} \rightarrow\rightarrow CH_3\text{—}CH_3\text{—}\overset{\displaystyle COOH}{\overset{|}{C}}H\text{—CO—SCoA}$$
$$\rightarrow CH_3\text{—}CH_2\text{—}CH_2\text{—CO—SCoA} + CO_2 \qquad (7)$$

In this scheme the saturated acyl-CoA were intermediates, but not the substituted derivatives of the β-oxidation sequence, which was in complete agreement with the experimental observations. On the other hand, from the standpoint of energetics, such a scheme could not take advantage of the decarboxylation of malonyl-CoA for a synthetic reaction. We felt this to be a serious objection to the scheme.

After considerable effort in collaboration with Ingrid Kessel and Hermann Eggerer,[20,21] we were able to gain insight into the secrets of the process. The first inkling of the direction which we were to follow came from the observation that fatty acid synthesis can be inhibited by sulfhydryl blocking agents, such as iodoacetamide or arsenite. At 0.01 molar concentration, the enzyme was inhibited completely by both agents, and even with 10^{-4} molar concentration, some inhibition was observed.[21] In view of this sensitivity, we routinely added thiols, such as cysteine or glutathione, to the incubation mixtures, which resulted in a stimulation of fatty acid synthesis. In order to explain the stimulation by thiols and the inhibition by sulfhydryl binding agents, we assumed that the enzyme itself contains SH-groups, which are indispensible for the catalytic effect. This fact, taken together with our repeated inability to demonstrate any free intermediates in the synthetic process, lead us to the working hypothesis that the actual mechanism involves a transfer of the acyl portion of the coenzyme A esters to sulfhydryl groups on the enzyme itself. The process was visualized as being initiated by the transfer of the malonyl residue from malonyl-CoA to the enzyme to form a malonyl-S-enzyme [equation (8)]

$$\overset{\displaystyle COOH}{\overset{|}{C}}H_2\text{—CO—SCoA} + \text{HS-enzyme} \rightarrow \overset{\displaystyle COOH}{\overset{|}{C}}H_2\text{—CO—S-enzyme} + \text{HSCoA} \qquad (8)$$

$$CH_3\text{—}\overset{*}{C}O\text{—SCoA} + \overset{\displaystyle COOH}{\overset{|}{C}}H_2\text{—CO—S-enzyme}$$
$$\rightleftharpoons CH_3\text{—}\overset{*}{C}O\text{—}CH_2\text{—CO—S-enzyme} + \text{HSCoA} + CO_2 \qquad (9)$$

The next step was the most distinguishing point in our scheme: the condensation of the saturated acyl-CoA with the malonyl-enzyme to form the β-ketoacyl enzyme [equation (9)]. The accompanying decarboxylation merely adheres to the principle discussed earlier.

After condensation, there follows a stepwise conversion involving first a reduction to the β-hydroxyacyl enzyme, then a dehydration to the dehydroacyl enzyme, and finally, a reduction to the saturated acyl enzyme, which can then enter into a new condensation with a malonyl group.

The first experimental support for this scheme was the identification of the intermediate formation of a β-keto-acyl-enzyme, namely acetoacetyl-enzyme.[20,21] When 1-^{14}C-labeled acetyl-CoA and malonyl-CoA were incubated with stoichiometric amounts of the purified yeast enzyme and the protein was precipitated by the addition of trichloroacetic acid, it contained bound radioactive acetoacetate, which could be released by mild alkaline hydrolysis and which yielded radioactive acetone upon decarboxylation. This result proved unequivocally that the condensation of acetyl-CoA and malonyl-CoA had occurred in the manner depicted in equations (8) and (9).

In order to demonstrate the manifold catalytic activities attributed to the enzyme, which was visualized as a multienzyme complex, we were able to utilize the earlier experience gained, by replacing the complicated coenzyme A residue by *N*-acetyl cysteamine or pantetheine. For the problem under discussion, this technique proved to be tremendously useful, so that we were able to demonstrate every single reaction step in our hypothetical sequence independent of preceding or subsequent reactions.[20] The affinity of these model compounds is generally rather low compared with the natural substrates. But this could be circumvented by employing high concentrations of the model substrates.

Further systematic studies undertaken on the separate enzyme functions indicated that two types of sulfhydryl groups participate in the overall process. If only the sulfhydryl-group fixing the malonyl residue would be involved in fatty acid synthesis, it would be predicted on the basis of our experience gained with thiolase that preincubation of fatty acid synthetase with malonyl-CoA would protect the catalytic functions from the attack of *N*-ethyl maleimide. When Kasper Kirschner performed the experiment, he found that

Preincubated 5 min. at 0°C with 0.28×10^{-3} M	Relative enzyme activity
—	5.8
malonyl CoA	4.8
acetyl CoA	84.6
butyryl CoA	74.1
decanoyl CoA	66.9
myristyl CoA	48.4
palmityl CoA	33.7
control without N-ethylmaleimide	100

FIG. 4. Protection of fatty acid synthetase from inactivation with *N*-ethylmaleimide by preincubation with acyl-CoA derivatives. For inactivation the enzyme was incubated with 5×10^{-3} M *N*-ethylmaleimide for 5 minutes at 0°C.

malonyl-CoA had no protective effect at all.[23] In contrast to this, saturated acyl-CoA derivatives, especially acetyl-CoA, had a marked protective effect against the sulfhydryl-reagent (Fig. 4). We were therefore forced to conclude that two different kinds of functional sulfhydryl groups participate in fatty acid synthesis. To one of them the malonyl residue and the intermediates of fatty acid synthesis are attached, whereas the other can only accept the saturated fatty acids. We called the two types of sulfhydryl groups "central" and "peripheral". By labeling with ^{14}C-iodoacetamide which reacts very preferentially with the latter Alexander Hagen[24] could identify it with protein-bound cysteine, located in the condensing enzyme component of the multienzyme complex.

The chemical nature of the "central" sulfhydryl group was then elucidated by the brilliant studies of Roy Vagelos on the "acyl carrier protein" isolated from *E. coli*.[25] In these studies it was discovered that the "central" group is represented by 4′-phosphopantetheine attached in phosphodiester linkage to a serine residue of the protein. In addition, the studies on the fatty acid synthesizing enzyme system from *E. coli* (cf. ref. 25), where the component enzymes fail to show any signs of physical interaction at the cell-free level, fully confirmed and at the same time refined our scheme of fatty acid biosynthesis as presented in 1961.[20]

In the meantime it was discovered that "acyl carrier proteins" with 4′-phosphopantetheine as functional group represent also essential constituents of multienzyme systems involved in the biosynthesis of other "polyacetate" compounds, like 6-methyl-salicylic acid,[26] in nonribosomal cyclic peptide biosynthesis[27] and in the enzymic cleavage of citrate by citrate lyase.[28] These observations opened a broad new field of research in sulfur biochemistry.

REFERENCES

1. Stern, J. R., Ochoa, S. and Lynen, F. (1952) Enzymatic synthesis of citric acid. V. Reaction of acetyl coenzyme A. *J. Biol. Chem.* **198,** 313–321.
2. Lynen, F. and Ochoa, S. (1953) Enzymes of fatty acid metabolism. *Biochim. Biophys. Acta,* **12,** 299–314.
3. Martius, C. and Lynen, F. (1950) Probleme des Citronensäurecyklus. *Adv. Enzymology,* **10,** 167–222
4. Lynen, F. (1942) Zum biologischen Abbau der Essigsäure. I. Über die "Induktionszeit" bei verarmter Hefe. *Liebigs Ann. Chem.* **552,** 270–306.
5. Lipmann, F. (1942) Acetyl phosphate. *Adv. Enzymology,* **6,** 231–267.
6. Lynen, F. (1940) Über die gemischten Anhydride aus Phosphorsäure und Essigsäure. *Ber. Deutsche Chem. Ges.* **73,** 367–375.
7. Lipmann, F. (1948/9) Biosynthetic mechanisms. *Harvey Lectures,* Ser. **44,** 99–123.
8. Stadtman, E. R. (1950) Coenzyme A-dependent transacetylation and transphosphorylation. *Fed. Proc.* **9,** 233.
9. Ochoa, S. (1951) Biological mechanisms of carboxylation and decarboxylation. *Physiol. Rev.* **31,** 56–106.
10. Lynen, F. and Reichert, E. (1951) Zur chemischen Struktur der "aktivierten Essigsäure". *Angew. Chem.* **63,** 47–48.
11. Lynen, F., Reichert, E. and Rueff, L. (1951) Zum biologischen Abbau der Essigsäure IV. "Aktivierte Essigsäure", ihre Isolierung aus Hefe und ihre chemische Natur. *Liebigs Ann. Chem.* **574,** 1–32.
12. Lynen, F. (1951) Quantitative Bestimmung von Acyl-mercaptanen mittels der Nitroprussid-Reaktion. *Liebigs Ann. Chem.* **574,** 33–37.
13. Lynen, F. (1952/3) Acetyl coenzyme A and the "fatty acid cycle". *Harvey Lectures,* Ser. **48,** 210–244.
14. Lynen, F. (1953) Functional group of coenzyme A and its metabolic relations, especially in the fatty acid cycle. *Fed. Proc.* **12,** 683–691.
15. Gehring, U., Riepertinger, C. and Lynen, F. (1968) Reinigung und Kristallisation der Thiolase, Untersuchungen zum Wirkungsmechanismus. *Eur. J. Biochem.* **6,** 264–280.
16. Gehring, U. and Harris, J. I. (1970) The active site cysteines of thiolase. *Eur. J. Biochem.* **16,** 492–498.
17. Gibson, D. M., Titchener, E. B. and Wakil, S. J. (1958) Studies on the mechanism of fatty acid synthesis. V. Bicarbonate requirement for the synthesis of long chain fatty acids. *Biochim. Biophys. Acta,* **30,** 376–383.

18. Cf. Lynen, F. (1959) Participation of acyl-CoA in carbon chain biosynthesis. *J. Cell Comp. Physiol.* **54,** Suppl. 1, 33–49.
19. Wakil, S. J. (1958) A malonic acid derivative as an intermediate in fatty acid synthesis. *Am. Chem. Soc.* **80,** 6465.
20. Lynen, F. (1961) Biosynthesis of saturated fatty acids. *Federation Proc.* **20,** 941–951.
21. Lynen, F., Hopper-Kessel, I. and Eggerer, H. (1964) Zur Biosynthese der Fettsäuren. III. Die Fettsäuresynthetase der Hefe und die Bildung enzymgebundener Acetessigsäure. *Biochem. Z.* **340,** 95–124.
22. Wakil, S. J. and Ganguly, J. (1959) On the mechanism of fatty acid synthesis. *J. Am. Chem. Soc.* **81,** 2597–2598.
23. Cf. Lynen, F. (1962) Biochemical mechanisms in lipid synthesis. The Robert A. Welch Foundation Conferences on Chemical Research. V. *Molecular Structure and Biochemical Reactions*, pp. 293–329, Houston, Texas.
24. Hagen, A. (1963) Untersuchungen zur Wirkungsweise und zur Struktur der Fettsäuresynthetase. Dr. Thesis University of Munich.
25. Prescott, D. J. and Vagelos, P. R. (1972) Acyl-carrier protein. *Adv. Enzymology*, **36,** 269–311.
26. Lynen, F. (1972) Structure and function of multienzyme complexes. *Enzymes: Structure and Function* **29,** 177–200. Proc. 8th FEBS-Meeting, Amsterdam (North-Holland Publ. Co., Amsterdam–London).
27. Lipmann, F. (1973) Nonribosomal Polypeptide Synthesis on Polyenzyme Templates. *Accounts Chem.* **6,** *Res.* 361.
28. Dimroth, P., Dittmar, W., Walther, G. and Eggerer, H. (1973) The acyl-carrier protein of citrate lyase. *Eur. J. Biochem.* **37,** 305–315.

THE *CLOSTRIDIUM KLUYVERI*–ACETYL CoA EPOCH

E. R. STADTMAN

Laboratory of Biochemistry, National Heart and Lung Institute,
National Institutes of Health, Bethesda, Maryland 20014

Very often in the course of scientific experimentation a totally unexpected observation is made that is either unrelated or only incidentally related to the problem under immediate investigation. When it captures the attention of an alert, inquisitive mind, this observation may open the door for an entirely new study that is often more fruitful than that of the original design. Such was the beginning, almost 40 years ago, of a timely and highly productive investigation in H. A. Barker's laboratory which was fundamental to the development of our knowledge concerning the biochemical functions of acetyl-CoA and especially to an understanding of the mechanism of fatty acid oxidation and synthesis.

This investigation was the outgrowth of studies initiated by Barker in 1935 while he was a visiting fellow working with Professor A. J. Kluyver in the Laboratorium voor Mikrobiologie in Delft. To obtain support for C. B. Van Niel's theory that in methane fermentations of organic compounds the methane is produced by reduction of CO_2, Barker(1) set up anaerobic enrichment cultures containing calcium carbonate, ethanol and a generous innoculum of Delft canal mud.

Chemical analysis showed that the vigorous gas production, which occurred after a few days, was accompanied, as expected, by the oxidation of ethanol to acetate and the reduction of CO_2 to methane according to the equation:

$$2CH_3CH_2OH + CaCO_3 \rightarrow (CH_3COO)_2Ca + CH_4 + H_2O$$

However, in some enrichment cultures the yield of acetic acid was much less than the amount of ethanol which had disappeared, and in these cultures considerable amounts of butyric and caproic acids accumulated. In a subsequent report dedicated to A. J. Kluyver,(2) Barker, in referring to the accumulation of caproic acid in these cultures, wrote: "Since I was not aware of any previous observations of microbial formation of a fatty acid of this sort in high yield, particularly from so simple a substrate as ethyl alcohol, I immediately realized that I had made a discovery of considerable interest". He acknowledged, however, that a "later search of the literature revealed that Bechamp, a student of Pasteur, had in 1868 observed the conversion of ethyl alcohol to butyric and caproic acids". In any case, microscopic examination disclosed that in those cultures which produced caproic acid, two kinds of bacteria were always present. Barker isolated both. One of these catalyzed the formation of methane as noted above and was given the name *Methanobacterium omelianskii.*(3) The other catalyzed the conversion of ethanol and acetate to butyrate and caproate and was given the name *Clostridium kluyveri.*(4)

An opportunity to investigate further the mechanism of butyrate synthesis in *C. kluyveri* presented itself in 1944, when for the first time small quantities of the long-lived radioactive isotope of carbon, [^{14}C] became available. In collaboration with M. D. Kamen and B. Bornstein, Barker demonstrated that the fermentation of [carboxyl-^{14}C] acetate and unlabeled ethanol led to the production of [^{14}C] butyrate that was almost equally labeled in the carboxyl and beta carbon atoms; moreover, during the fermentation the specific radioactivity of the added acetate was decreased by an amount equivalent to the amount of ethanol utilized.[5] This established that the formation of butyrate from ethanol and acetate is a coupled oxidation-reduction process in which ethanol is oxidized to acetate (or to a compound that is in equilibrium with acetate) which can condense with another equivalent of acetate to form a four carbon derivative (viz. acetoacetate) that is reduced to butyrate. In view of Lipmann's calculations[6] showing that the condensation of two molecules of acetate to form acetoacetate is highly endergonic ($\Delta G^\circ \approx +16{,}000$ cal) it appeared likely than an "energy-rich" form of acetate is involved in the condensation reaction to form the four carbon intermediate: accordingly, Barker proposed that acetyl-P rather than acetate might be the product of ethanol oxidation. Acetyl-P was a logical energy-rich intermediate since Lipmann had shown that acetyl-P produced in the oxidation of pyruvate by *Lactobacillus delbruckii*.

It was at this stage of development that, as a graduate student, Barker offered me the opportunity to continue this investigation for my thesis research. Initially, I had planned to utilize tracer techniques to investigate the capacity of various compounds to be metabolized by resting cell suspensions of *C. kluyveri*. This approach was abandoned, however, as a consequence of a brief trip to New York State to get married. During this trip Terry and I stopped for a visit at Cornell University, where she introduced me to one of her former professors, Dr. I. C. Gunsalus. In the course of this visit Gunsalus demonstrated to us his technique for making dried cell preparations by desiccation of cell pastes in an evacuated desiccator over anhydrous $CaCl_2$. Upon my return to Berkeley I utilized this technique to make dried cell preparations of *C. kluyveri*. I was elated to discover that such preparations had the capacity to catalyze rapid synthesis of butyrate and caproate from ethanol and acetate. Even more surprising was the fact that brief incubation of dried cell suspensions in phosphate buffer led to solubilization of all the enzymes needed for the synthesis and oxidation of these simple fatty acids.[7]

The discovery that soluble enzyme preparations could catalyze synthesis and oxidation of fatty acids was of especial significance because it dispelled once and for all the then commonly accepted belief, that the capacity to synthesize and oxidize fatty acids is the unique property of particulate organelles, infamously referred to by D. E. Green as the "cyclophorase" system and by others simply as mitochondria or particulate suspensions.

In the months that followed it was demonstrated that these soluble extracts catalyzed a number of interesting reactions which are as follows:*

A. *Reactions using* O_2 *as an electron acceptor* — Refs.

$$CH_3CH_2OH + \tfrac{1}{2}O_2 \rightarrow CH_3CHO + H_2O \quad (1)$$

$$CH_3CHO + HPO_4^{=}\ \tfrac{1}{2}O_2 \rightarrow CH_3COOPO_3^{=} + H_2O \quad (2)$$

$$CH_3CH_2CH_2COO^- + HPO_4^{=} + O_2 \rightarrow CH_3COOPO_3^{=} + CH_3COO^- + H_2O \quad (3)$$

$$2CH_3CH_2CH_2COO^- + 2O_2 \rightarrow CH_3COCH_2COO^- + 2CH_3COO^- + H_2O \quad (4)$$

$$CH_2{=}CHCH_2COO^- + HPO_4^{=} + \tfrac{1}{2}O_2 \rightarrow CH_3COOPO_3^{=} + CH_3COO^- \quad (5)$$

$$CH_2{=}CHCH_2COO^- + \tfrac{1}{2}O_2 \rightarrow CH_3COCH_2COO^- \quad (6)$$

* References to the original literature where these reactions are described can be found in the review articles by Barker[8] and Stadtman.[9]

B. *Reactions using* H_2 *as an electron donor*

$$CH_3COOPO_3^{=} + CH_3COO^- + 2H_2 \rightarrow CH_3CH_2CH_2COO^- + HPO_4^{=} + H_2O \quad (7)$$

$$RCOOPO_3^{=} + 2H_2 \rightarrow RCH_2OH + HPO_4^{=} \quad (8)$$

$$CH_3CHO + H_2 \rightarrow CH_3CH_2OH \quad (9)$$

$$CH_2 = CHCH_2COO^- + H_2 \rightarrow CH_3CH_2CH_2COO^- \quad (10)$$

$$CH_3COCH_2COO^- + H_2 \rightarrow CH_3CHOHCH_2COO^- \quad (11)$$

$$CH_2 = CHOO^- + H_2 \rightarrow CH_3CH_2COO^- \quad (12)$$

C. *Dismutation and anaerobic oxidation reactions*

$$2CH_3CHO + HPO_4^{=} \rightarrow CH_3CH_2OH + CH_3COOPO_3^{=} \quad (13)$$

$$2CH_2 = CHCH_2COO^- + H_2O + HPO_4^{=} \rightarrow CH_3CH_2CH_2COO^- + CH_3COOPO_3^{=} + CH_3COO^- \quad (14)$$

$$CH_2 = CHCH_2COO^- + H_2O + HPO_4^{=} \rightarrow CH_3COOPO_3^{=} + CH_3COO^- + H_2 \quad (15)$$

D. *Nonoxidative reactions*

$$CH_3COCH_2COO^- + HPO_4^{=} \rightarrow CH_3COOPO_3^{=} + CH_3COO^- \quad (16)$$

$$CH_3COOPO_3^{=} + CH_3CH_2CH_2COO^- \rightarrow CH_3COO^- + CH_3CH_2CH_2COOPO_3^{=} \quad (17)$$

$$CH_3COOPO_3^{=} + H_2O \xrightarrow{HA_sO_4^{=}} CH_3COO^- + HPO_4^{=} \quad (18)$$

$$CH_3COOPO_3^{=} + RNH_2 \xrightarrow{HCN} CH_3CONHR + HPO_4^{=} \quad (19)$$

$$CH_3COOPO_3^{=} + ADP \rightarrow CH_3COO^- + ATP \quad (20)$$

Among the more significant of these from the standpoint of fatty acid synthesis in *C. kluyveri* are reactions (1) and (2) which show that ethanol is oxidized via acetaldehyde to acetyl-P and reaction (7) which shows that acetyl-P and acetate are reduced to butyrate. These observations appeared to confirm Barker's suggestion that acetyl-P might be the "active acetate" involved in the oxidation of ethanol and in the synthesis of butyrate.[5] With the demonstration that cell-free extracts catalyze both reactions (3) and (7), we were in an excellent position to investigate the roles of postulated intermediates in butyrate metabolism. In these extracts which were devoid of permeability barriers, obligatory intermediates should be converted to acetyl-P and acetate in the presence of O_2 and should be reduced to butyrate in the presence of H_2. Of some fifteen different compounds tested,[8,9] only two compounds, acetoacetate and vinylacetic acid, were metabolized by the extracts. Although these compounds underwent many of the reactions expected [reactions (4), (5), (6), (10), (11), (16)], their roles as free intermediates in either butyrate synthesis or oxidation were excluded by a number of criteria,[9] most important of which was the fact that added pools of neither substance became appreciably labeled during the overall oxidation of [^{14}C] butyrate to [^{14}C] acetyl-P and [^{14}C] acetate, or during reduction of the latter labeled compounds to [^{14}C] butyrate. It was therefore concluded that 4-carbon carboxylic acids at the various states of oxidation do not normally occur as free intermediates, but are present only as activated derivatives or as enzyme complexes that do not readily equilibrate with the free acids.

Phosphotransacetylase

While this study was in progress, evidence was accumulating from studies in Lipmann's laboratory[10,11] for the involvement of a pantothenic acid containing coenzyme, so-called coenzyme A, in a number of ATP-dependent acetylation reactions. Although it was inferred that this coenzyme might serve as an acetyl group carrier,[11] direct support for such a role was lacking.

It was at this state of knowledge that an incidental observation with extracts of *C. kluyveri* led to a development which played a significant role in elucidating the mechanism of CoA action. It was observed that arsenate inhibited the ability of extracts to catalyze the oxidation of butyrate, as well as the reduction of acetyl-P and acetate to butyrate.[12] The inhibition was ultimately explained by the fact that arsenate stimulated the enzymic hydrolysis of acetyl-P [reaction (18)]. By analogy to the postulated role of arsenate in the hydrolysis of glucose-1-P by sucrose phosphorylase,[13] it was proposed [12] that the arsenolysis of acetyl-P could be the result of a reversible transfer of the acetyl moiety to an acetyl acceptor, "x", catalyzed by a phosphotransacetylase [reaction (21)].

$$\text{Acetyl-P} + \text{“}x\text{”} \rightleftarrows \text{acetyl-“}x\text{”} + \text{Pi} \tag{21}$$

$$\text{acetyl-“}x\text{”} + HAsO_4 \rightleftarrows \text{acetyl-}AsO_4^{=} + \text{“}x\text{”} \tag{22}$$

$$\text{acetyl-}AsO_4^{=} + H_2O \rightleftarrows \text{acetate} + HAsO_4^{=} \tag{23}$$

$$\text{Sum: Acetyl-P} + H_2O \rightleftarrows \text{acetate} + \text{Pi} \tag{24}$$

Substitution of arsenate for Pi in the reverse of reaction (21) would then lead to the formation of acetyl-arsenate [reaction (22)] which would undergo spontaneous hydrolysis. This explanation was supported by the fact that these preparations catalyzed the incorporation of [^{32}P] orthophosphate into acetyl-P as is predicted by the reversal of reaction (21).[12] Although we had first thought that "x" in reaction (21) might be the transacetylase itself, Barker and I discussed the possibility that it could in fact be CoA. An opportunity to test this hypothesis presented itself when, after completion of my thesis research in late 1949, I moved to Boston to continue my studies as a postdoctoral fellow in the laboratory of Dr. Fritz Lipmann. The year in Lipmann's laboratory was an unusually rich and satisfying experience for me, both from the personal and the scientific point of view. Barker had assured me that in Lipmann's laboratory I would find more than an opportunity to do stimulating research. Here there was an intangible benefit to be gained by close association with a man of unusual imagination and perception and one who possessed a gentle warmth of character and concern for others that endeared him to all of his associates. It was here I discovered that productive laboratories are not merely the reflection of good scientific discipline and expert direction but depend almost as much on the establishment of a congenial atmosphere in which science can flourish as a consequence of free thought, unguarded exchange of ideas, critical discussion and a respectful interaction among all of its personnel. Such was Lipmann's laboratory.

Unfortunately, I cannot discuss here the important contributions of other workers in Lipmann's laboratory which were important in establishing the acetyl carrier function of CoA. Few observations were more decisive, however, than the demonstration by G. D. Novelli and myself[14] that CoA is absolutely required for the phosphotransacetylase catalyzed arsenolysis of acetyl-P, as well as for the exchange of orthophosphate into acetyl-P. It seemed almost certain therefore that transacetylase catalyzed the reversible transfer of an acetyl group from acetyl-P to CoA [reaction (28)]; i.e. CoA was the hypothetical compound "*x*" in reaction (21).

$$\text{Acetyl-P} + \text{CoA} \rightarrow \text{Acetyl-CoA} + \text{Pi} \qquad (25)$$

The acyl carrier function of CoA was supported further by the demonstration that, when coupled with partially purified preparations of the sulfanilamide acetylase and the aceto-acetate synthetase from pigeon liver, purified preparations of transacetylase and acetyl-P could replace the ATP-dependent acetate activating enzyme as a source of active acetate for the acetylation of sulfanilamide[14,15] and for the synthesis of acetoacetate.[15] It was during my year in Boston that Professor Ochoa invited me to spend a few days in his laboratory at New York University to explore with J. Stern the possibility that transacetylase and acetyl-P could replace ATP + acetate and the acetate-activating enzyme, as a source of active acetate in the CoA-dependent synthesis of citrate by a crystalline preparation of citrate condensing enzyme from pig heart. This turned out to be a very rewarding experience for me, but one fraught with moments of extreme anxiety. In New York, my purified preparation of transacetylase failed to catalyze the CoA-dependent arsenolysis of acetyl-P. Convinced that the enzyme could not have lost activity in transit from Boston, I systematically replaced each component in the assay mixture with another sample from a different stock bottle. This led to the discovery that K^+ is required for transacetylase activity and that Na^+ is strongly inhibitory. By chance the sample of arsenate used in the preliminary experiment was the sodium salt and that used in the replacement study was the potassium salt. Having solved this mystery, we proceeded with the coupling experiments which showed that acetyl-CoA generated in the transacetylase system did serve as a source of active acetate for the synthesis of citrate by the crystalline citrate condensing enzyme.[16] The brief visit to Ochoa's laboratory was therefore a great success; it established the role of acetyl-CoA in citrate synthesis and it led to the discovery that K^+ is required for the activity of transacetylase.

Before leaving the subject of Lipmann's laboratory I should mention the results of two other studies that were key to an understanding of the mechanism of fatty acid metabolism. These studies were carried out in collaboration with Michael Douderoff who spent a few weeks in Lipmann's laboratory while I was there. To determine whether acetoacetate synthesis involved the condensation of two molecules of acetyl CoA or one equivalent each of acetyl-CoA and acetate, Douderoff and I studied the formation of acetoacetate in a coupled enzyme system composed of phosphotransacetylase and acetoacetate synthetase from pigeon liver. We found that the acetoacetate produced from 1-[^{14}C] acetyl-P and unlabeled acetate was equally labeled in the carboxyl and carboxyl carbons, whereas acetoacetate was produced from unlabeled acetyl-P and [^{14}C] acetate contained no ^{14}C.[17]

Therefore the synthesis of acetoacetate must involve the condensation of two molecules of acetyl CoA [reaction (26)].

$$\text{2Acetyl CoA} \rightarrow \text{acetoacetyl CoA} + \text{CoA} \qquad (26)$$

The other pertinent observation that Mike Douderoff and I made was that CoA is required for the synthesis of butyryl-P from acetyl-P and butyrate [reaction (17)].

More detailed studies which were carried out in Bethesda, showed that butyryl-P synthesis involves the intermediate formation of acetyl-CoA and butyryl-CoA as follows:

acetyl-P + CoA ⇄ acetyl-CoA + Pi (25)

acetyl-CoA + butyrate ⇄ butyryl-CoA + acetate (27)

butyryl-CoA + Pi ⇆ butyryl-P + CoA (28)

Sum: Acetyl-P + butyrate ⇄ butyryl-P + acetate (29)

Both reactions (25) and (28) are catalyzed by phosphotransacetylase, but the transfer of CoA from acetyl-CoA to butyrate [reaction (27)] is catalyzed by a specific CoA transferase.[18]

This finding was of unique biochemical interest because it represented a new type of energy transfer process (i.e. a thioalkyl group transfer) by means of which one activated carboxylic acid could be converted to another. Moreover, the coupling of reactions (25) and (27) offered an attractive explanation for the acetyl-P requirement in the oxidation of butyrate by dialyzed extracts of *C. kluyveri*,[9] and focused attention also on the possible role of CoA derivatives as the activated intermediates in fatty acid metabolism.

Barker's theory of fatty acid oxidation

At the first Symposium on Phosphorous Metabolism, held in Baltimore in the spring of 1951, Barker reviewed the results of studies with enzyme preparations of *C. kluyveri* as well as complimentary studies in the field of CoA metabolism.[8] Based on a most remarkable critical analysis of the available information, he proposed a mechanism for the oxidation of butyrate by soluble extracts of *C. kluyveri* (see Fig. 1) in which CoA derivatives of 4-carbon compounds at various states of oxidation were postulated as the "active" intermediates, and in which cleavage of acetoacetyl CoA to form two equivalents of acetyl-CoA) [the reverse of reaction (26)] was the final step. The observed stoichiometry of one mole each of acetyl-P and acetate as ultimate products of butyrate oxidation [reaction (3)] was explained by the obligatory coupling of butyrate oxidation with reaction (25), needed to regenerate free CoA, and with reaction (27), in which butyryl-CoA is formed.

It is noteworthy that when Barker first proposed this scheme of fatty acid oxidation, there was only inferential evidence for a role of CoA in butyrate oxidation in *C. kluyveri*, and there was no evidence of any kind to implicate CoA in the oxidation of fatty acids by animal enzyme systems. In fact it was not until one year later[19] that a soluble system capable of oxidizing fatty acids in the presence of added CoA was prepared from animal tissues. It is therefore a tribute to Barker's imagination and conceptual analysis that within a few years after his report, his hypothesis was shown to be correct in every significant detail,† not only in *C. kluyveri* but in higher animals as well.‡

The mechanisms of butyrate oxidation in *C. kluyveri* and in animals differ significantly

† In the scheme as originally postulated by Barker, vinylacetyl-SCoA rather than crotonyl-SCoA was proposed as the immediate end-product of reaction (II), Fig. 1, and -hydroxybutyryl-SCoA, though implied, was not specifically designated as an intermediate. Otherwise the original scheme was essentially the same as that given in Fig. 1.

‡ There can be little doubt that Barker's hypothesis served as a model for most if not all subsequent studies in the field of fatty acid oxidation. It is therefore regrettable that whether by conscious neglect or by careless oversight, reports of research confirming this hypothesis have failed to acknowledge Barker's important contributions to the development of the field.

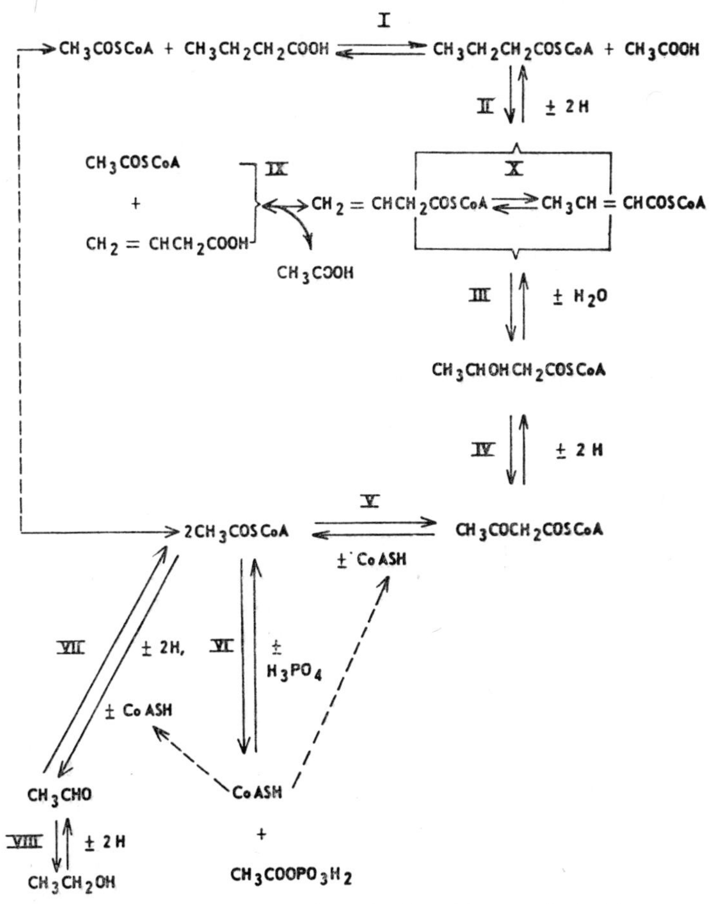

FIG. 1. Barker's scheme for the oxidation of butyrate in C. *kluyveri*

only in the manner of CoA regeneration and of butyrate activation. In animals the regeneration of CoA is obtained through the coupling of fatty acid oxidation with the oxidation of acetyl-CoA via the tricarboxylic acid cycle. Free CoA produced in the condensation of acetyl-CoA with oxalacetate to form citrate is needed for continued thiolytic cleavage of acetoacetyl CoA and also for the synthesis of butyryl-CoA by the ATP-CoA dependent fatty acid activating enzyme. However, since *C. kluyveri* is an obligate anaerobe, it catalyzes little if any oxidation of acetyl-CoA by the tricarboxylic acid cycle. In this organism free CoA regenerated from acetyl-CoA by means of the reaction catalyzed by phosphotransacetylase [reaction (25)], and the synthesis of butyryl CoA is achieved by the transfer of CoA from acetyl-CoA to butyrate [reaction (27)].

The net synthesis of acetyl CoA

Early in 1951, the scientific community was astounded by a preliminary communication from Lynen's laboratory[20] announcing the isolation of acetyl-CoA from yeast juice, and its characterization as a stable thiolester. In retrospect, it is ironic that many investigators were deterred from attempting to isolate acetyl-CoA because of the prevailing assumption that it would be too unstable to survive the rigors of isolation procedures; yet, it was just the opposite assumption that encouraged Lynen to undertake its direct isolation. He reasoned, rightly, that because of its central importance in metabolism it must be a relatively stable compound.

The demonstration that acetyl-CoA is a relatively stable thiolester as well as later reports from Lynen's laboratory showing, (a) that thiolester derivatives of saturated, unsaturated and β-ketoacids could be differentiated from one another by differences in their ultraviolet absorption spectra, and (b) that acyl-CoA analogs (i.e. thiolesters of *N*-acetylthioethanolamine and panthetheine) are catalytically active substrates for sheep liver enzymes, provided the much needed tools for more definitive studies on the role of acyl-CoA derivatives in intermediary metabolism. These studies were facilitated also by the eventual development of reliable methods for the isolation of substrate quantities of CoA of relatively high purity.§

Taking advantage of Lynen's spectrophotometric methods and the stability characteristics of thiolesters, it was readily shown that the incubation of phosphotransacetylase with acetyl-P and substrate levels of CoA led to the accumulation of a thiolester which was isolated and identified as acetyl-CoA.[22] This represented the first net synthesis of acetyl-CoA *in vitro* and established once and for all the mechanism of the phosphotransacetylase catalyzed reaction. From equilibrium measurements of reaction (25), it was calculated that the standard free energy of hydrolysis of acetyl-CoA is about the same as that of the pyrophosphate bond of ATP (i.e. -8 kcal). The "energy-rich" nature of acetyl CoA was thus firmly established.

Further studies in Bethesda

Extracts of *C. kluyveri* proved to be a fertile medium for further studies of fatty acid metabolism. As shown in Fig. 2, these studies led to the discovery of a number of reactions in which acetyl-CoA or other CoA derivatives are involved. Although lack of space precludes a detailed discussion of these studies, it may be noted that the DPN-and CoA-dependent oxidation of acetaldehyde to acetyl-CoA established as a new biological principle the fact that oxidation of aldehydes can be coupled with the esterification of CoA. This mechanism has since been shown to be widely utilized in intermediary metabolism.

It is also noteworthy that the acetylations of imidazole, HCN and mercaptans (steps 23, 22 and 27, respectively, in Fig. 2) represent different mechanisms by which metabolic energy in the form of acetyl-CoA can be transferred at the "energy-rich" level.[23] However, to date, the physiological function of the enzymes catalyzing these reactions is unknown. It seems likely that the capacities of imidazole, HCN and mercaptans to serve as acetyl group acceptors is the manifestation of nonspecificity of the respective transacylases for the acceptor substrate. The normal function of these enzymes may be to transfer acyl groups between acyl-CoA's and the functional groups (viz. imidazolium groups of histidine residues and sulfhydryl groups) at the catalytic sites of various enzymes.[23] The latter possibility is supported by the studies in Vagelos' laboratory[24] which show that *C. kluyveri* and *E. coli* catalyze thioltransacylations between acetyl-CoA (and malonyl-CoA) and the

§ Driven by a mutual desire to obtain substrate quantities of CoA for our independent research activities, Arthur Kornberg and I pooled our resources and developed a relatively simple method for the isolation of large amounts of CoA from yeast.[21]

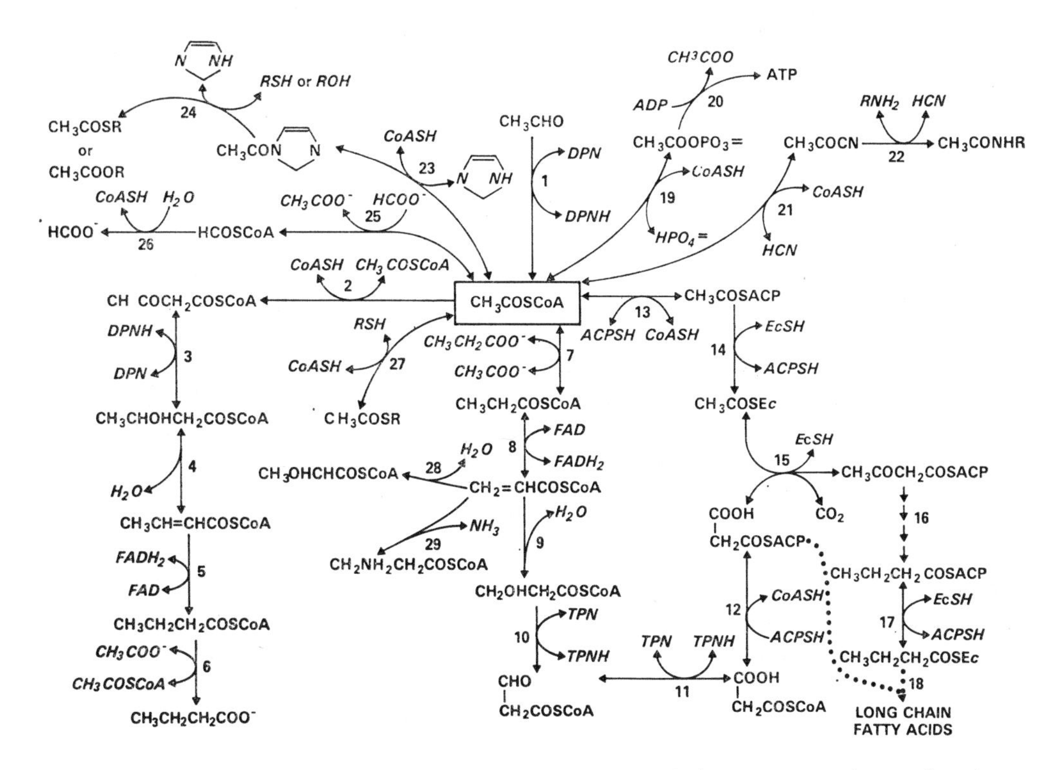

FIG. 2. The metabolism of acetyl CoA in cell-free extracts of *C. kluyveri:* Citations to the original literature describing these reactions can be found as follows: step 1, acetaldehyde oxidation;[23] steps 7–18, long chain fatty acid synthesis;[24,25] steps 19–20, synthesis of ATP by the coupling of phosphotransacetylase and acetate kinase;[12] steps 21–22, cyanide-dependent acetylation of amino groups;[23] steps 23–24, *N*-acetylimidazole synthesis and degradation;[28] steps 25–26, formate-dependent decomposition of acetyl-CoA by coupling of CoA-transferase and formyl-CoA thiolesterase activities;[23] step 27, thioltranacetylation;[23] step 28, acrylyl CoA hydratase (demonstrated in *Pseudomonas sp.* and heart muscle only;[26] step 29, acrylyl-CoA aminase, demonstrated in *C. propionium*, only.[26] *Abbreviations:* acylcarrier protein, ACPSH; acetoacetyl-S ACP synhetase (condensing enzyme), E_c.

sulfhydryl groups of acyl carrier protein (steps 12, 13, Fig. 2) and between the acyl carrier protein and the condensing enzyme (step 14, Fig. 2). Whether or not either of the latter transacylases is identical with the thioltransylases detected in our earlier investigation has not been determined.

Discovery of the acyl carrier protein

After the basic mechanism of fatty acid oxidation was established, it was generally assumed that fatty acid synthesis would occur by a reversal of the oxidative pathway, except of course for those steps concerned with the transfer of electrons to molecular oxygen. This assumption was readily verified in the case of *C. kluyveri* where the synthesis of butyrate occurs by reversal of those steps shown in Fig. 1. However, subsequent studies in the laboratories of Wakil, Lynen, Brady and Vagelos[24,25] demonstrated that the synthesis of long-chain fatty acids occurs by a different mechanism in which malonyl-CoA rather than acetyl-CoA is the immediate source of "active acetate" in the synthesis of β-keto thiolester intermediates, and in which thiolester derivatives of acyl carrier protein are activated intermediates. It is therefore of historical interest that studies with *C. kluyveri*, which had contributed so much to elucidation of basic mechanism of fatty acid oxidation, played a significant role also in elucidation of the mechanism of long-chain fatty acids synthesis. Whereas the energy metabolism of *C. kluyveri* is linked to the synthesis of butyrate (and caproate) by the reversal of steps illustrated in Fig. 1, *C. kluyveri* utilizes the malonyl CoA-acyl carrier protein mechanism for the synthesis of long-chain fatty acids. In fact, the involvement of the acyl carrier protein in fatty acid synthesis was discovered by P. R. Vagelos in the course of studies on the oxidation of propionate by extracts of *C. kluyveri*. In 1956 Vagelos came to NIH as a Clinical Associate to seek experience in the fields of enzymology and biochemistry. There he undertook a series of investigations to determine if acyl-CoA derivatives of 3-carbon compounds are intermediates in propionate metabolism.[26] Among other important results, shown in Fig. 2, he found that extracts of *C. kluyveri* catalyzed the oxidation of β-hydroxypropionyl CoA to malonyl-CoA (steps 10 and 11, Fig. 2).[27] Efforts to demonstrate the further conversion of malonyl-CoA to acetyl-CoA and CO_2 in extracts of *C. kluyveri* failed, but led to the curious observation that these extracts catalyzed rapid exchange of $^{14}CO_2$ with the carboxyl group of malonyl-CoA.[24,25] In a series of brilliant investigations, Vagelos and his associates showed that the malonyl CoA–CO_2 exchange reaction is catalyzed by a multi-component system consisting of a fatty acyl-CoA derivative, several enzymes, and a heat-stable protein. The CO_2-exchange reaction (ultimately explained by steps 12–15, Fig. 2) proved to be a part of the enzyme system concerned with the synthesis of long-chain fatty acids in *C. kluyveri*.[27] In comparative studies the CO_2-exchange and fatty acid synthetase systems were detected in extracts of *Escherichia coli* and for technical reasons *E. coli* was selected for more detailed investigations. The heat-stable protein required for the CO_2-exchange reaction was isolated in pure form from *E. coli* and its function as an acyl carrier protein (ACPSH) in steps 12–18 in Fig. 2 was firmly established.[24,25]

This is not the place to review in detail the elegant studies in Vagelos' laboratory nor those in the laboratories of Lynen and Wakil[24] that led to complete elucidation of the mechanism of long-chain fatty acid synthesis which is depicted in Fig. 2, steps 12–18. Needless to say, for the reasons summarized above, the studies with enzymes from *C. kluyveri* contributed richly to this effort.

Retrospection of prospect

When he first detected butyric and caproic acids in crude ethanol enrichment cultures, Barker was keenly aware of the fact that the responsible organism would offer a unique opportunity to investigate the mechanism of fatty acid synthesis. I wonder, however, if in his wildest imagination he could have forseen that the obscure, then nameless bacterium which he extracted from the murky waters of a Dutch canal was destined to contribute so much to our knowledge of the biochemistry of fatty acid metabolism and to the establishment of basic principles of metabolic energy transfer. The tremendous impact that studies with *C. kluyveri* have had on the development in a major branch of biochemistry should along with the fundamental contributions of bacterial metabolism in other fields of biochemistry, serve notice to those who would have us believe that research on procaryotic systems has little relevance to the solution of problems in higher life forms. I suspect that such short-sightedness often stems from the frustrations of some biochemists who, bewildered by the complexities of their experimental material, draw unconsciously on the basic principles gleaned from the pioneering studies with the simpler bacterial systems in order to interpret their results. Nevertheless, in their eagerness to disclaim the importance of these pioneering studies in the development of a particular problem some investigators call attention to minor differences between procaryotic and eucaryotic systems but fail to acknowledge the gross similarity in basic biochemical mechanisms and fundamental principles that govern all forms of life.

Surely, by whatever name, biochemistry, molecular biology, or biophysics, our base of knowledge is not yet so great as to justify the ill-conceived notion that studies of bacterial metabolism are irrelevant to the metabolism of higher organisms. Certainly, basic information that is more easily obtained through reasearch on carefully selected microorganisms that are uniquely endowed with exaggerated biochemical capacities, will continue to contribute significantly to better understanding of many of the major problems of cell biology.

REFERENCES

1. Barker, H. A. (1936) On the biochemistry of the methane fermentation. *Arch. Mikrobiol.* **7**, 404.
2. Barker, H. A. (1947) *Clostridium kluyveri Anthonie Van Leeuwenhoek*, **12**, 167.
3. Barker, H. A. (1939–40) Studies upon the methane fermentation. IV. The isolation and culture of *Methanobacterium Omellianskii. Anthonie Van Leewenhoek, J. Mikrobiol. Serol.* **6**, 201.
4. Barker, H. A. and Taha, S. M. (1942) *Clostridium kluyveri*, an organism concerned in the formation of caproic acid from ethyl alcohol. *J. Bacteriol.* **43**, 347.
5. Barker, H. A., Kamen, M. D. and Bornstein, B. T. (1946) The synthesis of butyric and caproic acids from ethanol and acetic acid by *Clostridium kluyveri. Proc. Nat. Acad. Sci. U.S.A.* **31**, 373.
6. Lipmann, F. (1946) Acetyl phosphates. *Adv Enzymol.* **6**, 231.
7. Stadtman, E. R. and Barker, H. A. (1949) Fatty acid synthesis by enzyme preparations of *Clostridium kluyveri*. I. Preparation of cell-free extracts that catalyze the conversion of ethanol and acetate to butyrate and caproate. *J. Biol. Chem.* **180**, 1095.
8. Barker, H. (1951) Recent investigations on the formation and utilization of active acetate. In *A Symposium on Phosphorous Metabolism*, vol. I (W. D. McElroy and B. Glass, eds.), p. 204, The Johns Hopkins Univ. Press, Baltimore.
9. Stadtman, E. R. (1954) Studies on the biochemical mechanism of fatty acid oxidation and synthesis. *Record of Chemical Progress*, **15**, 1.
10. Lipmann, F. (1945) Acetylation of sulfanilamide by liver homogenates and extracts. *J. Biol. Chem.* **160**, 173.
11. Lipmann, F. (1948) Biosynthetic mechanisms. *Harvey Lect.* **44**, 99.
12. Stadtman, E. R. and Barker, H. (1950) Fatty acid synthesis by enzyme preparations of *Clostridium kluyveri*. VI. Reactions of acyl phosphates. *J. Biol. Chem.* **184**, 769.
13. Douderoff, M. Barker, H. A. and Hassid, W. Z. (1947) Studies with bacterial sucrose phosphorylase. I. The mechanism of action of sucrose phosphorylase as a glucose-transferring enzyme (transglucosidase). *J. Biol. Chem.* **168**, 725.
14. Stadtman, E. R., Novelli G. D. and Lipmann, F. (1951) Coenzyme A function in acetyl transfer by the phosphotransacetylase system. *J. Biol. Chem.* **191**, 365.
15. Chou, T. C., Novelli, G. D., Stadtman, E. R. and Lipmann, F. (1950) Fractionation of coenzyme A dependent transfer reactions. *Federation Proc.* **9**, 160.
16. Stern, J. R., Shapiro, B., Stadtman, E. R. and Ochoa, S. (1951) Enzymatic synthesis of citric acid. III. Reversibility and mechanism. *J. Biol. Chem.* **193**, 703.
17. Stadtamn, E. R. Douderoff, M. and Lipmann, F. (1951) The mechanism of acetoacetate synthesis. *J. Biol. Chem.* **191**, 377.
18. Stadtman, E. R. (1953) The coenzyme A transphorase system in *Clostridium kluyveri. J. Biol. Chem.* **203**, 501.

19. DRYSDALE, G. R. and LARDY, H. A. (1951) Fatty acid oxidation by a soluble enzyme system from mitochondria. In *A Symposium on Phosphorus Metabolism*, vol. II (W. D. McELROY and B. GLASS, eds.), p. 281, The Johns Hopkins Univ. Press, Baltimore.
20. LYNEN, F. and REICHERT, E. (1951) Zur Chemischen Struktur der Aktiverten Essigsaure. *Angew. Chem.* **63**, 47 .
21. STADTMAN, E. R. and KORNBERG, A. (1953) The purification of coenzyme A by ion exchange chromatography. *J. Biol. Chem.* **203**, 47.
22. STADTMAN, E. R. (1952) The net enzymatic synthesis of acetyl coenzyme A. *J. Biol. Chem.* **196**, 535.
23. STADTMAN, E. R. (1966) Some considerations of the energy metabolism of anaerobic bacteria. In *Current Aspects of Biochemical Energetics* (N. O. KAPLAN and E. P. KENNEDY, eds.), p. 39, Academic Press, New York.
24. VAGELOS, P. R. (1964) Lipid metabolism. *Ann. Rev. Biochem.* **33**, 139.
25. VAGELOS, P. R. (1973) Acyl group transfer (acyl carrier protein). In *The Enzymes*, Vol. VIII, 3rd ed. (P. D. BOYER, ed.), p. 155, Academic Press, New York.
26. STADTMAN, E. R. and VAGELOS, P. R. (1958) Propionic acid metabolism. *International Symposium on Enzyme Chemistry*, p. 86, Tokyo and Kyoto, 1957, Pan-Pacific Press, Tokyo.
27. VAGELOS, P. R. and EARL, J. M. (1959) Propionic acid metabolism. III. β-Hydroxpropionyl coenzyme A and malonyl semialdehyde coenzyme A, intermediates in propionate oxidation by *Clostridium kluyveri*. *J. Biol. Chem.* **234**, 2272.
28. ALBERTS, A. W., GOLDMAN, P. and VAGELOS, P. R. (1963) The condensation reaction of fatty acid synthesis. I. Separation and properties of the enzymes. *J. Biol. Chem.* **238**, 557.

BIOTIN UNMASKED

M. DANIEL LANE

Department of Physiological Chemistry, Johns Hopkins University Medical School,
Baltimore, Maryland 21205

The hen's egg has played an important part in both the discovery of the vitamin, biotin and its biochemical role. It is curious that nature has brought together within the egg one of the richest sources of biotin in the yolk and a "toxic" factor in the white that causes biotin deficiency when fed to animals. By the turn of the century[1] it was recognized that uncooked egg white was toxic to animals. The basis for this toxicity—revealed only later—was that egg-white contains avidin, a glycoprotein which binds biotin with extraordinary affinity rendering it unavailable for absorption from the intestine. Around 1930 several research groups[2] launched efforts to isolate and purify a vitamin-like factor found widely distributed in nature that cured "egg-white injury" to animals. In particular, egg yolk and liver were rich in this factor. By coincidence, work was also under way in Kögl's laboratory on the purification of a yeast growth factor from egg yolk. The rapid progress in purifying this factor can be attributed to the rapid yeast-growth assay developed by Kögl to monitor fractionation. In 1936, after a heroic fractionation effort in which over 500 pounds of dried egg yolk were processed, Kögl and Tönnis obtained 1.1 mg of the pure methyl ester of biotin.[3] Unfortunately, progress in the purification of the anti-egg-white injury factor (or vitamin H) was delayed by the slow and sometimes inaccurate animal curative assay. Nevertheless, it became apparent to Paul György that the distribution, fractionation properties, and chemical behavior of this factor were similar to those of Kögl's yeast growth factor. When it became available, a sample of Kögl's pure biotin methyl ester was found to be extremely potent in protecting rats against "egg-white injury";[2] thus, it became apparent that biotin and the egg-white injury factor were identical. Within a few years, Vincent Du Vigneaud and his colleagues[2] had determined the correct chemical structure of biotin (shown below), clearing the way for an attack on the role of the vitamin at the molecular level.

By 1950 biotin had been implicated[2] in a number of seemingly unrelated enzymatic processes, including: (1) the decarboxylation of oxaloacetate and succinate; (2) the "Wood–Werkman reaction", that is, the β-carboxylation of pyruvate; (3) the biosynthesis of aspartate; (4) the biosynthesis of unsaturated fatty acids; (5) the deamination of certain amino acids; (6) the synthesis of certain biotin-independent enzymes. In each case the enzymatic or metabolic basis for these observations can now be explained in terms of the unequivocally established role of biotin as an enzymatic "CO_2 carrier" or through indirect effects.

The long-sought link between biotin and a specific enzyme function was provided by Henry Lardy at the University of Wisconsin.[4] He and Peanasky demonstrated that soluble

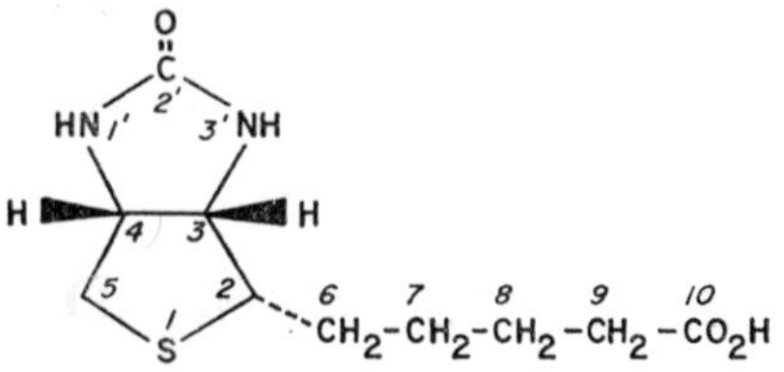

FIG. 1. The structure of biotin.

extracts of acetone-dried liver mitochondria catalyzed the ATP- and divalent cation-dependent carboxylation of propionate (or propionyl-CoA) to form succinate. Later, work in Severo Ochoa's laboratory showed[5] the initial carboxylation product to be methylmalonate, an intermediate *en route* to succinate. Most importantly, Lardy found the carboxylating activity to be almost completely lacking in mitochondrial enzyme extracts from rats rendered biotin deficient by feeding them dried egg white. Injection of biotin quickly brought the carboxylase activity back to normal.

I remember being particularly intrigued by the papers of Lardy's group on the carboxylation of propionate because I had just initiated efforts to determine the metabolic fate of propionate in ruminant liver. This interest derived from the fact that propionate, a major fermentation product of the rumen or "fore-stomach" of ruminants, is the principal gluconeogenic precursor in ruminants. After absorption from the rumen, it is carried by portal blood directly to the liver. Little was known at that time about how propionate was metabolized. In 1957 I wrote to Henry Lardy to determine whether they had progressed substantially beyond what their published reports indicated. To my surprise, he indicated that they had dropped the problem because Severo Ochoa's lab was way ahead having isolated and substantially purified the propionyl-CoA carboxylation enzyme system from heart tissue. This was confirmed in a cordial letter from Severo. Although discouraged by this news, my students and I decided to push ahead with the purification of the propionyl-CoA carboxylase from bovine liver mitochondria. By 1961, both the pig heart and bovine liver carboxylases were essentially homogeneous.[2]

Two other enzymes discovered at about the same time,[2] acetyl-CoA carboxylase by Salih Wakil in David Green's lab and β-methylcrotonyl-CoA carboxylase by Feodor Lynen, played important parts in linking biotin function to enzymatic carboxylation. It was reported by Green at a Gordon Conference in 1958 that although bicarbonate was required for fatty acid synthesis from acetyl-CoA (catalyzed by two enzyme components (R_1 and R_2) from avian liver), [^{14}C] bicarbonate did not enter the fatty acid product. In the ensuing discussion, Lynen suggested[6] that the action of bicarbonate could be explained by the intermediate carboxylation of acetyl-CoA to form malonyl-CoA followed by decarboxylation when malonyl-CoA condensed with acetyl-CoA (in reality, acetyl-S-enzyme) for chain elongation. The analogy to the carboxylation of propionyl-CoA to form methylmalonyl-CoA was pointed out. Lynen's suggestion proved to be correct. Roscoe Brady, who was present at the Conference, as well as Wakil in Green's laboratory, shortly thereafter published communications[2] in which malonyl-CoA was found to be an intermediate in the conversion of acetyl-CoA to palmitic acid.

Thus, it became evident that acetyl-CoA carboxylase and propionyl-CoA carboxylase catalyzed analogous reactions; moreover, the homology of reaction could be stretched*

* Since β-methylcrotonyl-CoA has an α,β-double bond, the electrophilic site—hence the site of carboxylation—is displaced to the α-position by conjugation.

to β-methylcrotonyl-CoA carboxylase. Lardy's earlier work showing that biotin deficiency led to a decreased propionyl-CoA carboxylase activity in liver suggested that perhaps all three acyl-CoA carboxylases involved biotin. Similar experiments in Minor Coon's laboratory in Michigan revealed[(2)] that β-methylcrotonyl-CoA carboxylase activity was almost completely lacking in the livers of rats fed a biotin deficient (avidin-containing) diet. Finally Wakil provided the first compelling evidence that the carboxylase per se contained functional biotin using a technique introduced earlier by Wessman and Werkman.[(2)] It was found that incubation of avidin with an acetyl-CoA carboxylase preparation irreversibly blocked carboxylase activity. That this inhibition is specific was indicated by the fact that prior treatment of avidin with free biotin masked its inhibitory action. Similar experiments in our lab.[(2)] revealed that, contrary to an earlier erroneous report, propionyl-CoA carboxylase was also a biotin enzyme. Since that time, avidin has become a routine diagnostic agent for the identification and study of biotin-dependent enzymes.

At this juncture, a major breakthrough was made by Feodor Lynen and his colleagues at the Max Planck-Institut für Zellchemie in Munich[(7)] which ultimately defined the function of biotin in enzymatic carboxylation. They had shown that β-methylcrotonyl-CoA carboxylase isolated from a Mycobacterium carried out the following reaction:

$$\mathrm{ATP} + \mathrm{HCO_3^-} + \begin{matrix}\beta\text{-methyl}\\ \text{crotonyl-CoA}\end{matrix} \underset{}{\overset{Mg^{2+}}{\rightleftharpoons}} \mathrm{ADP} + \mathrm{P_i} + \begin{matrix}\beta\text{-methyl}\\ \text{glutaconyl-CoA}\end{matrix} \quad (1)$$

Since biotin was thought to be covalently bound to protein in biological materials, perhaps through its carboxyl group, Lynen presumed that the vitamin functioned as a prosthetic group. This belief was supported by the observation of Woessner and Coon[(2)] that hepatic β-methylcrotonyl-CoA carboxylase activity was markedly depressed by biotin deficiency. Lynen reasoned that *free* biotin might mimic the biotin prosthetic group at the active site of the enzyme and undergo reaction. This in fact, turned out to be the case. Free (+)-biotin behaved as a "substrate", or model prosthetic group, in a carboxylase-catalyzed reaction which led to the formation of an unstable free carboxybiotin derivative. The carboxylation product, after stabilization by methylation with diazomethane, was conclusively identified as 1′-*N*-carboxybiotin as illustrated in Fig. 2.

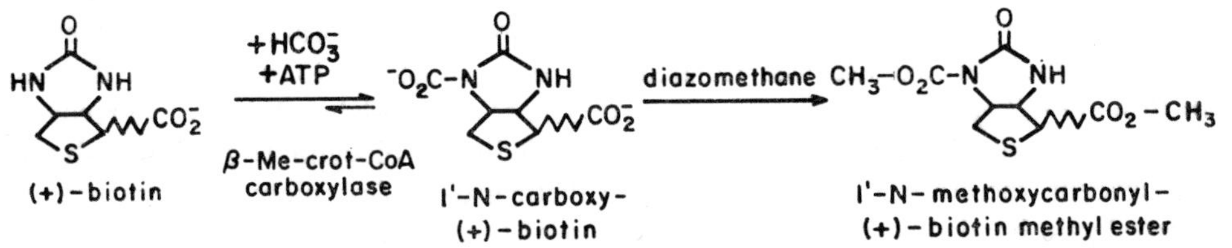

(+)-biotin
+HCO$_3^-$
+ATP
β-Me-crot-CoA carboxylase
1′-N-carboxy-(+)-biotin
diazomethane
1′-N-methoxycarbonyl-(+)-biotin methyl ester

FIG. 2. Site of carboxylation on free biotin.

Thus, Lynen proposed that—like free biotin—the biotinyl prosthetic group of the enzyme undergoes carboxylation on its 1′-nitrogen after which carboxyl transfer to the acyl-CoA substrate occurs. This and the results of isotopic exchange experiments carried out in several laboratories, including our own,[(2)] with various biotin-dependent carboxylases led to the well-known 2-step reaction sequence.

$$\mathrm{ATP} + \mathrm{HCO_3^-} + \text{enz-biotin} \overset{Mg^{2+}}{\rightleftharpoons} \text{enz-biotin-}\mathrm{CO_2^-} + \mathrm{ADP} + \mathrm{P_i} \quad (2)$$

$$\text{Enz-biotin-}\mathrm{Co_2^-} + \text{acceptor} \rightleftharpoons \begin{matrix}\text{carboxylated}\\ \text{acceptor}\end{matrix} + \text{enz-biotin} \quad (3)$$

Proof that a carboxylated enzyme intermediate (enzyme-CO_2^-) actually participates in biotin-dependent carboxylations was provided by Yoshito Kaziro in Severo Ochoa's laboratory at New York University.[8] They were able to isolate enzyme-CO_2^- and show an exact stoichiometry between bound biotin and the "active carboxy" group. Importantly, the isolated enzyme-CO_2^- transferred its labile carboxy group to propionyl-CoA yielding methylmalonyl-CoA [reaction (3)] or underwent quantitative decarboxylation upon exposure to ADP, P_i and Mg^{2+} [reverse of reaction (2)].

Although Knappe and Lynen's experiments indicated that free biotin mimicked the enzyme-bound biotin prosthetic group, some doubt was raised by the apparent inability* of free 1'-*N*-carboxybiotin (chemically or enzymatically synthesized) to serve as carboxyl donor to β-methylcrotonyl-CoA in the presence of the carboxylase. It became apparent that to ascertain whether enzyme-bound biotin was carboxylated in the same manner as free biotin, it would first be necessary to determine the nature of the chemical linkage of the prosthetic group to the enzyme. Only then could a logical strategy be developed to isolate the putative biotin-containing carboxylated fragment. Dave Kosow in my lab. approached the problem in the following way. We assumed that the terminal step in the synthesis of biotin enzymes was the covalent attachment of the prosthetic group to the apoenzyme to form an active holoenzyme. The idea was to get out the apocarboxylase and the enzyme which catalyzed the "loading" reaction. Then using [^{14}C] biotin it would be possible to obtain [^{14}C]biotinylated holocarboxylase which could be degraded and the fragment identified. Dave spent the better part of a year trying to demonstrate the *in vitro* enzymatic attachment of labeled biotin to propionyl apocarboxylase obtained from liver mitochondria of biotin-deficient rats. Finally, an enzyme was obtained[2] which carried out the following reaction:

$$\text{Biotin} + \epsilon\text{-NH}_2\text{-lysyl-apocarboxylase} \xrightarrow[\text{Synthetase}]{\text{ATP} \searrow\ \text{Mg}^{2+}\ \nearrow \text{AMP, PP}_i} \text{biotinyl-NH}_2\text{-lysyl-holocarboxylase} \qquad (4)$$

The rest was easy.

Following the incubation of [^{14}C]biotin with the propionyl holocarboxylase-synthesizing system, [^{14}C]holocarboxylase was isolated and then subjected to proteolytic digestion with Pronase (*Streptomyces griseus* protease). The sole [^{14}C]-labeled derivative recovered from the Pronase digest was isolated and identified[9] as biocytin (ϵ-*N*-biotinyl-lysine). It was evident, therefore, that the biotinyl prosthetic group of propionyl-CoA carboxylase was covalently bonded to a lysyl ϵ-amino group of the apoenzyme. This type of linkage was subsequently detected in several other biotin enzymes.[2] Some years later[2] it became apparent that the long (14 Å) side-chain was necessary to allow the functional bicyclic ring to flip to and fro between remote active sites which carry out the two half-reactions during catalysis.

We were quite pleased that the completion of these experiments coincided with a visit by Severo Ochoa to our outpost in Blacksburg, Virginia (late, 1961). News had reached us via the grapevine that his group had attempted unsuccessfully to answer the same question using another approach. Severo seemed genuinely pleased with our findings and took some Pronase and the method back to New York with him. At the Federation

* Only recently[11] was it possible to demonstrate that free 1'-*N* carboxybiotin derivatives are enzymatically active carboxyl donors.

Meeting in Atlantic City (April 1962), Dave Kosow presented our results at a session chaired by Severo. In the discussion that followed Dave's paper, Al Grossman from the NYU group indicated that they had confirmed our results. This really boosted our collective egos.

That summer (1962) before leaving for Munich to spend a year in Lynen's institute, I became aware of some results which were subsequently published and which generated a great deal of controversy. It had been found[2] that the ureido carbon atom of biotin,

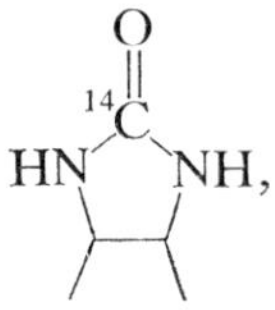

on acetyl-CoA carboxylase became labeled when the enzyme was incubated with ATP, Mg and [^{14}C]bicarbonate. Moreover, it appeared that the ureido [^{14}C] label could be enzymatically transferred to acetyl-CoA to form [^{14}C]-malonyl-CoA. This, of course, conflicted with the results of Lynen who had obtained evidence that enzyme-bound 1′-*N*-carboxybiotin was involved. Lynen's compound was acid-labile, whereas the ureido carbon atom of biotin was acid-stable. As pointed out above, one of the criticisms leveled at Lynen's approach was that the 1′-*N*-carboxy group had not been proven to be on the bound prosthetic group, but rather on the *free* carboxybiotin product. That spring and summer before leaving for Lynen's lab. I had shown that under appropriate conditions, enzyme-$^{14}CO_2^-$ (propionyl-CoA carboxylase) could be stabilized by methylation with diazomethane; the enzyme-$^{14}CO_2^-$ was labile to acid before, but not after methylation. Moreover, by digesting the methylated enzyme-$^{14}CO_2^-$ with Pronase, a single radioactive derivative, presumably 1′-*N*-methoxy [^{14}C]carboxyl-ε-*N*-(+)-biotinyl lysine which had chromatographic properties similar to, but not identical with ε-*N*-biotinyl lysine, was isolated. Since I did not have the authentic compound for comparison, these experiments could not be completed at that time. Fortunately, Joachim Knappe, then at the University of Heidelberg, had synthesized the derivative and provided Lynen with a sample. Therefore, shortly after arriving in Munich, I was able to confirm the presumptive identification.[10] This indicated that, like free biotin, the covalently-bound biotinyl prosthetic group was carboxylated at the 1′-*N* position.

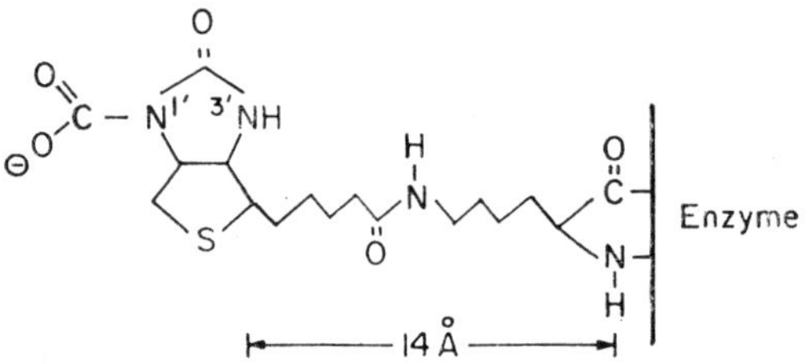

FIG. 3. Carboxylated active site of propionyl-CoA carboxylase.

In short order, Harland Wood, who was also on sabbatical leave in Munich, showed by the same technique[2] that the carboxybiotin prosthetic group of transcarboxylase had the identical structure. This more or less laid to rest the notion that the ureido C per se of biotin was generated from bicarbonate or underwent transfer. The idea was later retracted.

Only within the last few years has it become apparent[11] that the biotin group on enzymes, although covalently bound, is in a dynamic state and shuttles between remote active sites during catalysis. This was proven most directly by work in Roy Vagelos' and my laboratories[11] with the unique *E. coli* acetyl-CoA carboxylase system. Unlike its counterpart in animal tissues, the carboxylase from *E. coli* dissociates readily into three protein components all of which are essential foı the overall reaction. Two of these, i.e. biotin carboxylase (BC) and carboxyl transferase (CT), possess catalytic centers for the first [reaction (5)] and second [reaction (6)] half-reactions, respectively, of acetyl-CoA carboxylation.

$$\text{CCP—biotin} + HCO_3^- + \text{ATP} \underset{\text{BC}}{\overset{Mg^{2+}}{\rightleftharpoons}} \text{CCP—biotin—}CO_2^- + \text{ADP} + P_i \qquad (5)$$

$$\text{CCP—biotin—}CO_2^- + \text{acetyl-CoA} \underset{\text{CT}}{\rightleftharpoons} \text{CCP—biotin} + \text{malonyl-CoA} \qquad (6)$$

Neither catalytic component contains a trace of bound biotin. The biotin prosthetic group is covalently linked to the third component, carboxyl carrier protein (CCP—biotin). As with other biotin enzymes, the bicyclic ring of the prosthetic group resides at the distal end of a flexible 14 Å side chain which allows it to act as a "mobile carboxyl carrier" between the two catalytic centers as illustrated below. A large number of biotin-dependent enzymes which carry out diverse reaction types are now known.[2] All of these reactions proceed through a carboxylated intermediate with the carboxybiotinyl prosthetic group functioning as a "mobile carboxyl carrier".

In retrospect, the number of people and laboratories contributing to the solution of biotin function were many. There was much cooperation and encouragement as well as

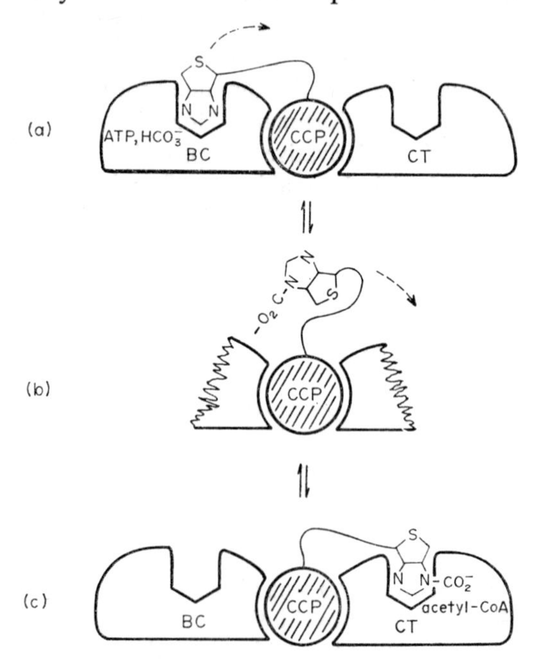

Fig. 4. "Ping-pong" action of the mobile biotin group of acetyl-CoA carboxylase.

competition. It is one more area in which Severo Ochoa and his laboratory played an important role. In 1964 I joined his department at NYU as a faculty member, and had the opportunity to appreciate even more his scientific interests, enthusiasm, expertise and his friendship.

REFERENCES

1. STEINITZ, F. (1898) Ueber das Verhalten Phosphorhaltiger Eiweisskörper in Stoffwechsel. *Pflüger's Archiv. für Physiologie*, **72**, 75.
2. MOSS, J. and LANE, M. D. (1971) The biotin dependent enzymes. *Adv. Enzymol.* **35**, 321.
3. KÖGL, F. and TÖNNIS, B. (1936) Uber das Bios-Problem. Darstellung von Krystallisiertem Biotin aus Eigelb. *Z. Physiol. Chem.* **242**, 43.
4. LARDY, H. A. and PEANASKY, R. (1953) Metabolic functions of biotin. *Phys. Rev.* **33, 560.**
5. FLAVIN, M., ORTIZ, P. J. and OCHOA, S. (1955) Metabolism of propionic acid in animal tissues. *Nature*, **176**, 823.
6. LYNEN, F. (1959) Participation of acyl-CoA in carbon chain biosynthesis. In Symposium on Enzyme Reaction Mechanisms. *J. Cell. Comp. Physiol.* **54**, Suppl. 1, 33.
7. LYNEN, F., KNAPPE, J., LORCH, E., JÜTTING, G. and RINGELMANN, E. (1959) Die Biochemische Funktion des Biotins. *Angewandte Chemie*, **71**, 481.
8. KAZIRO, Y. and OCHOA, S. (1961) Mechanism of the propionyl carboxylase reaction, I. Carboxylation and decarboxylation of the enzyme. *J. Biol. Chem.* **236**, 3131.
9. KOSOW, D. P. and LANE, M. D. (1962) Propionyl holocarboxylase formation: covalent bonding of biotin to apocarboxylase lysyl ε-amino groups. *Biochem. Biophys. Res. Commun.* **7**, 439.
10. LANE, M. D. and LYNEN, F. (1963) The biochemical function of biotin. VI. Chemical structure of the carboxylated active site of propionyl carboxylase. *Proc. Nat. Acad. Sci., U.S.A.* **49**, 379.
11. LANE, M. D., MOSS, J. and POLAKIS, S. E. (1974) Acetyl coenzyme A carboxylase. *Current Topics in Cellular Regulation*, vol. 8, p. 139, Academic Press.

THE SEVEN VEILS OF DIPICOLINIC ACID

CHARLES GILVARG
Department of Biochemical Sciences, Princeton University, Princeton, New Jersey 08540

Many articles that deal with dipicolinic acid (pyridine 2,6 dicarboxylate) (DPA), at some point are sure to mention that this single substance constitutes 10–15% of the dry weight of the bacterial endospore. This is an arresting fact that immediately draws one's attention to an exceptional aspect of what is already an unusual biological material. Bacterial endospores are indeed a dramatic representative of the cryptobiotic state. Their indetectable metabolic rate makes it possible for spores to remain in a state of suspended animation for periods of at least several centuries. This remarkable dormancy is complemented by extraordinary levels of resistance to many environmental hazards. Though one usually finds this point illustrated with a reference to the ability of clostridial spores to withstand boiling water (indeed, "activation" of spores for germination, and a return to vegetative growth, is often best achieved by a heat shock that would rapidly dispatch the vegetative cell) my own appreciation of spore resistance was enhanced when I discovered that spores of *Bacillus megaterium* could remain viable after spending a week in Bray's solution.

In view of the very high DPA concentrations in spores, it is not surprising that soon after the compound's discovery, many authors proposed rather direct relationships between DPA and spore dormancy and resistance. However, at this moment, many years later, the role of DPA in the spore is still a matter of conjecture. Curiously, even less demanding questions than the function of the compound have been sources of considerable confusion. Perhaps the simplicity of the molecule made it all look too easy and the investigators were put off their guard. As one surveys the literature, there is a definite feeling that there is an analogy to be drawn between the multilayered ultrastructure of the spore and the successive veils of secrecy that have enshrouded the mysteries of dipicolinate. As we shall see below, some of these veils have been lifted, some have only been slightly parted, and the last veil, that guarding the *raison d'être*, is still quite secure.

1. DISCOVERY

Though major reviews[1,2] date the discovery of DPA to Powell's classic 1953 paper, in which she demonstrated the occurrence of the compound in germination exudates, the molecule had been established as a natural product some 17 years earlier. Udo in 1936 had described the isolation of DPA from "natto", a Japanese foodstuff.[3] Retrospectively, it is clear that the involvement of *B. natto* in the fermentation of the soy beans is the reason that DPA occurs in the condiment. As an organic chemist, though, it was not a concern of Udo to ascertain the link between the microorganism and the compound and that major point is really the contribution of Powell.

It is an interesting coincidence, one that does not appear in the journals, but serves to enrich the subject, that an undergraduate who had studied organic chemistry under Udo in Tokyo should, on coming to the United States, quite adventitiously do his Ph.D. thesis on the biosynthesis of DPA and provide the first solid genetic evidence on the relationship between DPA and the bacterial lysine pathway.[4] But more of that below.

2. DISTRIBUTION

Though one cannot be sure that DPA, in smaller amounts that have escaped detection, might not have a more widespread distribution in nature, it presently appears that its biosynthesis in massive concentrations is restricted to sporulating bacteria. Among the bacteria that form endospores (Bacilli and Clostridia), it is a ubiquitous constituent so much so that *Sporosarcina ureae*, the one exceptional example of a coccus that forms an endospore, is unexceptional in regard to the presence of DPA. The firmness of the association of the endospore and DPA is further strengthened by noting that the compound has never been detected during the vegetative phase of growth of these microorganisms but appears only after the sporulation sequence has been well advanced.

At one point, this inviolate association was somewhat threatened by the finding that DPA occurred in the medium in which *Penicillium citreo-viride* had been grown.[5] However, it now appears quite likely that the DPA, in this instance, arises as a result of the non-enzymatic interaction of α,ε-diketopimelic acid, a fermentation product of the mold, and ammonia. It does not seem quite fair to use this atavistic retention of pre-biotic chemistry as a basis for including the mold in the select list of DPA producers and, more discriminatorily, users. However, it is perhaps worth noting that commerce, which looks at matters from a more pragmatic view than academicians, has chosen to take out a patent on the biological production of DPA using a mold rather than a bacterial fermentation.[6] In this process, n-hexadecane is used as a carbon source which is some measure of how far removed the carbon source can be from the final product. This is perhaps a particularly fitting introduction to the next section.

3. BIOSYNTHETIC ORIGIN

The common possession of seven carbon atoms and two carboxyl groups by diaminopimelic acid and DPA appeared too irresistible to ascribe to mere coincidence and Perry and Foster did succumb to the temptation of adding labeled diaminopimelate to sporulating cultures of *B. cereus*.[7] The low levels of incorporated radioactivity should have led to the conclusion that there must be something wrong with diaminopimelate transport in the sporulation phase of growth (as was indeed confirmed by others many years later[8]). However, the few counts over background in the DPA was instead advanced as a verification of the biosynthetic relationship of the amino acid and DPA. This conclusion was all the more remarkable because the authors took the trouble to isolate several other compounds from the spores. One of these was lysine, a known product of diaminopimelate metabolism, and it was devoid of any radioactivity.

A careful repetition of this early experiment on biosynthetic origins by Finalayson and Simpson[9] served to confirm that the wish had been father to the thought. Actually, the thought did have considerable merit since as it turned out, the lysine pathway does supply the intermediate for DPA synthesis. It was simply that the intermediate is at the beginning rather than at the end of the pathway. Before this point was established, however, a theory of a C_2 plus C_5 condensation did enjoy a brief vogue.[10] The arguments for this hypothesis, i.e. the central role of glutamate in metabolism during sporulation, the promotion of spore formation by certain C_2 compounds and the elevated DPA levels with glycolate and glyoxylate grown cells, seem less than compelling in view of Schönheimer's dictum, "because the insertion of a penny results in the output of a stick of chewing gum, the penny is not to be regarded as the precursor of the chewing gum."

My own involvement in the question of the biosynthetic origin of DPA arose through an interest not in that molecule but in the intermediates in the lysine pathway. The very naming of the first pathway specific intermediate, dihydrodipicolinate, which results from the condensation of pyruvate and aspartic semialdehyde immediately drew our attention to the possibility that the lysine pathway might be supplying the precursor of DPA. As all of our work on the lysine pathway had been done with *E. coli*, an organism that does not sporulate, we were forced to become acquainted with *B. megaterium* in order to evaluate this attractive possibility. It was possible to show that extracts prepared from the sporulation phase but *not* from vegetative cells would markedly accelerate the conversion of enzymatically generated dihydrodipicolinate to DPA.[11] However, in the course of these experiments, we became painfully aware of how easily chemically synthesized dihydrodipicolinate would spontaneously convert to DPA. It was not until the experiments of Fukuda,[4] who showed that mutants lacking the pyruvate–aspartic semialdehyde condensing enzyme, but not those blocked at a later point in the lysine pathway, would fail to sporulate unless supplied with an exogenous source of DPA, that the proof of the involvement of the lysine pathway in DPA synthesis could be regarded as secure. The lysine pathway represents one of the few exceptions to the "Unity of Biochemistry". The intermediates of the bacterial pathway, which also operates in blue-green algae, water molds and higher plants, are entirely different from those that participate in the formation of lysine in many fungi. It is not without interest therefore that in bacteria the lysine pathway should, in addition to its primary function, serve as a source of two materials that occur in and are so characteristic of the schizomycetes, i.e. DPA and diaminopimelate.

4. ENZYMATIC SYNTHESIS

The observation that extracts prepared from sporulating organisms can catalyze the conversion of dihydrodipicolinate to DPA[11] was readily confirmed in several laboratories.[12,13] However, the path from that position to a highly purified well-characterized enzyme appears to be a thorny one. Not only is the precursor exceedingly labile, but the enzyme appears to share this attribute as well (though it may be the rich content of proteases, with which sporulating extracts abound, rather than intrinsic lability that is confounding the investigator here). In view of these difficulties, it is not too surprising that one laboratory has presented evidence for a stimulation of the synthesis by DPN,[12] while another group swears by flavin.[13]

Quite recently we have succeeded in securing a 100-fold purified preparation of the DPA synthetase. This material rather than being a source of unalloyed joy has instead been greeted with dark suspicion and is being examined as though it could only be some subtle artifact. Clearly, this veil has not really lifted and though the promise of insights into such questions as the electron acceptor in dipicolinate synthesis and the mechanism of regulation heightens interest, at the moment, it is more a source of frustration.

5. LOCALIZATION

At first glance, the question of deciding where a material that comprises 10–15% of the object under study is situated, would not appear to be particularly difficult. However, when the object is on the order of 1 μm in diameter and sectioning the specimen facilitates solubilization of the material, the dimensions of the problem begin to take on an awesome aspect. Nonetheless, it is clear that any speculations on the function of DPA have to start with a discussion of its relation to its immediate environment be that the core, cortex or coat regions of the spore. (We can dismiss the fourth major anatomical region of the spore, the exosporium, as a likely site since all endospores contain DPA though many lack an exosporium.)

When we became interested in the problem of localization and examined the literature in this area, the difficulty was not that there was a lack of papers dealing with the subject, it was rather that there were too many solutions. The core, cortex or coat localization each appeared to have near equal numbers of adherents though the cortex was emerging as the popular favorite.[14] Entering this arena seemed a thankless task since at best one would be proving correct those that had prematurely staked their claims. However, we needed to know the answer and there was at least one hole in every argument in the literature.

We decided at the outset that we would not use a method that would require first dismembering the spore and this stricture forced us to invent the technique of β-attenuation. The method depends on the expectation that the closer a tritium-labeled material is to the center of the spore, the less likely that the β-particles will escape the dimensions of the spore and be scored by the surrounding scintillation fluid. It was possible to empirically prove the correctness of this assumption by preparing spores labeled with tritium-marked uracil, diaminopimelate or lysine, added late in the sporulation sequence, to selectively label the core, cortex and coat respectively. The observation of increasing attenuations of 3, 18 and 26% as one placed the marker at positions nearer the spore center, validated the procedure.[15,16]

It was with some delight that we observed the value of 33% attenuation for DPA. This placed the bulk of the DPA unequivocally in the core. Since this is where the nucleic acids and cytoplasmic proteins reside, it puts the DPA in the place where it is most likely to have an exciting function. We hope that this particular aspect of DPA biology has been settled and have noted with increasing satisfaction that the most recent papers have come in exclusively on the core localization side.[17,18] And yet, has this issue really been resolved? What if a minor portion of the DPA and not the bulk of the compound is involved in a critical function?

6. STRUCTURE

The invariant association of approximately equimolar amounts of calcium with DPA in the endospore[19] has led to the automatic expectation that DPA would exist in the spore as the calcium salt. It was therefore a welcome addition to the armamentarium of information on DPA that X-ray analysis of the crystal structure of the calcium salt should have been undertaken and indeed led to a solution of the crystal structure of the trihydrate.[20] While analogous work with proteins and nucleic acids has enormously enhanced our understanding of how these molecules function, to date the detailed knowledge of the structure of CaDPA has provided no clues to biological function. In fact, at the moment new evidence has raised the disturbing and yet intriguing spectre of a novel structure for spore DPA. Laser Raman spectroscopy of intact spores done in collaboration with W. Woodruff and T. Spiro[21] indicates that calcium dipicolinate crystals are not a useful model for the state of DPA within the spore. In its biological environment the carboxyl groups of the DPA are partially polarized and neither the dianion nor the fully protonated acid form is an adequate exemplar for what is going on in the depths of the core.

7. FUNCTION

It is not unexpected that the last veil should hide the most intriguing aspects of DPA physiology, those concerning function. Despite general agreement that we do not know, there is no dearth of suggestions as to what the DPA might be doing.[22] The major proposals center on the two key attributes of spores, heat resistance and dormancy. They include direct theories that require specific interaction with individual spore proteins to yield resistant and inactive conformations and postulates of indirect action in which DPA participates in creating a relatively anhydrous environment that would automatically confer the desired properties on the spore components. Minor proposals have also been advanced such as the one that invokes a role for DPA in electron transport. It is possible that the compound has multiple functions and that, when the dust clears, there will have been merit in many of the proposals.

The finding that DPA is relegated to the core confronts the investigator with the calculation that if this compound sometimes amounts to 15% of the dry weight of the total spore, it may represent nearly half of the material present in the spore core. It appears to me that by narrowing one's objective to encompass only those functions of DPA that are dependent on its spectacular concentration, one can gain in focus for what might be lost in scope. Under such a guideline, it is possible to put aside the observation of the effect on electron transport since clearly the activation can be achieved with minute amounts of DPA. We are left to conclude that what can be accomplished only with high concentrations of DPA is the provision of a unique environment for the spore cytoplasm. The observations, made with the laser Raman spectroscopy, of a special form of DPA, offer encouragement for this thought but no concrete model.

One wishes that the last veil were just a little transparent so that one could have a glimmering of what is to be seen when it is lifted.

BIBLIOGRAPHY

1. MURRELL, W. G. (1967) The biochemistry of the bacterial endospore. *Adv. Microbial Physiol.* **1,** 133–251.
2. GOULD, G. W. and HURST, A. (1969) *The Bacterial Spore*, Academic Press, New York.
3. UDO, S. (1936) Chemical constituents of "Natto", fermented soy-beans. *J. Agric. Chem. Soc. Japan*, **12,** 386–94.
4. FUKUDA, A. and GILVARG, C. (1968) The relationship of dipicolinate and lysine biosynthesis in *Bacillus megaterium*. *J. Biol Chem.* **243,** 3871–3876.
5. TANENBAUM, S. W. and KANEKO, K. (1964) Biosynthesis of dipicolinic acid and of lysine in *Penicillium citero-vivide*. *Biochem. A.C.S.* **3,** 1314–1322.

6. Hudson, P. H. and Darlington, W. A., Process for preparing dipicolinic acid. U.S. Pat. 3 334 021.
7. Perry, J. J. and Foster, J. W. (1955) The biosynthesis of dipicolinic acid by spores of *Bacillus cereus* var. *mycoides*. *J. Bacteriol.* **69**, 337–346.
8. Pitel, D. W. and Gilvarg, C. (1970) Mucopeptide metabolism during growth and sporulation in *Bacillus megaterium*. *J. Biol. Chem.* **245**, 6711–6717.
9. Finlayson, A. J. and Simpson, F. J. (1961) The conversion of 2,6-diaminopimelic acid-1,7-^{14}C to lysine-1-^{14}C by certain bacteria. *Can. J. Biochem. Physiol.* **39**, 1551–1558.
10. Benger, H. (1962) Zur Biosynthese der Dipicolinsaure. *Z. Hyg. Infektskr.* **148**, 318–344.
11. Bach, M. L. and Gilvarg, C. (1966) Biosynthesis of dipicolinic acid in sporulating *Bacillus megaterium*. *J. Biol. Chem.* **241**, 4563–4564.
12. Chasin, L. A. and Szulmajster, J. (1967) Biosynthesis of dipicolinic acid in *Bacillus subtilis*. *Biochem. Biophys. Res. Commun.* **29**, 648–654.
13. Zytkovicz, T. H. and Halvorson, H. O. (1972) Some characteristics of dipicolinic acid-less mutant spores of *Bacillus cereus*, *Bacillus megaterium* and *Bacillus subtilis*. In *Spores*, V (H. O. Halvorson, R. Hanson and L. L. Campbell, eds.), Amer. Soc. Microbiol, pp. 49–52.
14. Murrell, W. G., Ohye, D. F. and Gordon, R. A. (1969) Cytological and chemical structure of the spore. In *Spores*, IV (L. L. Campbell, ed.), Amer. Soc. Microbiol., pp. 1–19.
15. Leanz, G. F. and Gilvarg, C. (1972) Localization of bacterial spore components by beta-attenuation analysis. In *Spores*, V, *ibid*, pp. 45–48.
16. Leanz, G. F. and Gilvarg, C. (1973) Dipicolinic acid location in intact spores of *Bacillus megaterium*. *J. Bact.* **114**, 455–456.
17. Hitchins, A. D., Greene, R. A. and Slepecky, R. A. (1972) Effect of carbon source on size and associated properties of *Bacillus megaterium* spores. *J. Bact.* **110**, 392–401.
18. Germaine, G. R. and Murrell, W. G. (1974) Use of ultraviolet radiation to locate dipicolinic acid in *Bacillus cereus* spores. *J. Bact.* **118**, 202–208.
19. Walker, H. W., Matches, J. R. and Ayres, J. C. (1961) Chemical composition and heat resistance of some aerobic bacterial spores. *J. Bact.* **82**, 960–966.
20. Strahs, G. and Dickerson, R. E. (1968) The crystal structure of calcium dipicolinate trihydrate. *Acta Cryst.* **4**, 571–578.
21. Woodruff, W. H., Spiro, T. G. and Gilvarg, C. (1974) Raman spectroscopy *in vivo*: evidence on the structure of dipicolinate in intact spores of *Bacillus megaterium*. *Biochem. Biophys. Res. Commun.* **58**, 197–203.
22. Halvorson, H. and Howitt, C. (1961) The role of DPA in bacterial spores. In *Spores*, II (H. O. Halvorson, ed.), Burgess Publ. Co., Minneapolis.

ENZYMIC MECHANISMS INVOLVED IN THE SYNTHESIS OF YEAST CELL WALL

R. SENTANDREU AND J. R. VILLANUEVA

Departmento de Microbiologia
Facultad de Ciencias, C.S.I.C.
Universidad de Salamanca, Spain

Bacteria, fungi and plant cells are enclosed in a rigid envelope which protects the protoplast and which is called "cell wall". The form and even the cell viability are dependent upon the "cell wall".

Chemical, cytological, genetic and other studies carried out in the last few years have brought important changes in the concept of the biological role of the cell wall. Today we do not regard cell walls as mere inert structures; they are organelles which show specific turnover and support certain catalytic processes related to cell economy and morphogenesis.

STRUCTURE OF *SACCHAROMYCES CEREVISIAE* WALL

The structure of cell walls is similar at different biological levels. They contain structural polymers which, even though they may represent a small percentage of the whole wall, are responsible for its mechanical strength and also for the cell morphology. These polymers form the framework upon which the soluble polymers are accreted.

In *Saccharomyces cerevisiae* the most abundant structural polymer consists of glucose units joined mainly by β $(1 \rightarrow 3)$ linkages. This compound was wrongly called "yeast cellulose" by Salkowski as early as 1894. Manners and colleagues[1] have proposed the structure shown in Fig. 1. The main component—about 85%—is a branched β $(1 \rightarrow 3)$ glucan of about 240,000 daltons containing 3% of β $(1 \rightarrow 6)$ glucosidic interchain linkages. The minor component—about 15%—is a branched β $(1 \rightarrow 6)$ glucan.

The β $(1 \rightarrow 3)$ glucan is very insoluble and forms a net resistant to physical and chemical agents due to cross-linking of its molecules. Evidence that the β-glucan is the structural polymer of *Sacch. cerevisiae* walls was obtained when it was found that after extraction of the soluble compounds of wall by alkaline and acid treatment the residue was the β-glucan.

Some structural polymers such as the true cellulose of plant cells show a crystalline structure but the yeast β-glucan does not because branching distorts the regularity of the structure.

A structural polymer also present in small quantities is chitin, formed by repeating units of *N*-acetyl-glucosamine, localized at the neck between mother and bud cells. The synthesis and control of this polymer has been thoroughly studied by Cabib and his colleagues.[2]

The β-glucan is embedded into a gel of high viscosity formed by different substances. The main component of this gel is glycoprotein(s) which is called "yeast mannan". Though

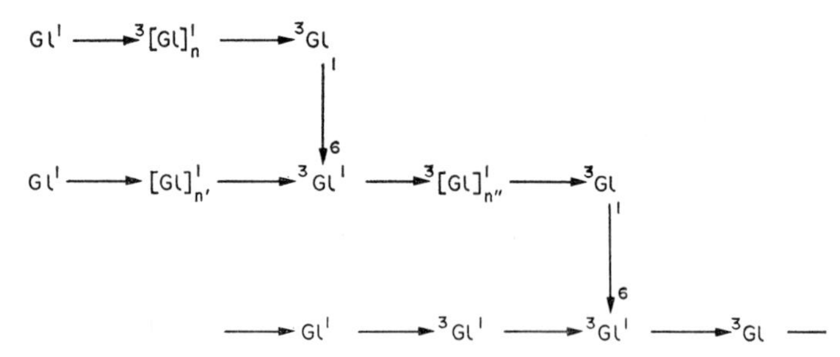

FIG. 1. Structure of the yeast wall β (1 → 3) glucan. n + n′ + n″ represent about sixty glucose molecules.

the glycosidic moiety of the molecule has been studied by several workers (for a review see ref. 3), the structure of this macromolecule was first proposed in 1968[4] (Fig. 2).

The peptide moiety is rich in threonine and serine and one type of link present is *O*-glycosidic between mannose and oligosaccharides and those aminoacids. Another type of link connects the high molecular weight highly branched mannan and the protein probably via an *N*-glycosidic bridge between *N*-acetylglucosamine and aspartamide. Recently Ballou and coworkers have shown that this last linkage involves an *N*-acetyl chitobiose, a disaccharide also often found in glycoproteins of animal origin.

The protein moiety of this molecule may act as a bridge between several mannan subunits giving rise to a polymer with a very high molecular weight.[5]

In yeast walls there are several exoenzymes which are also glycoproteins with mannan as a part of their molecule. Are all the glycoproteins of the wall enzymes? We do not really know the answer to this question. We know that some are inducible enzymes which are synthesized only when the cells need them. When *Sacch. cerevisiae* is grown in the presence of glucose and sucrose, the cells do not produce invertase, the enzyme that breaks down sucrose into its components, glucose and fructose. When the glucose concentration goes down, the cells unable to utilize sucrose synthesize invertase and the glucose and fructose produced by the hydrolysis of sucrose allow them to go on growing.

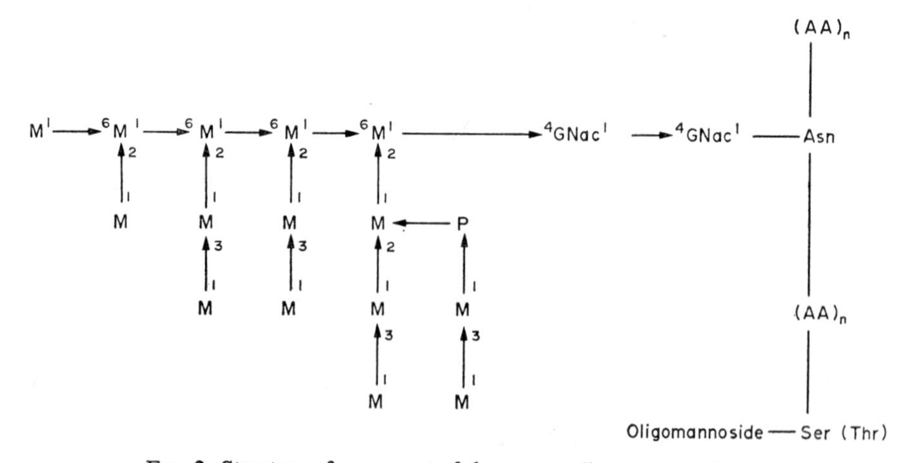

FIG. 2. Structure of a segment of the yeast wall mannoproteins.

BIOSYNTHESIS OF *SACCHAROMYCES CEREVISIAE* WALL: FORMATION OF WALL GLYCOPROTEIN(S)

The initial studies carried out on the biosynthesis of yeast wall were directed towards learning the mechanisms of sugar polymerization *in vivo*. Bretthauer *et al.*[6] fed *Hansenula holstii* cells with radioactive glucose or mannose and found that the extracellular mannan was produced from either sugar.

Algranati *et al.*[7] demonstrated that a membrane preparation obtained from lysed protoplasts was able to catalyze the formation of a substance that behaved chemically like yeast mannan. The immediate precursor was shown to be an activated sugar molecule, guanosine diphosphate-mannose.

Once the structure of mannan was established to be manno-protein(s) we studied its biosynthesis by following the formation of both the peptide and carbohydrate moieties. In these studies one of the most powerful techniques is the dissection of a metabolic route by inhibiting specific reactions with antimetabolites. In the case of mannoproteins the two moieties are quite different and are polymerized by different biosynthetic routes. We inhibited protein formation with the antibiotic cycloheximide which "freezes" translation of the genetic message at the level of polysomes.

Formation of the carbohydrate moiety can be blocked with 2-deoxyglucose. Invertase-synthesizing cells, after addition of 2-deoxyglucose, pile up molecular forms of the enzyme with a low content of mannan. These results indicate that an inhibition of glycosylation and secretion processes occurs in the presence of the drug.[8]

Addition of cycloheximide to a yeast culture results in almost complete inhibition of protein synthesis and after a short period incorporation of glycoproteins into the cell wall is also inhibited. Synthesis of yeast glucan takes place normally. These results suggest that the inhibition of protein synthesis by the antibiotic leads to a decrease in the concentıation of mannose acceptor molecules, but glycosylation of those already present takes place normally. That explains the continued incorporation of mannoproteins into the wall after inhibition of protein synthesis.

We can then conclude that the biosynthesis of wall glycoproteins requires as a primary step protein formation and specifically synthesis of the mannose acceptor proteins. It also requires formation of the enzymic apparatus which catalyzes polymerization of aminoacids and carbohydrates.

FORMATION OF THE CARBOHYDRATE MOIETY OF WALL GLYCOPROTEIN(S)

Polysaccharide formation is carried out by sequential addition of glycosidic residues (one or more monosaccharide molecules) at a time to different acceptors. Glycoprotein formation is also carried out similarly, but how is the first sugar resiaue introduced? It is thought that a spatial and temporal union between protein and glycosylation mechanisms must be established.

To obtain information about the mechanism of incorporation of the first sugar residue many research workers have followed the kinetics of sugar incorporation *in vivo*. After incubation the cells are broken, fractionated and the precursor–product relationship between the fractions determined.

We incubated yeast cells with short pulses of radioactive sugars, then the cells were broken by shaking them with glass beads in a Braun homogenizer and the suspension fractionated as indicated in Fig. 3. Radioactivity was found mainly in the particulate preparation sedimented between 10,000 g and 40,000 g. When longer incubation times were used radioactivity was found in other fractions, but walls were preferentially labeled. This procedure does not permit studying incorporation of sugars at the level of polysomes because they are destroyed during cell breakage.

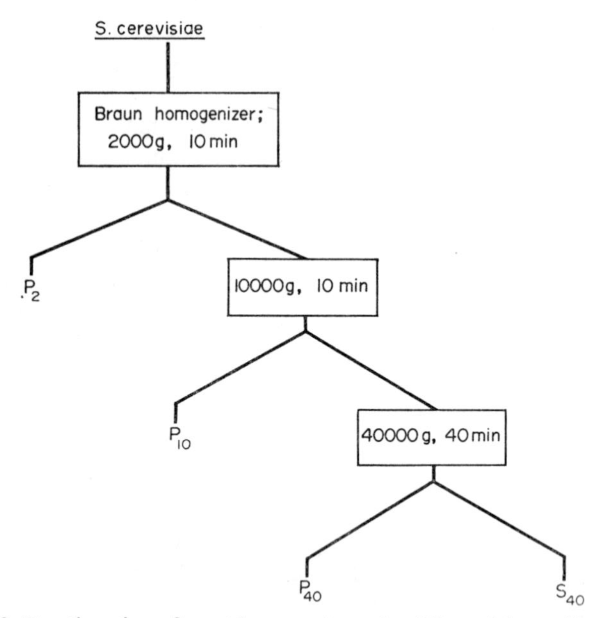

FIG. 3. Fractionation of yeast homogenizates by differential centrifugation.

To study incorporation of sugar at the polysome level—in collaboration with Dr. J. Ruiz-Herrera of the Instituto Politécnico Nacional of Mexico, D. F.—we repeated the experiments using protoplasts. Protoplasts are cells whose walls have been removed by solubilization of wall components with lytic enzymes. Protoplasts are stable only when the osmotic pressure of the surrounding medium is the same as that of the cytoplasm (for a review see ref. 9.)

We incubated protoplasts with radioactive mannose for 30 seconds, then they were lysed and fractions isolated by differential centrifugation in a discontinuous sucrose gradient. The smooth endoplasmic reticulum incorporated most of the radioactivity whereas only a small amount appeared in the rough endoplasmic reticulum.[22]

To determine if glycosylation occurs on ribosomes and at the level of nascent polypeptides, we pulse-labeled whole cells with radioactive mannose, stopped the labeling with cycloheximide to "freeze" the polysomes and then prepared protoplasts from the labeled cells. A significant amount of radioactivity was found in the isolated ribosomal fraction.

Evidence that sugars were incorporated into nascent polypeptide chains was obtained by use of puromycin. Puromycin blocks protein synthesis by functioning as an analogue of amino acylated-tRNA and substitutes for the incoming amino acyl-tRNA as the acceptor of the activated peptide. Addition of puromycin to a cell-free system leads to the release of the nascent polypeptides.

Protoplasts obtained from cells pulse-labeled with either [U-^{14}C]glucose or [U-^{14}C] mannose were lysed and treated with puromycin. The antibiotic-treated cells showed a drop in the radioactivity associated with the polysomes and the absorbancy at 254 nm was also lower than the control.[22]

These results indicate that the initial glycosylation of the glycoprotein(s) of *Sacch. cerevisiae* wall takes place at the level of the nascent polypeptide.

Surprisingly we found that our polysomes contained radioactive glucose and at the time we did not know whether this result was an indication that there are glucose-containing proteins in yeast. This now seems to be the case since recently Eugenio Santos, an undergraduate student of our laboratory, working with a wall material called Fraction B[10] which is alkali-soluble, water-insoluble has found that treatment with 0.1 *M* NaOH results in its almost instantaneous solubilization and the extinction at 241 nm of the solution increases with time.

The increase in the extinction at this wavelength is related to the formation of α-aminoacrylic acid. The ultraviolet spectrum of α-aminoacrylic acid has been shown to have a maximum at 241 nm. This result could be explained if it is assumed that a nucleophilic attack (β-elimination) of the alkoxide moiety (sugar residue) of Fraction B occurred to give a dehydro derivative (Fig. 4). The fact that glucose is released by alkaline treatment sug-

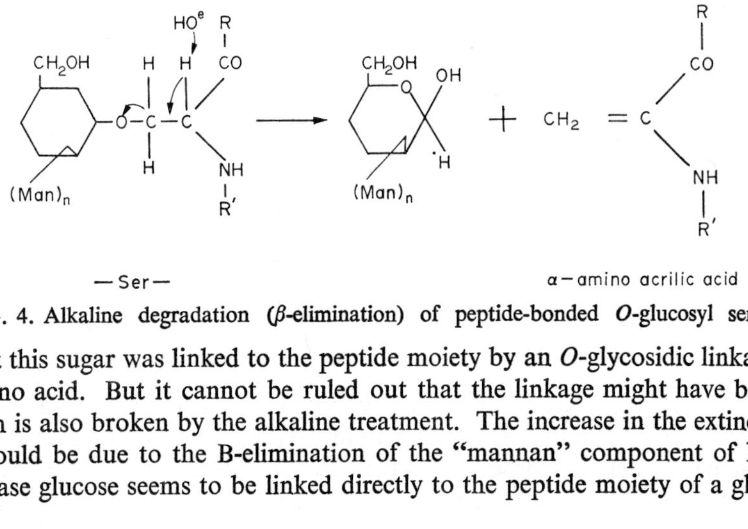

FIG. 4. Alkaline degradation (β-elimination) of peptide-bonded *O*-glucosyl serine.

gested that this sugar was linked to the peptide moiety by an *O*-glycosidic linkage to a hydroxy amino acid. But it cannot be ruled out that the linkage might have been an acyl ester which is also broken by the alkaline treatment. The increase in the extinction of the solution could be due to the B-elimination of the "mannan" component of Fraction B. In either case glucose seems to be linked directly to the peptide moiety of a glycoprotein.

INCORPORATION OF MANNOSE IN THE SMOOTH ENDOPLASMIC RETICULUM

As discussed previously, our experiments *in vivo* suggested that the microsomal membranes are the sites of bulk mannose incorporation. To check this point we decided to determine the location of the mannosyltransferases by following the incorporation of mannose from GDP-mannose in fractions obtained by differential centrifugation of the broken cell (Fig. 3). The largest amount of radioactivity appeared in the Fraction P_{40} which was also the fraction most heavily labeled *in vivo*. However, as only endogenous acceptors were present in these experiments, the presence or absence of enzymes in other fractions could not be evaluated with certainty. The origin of the enzymatic particulate preparation seems to be intracytoplasmic.

ROLE OF LIPIDS IN GLYCOPROTEIN SYNTHESIS

Mannose incorporation into the glycoprotein takes place initially on polypeptide acceptors at the level of ribosomes but the massive incorporation occurs in a cytoplasmic particulate preparation. The fact that the transfer activity is found in a particulate system was the basis for the discovery that membrane lipids play an essential role in glycoprotein synthesis.

There are two mechanisms by which lipids may participate in enzymic systems. Glycerophosphatides act as enzymic cofactors modifying protein structure but they do not participate directly in catalytic processes. Other lipids are acceptors of a substrate in a reaction and act as donors of it in a second one. These lipids are called "lipid carriers" and have been characterized as phosphates or pyrophosphates of isoprenoid alcohol. The fact that these lipids are found in particulate preparations suggests that they may give the sugars the lipophilic characteristics needed to pass through membranes.

The participation of polyprenol sugar phosphates as intermediates in the biosynthesis of several types of complex carbohydrate has been clearly established in bacteria (for a review see ref. 11). Tanner in 1969[5a] presented the first evidence of the possible participation of lipid carriers in yeast. This was confirmed by us and the carrier was found to be a phosphodiester of mannose and an isoprenoid alcohol.[12] Finally, Jung and Tanner[13] purified a lipid which was able to accept mannose from GDP-mannose when added exogenously to a particulate preparation. This lipid was characterized as a dolichol phosphate.

In a very elegant study Anatalis and collaborators[14] have demonstrated by genetic analysis that yeast mannan is built up by the combined activity of specific enzymes. In other words, mannan is made by stepwise addition of single mannose residues to the growing chain and each addition is carried out by different transfer enzyme. This mechanism of synthesis has also been confirmed in our laboratory and others.

SECRETION OF GLYCOPROTEINS

Our studies *in vivo* showed that labeled glycoproteins appeared in the wall without any perceptible time lag. When the cells are grown in the presence of cycloheximide, a decline in the radioactivity which is incorporated in the wall takes place immediately and ultimately the incorporation is brought to a halt. The time between complete inhibition of protein synthesis and mannose incorporation into the wall is thought to be the time needed to build up the sugar chains and to secrete the glycoproteins.

The secretion takes place by formation of small vesicles probably derived by proliferation of the endoplasmic reticulum.[15] This has been confirmed particularly during cell division, at which time the requirements for wall materials are very large.[16]

SYNTHESIS OF WALL GLUCAN

For many years, people interested in yeast have been addressing themselves to the possible mechanisms of glucan synthesis. We tried several times using all kinds of tricks, but failed. One day, we came across a paper from Prof. Sols' laboratory[17] in which it was proposed that studies of regulatory mechanisms of proteins in yeast might be carried out *in situ*. What they did was to destroy the permeability barrier by shaking cells in the presence of small amounts of toluene-ethanol. This treatment produces small openings in the plasmalemma which allow free access of small molecules into the cytoplasm. The enzymes are too large to escape and remain inside.

We talked to the people in the laboratory and Dr. M. V. Elorza decided to try incorporation of glucose from UDP-glucose by toluenized cells. Everybody was rather reluctant to do such as experiment but when it was tried we found that the cells were labeled. At first it was thought that the label was in yeast glycogen, but after extraction of the preparation with alkali and acid and digestion with amylase, the label was still insoluble. Partial hydrolysis of the preparation revealed the presence of radioactive substances which co-chromatographed with glucose, laminaribiose (a β $(1 \rightarrow 3)$ linked glucose disaccharide) and possibly higher lamina-saccharides. These results suggested that at least part of the glucose transferred to endogenous acceptor(s) *in situ* had been incorporated into β $(1 \rightarrow 3)$ glucan found in the yeast wall or into its precursors.[18]

The subcellular location of the radioactive material incorporated was studied by differential centrifugation. We found that the radioactivity was mainly found in the particulate preparation $P_{10} + P_{40}$ (Fig. 2). This particulate preparation ($P_{10} + P_{40}$) when incubated with UDP—[U—^{14}C]glucose in an *in vitro* system did not incorporate label into glucan. The reason for this negative result is unknown but during breakage of the cells some cofactor may be lost or the structure of the enzymic system may be changed.

We also carried out a search for a lipid carrier which might be involved in the synthesis of glucan but no traces of it could be found, either in the *in situ* system or in the particles *in vitro*.

We postulate that the β-glucan of the yeast wall may be synthesized as follows: glucan subunits are built up in the endoplasmic reticulum and/or on vesicles in the cytoplasm and transported to the sites of wall growth. These vesicles fuse with the plasmalemma by reverse pinocytosis to produce external glucan subunits. These subunits, once on the outside of the cell, give rise to the three-dimensional framework of the wall. Activity of hydrolytic enzymes may be needed to produce openings in the structural glucan mesh. Once these have been made, they may be widened by turgor pressure. The subunits may now be translocated to the gap producing an extension of the wall without any increase in thickness.

REGULATION OF WALL FORMATION

The yeast envelope is a rigid structure and can be considered as one of the largest molecules of the cell. At the same time, it is a dynamic structure. Progress toward an understanding of the biochemistry and biology of the wall is crucial: we cannot study the structure and biosynthesis independently of the control mechanisms involved in the polymerization of the components and also of the interactions which take place during assembly of the cell wall.

Cell division in yeast is a polarized process so it is not only a problem of quantitative synthesis but also of topological location and coordination of all kinds of related processes.

These considerations emphasize the importance of the regulation of enzyme activities, formation and location of glucan and mannan synthesis, and their interrelationships with other metabolic processes of the cell. We now consider different levels at which the cells may control wall morphogenesis.

(a) Genetic level

In synchronized cultures glucan and mannan are exponentially synthesized during the cell division cycle.[19] Halvorson and collaborators[20] have proposed that the genes of each chromosome are transcribed in an orderly and linear way. This hypothesis suggests that in *Sacch. cerevisiae* every gene is transcribed only once every cell cycle. If this is true, the continuous synthesis of wall polymers could be due to: (a) the presence of several duplicated copies of the gene involved, distributed throughout the cell genome. This could result in the continuous transcription of these genes. (b) The genes being present in small number but transcribed constantly, without "linear reading". (c) The RNA messengers and (or) the glucan and mannan synthetases showing slow turnover.

Recent studies carried out in our laboratory suggest that the mRNA for the mannoprotein might have a longer half-life than the average mRNA for cellular proteins. On the other hand, inhibition of protein synthesis with cycloheximide does not interfere with glucan formation indicating that glucan synthetase(s) show a slow turnover.

Independently of any control at the level of transcription, the existence of RNA messengers and synthetases of slow turnover would result in the continuous synthesis of both glucan and mannan during the *Sacch. cerevisiae* cell cycle.

(b) Control of enzyme activity

The massive accumulation of GDP-mannose by cells incubated in the presence of cycloheximide suggests that there is no regulation by a feed-back mechanism on the metabolic route that leads to the formation of nucleoside-diphosphate mannose.[21]

(c) Regulation at the lipid level

The existence of several mannolipids in yeast cells points to the possibility that they might control dolichol phosphomannose concentration in a competitive manner or by altering the structure of the environment needed to produce active enzymes.

(d) At the cellular level

As previously indicated, the materials formed in the cytoplasm must be transported to their location in the walls. This transport, at least during cell division, is orientated. The control might be carried out by microfilaments or by changes in the electric potential which would result in a channeling of the vesicles towards specific points on the cell surface. An interesting fact intimately related is that in the presence of inhibitors of protein and RNA synthesis the overall rate of glucan synthesis is virtually unchanged but the topology of the deposition process may be altered. Perhaps this is a consequence of uncoordinated synthesis of glucan and manno-proteins. The relationship between the structural glucan and the glycoproteins of the wall is unknown but it may be that other substances present in low concentration in the wall act as bridges between them. These substances could be responsible for the vectorial deposition of both glucan and mannoproteins and hence for the morphogenesis of the *Sacch. cerevisiae* cell wall.

REFERENCES

1. MANNERS, D. J., MASSON, A. J. and PATTERSON, J. C. (1973) The structure of a β (1 $\rightarrow$ 3)-D-glucan from yeast cell walls. *Biochem. J.* **135,** 19–30.
2. CABIB, E., ULANE, R. and BOWERS, B. (1974) A molecular model for morphogenesis. The primary septum of yeast. In *Current Topics in Cellular Regulation* (B. L. HORECKER and E. R. STADTMAN, eds.), vol. **8,** pp. 2–32, Academic Press, New York and London.
3. PHAFF, H. J. (1971) Structure and biosynthesis of the yeast cell envelope. In *The Yeast* (A. H. ROSE and J. S. HARRISON, eds.), vol. **2,** pp. 135–210, Academic Press, London.
4. SENTANDREU, R. and NORTHCOTE, D. H. (1968) The structure of a glycopeptide isolated from the yeast cell wall. *Biochem. J.* **109,** 419–432.
5. THIEME, T. R. and BALLOU, C. E. (1972) Subunit structure of the phosphomannan from *Kloeckera brevis* yeast cell wall. *Biochemistry*, **11,** 1115–1120.

5a. TANNER, W. (1969) A lipid intermediate in mannan biosynthesis in yeast. *Biochem. Biophys. Res. Commun.* **35,** 144–150.

6. BRETTHAUER, R. K., VILKEN, D. R. and HANSEN, R. G. (1963) The biosynthesis of guanosine-diphosphate mannose and phosphomannan by *Hansenula holstii*. *Biochem. Biophys. Acta*, **78,** 420–429.
7. ALGRANATI, I. D., CARMINATTI, H. and CABIB, E. (1963) The enzymatic synthesis of yeast mannan. *Biochem. Biophys. Res. Commun.* **12,** 504–509.
8. MORENO, F., OCHOA, A. G., GASCÓN, S. and VILLANUEVA, J. R. (1975) Molecular forms of yeast invertase. *Eur. J. Biochem.* **50,** 571–579.
9. VILLANUEVA, J. R. and GARCÍA ACHA, I. (1971) Production and use of fungal protoplasts. In *Methods in Microbiology* (BOOTH, C. ed.), vol. **4,** pp. 665–718, Academic Press, London.
10. KORN, E. D. and NORTHCOTE, D. H. (1960) Physical and chemical properties of polysaccharides and glycoproteins of the yeast-cell wall. *Biochem. J.* **75,** 12–17.
11. HEATH, E. C. (1971) Complex polysaccharides. *Ann. Rev. Biochem.* **40,** 29–56.
12. SENTANDREU, R. and LAMPEN, J. O. (1972) Biosynthesis of mannan in *Saccharomyces cerevisiae*. Isolation of a lipid intermediate and its identification as a mannosyl-1-phosphoryl-polyprenol. *FEBS Lett.* **27,** 331–334.
13. JUNG, P. and TANNER, W. (1973) Identification of the lipid intermediate in yeast mannan biosynthesis. *Eur. J. Biochem.* **37,** 1–16.
14. ANTALIS, C., FOGEL, S. and BALLOU, C. E. (1973) Genetic control of yeast mannan structure. Mapping the first gene concerned with mannan biosynthesis. *J. Biol. Chem.* **248,** 4655–4659.
15. MOOR, H. (1967) Endoplasmic reticulum as the initiator and bud formation in yeast (*S. cerevisiae*). *Archiv. Mikrobiol.* **57,** 135–146.
16. SENTANDREU, R. and NORTHCOTE, D. H. (1969) The formation of buds in yeast. *J. Gen. Microbiol.* **55,** 393–398.
17. SERRANO, R. GANCEDO, J. M. and GANCEDO, C. (1973) Assay of yeast enzymes "in situ": A potential tool in regulation studies. *Eur. J. Biochem.* **34,** 479–482.
18. SENTANDREU, R., ELORZA, M. V. and VILLANUEVA, J. R. (1975) Synthesis of yeast wall glucan. *J. Gen. Microbiol.* **89** (in press).
19. SIERRA, J. M., SENTANDREU, R. and VILLANUEVA, J. R. (1973) Regulation of wall synthesis during *Saccharomyces cerevisiae* cell cycle. *FEBS Lett.* **34,** 285–290.
20. HALVORSON, H. O., CARTER, B. L. A. and TAURO, P. (1971) Synthesis of enzymes during the cell cycle. In *Advances in Microbial Physiology* (A. H. ROSE, and F. J. WILKINSON., eds.), vol. **6,** pp. 47–106, Academic Press, London.
21. SENTANDREU, R. and LAMPEN, J. O. (1970) Biosynthesis of yeast mannan: Inhibition of synthesis of mannose acceptor by cycloheximide. *FEBS Lett.* **11,** 95–99.
22. RUIZ-HERRERA, J. and SENTANDREU, R. (1975) Site of initial glycosylation of mannoproteins from *Saccharomyces cerevisiae*. *J. Bacteriol.* **124,** 127–133.

III. REGULATION

THE PASTEUR EFFECT IN THE ALLOSTERIC ERA

ALBERTO SOLS*

"Une fois engagé dans la voie de l'erreur, il est malaisé d'en sortir."

Louis Pasteur

"The certainty with which various authors claim to have explained the Pasteur effect is almost as general a phenomenon as the effect itself."

Frank Dickens, paraphrasing Dean Burk

The inhibition of glycolysis by respiration, generally known as the Pasteur effect, has been perhaps the longest-standing puzzle in metabolic regulation. It was glimpsed by Pasteur just over a century ago,(1) systematically measured by Otto Meyerhof and christened by Otto Warburg in the 1920s(2) and has been the subject of long and frequently heated controversies from the 1930s through the 1960s. On its mechanism there is no clear agreement even now. Tentative interpretations have been almost as varied as the backgrounds of those who have aimed at its unraveling. Outstanding claims along time have tended to reflect the prevailing trends in the development of biochemistry. Hence it is now not surprising that in a complex problem of metabolic regulation no definite identification of the real mechanism could be attained before the emergence of the allosteric era. The allosteric concept, i.e. the specifically induced conformational changes which enable certain enzymes to act as chemical transducers, has opened a third dimension in enzymology and definitely involves a change of paradigm for the interpretation of metabolic regulation at the enzyme level. As Pasteur himself said "la science vit des solutions successives données à des *pourquoi* de plus en plus subtils, de plus en plus rapprochés de l'essence même des phenomènes".

A correct elucidation of the mechanism of the Pasteur effect has been hindered by the fact that there are significant differences in the pathways of glucose degradation among the main organisms studied and because there has been some confusion on the meaning of the concept itself. In the literature, and frequently in the minds of individual biochemists, there have been at least two different but not clearly distinguished and partially overlapping concepts of the "Pasteur effect". A coarse concept refers to the change from CO_2 producing fermentation, or lactic acid formation, to oxidation, i.e. it refers to the *decreased formation* of the anaerobic carbon by-product. This was basically what Pasteur initially discovered. A more subtle concept emphasizes the fact that in aerobiosis there is a *decreased consumption* of the sugar source. The latter, and harder to explain concept, was clearly expressed, among other distinguished biochemists, by Severo Ochoa in the following terms (1947): "It is well known that, under anaerobic conditions, the rate of consumption of carbohydrate by cells is six to eight times faster than under aerobic conditions. This decreased rate of utilization in the latter case is known as the *Pasteur effect*".

* Instituto de Enzimología del CSIC, Facultad de Medicina de la Universidad Autónoma, Madrid-34, Spain.

In anaerobic glycolysis much of the chemical energy of the sugar is wasted by the cell with the carbon by-product(s). Obviously, if a cell can oxidize glucose completely to carbon dioxide and water it can get much more energetic profit per unit consumed. Hence, it can grow faster, as observed and emphasized by Pasteur. Then some other limitation usually sets in so that the possibility of energy supply when there is plenty of both sugar and oxygen exceeds requirements. As a consequence it is reasonable that whenever oxidation is possible the cell tends to use glucose at a lower rate. In contrast with this obvious reasonableness of the Pasteur effect, understanding of the mechanism through which it is accomplished has been most elusive indeed.

Of the many review articles on the Pasteur effect there are four that deserve particular consideration, two as historical landmarks: those of Burk in 1939[(3)] and Dickens in 1951,[(4)] and two as representatives of the currently prevailing mixture of old and new: those of Hans Krebs[(5)] and Abburi Ramaiah,[(6)] The latter is focused on phosphofructokinase, while that of Krebs is a general treatment full of insights. After the last two reviews two interesting recollections have been published by such distinguished old-timers of the Pasteur effect as Vladimir Engelhardt[(7)] and Efraim Racker.[(8)] These two articles, published together in the same issue of a journal, speak dramatically different languages. Realization of this fact should have a sobering effect on the dangers of overconfidence in this triply complex subject: complex in any given organism, in the variations among organisms, and in its historical precedents (frequently hard to shed: see the opening quotation). If the experts can be so wide apart, it is not surprising that the second-hand treatment of the subject in current textbooks tends to be an undigested mixture of some old fancies and some solid facts, or simply does not attempt an explanation. Two examples taken from first class books are illustrative:

"... despite years and years of many biochemists' careers, no one really knows how glucose consumption is regulated". (1970)

"... glycolysis requires a supply of ADP ... and respiration will keep the ADP level low.... However, the Pasteur effect may not depend entirely on substrate limitations, since numerous interactions between metabolic pathways are now known to involve specific regulatory effects of metabolites on the activity of allosteric enzymes". (1973)

In 1941 Feodor Lynen in Munich[(9)] and Marvin Johnson in Madison[(10)] independently made the most significant proposal among the early attempts to account for the Pasteur effect. They proposed the theory that in aerobiosis there was a lack of phosphate because of the efficient competition of oxidative phosphorylation for the inorganic phosphate required for glycolysis. This phosphate competition hypothesis has been most vigorously championed by Racker.[(8)] A related alternative possibility, competition for the nucleotide phosphate acceptor, was implicit in Johnson's paper, explicitly formulated simultaneously by Fritz Lipmann[(11)] and vigorously supported by Britton Chance on the basis of the great affinity of the oxidative phosphorylation for ADP.

At the time of the earlier attempts to ascertain the mechanism of the Pasteur effect most of the enzymes involved in the utilization of glucose had not yet been identified. For a time it was not even known whether the aerobic utilization of glucose involved a pathway parallel to or subsequent to the anaerobic one. With the full unraveling of the glycolytic pathway, the discovery of the citric acid cycle and the connecting pyruvate dehydrogenase, it became abundantly clear that aerobic utilization of glucose when oxygen is available involves a metabolic branching at the level of pyruvate in the more primitive fermentative pathway, as

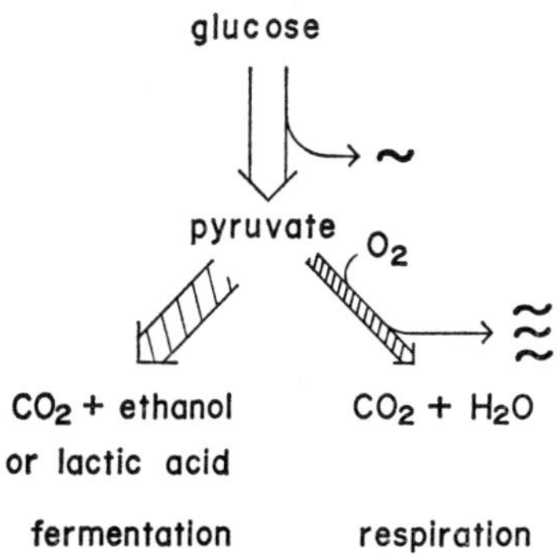

glucose

~

pyruvate O_2

$\approx$

CO_2 + ethanol
or lactic acid

$CO_2 + H_2O$

fermentation respiration

FIG. 1. Metabolic branching at the level of pyruvate when availability of oxygen allows its oxidation. At the branching the width of the arrows indicates relative maximal rates, while the intensity of the shadowing indicates relative apparent affinities for the common substrate. Generation of energy-rich bonds (~) is small in glycolysis to pyruvate and large in the oxidation of the latter.

illustrated in Fig. 1. Helmut Holzer clearly showed in 1961[12] that in yeast simple enzyme competition controls this metabolic branching, the O_2-dependent oxidative pathway having preference because it has much greater affinity for pyruvate than the alternative pyruvate decarboxylase. This enzyme competition at a metabolic branching leads of course to an automatic diminution in aerobiosis of the formation of the anaerobic carbon by-products, but does not affect, by itself, the rate of glucose utilization. In other words, competition could nearly account for the coarse Pasteur effect of decreased formation of carbon by-product, but not for the more basic one of regulation of the rate of glucose utilization.

Along the 1930s and 1940s the redox hypothesis of oxidative inhibition of the key glycolytic enzyme formulated by Lipmann, was apparently confirmed by Engelhardt at the level of phosphofructokinase, but killed by the fact that uncouplers of oxidative phosphorylation can prevent the Pasteur effect leading to aerobic fermentation, as shown by the results of Warburg, Lipmann and others. Oxidation of phosphofructokinase was an artefact, although phosphofructokinase is indeed the key enzyme in the regulation of glycolysis and the Pasteur effect, as we will see below.

Estimation, in the late 1950s, of metabolite levels by Lynen and others gave unequivocal evidence that phosphofructokinase was critically involved in the Pasteur effect. Gratuitous compartmentation hypotheses for ATP were advanced by some investigators in an attempt to account for the decrease in phosphofructokinase activity. Then in the early 1960s an inhibitory site for ATP in phosphofructokinase was postulated by Janet Passonneau and Oliver Lowry[13] and identified in our laboratory as an allosteric site for feedback inhibition of glycolysis[14] (Fig. 2). Nevertheless, in resting cells in the presence of glucose the level of ATP does not consistently change from anaerobiosis to aerobiosis, as reported by Holzer and others. Some interesting experiments by Van Potter and coworkers as early as 1957 had given evidence strongly suggestive of inhibition of phosphofructokinase by an unidentified mitochondrial product thought to be an intermediate of oxidative phosphorylation. In 1965 we identified citrate as the mitochondrial product that increases in aerobiosis (Fig. 3) and inhibits phosphofructokinase in yeast[15] (Fig. 2). Lowry and Philip Randle also found inhibition of the enzyme from animal tissues by physiological concentrations of citrate.

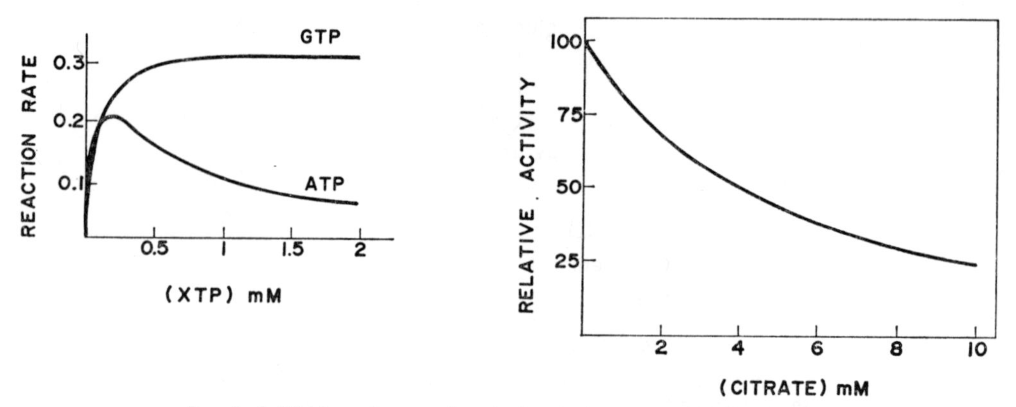

FIG. 2. Inhibition of yeast phosphofructokinase by ATP and citrate.

Citrate is the first product of the aerobic utilization of pyruvate that can permeate out of the mitochondria, and its utilization is feedback controlled by the dependence of the first irreversible enzyme, the NAD-linked isocitrate dehydrogenase on activation by AMP in yeast (as observed by Arthur Kornberg before the allosteric era and later characterized by Daniel Atkinson), and by ADP in animal tissues (as shown by Gerhard Plaut). Thus there is a link between the Krebs cycle, which coupled to the oxidative phosphorylation is the major energy "producer" in the cell, and the pacemaker of the common glycolytic pathway, through the operation of a sequential series of two feedbacks acting through allosteric mechanisms.

Regulation of phosphofructokinase is not enough to account for the inhibition of glucose utilization in aerobiosis. An additional link was needed. And it could not be understood without the realization that there is a glucose phosphorylation pathway that involves two steps: a catalysed transport and an irreversible phosphorylation. Now, control of a pathway can take place either at the level of the first step or at the first irreversible step. Of the organisms more used in studies on the Pasteur effect, in animal cells the glucose phosphorylation pathway is feedback controlled by allosteric inhibition of hexokinase by glucose-6-P,

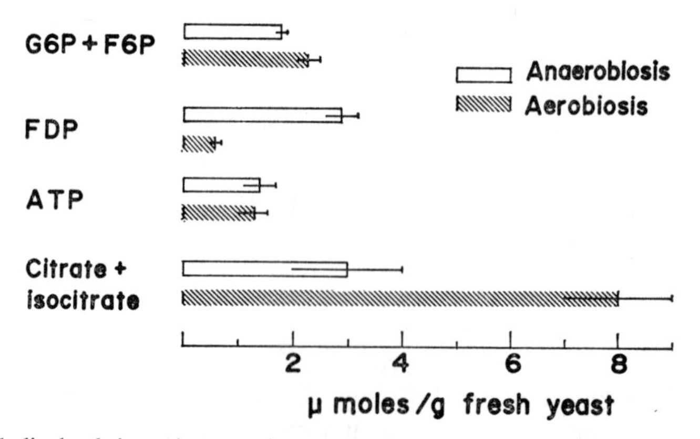

FIG. 3. Metabolite levels in resting yeast in the presence of glucose in anaerobiosis and aerobiosis.

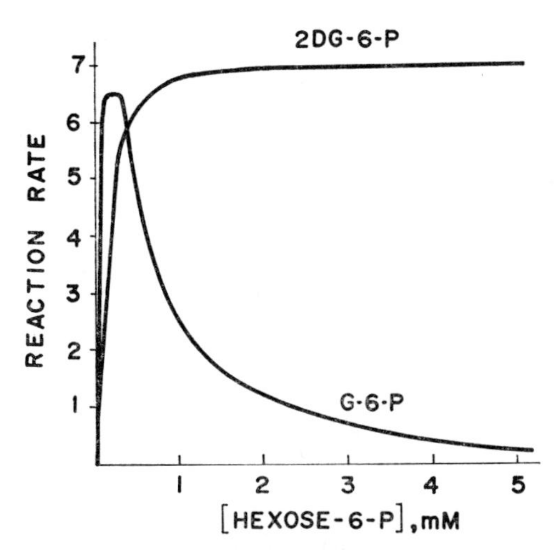

FIG. 4. Allosteric inhibition of an animal hexokinase by glucose-6-P, as shown by the kinetics of the reverse reaction with glucose-6-P and with the analogue 2-deoxyglucose-6-P, which is a good substrate but not an inhibitor.

as shown already in the early 1950s by Robert Crane and myself[16] (Fig. 4), while in the free living yeast hexokinase is not similarly inhibited. In the early sixties, on the basis of indirect evidence I postulated that the constitutive hexose transport in yeast is feedback controlled by the intracellular level of glucose-6-P.† Some direct evidence in support of this hypothesis has been obtained by Arnost Kotyk in Prague. And recently Ramón Serrano and Gertrudis DelaFuente in Madrid have found[18] that in *Saccharomyces cerevisiae* aerobiosis raises the Km for glucose of the constitutive hexose transport, which apparently can occur in two functional states with different affinities for glucose. This would account for the repeated observation that the Pasteur effect is decreased at high concentrations of glucose in short-term experiments; in cultures this inhibition by high glucose is followed by catabolite repression of respiration capability. The fact that it disappears entirely at very low concentrations of sugar is an obvious consequence of the feedback series nature of the Pasteur effect: there is no place for significant feedbacks when the rate of entrance into the system is too low, as happens when the concentration of sugar is well below saturation of the transport.

The Pasteur effect can then now be explained as a consequence of basic allosteric regulatory mechanisms involving different key enzymes of glucose metabolism. These mechanisms make possible a precise and immediate adjustment of the rate of utilization of glucose to the metabolic requirements of the cell for energy and carbon skeletons. Recognition of the fact that overall *glucose degradation generally involves a sequence of two individual pathways in anaerobiosis and three in aerobiosis* is crucial for the understanding of the metabolic regulation of the whole system. These pathways (see Fig. 5) are: (i) from extracellular glucose to glucose-6-P, a major metabolic crossroads, (ii) from glucose-6-P to ATP and a carbon

† The idea that the glucose phosphorylation pathway in yeast was likely to be feedback controlled by glucose-6-P at the level of glucose transport came to me in the spring of 1961 and was advanced for the first time in a Colloquium of Spanish Biochemists chaired by Ochoa in Santander. Afterwards, it took years of work on both hexose transport and hexokinase to allow the move from working hypothesis to formal proposal, which I made in a Symposium on Yeast Metabolism held in Dublin in 1965.[17]

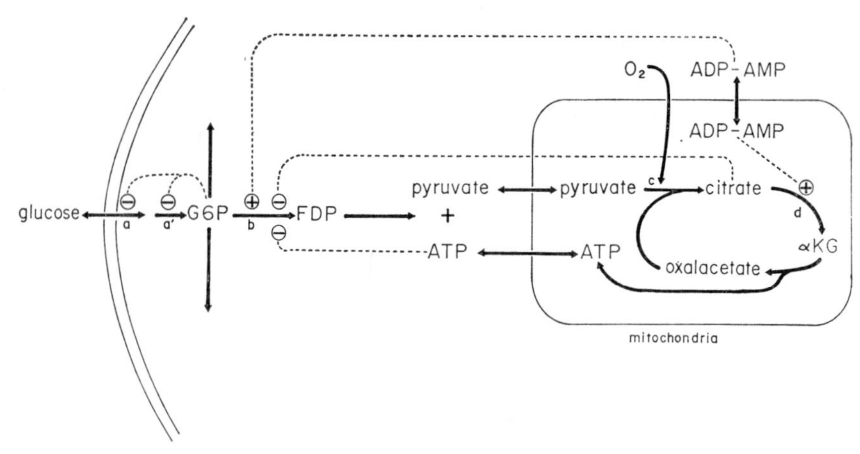

FIG. 5. Allosteric mechanisms for the feedback regulation of glucose degradation and the Pasteur effect. The key steps are: a, glucose transport, feedback regulated in yeast; a′, hexokinase, feedback regulated in animal tissues; b, phosphofructokinase; c, pyruvate dehydrogenase; d, isocitrate dehydrogenase.

by-product in anaerobiosis, or to ATP *and* citrate in aerobiosis, followed in the latter case by (iii) from citrate to more ATP and CO_2 plus water. *Each of these individual pathways can be controlled by allosteric feedback mechanisms affecting either its first step or its first irreversible step. The Pasteur effect involves the integration of a sequential series of feedback mechanisms.*

Within the general principle, the specific regulatory mechanisms implicated in the Pasteur effect in the historically prominent baker's yeast are as follows:[19]

(I) Feedback control of the NAD-dependent isocitrate dehydrogenase, the first irreversible step in the pathway of oxidation of citrate through the citric acid cycle, by its dependence on allosteric activation by AMP for activity in physiological conditions. When the energy level of the cell is high, with a very low AMP/ATP ratio, the slowing down of isocitrate dehydrogenase activity allows the concentration of citrate to increase and facilitates the building up of storage lipids and/or increases the feedback inhibition of the preceding pathway.

(II) Feedback inhibition of phosphofructokinase, the first irreversible step of the common glycolytic pathway from glucose-6-P, by the physiological products of this pathway: ATP in anaerobiosis, and ATP *and* citrate in aerobiosis. The feedback regulation of phosphofructokinase by ATP is admirably reinforced by a strong allosteric activation by AMP. Feedback inhibition of phosphofructokinase raises the concentration of glucose-6-P, favoring the formation of storage polysaccharides and/or feedback inhibiting the preceding pathway.

(III) Allosteric feedback inhibition of the catalysed transport of glucose across the cell membrane, the first step in the glucose phosphorylation pathway, by glucose-6-P, the end-product of this pathway.

Basically irrelevant for the Pasteur effect is the allosteric regulation of the pyruvate kinases of yeast and certain animal tissues, which is not intrinsic to glycolysis, but rather is a device for the benefit of the potential switch over to gluconeogenesis. So is the interconversion system of the pyruvate dehydrogenase of animal tissues, which probably is not intrinsic to cellular metabolism but rather a device for the diversion of pyruvate glycolytically generated in certain tissues for oxidation in other tissues. The very marked glycolytic oscillations observed by Chance and Benno Hess in certain conditions are related to the long feedbacks involved and simulate a multisite control of the common glycolytic pathway that could be physiologically misleading.

The classical hypothesis of substrate level competition for phosphate should be shelved, displaced by the theory of the sequence of allosteric feedbacks, involving the adenylic nucleotides ATP and AMP (or ADP), citrate, and glucose-6-P. The advance to the allosteric era is an irreversible step. The present interpretation of the mechanism of the Pasteur effect is a coherent theory based on specific regulatory mechanisms and represents a decisive achievement that can certainly be subject to refinements but essentially is likely to withstand the test of time. Among the possible refinements it is interesting that very recent findings of Carlos and Juana Mª. Gancedo in Madrid could bring back inorganic phosphate into the Pasteur effect in yeast, although now as a potent allosteric activator of phosphofructokinase in physiological conditions that could contribute to the aerobic inhibition of fermentation when oxidative phosphorylation decreases its level.

An important factor in the maturation of Biochemistry as a modern science was the realization that there is a basic unity in glycolysis, despite variations as obvious as the formation of CO_2 plus ethanol, lactic acid, or glycerol. There is also a basic unity in the regulation of glycolysis, centered in allosteric feedback inhibition of key steps in each of the succeeding pathways, delineated by major metabolic branchings, by its physiological end-products. It appears that Nature has answered to the pressure for economical order with the selection of a coordinable set of functionally equivalent, even if distinctly different, regulatory mechanisms. For these reasons it is possible to make the generalization that there is indeed a mechanism of the Pasteur effect despite the existence of secondary differences: the sequential integration of allosteric feedbacks along the series of pathways involved in glucose degradation.

REFERENCES

1. PASTEUR, L. (1861) Expériences et vues nouvelles sur la nature des fermentations. *Compt. Rend. Acad. Sci.* **52,** 1260–1264.
2. WARBURG, O. (1926) Über die Wirkung von Blausäureäthylester (Äthylcarbylamin) auf die Pasteursche Reaktion. *Biochem. Z.* **172,** 432–441.
3. BURK, D. (1939) A colloquial consideration of the Pasteur and neo-Pasteur effects, *Cold Spring Harbor Symp. Quant. Biol.* **7,** 420–455.
4. DICKENS, F. (1951) Aerobic glycolysis, respiration, and the Pasteur effect. In *The Enzymes* (J. B. SUMNER and K. MYRBACK, eds), vol. 2, part 1, pp. 624–683. Academic Press, New York.
5. KREBS, H. A. (1972) The Pasteur effect and the relations between respiration and fermentation. In *Essays in Biochemistry* (P. N. CAMPBELL and F. DICKENS, eds), vol. 8, pp. 1–34. Academic Press, London.
6. RAMAIAH, A. (1974) Pasteur effect and phosphofructokinase. In *Current Topics in Cellular Regulation* (B. L. HORECKER and E. R. STADTMAN, eds), vol. 8, p. 297–345. Academic Press, New York.
7. ENGELHARDT, W. A. (1974) On the dual role of respiration. *Molec. Cell. Biochem.* **5,** 25–33.
8. RACKER, E. (1974) History of the Pasteur effect and its pathobiology. *Molec. Cell. Biochem.* **5,** 17–23.
9. LYNEN, F. (1941) Über den aeroben Phosphatbedarf der Hefe. Ein Beitrag zur kenntnis des Pasteur'schen Reaktion. *Justus Liebigs Ann. Chem.* **546,** 120–141.
10. JOHNSON, M. J. (1941) The role of aerobic phosphorylation in the Pasteur effect. *Science*, **94,** 200–202.
11. LIPMANN, F. (1941) Metabolic generation and utilization of phosphate bond energy. *Adv. Enzymol.* **1,** 99–162.
12. HOLZER, H. (1961) Regulation of carbohydrate metabolism by enzyme competition. *Cold Spring Harbor Symp. Quant. Biol.* **26,** 227–288.

13. Passonneau, J. V. and Lowry, O. H. (1962) Phosphofructokinase and the Pasteur effect. *Biochem. Biophys. Res. Commun.* **7**, 10–15.
14. Viñuela, E., Salas, M. L. and Sols, A. (1963) End-product inhibition of yeast phosphofructokinase by ATP. *Biochem. Biophys. Res. Commun.* **12**, 140–145.
15. Salas, M. L., Viñuela, E., Salas, M. and Sols, A. (1965) Citrate inhibition of phosphofructokinase and the Pasteur effect. *Biochem. Biophys. Res. Commun.* **19**, 371–376.
16. Crane, R. K. and Sols, A. (1954) The non-competitive inhibition of brain hexokinase by glucose-6-phosphate and related compounds. *J. Biol. Chem.* **210**, 597–606.
17. Sols, A. (1967) Regulation of carbohydrate transport and metabolism in yeast. In *Aspects of Yeast Metabolism* (A. K. Mills and H. A. Krebs, eds), pp. 47–66. Blackwell Scientific Publications, Oxford.
18. Serrano, R. and DelaFuente, G. (1974) Regulatory properties of the constitutive hexose transport in *Saccharomyces cerevisiae. Molec. Cell. Biochem.* **5**, 161–171.
19. Sols, A., Gancedo, C. and DelaFuente, G. (1971) Energy-yielding metabolism in yeast. In *The Yeasts* (A. H. Rose and J. S. Harrison, eds), vol. 2, pp. 271–307. Academic Press, London.

COMPARATIVE ASPECTS OF HORMONAL REGULATION OF ADIPOSE TISSUE LIPOLYSIS

FRANCISCO GRANDE

Instituto de Bioquímica y Nutrición,
Fundación F. Cuenca Villoro, Zaragoza, 6, Spain

In contrast with the traditional view of adipose tissue as a passive deposit of fat, it is now recognized that this tissue is the site of considerable metabolic activity which is influenced by a number of hormones.

The metabolic importance of the adipose tissue was noted in 1928 by the Viennese clinicians Schur and Löw[42] who wrote: "Dass Fettegewebe ist nicht, wie es bisher meist geschieht, als reines and den Stoffwechselvorgängen wenig beteiligtes Depot zu betrachten, sondern steht als Durchgangstation für die aufgenomenen Nahrungsmittel in Zentrum der Stoffwechselvorgänge".* These authors considered that the function of insulin was related to the formation and maintenance of the fat deposit; but it was not until 1958 that the remarkable sensitivity of adipose tissue to insulin action was demonstrated.[49]

Current interest in the study of adipose tissue metabolism has been stimulated by the work of many researchers in various countries. It is not my purpose to give a full account of the historical development of the subject, but a few highlights should be mentioned. Among these: (a) the demonstration of the rapid turnover of stored fat;[41] (b) the publication in 1948 of the classic review on the physiology of adipose tissue by Wertheimer and Shapiro;[47] (c) the identification of the plasma free fatty acids (FFA) as the form in which the triglycerides stored in the adipose tissue are mobilized and made available as fuel to the body tissues,[7,12,29] and the introduction of the concept of "caloric homeostasis".[9]

In 1965 the Section 5, Adipose tissue, of the American Physiological Society *Handbook of Physiology* was published[37] under the editorship of A. E. Renold and G. F. Cahill. This volume gives proof of the enormous progress made by the physiology of the adipose tissue. It contains more than 4000 references, most of them from work done after 1948, and shows that the adipose tissue of young rats performs four main metabolic functions: (1) synthesis of fatty acids and alpha-glycerophosphate from glucose, (2) uptake of fatty acids from extracellular triglycerides, (3) esterification of fatty acids with alpha-glycerophosphate and storage of the triglycerides formed, (4) breakdown of the stored triglycerides with release of FFA and glycerol.

Glucose transport into the fat cell is stimulated by insulin, which has also an antilipolytic effect, thus favoring the accumulation of fat in the adipose tissue. The activity of

* The adipose tissue is not to be considered, as it has been up to now, as a pure deposit of fat with little participation in the metabolic processes, but as a transit station for the ingested foodstuffs which is at the center of the metabolic activity.

the lipase which hydrolyzes the triglycerides, the so-called hormone-sensitive lipase, is stimulated by a number of hormones such as catecholamines, glucagon, TSH, ACTH, and other pituitary peptides.

The question is whether the results obtained in the rat adipose tissue apply to the adipose tissue of other animals. In the introduction to the 1965 volume, Wertheimer showed his awareness of this problem by the following statement: "What may well be an exaggeration is the overemphasis placed on studies of rat adipose tissue. There is enough experimental evidence today to suggest that conclusions based on rat adipose tissue must be interpreted with caution, because such results do not necessarily apply to other mammals. Moreover, epididymal fat, which is used in so many experiments, may have a unique metabolic pattern and cannot be considered as a 'typical' adipose tissue".

In 1967 Rudman and Di Girolamo[40] published a comprehensive comparative study of the physiology of adipose tissue which clearly shows that there are remarkable differences between the adipose tissues from different mammalian species. There is less information in the review regarding the adipose tissue of birds. This class of vertebrates, however, is of interest because birds share with mammals the common property of having a well-developed subcutaneous adipose tissue and that of being homeotherms.

My interest in avian adipose tissue developed in a somewhat devious way, as a consequence of observations about the effect of glucagon on the serum lipids of hyperlipemic human subjects. We found that daily repeated glucagon injection caused a marked decrease of the various lipid fractions (cholesterol, phospholipids, total fatty acids), followed by an elevation when the administration of the hormone was discontinued.[1] This and other observations[3,34] suggested a role of glucagon in the regulation of lipid metabolism, and I decided to look into the question.

In searching for a suitable animal I followed the advice of my teacher August Krogh, who once remarked that if you look carefully you will find that nature has provided an animal particularly well suited for the experiment that you have in mind. Review of the literature indicated that birds could be ideal animals to study the role of glucagon for the following reasons. (a) The glucagon-producing alpha cells of the Langerhans islets are particularly abundant in the splenic lobe of the avian pancreas.[33,34] (b) The glucagon content of the avian pancreas is higher than that of the mammalian pancreas.[45] (c) Early work by Minkowski and his co-workers indicated that in some avian species pancreatectomy is not followed by hyperglycemia, as it is in mammals.[26,32] More recent work by Miahle[30] and others[31] showed that in certain birds pancreatectomy produces hypoglycemia and death in a few hours. This, and the fact that administration of glucagon prolonged the life of the animals, suggested that glucagon is an indispensable hormone for these avian species.

My first experiments with pancreatectomized ducks[2] gave results which were basically in agreement with those reported by Miahle[30] and in view of the limited information about the effect of glucagon on the serum lipids of birds available at the time, I turned my attention to this question. It was then observed that injection of glucagon caused a marked and rapid elevation of plasma FFA in all of the avian species examined and this prompted a study of the effects in birds of other hormones known to affect fatty acid mobilization in mammals.

In this essay I intend to review briefly the results of these experiments, in order to illustrate the differences in the patterns of hormonal control of adipose tissue lipolysis noted among the various species of mammals and birds used. For complete information about methods and original data the reader is referred to our previous publications.[2,10,13–22,36]

EFFECT OF VARIOUS HORMONES ON FFA MOBILIZATION IN THE INTACT ANIMAL

Effect of glucagon

As already noted intravenous injection of glucagon in fasting birds produces an elevation of plasma FFA which reaches a maximum between 5 and 15 minutes, depending on the dose, and is followed by a slower return towards the preinjection level. In adult geese the dose of 100 μg/kg causes a 3- to 4-fold elevation of plasma FFA in 15 minutes, and the level is still twice the preinjection level 2 hours after injection. With the dose of 1.0 μg/kg the maximum FFA level 5 minutes after injection is nearly twice the preinjection level. Two hours after injection the FFA level reaches the preinjection level. There was a straight-line relationship between $\log_{10}$ of the glucagon dose and the FFA response for the doses used[10,13] Comparable results were obtained in other avian species such as ducks (Mallards and Peking), chicken, turkeys and owls.[13,16,17] Continuous intravenous infusion of glucagon caused a prompt elevation of plasma FFA reaching in 10 to 30 minutes a level that remains constant until the end of the infusion, for infusion periods up to 8 hours. The FFA level quickly decreases upon discontinuing the infusion.

The lipolytic activity of the subcutaneous abdominal fat pads of geese, removed before and 5 minutes after glucagon injection, was tested *in vitro*. The lipolytic activity of the fat pads removed after the injection was significantly higher than that of the pads removed before the injection.

The injection of glucagon was followed by an elevation of plasma triglycerides. Infusion of glucagon for 2 hours (0.5 μg/kg/min.) caused the development of fatty liver in ducks and geese. Analysis of the liver lipids demonstrated an increased triglyceride content. Fatty acid analysis showed that the triglyceride deposited in the liver during the glucagon infusion was comparable in fatty acid composition to the triglycerides of the animal's adipose tissue.[19]

Injection of glucagon in the dog caused only a very small and transient elevation of plasma FFA, followed by a marked decrease. This decrease has been attributed to insulin release induced by glucagon, but we and others[18,48] have shown that the effect is not abolished by removal of the pancreas.

Intravenous injection of glucagon (20.0 μg/kg) caused a significant elevation of plasma FFA in the rabbit. This animal shows also an elevation of plasma FFA when glucagon was infused intravenously at rates between 0.1 and 0.5 μg/kg/min.[22,36]

Effect of insulin

It is well known that insulin injection causes a decrease of plasma FFA concentration in mammals like man, dog and rat[37] and this decrease can be obtained with insulin doses too small to affect the blood sugar level. Intravenous injection of insulin at doses up to 2.0 U/kg had no effect on the FFA of geese and owls and did not prevent the elevation of plasma FFA produced by glucagon injection.[14,16] Since the insulin used was crystalline beef–pork insulin the question arises of whether this negative result could be ascribed to the use of heterologous insulin, in spite of the fact that all of the insulin doses used caused significant decrease of blood sugar. However, experiments with chicken insulin kindly donated by Dr. J. Kimmel of Kansas University gave the same results. It was therefore concluded that insulin is devoid of antilipolytic effect in intact birds, and that it does not inhibit the adipokinetic effect of glucagon in these animals.

Effect of catecholamines

The extensive literature about the lipolytic effect of the catecholamines[37,43] should not be reviewed here. Our experiments showed that epinephrine and norepinephrine at doses higher than those producing maximal effects in mammals do not elevate plasma FFA in the duck. Continuous infusion of either epinephrine or norepinephrine failed to produce fatty liver in the duck, in contrast with their effect in dogs.[8,19]

In the intact goose intravenous injection of epinephrine caused a modest, but statistically significant elevation of plasma FFA. Norepinephrine also caused an elevation of FFA in the goose, but the effect was smaller than that of epinephrine and much smaller than the effects reported in the dog. Continuous infusion of norepinephrine caused elevation of plasma FFA in the goose, but the effect was much smaller than that observed in dogs under comparable conditions.

The adipokinetic effect of norepinephrine in the goose was much smaller than that of glucagon. My calculations indicate that, on a molar basis, glucagon is some 1000 times more effective than norepinephrine in elevating the plasma FFA of the goose.[15]

Our experiments and other observations in the literature[5] indicate therefore that catecholamines have no adipokinetic effect in some birds such as duck, turkey and chicken. In other birds like the goose, catecholamines have some adipokinetic effect but their ability to elevate plasma FFA seems to be much smaller than that of glucagon.

HORMONAL EFFECTS ON THE LIPOLYTIC ACTIVITY OF ADIPOSE TISSUE INCUBATED *IN VITRO*

The *in vitro* lipolytic effects of glucagon and epinephrine were tested in tissues from three avian (goose, duck and owl) and three mammalian species (dog, rat, rabbit). In addition, the effects of insulin and nicotinic acid on the glucagon-induced lipolysis were also tested. The concentrations of the various compounds tested were those known to produce maximal responses in rat epididymal adipose tissue. The experiments provide a comparison of the responses of the adipose tissue from various species with each other, and with that elicited in the rat adipose tissue by hormone concentrations capable of producing the maximal response in the latter tissue (see ref. 36).

The results of the *in vitro* experiments were in good agreement with the observations *in vivo*. Glucagon stimulated lipolysis in all the tissues examined except that of the dog. The greater effects were observed with the tissues of goose and owl. Epinephrine stimulated the lipolytic activity of rat and dog adipose tissues, and had a small effect on the tissues from goose and owl, but no detectable effect on the tissues from rabbit and duck. Insulin had no antilipolytic effect in any of the avian tissues, in agreement with other observations,[5,11,28] and did not inhibit the lipolytic effect of glucagon in these species or in the rabbit. Insulin, however, inhibited the effect of glucagon in rat adipose tissue as reported by others.[4,24,39]

Results with nicotinic acid were parallel to those obtained with insulin. Nicotinic acid inhibited glucagon-stimulated lipolysis in rat adipose tissue, but not in the adipose tissue of the other species responsive to the lipolytic action of glucagon (goose, duck, owl, and rabbit). Again, these results *in vitro* are in good agreement with the results *in vivo*. We were unable to prevent in the goose the glucagon-induced elevation of plasma FFA by either pretreatment or simultaneous injection of large doses of nicotinic acid.[21] Similar negative results have been reported by Hoak *et al.*[25]

PATTERNS OF CONTROL OF ADIPOSE TISSUE LIPOLYSIS

The results of our *in vivo* and *in vitro* experiments, together with other observations in the literature,[4,11,40] strongly suggest the existence of different patterns of hormonal control of adipose tissue lipolysis among the species studied.

Table 1 summarizes our findings *in vitro* which, it should be remembered, are in general agreement with the observations *in vivo*.

TABLE 1.

Lipolytic Effects of Glucagon and Epinephrine and the Effects of Insulin and Nicotinic Acid on Glucagon-induced Lipolysis Adipose tissue *in vitro* from various mammals and birds[a]

Lipolytic effect		Inhibition of glucagon-induced lipolysis		Species
Glucagon	Epinephrine	Insulin	Nicotinic acid	
0	+ +			Dog
+ +	0	0	0	Rabbit, Duck
+ + +	+	0	0	Goose, Owl
+ +	+ + +	+	+	Rat

[a] Concentrations of the compounds tested, per ml of incubation fluid, were as follows: glucagon 5.0 μg, epinephrine 1.0 μg, insulin 0.1 U., nicotinic acid 12.0 μg.

As shown in Table 1 it is possible to distinguish four different patterns among the six species studied. At one extreme, the dog is responsive to the lipolytic effect of epinephrine, but not to that of glucagon. At the other, the rabbit and the duck respond to glucagon but not to epinephrine. The other three species (goose, owl, rat) are responsive to the lipolytic actions of glucagon and epinephrine, both *in vivo* and *in vitro*. The information available indicates that the lipolytic effect of glucagon is greater than that of epinephrine in the goose and the owl and in these two species the lipolytic effect of glucagon is not inhibited by either insulin or nicotinic acid. Rat adipose tissue, on the other hand, showed in our *in vitro* experiments greater lipolytic response to epinephrine than to glucagon, in agreement with other reports.[23,24] Moreover, the lipolytic effect of glucagon in the adipose tissue of the rat was inhibited by both insulin and nicotinic acid, as reported by others.[4,24,39] The pattern of response observed in our experiments for the rat adipose tissue is therefore similar to that described by other workers, and different from that of the other species studied in our laboratory.

In their excellent study on the comparative physiology of adipose tissue, Rudman and Di Girolamo[40] have also noted remarkable differences between the rat and other species. The study includes an examination of the lipolytic effects of other hormones, in particular a number of pituitary peptides. Of great importance is the critical analysis made by the authors of the limitations of the experimental data. These limitations should be clearly kept in mind.

The data reported by Rudman and Di Girolamo[40] indicate that mouse adipose tissue compares with that of rat regarding the effects of the catecholamines and insulin. In addition, experiments *in vivo* indicate that glucagon is also adipokinetic in the mouse.[35]

The adipose tissues of pig and guinea pig do not respond to the lipolytic action of the catecholamines and, in this respect, are comparable to those of duck and rabbit. Hamster adipose tissue is responsive to the lipolytic action of the catecholamines, but not to that of glucagon. In this respect the hamster pattern of response compares with that of the dog observed in our experiments, and with the pattern of human adipose tissue described by Rudman and Di Girolamo.[40]

COMMENTS

Since the lipolytic effects of catecholamines and glucagon and the antilipolytic effect of insulin are believed to be mediated by Sutherland's "second messenger" mechanism, the species differences in response to these hormones pose a most intriguing problem. This problem was recognized by Robison, Butcher and Sutherland[38] when commenting on some of our findings as follows: "The impression created by these and other studies is that the role of cyclic AMP in a given system may be relatively constant from one species to another, but that the hormonal mechanisms regulating cyclic AMP may differ considerably . . . the situation seems to offer a specially interesting challenge for the comparative endocrinologist".

The puzzling question is how to reconcile the variety of responses to the lipolytic and antilipolytic hormones, with a single common mechanism of action.

I know that Severo Ochoa is a firm believer in the universality of the basic mechanisms of life phenomena, and Sir Hans Krebs once commented on the essential simplicity of the mechanisms involved in the energy transformations in the living matter by quoting Newton's statement: "Natura enim simplex est".[27]

It seems to me that the diversity of patterns of hormonal control of adipose tissue lipolysis which I have attempted to illustrate, provides just another example of the fact that, in Sir Hans words: "The amazing variety of living forms . . . tends to obscure the fact that relatively simple principles may hide behind complex phenomena".

It is tempting to speculate about the possible physiological significance of the species differences just described; but I have some hesitation to enter into a lengthy teleological argument. Moreover, I am reminded of von Bruecke's famous remark "Teleology is a lady without whom no biologist can live. Yet he is ashamed to show himself with her in public".

The evidence at hand suggests that glucagon is the main hormonal factor controlling fatty acid mobilization in birds, and perhaps in some mammals. It seems also evident that insulin is not antilipolytic in birds. Thus the elevation of plasma FFA induced by fasting in the bird, which we have observed, can not be ascribed to decrease of insulin secretion, and is likely to be due to increased production of glucagon. The importance of glucagon for the control of fat mobilization in birds is illustrated by the fact that pancreatectomy causes in ducks a decrease of plasma FFA[6] in contrast with the elevation of plasma FFA observed in the pancreatectomized dog.[18] Birds depend on fatty acid mobilization for their migratory flight and for the formation of the egg. If glucagon is indeed the main hormonal factor controlling fat mobilization in these animals, one has to conclude that this hormone seems to fulfill that role with admirable efficiency.

In presenting this essay to honor Severo Ochoa, I am well aware of the fact that our data pose many questions and give only few answers. Perhaps the main value of the data is that they give some more proof for the existence of a variety of patterns of hormonal control of adipose tissue lipolysis, and the number of questions they pose regarding the biochemical mechanisms of hormone action and the physiological significance of the diversity of patterns of control.

If I may be allowed to end this essay with a personal note, I would like to say that my feelings regarding the work summarized here can be described with the following words, which I borrow from Sir Isaac Newton: "... to myself I seem to have been only a boy playing on the sea-shore, and diverting myself in now and then finding a smoother pebble or a prettier shell than ordinary, whilst the great ocean of truth lay all undiscovered before me".

REFERENCES

1. AMATUZIO, D. S., GRANDE, F. and WADA, S. (1962). Effect of Glucagon on the serum lipids in essential hyperlipemia and in hypercholesterolemia. *Metabolism*, **11**, 1240.
2. ASSIMACOPOULOS, C. A., PRIGGE, W. F., ANDERSON, W. R. and GRANDE, F. (1966). Effects of pancreatectomy and of $CoCl_2$ in the duck. *Federation Proc.* **25**, 442 (Abstract).
3. CAREN. R, and CORBO, L. (1960). Glucagon and cholesterol metabolism. *Metabolism*, **9**, 938.
4. CARLSON, L. A. and BALLY, P. R. (1965). Inhibition of lipid mobilization. In *Handbook of Physiology*, Section 5, Adipose Tissue. (A. E. RENOLD and G. F. CAHILL, eds. Washington, D.C., Am. Physiol. Soc. (p. 557)).
5. CARLSON, L. A., LILJEDAHL, S. O., VERDY M. and WIRSEN C. (1964). Unresponsiveness to the lipid mobilizing action of catecholamines *in vivo* and *in vitro* in the domestic fowl. *Metabolism*, **13**, 227.
6. DESBALS, P. (1972). Effects de la pancréatectomie et de l'hypophysectomie sur la circulation des lipides chez le canard. Thesis University of Toulouse (France).
7. DOLE, V. P. (1956). A relation between non-esterified fatty acids in plasma and the metabolism of glucose. *J. Clin Invest.* **35**, 150.
8. FEIGELSON, E. B., PFAFF, W. W., KARMEN, A. and STEINBERG, D. (1961). The role of plasma free fatty acids in development of fatty liver. *J. Clin. Invest.* **40**, 2171.
9. FREDRICKSON, D. S. and GORDON, R. S. (1958). Transport of fatty acids. *Physiol. Rev.* **38**, 585.
10. GONZALEZ-SANTOS, P. and GRANDE, F. (1975). Age influence on the lipolytic effect of glucagon in geese. *Proc. Soc. Exptl. Biol. Med.*
11. GOODRIDGE, A. G. and BALL, E. G. (1965). Studies on the metabolism of adipose tissue, XVIII. *In vitro* effects of insulin, epinephrine, and glucagon on lipolysis and glycolysis in pigeon's adipose tissue. *Comp. Biochem. Physiol.* **16**, 367.
12. GORDON, R. S. and CHERKES, A. (1956). Unesterified fatty acids in human blood plasma. *J. Clin. Invest.* **35**, 206.
13. GRANDE, F. (1968). Effect of glucagon on plasma free fatty acids and blood sugar in birds. *Proc. Soc. Exptl. Biol. Med.* **128**, 532.
14. GRANDE, F. (1969). Lack of insulin effect on free fatty acid mobilization produced by glucagon in birds. *Proc. Soc. Exptl. Biol. Med.* **130**, 711.
15. GRANDE, F. (1969). Effects of catecholamines on plasma free fatty acids and blood sugar in birds. *Proc. Soc. Exptl. Biol. Med.* **131**, 740.
16. GRANDE, F. (1970). Effects of glucagon and insulin on plasma free fatty acids and blood sugar in owls. *Proc. Soc. Exptl. Biol. Med.* **133**, 540.
17. GRANDE, F., GRISOLIA, S. and DIEDERICH, D. (1972). On the biological and chemical reactivity of carbamylated glucagon. *Proc. Soc. Exptl. Biol. Med.* **139**, 855.
18. GRANDE, F. and PRIGGE, W. F. (1968). Effect of glucagon and of cysteine-treated glucagon in dogs before and after pancreatecomy. *Federation Proc.* **27**, 331 (Abstract).

19. GRANDE, F. and PRIGGE, W. F. (1970). Glucagon infusion, plasma FFA and triglycerides, blood sugar, and liver lipids in birds. *Am. J. Physiol.* **218,** 1406.
20. GRANDE, F. and PRIGGE, W. F. (1972). Influence of prostaglandin E_1 on the adipokinetic effect of glucagon in birds. *Proc. Soc. Exptl. Biol. Med.* **140,** 999.
21. GRANDE, F., PRIGGE, W. F. and ASSIMACOPOULOS, C. A. (1967). Fat-mobilizing effect of glucagon in birds. *Federation Proc.* **26,** 436 (Abstract).
22. GRANDE, F., PRIGGE, W. F. and de OYA M. (1972). Influence of theophylline on the adipokinetic effect of glucagon *in vivo*. *Proc. Soc. Exptl. Biol. Med.* **141,** 774.
23. Hagen, J. H. (1961). Effect of glucagon on the metabolism of adipose tissue. *J. Biol. Chem.* **236,** 1023.
24. HEPP, K. D., MENAHAN, L. A., WIELAND, O. and WILLIAMS, R. H. (1969). Studies on the action of insulin in isolated fat cells, II 3'5-nucelotide phosphodiesterase and antilipolysis. *Biochim. Biophys. Acta,* **184,** 554.
25. HOAK, J. C., CONNOR, W. E. and WARNER, E. D. (1968). Toxic effects of glucagon induced acute lipid mobilization in geese. *J. Clin. Invest.* **47,** 2701.
26. KAUSCH, W. (1896). Ueber den Diabetes mellitus der Voegel nach Pankreasextirpation. *Arch. Exp. Path. Pharmakol.* **37,** 274.
27. KREBS, H. A. (1953). Some aspects of the energy transformations in living matter. *Brit. Med. Bull.* **9,** 97.
28. LANGSLOW, D. R. and HALES, C. N. (1970). Lipolysis in chicken adipose tissue *in vitro*. *J. Endocrinol.* **46,** 243.
29. LAURELL, S. (1957) Turnover rate of unesterified fatty acids in human plasma. *Acta Physiol. Scand.* **41,** 158.
30. MIAHLE, P. (1958). Glucagon, insuline et regulation endocrine de la glycemie chez le canard. *Acta Endocrinol.* (Copenhagen) Suppl. **36.**
31. MIKAMI, S. I. and ONO K. (1962). Glucagon deficiency induced by extirpation of alpha cell islets of the fowl pancreas. *Endocrinology,* **71,** 464.
32. MINKOWSKI, O. (1893). Untersuchungen ueber den Diabetes mellitus nach Extirpation des Pankreas. *Arch. Exp. Pathol. Pharmakol.* **31,** 85.
33. OAKBERG, E. F. (1949). Quantitative studies of pancreas and islands of Langerhans in relation to age, sex and body weight in white Leghorn chicken. *Am. J. Anat.* **84,** 279.
34. PALOYAN, E. and HARPER, P. V. (1961) Glucagon as a regulating factor of plasma lipids. *Metabolism,* **10,** 315.
35. PAYNE, R. W. (1954). Fat-mobilizing properties of glucagon. *Federation Proc.* **24,** 438 (Abstract).
36. PRIGGE, W. F. and GRANDE, F. (1971). Effects of glucagon, epinephrine and insulin on *in vitro* lipolysis of adipose tissue from mammals and birds. *Comp. Biochem. Physiol.* **39,** B, 69.
37. RENOLD, A. E. and CAHILL, G. F. ed. (1965). Adipose tissue, Section 5 *Handbook of Physiology*. Washington D.C., *Am. Physiol. Soc.*
38. ROBISON, G. A., BUTCHER, R. W. and SUTHERLAND, E. W. (1971). *Cyclic AMP*. New York, Academic Press.
39. RODBELL, M. and JONES, A. B. (1966). Metabolism of isolated fat cells. III. The similar inhibitory effect of phospholipase C (*Clostridium perfringens* alpha toxin) and of insulin on lipolysis stimulated by lipolytic hormones and theophylline. *J. Biol. Chem.* **241,** 140.
40. RUDMAN, D. and DI GIROLAMO, M. (1967). Comparative studies on the physiology of adipose tissue. *Adv. Lipid Res.* **5,** 36.
41. SCHOENHEIMER, R. (1942). *The Dynamic State of Body Constituents.* Cambridge, Mass., Harvard University Press.
42. SCHUR, H. and LÖW, A. (1928). Studien ueber den Kohlenhydratstoffwechsel. II, Fettdepots und Kohlenhydratstoff-wechsel. *Wiener klin. Wochenschrift,* **41,** 261.
43. STEINBERG, D. (1966). Catecholamine stimulation of fat mobilization and its metabolic consequences. *Pharmacol. Rev.* **18** (Part I), 217.
44. VAN CAMPENHOUT, E. and CORNELIS G. (1954). Les ilots endocrines du pancreas des oiseaux. *Compt. Rend. Ass. Anat.* **79,** 422.
45. VUYLTEKE, C. A. and De DUVE, C. (1953). Le contenu en glucagon du pancreas aviaire. *Arch. Int. Physiol.* **61,** 273.
46. WERTHEIMER, H. E. (1965). Introduction—a perspective. In *Handbook of Physiology,* Section 5, Adipose tissue. (A. E. RENOLD and G. F. CAHILL, eds.). Washington, D.C., *Am. Physiol. Soc.* p. 5.
47. WERTHEIMER, E. and SHAPIRO, B. (1948). The physiology of adipose tissue. *Physiol. Rev.* **28,** 451.
48. WHITTY, A. J., SHIMA, K., TRUBOW, M. and FOA, P. (1969). Effect of glucagon and of insulin on serum free fatty acids in normal and pancreatized dogs. *Proc. Soc. Exptl. Biol. Med.* **130,** 55.
49. WINEGRAD, A. I. and RENOLD, A. E. (1958). Studies on rat adipose tissue *in vitro*. I. Effects of insulin on the metabolism of glucose, pyruvate and acetate. II. Effects of insulin on the metabolism of specifically labeled glucose. *J. Biol. Chem.* **233,** 267 and 273.

REFLECTIONS ON THE RELEVANCE OF BIOCHEMISTRY FOR THE UNDERSTANDING OF LIVING SYSTEMS

ZACHARIAS DISCHE
Department of Ophthalmology, College of Physicians and Surgeons, Columbia University, New York

When this writer was asked to contribute to the volume of essays dedicated to Severo Ochoa on his 70th birthday, he started to think about a subject suitable for such an occasion. Severo and this writer have been close friends for almost 40 years. This is in spite of the fact that our interpretation of the significance of science for the human life does not always appear to be identical. But one of the basic elements of our friendship seems to be our sharing of a deep respect for the variety of men. This short essay which attempts to deal with the relevance of biochemistry, as we know it today, for the understanding of the variety of life seems to be a suitable contribution in honor of a very eminent scientist and a dear friend.

I. STRUCTURAL AND METABOLIC FACTORS IN THE ORGANIZATION OF LIVING CELLS

Between 1933 and 1963[1] the biochemical landscape was radically changed by two waves of discoveries of new biochemical phenomena and elaboration of new concepts. These developments represent a scientific revolution in biology insofar as they show that catabolic and anabolic processes in living systems and even their fully developed structural characteristics can be understood, at least potentially, in chemical and physical terms. This was achieved by isolating and studying *in vitro* chemical processes of the cellular metabolism and establishing certain principles of the organization of these processes in living cells. These basic phenomena are:

1. The structure of the enzymes.
2. Flexible organization of metabolic processes which may consist of long sequential chains of often reversible reactions in which the product of one reaction becomes the substrate of the following one.
3. Crossing over of intermediates in processes of degradation and synthesis, of coenzymes and compounds for storing and dispensing of energy.
4. The use of these latter compounds as devices either for activation of substrates or of energy transformations, particularly chemical energy into mechanical, electrical and photochemical energy or for regulatory purposes.
5. The role of noncovalent bonds in addition to S—S bonds in determining the conformation and interactions of proteins in living systems.

6. The role of such conformations and interactions in the regulation of the cellular metabolism and as structural factors in morphological characteristics of cells and intercellular matrices.
7. The role of nucleotide sequences in nucleic acids as a source of information for the synthesis of proteins and cell reduplication.

In the following I will briefly discuss the potential significance of the most important steps in this historic development for the interpretation of life phenomena on the level of eukaryotic cells and certain cell assemblies.

1. Structure of enzymes

In 1936 H. Theorell demonstrated in the case of one of the so-called yellow enzymes that it consists of a specific protein bound in a reversible way to an active group, namely riboflavin. This discovery was followed by the demonstration by Warburg and his associates that dehydrogenases of the glycolytic and oxidative systems of living cells consist of specific proteins combining with DPN or TPN as active groups in a reversible equilibrium. The role of these coenzymes as H donors and acceptors was elucidated. This development led to the concept of an enzyme as a specific protein carrier for the active group that catalyzes the specific enzyme reaction.

2. Chains of reversible reactions of the anaerobic and aerobic cellular metabolism

Since the discovery of Embeden in 1933 that the Harden–Young ester is converted in muscle extracts to glycerol and glycerate phosphate and the pyruvate formed from the latter is reduced to lactic acid in a subsequent reaction with glycerol phosphate, the extensive work in the laboratories of Meyerhof, Warburg and others demonstrated the stepwise character of the anaerobic glycolysis and determined the nature of all the intermediates and the role and nature of coenzymes involved in these reactions. Characteristic features of these reaction chains are the reversibility of the individual steps of the chain and the fact that all the intermediates except the two final products are phosphate esters. A second major anaerobic metabolic system which contains chains of reversible reactions is the pentose phosphate cycle. Investigations of this reaction chain started with the demonstration by this writer that adensine is phosphorylated in human hemelysates to ribose 5 phosphate, 3 molecules of which are then converted in an equilibrium reaction to 2 molecules of hexose-6-P and 1 molecule of triosephosphate. The reaction chain leading from ribose-5-phosphate to glucose-6-phosphate and triosephosphate was then explored in depth in the comprehensive work of B. Horecker and E. Racker and associates. The energy-producing oxidative metabolic pathway consists of the reversible citric acid cycle discovered in 1938 by H. Krebs and the irreversible electron transfer system localized first in mitochondria by Lehninger.[2] The material to be oxidized is fed into the cycle in form of active acetyl[3] which can be derived from pyruvic acid or beta-oxidation of fatty acids and condenses with oxaloacetate to form isocitrate.

3. Coordination of cell metabolism

The metabolic processes in cells appear to be maintained in a dynamic equilibrium and organized as a unit by the operation of three factors.

(a) Reversibility of intermediate reactions in reaction chains.

(b) A characteristic which can be called the crossing over of substrates, intermediates and coenzymes between different reaction chains. Thus pyruvic acid, a key intermediate in the glycolytic system, enters the citric acid cycle by being oxidized in mitochondria to active acetyl which reacts with the oxaloacetic acid to form isocitric acid. Active acetyl is also a key component in the metabolism of fatty acids, and pyruvate is furthermore condensed with mannosamine to neuraminic acid, a key component of hexosamineglycans of glycoproteins. The hexose-and triosemonophosphates are crossing over as substrates and intermediates from the glycolytic cycle into the pentose phosphate cycle in which they are converted into ribose, ribulose and xylulose phosphates. These esters are important reactants in the synthesis of purines and nucleic acids, in photosynthesis through ribulose diphosphate and in hexuronic acid metabolism through xylulose phosphate. Ribose-5-phosphate can also be formed from glucose-6-phosphate by the action of glucose-6-phosphate dehydrogenase, at the same time reducing TPN to TPNH necessary for the synthesis of fatty acids; the participation of DPN and DPNH in cellular oxidoreductions is widespread. DPNH and TPNH are furthermore interconvertible by transhydrogenases.

(c) A further unifying factor in intracellular metabolism is the activation of substrates through esterification with phosphate by transphosphorylation with ATP; thus energy produced in metabolic reactions is stored to the greatest part in a single compound, the ATP formed by transphosphorylation between high energy intermediates and ADP.

The function of ATP as general dispenser and storage point of energy was first clearly formulated in 1941 by F. Lipmann.[4] The central role of ATP as dispenser of activation energy is not restricted to metabolic reactions providing energy for cellular operations, but also for reversible activation of cellular structural elements involved in numerous cell functions like muscle contraction,[5] propagation of nerve impulses,[6] and others. But metabolic substrates are also partly activated by phosphorylation with inorganic P, whereas dephosphorylation of end products to achieve storage of energy of the reaction must involve transphosphorylation from an end product to ADP. This results in a coupling of esterification which reaches a maximum ratio of 3 P per atom of O, first determined by S. Ochoa in 1943[7] in mitochondrial oxidations. There must, therefore, be a continuous dephosphorylation of ATP by ATPases synchronized with certain metabolic processes. This was clearly recognized by this writer in 1936[8] in red cell glycolysis and generalized for other tissues. ATP and ATPases are thus also a central element of self-regulation of the cellular metabolism. The unifying character of this central position of ATP resides in the fact that ATP and ADP can cross over from one metabolic system to another, synchronizing their activities in a meaningful way.

4. Regulation of the cellular metabolism

The reversibility of the metabolic reaction chains, the crossing over of intermediates and the role of ATP as central store of activating energy and as a factor in regulation of metabolic rates by synchronization of the rate of ATPase activities with those of certain intermediate reaction steps, all seem to contribute to the functional unification and regulation of cellular metabolism. But an adequate regulation of systems consisting of numerous subunits requires the possibility of shutting off certain units at certain stages of operation. This can be achieved by substrate inhibition of the first reaction of a chain or by feedback inhibitions by end products of reactions. The first mechanism leads to an accumulation of the first reaction product of the chain which may intervene as substrate in other chains in a disturbing way. The first observed example of the second mechanism was the inhibition of the phosphorylation of glucose to glucose-6-phosphate in human red cells by phosphoglycerate and phosphopyruvate described by this writer in 1941[9] and interpreted as a regulatory process by feedback inhibition of a reaction chain by an end product. Uyeda and Racker found in 1965[10] a similar feedback inhibition of the purified phosphofructokinase from ascites tumor cells and rabbit muscle by phosphopyruvate and creatine phosphate. Numerous such inhibitions were since observed in bacteria.[11] Most important among them as organizational factors are those of branched metabolic pathways which direct the further transformation of an end product in an initial reaction chain into one of several reaction chains, starting with this intermediate, by inhibition of the alternate branch by its final reaction product. Such mechanisms can explain a re-routing of certain metabolic intermediates from one metabolic process into another, thus achieving a coordination of such processes. Monod *et al.*[12] have shown that such feedback inhibitions can be due to conformational changes of the respective enzymes as a result of allosteric attachment of the inhibitor to the inhibited enzyme. But it is not yet established whether this is the case in all such feedback inhibitions, particularly in metazoa.

5. Factors controlling the conformation of proteins and their interactions with other macromolecules[13]

In 1936 Mirsky and Pauling recognized the role of hydrogen bonds as important determinants of the secondary and tertiary structure of the proteins, in addition to intra- and interchain S—S linkages and ionic forces. More than a decade later hydrophobic interactions between certain amino acids were added as important conformational factors. That non-covalent linkages between proteins and non-protein compounds can determine the quaternary structure of a protein was shown in 1957 by the reconstitution of the tobacco mosaic virus from its isolated RNA and protein parts by Fraenkel-Conrat.[14] On the other hand, Anfinsen[15] could show that when the tertiary structure and activity of the bovine pancreas RNase are destroyed by reduction of its S—S linkages, re-oxidation of the SH groups finally leads to complete restitution of the enzymatic activity. Further structural studies of the reconstituted active enzyme support the assumption that in this case initial tertiary structure of the protein was restored. This supports the assumption that the tertiary structure of a protein is defined by its amino acid sequence.

6. Glycoproteins and intercellular matrices

Between 1930 and 1950 two other types of polymers with a very broad spectrum of molecular sizes were discovered or their structure elucidated. These polymers are hexosaminoglycans of two types: (1) hexosaminohexuronides (acid mucopolysaccharides) and

(2) carbohydrates in glycoproteins, chains of hexose linked to hexosamine in sometimes highly branched structures carrying on the periphery of the molecule two characteristic compounds, 1-fucose and various derivatives of neuraminic acid. These carbohydrate chains are linked covalently to the polypeptide carrier. The structure of acid mucopolysaccharides was elucidated to a large extent by the work of K. Meyer *et al.*[(16)] Structural chemistry of the glycoproteins is still much less developed. These two types of compounds have in common that they seem to play a fundamental role in determining the tissue structures of metazoa on a supracellular level. Both types seem to play a role in the organization of the intercellular matrices in metazoa. Glycoproteins also appear to be essential constituents of cell membranes.

Hexosaminohexuronides can form assemblies of molecules of highly branched character and very high molecular weight, up to several millions, as has been recently demonstrated in the case of cartilage, and may determine physical properties of the intercellular matrix. Glycoproteins seem to mediate cell contacts in metazoan tissues. The latter function seems to depend to a large extent on the presence of derivatives of neuraminic acid and fucose which by differences in their affinity to water and in their electrical charges may determine the conformation of the cell surface and thus the nature of the contact of the cells. The operation of such mechanisms appears supported by species and organ-specific reaggregation of cells of desaggregated embryonal organs in tissue cultures and by species-specific reaggregation *in viro* of disaggregated spongae after addition of glycoproteins isolated from these organisms.[(17)] Further support is brought by the findings of Ashwell *et al.* that desialized blood glycoproteins and hormones form recognition sites with the neuraminic acid groups of glycoproteins of liver cell membranes.

7. Nucleic acids

It is remarkable from the point of view of general scientific methodology that a decisive role of nucleic acids was not recognized in this field in spite of the fact that genetic studies in the first two decades of the twentieth century clearly demonstrated the role of heredity determinants localized on chromosomes which were known to consist to a large extent of nucleic acids. This was due partly to the dominance of incorrect ideas about the structure of nucleic acids introduced by Levene according to which the chain of DNA and RNA consisted of tetranucleotides each containing all four different nucleotides. Such a structure was not suitable for the transferring of the extraordinary amount of information in reduplication of cellular material. The situation changed radically in 1948 when Chargaff and his associates showed that although the sum of the two purines is identical to the sum of the two pyrimidines in DNA, the ratio between the two purines and the two pyrimidines can vary in various parts of the DNA. Chargaff suggested in 1950[(18)] that this may be the basis of the transfer of information for the synthesis of polypeptide chains in proteins. On the basis of this discovery, and results of X-ray diffraction studies on DNA by Wilkins, Watson and Crick demonstrated the double helix form of the DNA molecule in which the two chains are held together by hydrogen bonds. It became clear that the reduplication of DNA and the coding of the amino acid sequences can take place after the separation of the two chains, each of which can serve as a template either directly for the synthesis of DNA or indirectly for the synthesis of polypeptide chains. The involvement of RNA as an intermediate in the coding of amino acid sequences in proteins was elucidated in the following few years as consisting of sequences of interactions by transcription of nucleotides from single DNA chains to form the messenger RNA (predicted by Monod) which carries this information to the cytoplasm where it is translated in the ribosomes into sequence of amino acid polypeptide chains by interaction with specific amino acyl compounds. The fit between isolated messenger RNA and denatured DNA chains was demonstrated by Spiegelman[19] in annealing experiments. The recognition of the alphabet represented by the messenger RNA and a series of trinucleotides, one of the greatest achievements in biochemistry in our times, was made possible by the intervening discovery of methods of synthesis of artificial polynucleotides synthetized by a bacterial enzyme discovered by Manago and Ochoa and by the discovery of Nirenberg and Matthaei of the synthesis of polyphenylamines on an artificial polynucleotide template and the development by the laboratories of Nirenberg[20] and Ochoa[21] of the major structures of the translational alphabet. Since then a series of additional factors necessary for the initiation and termination of the polypeptide chain synthesis have been discovered in various laboratories.

II. DOMINANCE OF THE OVERALL DESIGN IN THE ORGANIZATION OF LIVING SYSTEMS

The molecular structures of constituents of living systems and their metabolism and the forces with which they interact appear adequate to explain, at least, in principle, the characteristic structure and stable dynamic equilibrium of individual cells and unicellular organisms in their final state of organization. These structures can be interpreted in terms of cellular proteins whose interactions may determine unequivocally the relative position of every cytoplasmic molecule in a stable dynamic equilibrium.*

Even patterns of spatial distribution of discrete structural characteristics can be explained by assuming a quasi-fluid crystalline structure of the cytoplasm and temporal patterns of synthetic processes by mass action. Certain old experiments like that of Lillie in which a complete dislocation of cytoplasmic particles, except the cell membrane, and the yoke in a fertilized egg of an echinoderm did not prevent the recovery of the ability to produce initial stages of embryonal development seem to support such a concept of the structure of the cytoplasm, although there is some difficulty in reconciling this concept with such facts as the disappearance of the regular distribution of the cilia in the sucterian *Podophrya fixa* during the preparation for cell division (Lwoff, 1965).

But in metazoa we find a spatial organization of diverse morphological characteristics which is related to variations in functional properties. Furthermore, these systems consist either of cells separated by semipermeable membranes or connected only by very thin specialized cytoplasmic bridges. A temporal or spatial coordination requires in this case an elaborate manifold signalling system. The latter must again display the characteristics of a spatial and temporal organization. In this case, the activity and morphology of a certain region of the system is coordinated spatially and temporally with activities in other parts of the system. Thus the system displays an overall design of spatial and temporal distributions of biological entities and characteristics. Such patterns, furthermore, result from involved developmental processes with sometimes pronounced directive or anticipatory character. Examples of latter phenomena are the bending of the vertebrate column as early

* Recently Keates and Hall (*Nature*, **257**, 418, 1975) have demonstrated self-assembly of tubulin from brain to microtubules if an accessory protein of M.W. 360,000 is present in solution.

as the second month of the human embryo related to the future upright posture or the localization in the pupae of *Drosophila* of the imaginal discs which contain the anlage of appendices of the insect which appear later in the imago. They are located in the third final instar of the pupa at places which correspond to the localion of the appendices in the image.

Thus, patterns of specific spatial distribution of disparate discrete elements and meaningfully coordinated temporal sequences of events during the development prepare the final design. The latter unequivocally determines the morphological and functional interrelations of the subunits. This seemingly in spite of lack of adequate information transfer by direct contact between cellular macromolecules or hormonal signals not requiring prior spatio-temporal determination. These considerations are illustrated in the following by two examples of the dominance of the overall design taken from the embryonal development of poikilotherm vertebrates and one example of such dominance taken from the organization of social insects.

1. Development of retino–tectal connections in fish and amphibia

The difficulty in the interpretation of phenomena of embryonal development in quantitative and qualitative terms within the conceptual framework of present biochemical concepts is illustrated by the relatively simple process of establishing connections between retinal photoreceptors and the visual areas of the brain in fish and amphibians during embryogenesis and regeneration following the severing of the optic nerve (retino-tectal connection). R. W. Sperry[22] first showed in a newt that transsection of the optic nerve and rotation of the eye by 180° is followed after a certain time by an accurate restitution of the field of vision which, however, is permanently rotated by 180° in relation to the dorsoventral and nasotemporal body axis. This in spite of the complete disorganization of the original spatial arrangement of nerve fibers in the scar. Each photoreceptor of the retina must have acquired during regeneration its connection with its original neuron in the visual cortex of the tectum. More recent experiments by Sharma and Gaze,[23] in which a certain area in the visual cortex of the goldfish was excised and rotated by 90° showed also restitution of the right connection between the severed optic nerve and the cortical neurons. In this experiment the correspondence of the retinal receptors and the tectal neurons was determined by establishing with microelectrodes the correspondence of certain elements of the retina and the tectum before and after the operation. Similar results were obtained recently in the frog *Xenopus laevis* by Jacobsen and Levine (1975), after mutual translocation of ventral and temporal areas of the retina to the contralateral dorsal areas.

These results indicate that each nerve ending of a retinal ganglion cell linked to a photoreceptor carries a marker which fits a specific neuron in the visual certex. It seems necessary to assume that the marking of the individual nerve endings results from a rank order within the nerve endings and corresponding receptors. This order could be the result of some gradients across the retina and the visual cortex. The size and the position of the gradient, which would be two-dimensional, can hardly be coded purely by linear transcription.

The dominance of the overall design is even more striking in the embryonal development of retino–tectal connections. In all poikilotherms so far investigated, the retinal growth procedes in concentrical direction, whereas that of the visual cortex in ventro-caudal direction. The retino–tectal connections of the most peripheral photoreceptors are at every stage located in the most ventral parts of the cortex. When new photoreceptors are formed, they displace the already existing connections which were previously established and push them towards the newly formed, more caudally located neurons of the visual cortex.[24] This type of developmental and regenerative process illustrates the dominance of the overall design of the final organization of the system.

2. Effects of temperature on early embryonal development of amphibia

The directiveness towards the final organizational pattern of a living system is strikingly apparent in the influence of temperature on the amphibian development.

The embryonal development of poikilotherms at different environmental temperatures yields organisms of identical structure and morphology. If the morphogenetic process were solely the result of the operation of the known metabolic processes in the living cells and their regulatory mechanisms as we know them, the ratios of the rates of the numerous metabolic processes would have to be identical in temperature ranges, in which morphogenesis remains unchanged. But enzymatic reactions, particularly those which lead to an equilibrium far from the endpoint, will in general, have different temperature coefficients because of differences in activation energies. The effect of this factor could be corrected in individual reaction chains by appropriate changes in the concentration of intermediates. But because of the crossing over of certain intermediates from one reaction chain into a different one, the ratios of reaction rates of different chains could not be maintained without additional specific regulatory processes. The latter would have to belong to factors responsible for the dominance of the overall design of the process. This dominance is illustrated by the following three experiments on temperature effects on biological processes. In 1928 Twitty[25] compared the temperature coefficient of the time of determination of the direction of the ciliary beat of amblystoma larvae with the coefficient of the overall development of the larva between 15° and 25°. He showed that the latter is significantly larger than the first. Nevertheless, at both temperatures the larval development was identical. The response of a partial developmental event to temperature change was subordinated by some unknown mechanism to the design of the developmental process of the whole organism.

The invariance of the developmental design is illustrated by the influence of temperature gradients of varying orientation on the early development of a fertilized frog egg by J. Huxley.[26] In the frog egg initial cell divisions produce significantly larger cells at the vegetative pole adjacent to the yoke than those at the opposite animal cells. When a temperature gradient descending from the animal to the vegetative pole was applied before the first division, the difference in size of the cells at the two poles was reversed. Nevertheless, the first stage of development, up to the stage at which differentiation started, was completely normal.

Finally, in 1927[27] Vogt applied a lateral temperature gradient of 10° to a fertilized newt egg after the first division and followed the early development of each half up to 5 days. There was a retardation of the development in the half exposed to the lower temperature, but without change in the nature of the developmental process. After the removal of the gradient, the retarded half caught up with the other half in its development.

3. Construction of complex mound structures by termites

Termites are social insects which evolved from cockroaches, a rather lowly evolutionary parentage. They are conspicuous by the complexity and variety of their mounds which house their tightly organized societies. The fundamental design of these structures is very similar and often identical for the numerous species as far as the number and function of the various functional subunits is concerned as they all serve the same essential needs of the species. They differ radically, however, as far as the species specific position in the landscape, their form as a whole and the design of the elaborate ventilating and water vapor distributing systems are concerned. An example of such highly specialized design is the mound of the genus *Amitermes*; a large asymmetric block positioned in such a way that two of its narrow sides are directed towards the east and west and the broad sides towards the north and south to minimize the insolation effects. The mound is constructed in such a way that each worker of the population, which can consist of more than a million of individuals, constructs only a tiny part of the whole system. The latter, nevertheless, finally represents a highly elaborate, purposeful inner structure and a characteristic outer form adapted to environmental requirements. The problem arises, how does every one individual in the population building a tiny segment of the whole know to do the right thing so that the elaborate structure results from the specialized activities of the huge number of individual termites? There is no indication whatsoever that termites possess a symbolic language capable of transmitting this kind of information. In fact, their brains do not seem to differ significantly in structure of their non-social relatives.

III. THE GENOME AS A COORDINATED ASSEMBLY OF INFORMATION-PROCESSING DEVICES

The few examples of the phenomenon of the dominance of the overall design in the organization of developmental and behavioral processes make it appear most problematic that the transmission of information from the genome by simple readout of DNA templates can be considered an adequate explanation of the appearance of complex spatio-temporal patterns. To introduce the concepts of programming as an explanation is not helpful as long as nothing is known about the mechanism and origin of the programming. Some geneticists, therefore, have recently tended to conceive the genome as an organized assembly of gene subunits which are information processing devices (Wolpert,[28] Kaufman[29]) through interactions between DNA and nuclear proteins. The phenomenon of transdetermination in imaginal discs of *Drosophila* (Kaufman) which produces a well-developed antenna in the place where a leg should appear illustrates this concept. Here again we are confronted with seemingly unsurmountable difficulties in any attempt to interpret the structure of such an informational process on the basis of our present biochemical conceptual framework. In 1938 Dobzhansky and Sturtevant[30] found that regional subspecies of two North American species of *Drosophila* (*melanogaster* and *azteca*) often differ widely from each other in the conformation and relative position of their chromosomes. The differences are due to numerous inversions and translocations of chromosomal segments. These changes, however, do not affect at all any of the morphological or functional characteristics of these insects. As only a fraction of chromosomal nucleotides are involved in coding for structural and functional characteristics in

Drosophila, it seems possible that the intraspecific variations in chromosomal structure occur only in the non-functional parts of the chromosomes and therefore do not affect species-specific characteristics. If, on the other hand, nuclear proteins are essential participants of the informational apparatus of the genome, the latter must be rebuilt in every cell division, as the nuclear proteins apparently are synthetized in ribosomes of the cytoplasm. The difficulty in forming any reasonable, adequate concept of such a process is evident.

Those of us, who like this writer, were participants, to a lesser or major degree, in this magnificent effort, which in a single generation resulted in the founding of the new biochemistry, can only be elated at the thought that they were witnesses and contributors to a major event in the history of human culture. One thinks of the enthusiastic exclamation of the German humanist of the fifteenth century who, impressed by the intellectual life of his time, wrote to a friend, "What a century, what arts and sciences; it is a joy to be alive!" And yet, after this successful creative era, how little we know about the nature of the forces which determine some fundamental aspects of the life of metazoa. There is no doubt that increasing complexity of organisms brings new qualities of correlation and behavior that we cannot adequately explain. Why is the anticipatory, projecting and creative human mind dependent on a certain size and organization of the brain? Thinking about it one is sometimes overcome by a feeling of envy of those who will come after us and who will succeed in unraveling those mysterious, unsuspected forces that dominate life, as the physicists of our century unraveled the unsuspected, unimaginable forces and forms of the material universe.

REFERENCES

1. Horecker, B. L. (1961) *Harvey Lectures*, **57**, 35; Racker, G. (1954) *Adv. Enzym.* **15**, 141.
2. Lehninger, A. L. (1962) *Am. Rev. Biochem.* **31**, 47.
3. Lynen, F., Reićhert, E. and Ruef, L. (1961) *Liebig's Ann.* **574**, 1.
4. Lipmann, F. (1941) *Adv. Enzym.* 1, 99.
5. Szent-Gyergyi, A. (1960) In *Proteins of the Myofibril*, (G. M. Bourne, ed.), Academic Press, N.Y.
6. Nachmansohn, D. (1975) *Chemical and Molecular Basis of Nerve Activity*, p. 373, Academic Press, N.Y.
7. Ochoa, S. (1943) *J. Biol. Chem.* **151**, 493.
8. Dische, Z. (1936) *Enzymol.* **1**, 288.
9. Dische, Z. (1941) *Bull. Soc. Chim. Biol.* **23**, 1140.
10. Uyeda, K. and Racker, E. (1965) *J. Biol. Chem.* **240**, 4082.
11. Stadtman, E. R. (1970) in *Enzymes* **1**, 444–448.
12. Monod, J., Changeux, J. P. and Jacob, F. (1963) *J. Molec. Biol.* **6**, 306.
13. Sheraga, H. A. (1963) in *The Proteins*, 2nd ed., p. 477 (G. Neurath, ed.), Academic Press.
14. Fraenkel-Conrat, H., Singer, B. and Williams, R. C. (1975) *Biochim. Biophys. Acta*, **25**, 87.
15. Anfinsen, C. B., Haber, E., Sela, M. and White, J. R (1961) *Proc. Nat. Acad. Sci. U.S.A.* **47**, 1309.
16. Meyer, K. (1970) in *Chem. and Mol. Biol. of Int. Matrix*, vol. 1, p. 5, (E. A. Balazs, ed.), Academic Press.
17. Moscona, A. (1952) *Proc. Soc. Exp. Biol. Mod.* **92**, 410.
18. Chargaff, E. (1950) *Experientia*, **6**, 201.
19. Spiegelman, S. (1963) *Fed. Proc.* **22**, 36.
20. Nirenberg, M. W. and Matthaei, J. H. (1961) *Proc. Nat. Acad. Sci. U.S.A.* **47**, 1588.
21. Lengyel, P., Speyer, J. F. and Ochoa, S. (1961) *Proc. Nat. Acad. Sci. U.S.A.* **47**, 1961.
22. Sperry, R. W. (1944) *J. Neurophysiol.* **7**, 57.
23. Sharma, S. C. and Gaze, R. M. (1971) *Arch. Ital. de Biol.* **109**, 357.
24. Gaze, R. M., Keating, M. J. and Chung, S. H. (1974) *Proc. Roy. Soc. Lond.* B, **185**, 301.

25. Twitty, V. C. (1938) *J. Exp. Zool.* **50**, 319.
26. Huxley, J. S. (1927) *Roux Arch. f. Entwickl. Mech.* **112**, 480.
27. Vogt, (1927) *Anatomischer Anzeigert*, Erg. Haft 126.
28. Wolpert, L. and Lewis, J. H. (1975) *Fed. Proc.* **34**, 74.
29. Kaufman, S. (1973) *Science* **181**, 310.
30. Dobzhansky, Th. and Sturtevant, A. H. (1938) *Gen.* **23**, 28.

FORMATION AND CLEAVAGE OF C—N BONDS IN ARGININE BIOSYNTHESIS AND UREA FORMATION

SARAH RATNER

The Public Health Research Institute of the City of New York, Inc.
New York, New York 10016, U.S.A.

In the prevailing spirit of retrospective inquiry it is tempting to recall the lively climate of 30 years ago when biochemistry was bursting with new activities. The stimulation of advances in cofactor structures on enzymatic catalysis and the freshly realized advantages of working with broken or soluble cell preparations gave biochemists an open sesame into the mysteries of intermediary metabolism. Naturally the chief question addressed to every area and metabolic pathway then became: how does it work; what is the mechanism? This questioning atmosphere was reinforced in Severo Ochoa's department, which I joined in 1946, by continuous shop talk on experiments underway, or on papers we had read the night before, or the week before. The pool size kept changing from within and from without through a stream of visiting biochemists, many of whom are contributors to this volume. Communications from E. Racker, located on the floor below, on glycolysis and fumarase extended our spectrum. His frequent duets with Severo on biochemical problems then resisting solution provided an obligato.

At that time my own interests were already polarized around nitrogen metabolism and I had begun to address my questions to urea synthesis.

In 1932 Krebs had proposed, as the result of his well-known studies, the scheme for urea synthesis shown in Fig. 1. Accordingly NH_3 and CO_2 react with ornithine to form citrulline (step I); citrulline then condenses with a second molecule of NH_3 to form arginine (step II). The cycle is completed by arginase (step III). Ornithine occupied the role of a carrier compound in a series of reactions that brought about its regeneration. Implicit in the proposed mechanism was the novel concept that a metabolic synthesis can be accomplished by the repeated turnovers of a cyclic process. The experiments of Krebs and Henseleit[1] were carried out with respiring liver slices in an oxygen atmosphere and the presence of a respiratory substrate such as lactate was necessary for urea formation.

The specificity for NH_3 was reexamined in 1946, when Cohen and Hayano[2] succeeded in obtaining urea synthesis in respiring liver homogenates starting from citrulline (step II). They found that glutamate was a much better nitrogen donor than NH_3.

I had been puzzled by the requirement for oxygen and my experiments were taking a somewhat different direction. The first inkling of an answer to my questions came from a relatively simple experiment[3] with a soluble extract of liver not capable of respiration and free of pyridine nucleotides, but still containing transaminase, fumarase and arginase and arginine-synthesizing enzymes. Arginine (measured as urea) was formed from citrulline and aspartate in the presence of Mg^{++}, ATP and an ATP generating system. Glutamate

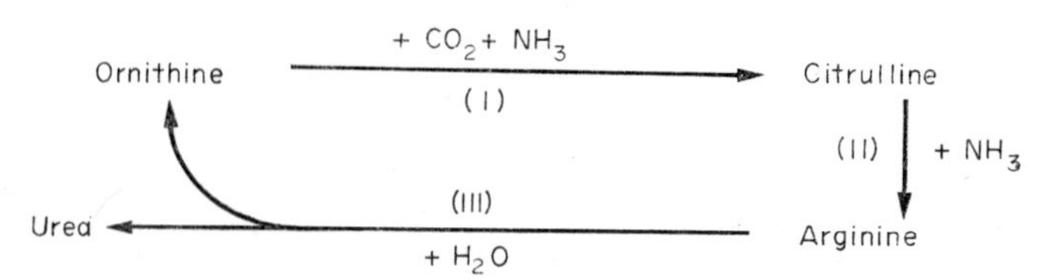

Fig. 1. Krebs–Henseleit ornithine cycle.

could not replace aspartate, but when glutamate and oxaloacetate were present together, so that aspartate could be formed by transamination, an almost equal amount of urea was formed. The elimination of oxidative processes allowed one more finding, that fumarate (originally found as malate) was formed only when arginine was formed.

This experiment told us some new things about arginine and urea synthesis: (a) that half of the nitrogen of the urea molecule originated specifically in aspartic acid, (b) that fumarate was a product of the reaction and (c) that phosphate bond energy, specifically in the form of ATP, was required for the new carbon to nitrogen bond formed.

New developments in intermediary metabolism in glycolysis, transamination and oxidative phosphorylation had begun to reveal the complexity of intracellular processes. Severo Ochoa had found earlier that α-ketoglutarate oxidation in respiring muscle homogenates was associated with ATP generation. This placed the suggested role of glutamate as nitrogen donor for urea in a different light. Our own experiments indeed showed[4] but with liver tissue, that when glutamate was used as a nitrogen donor in respiring homogenates, the specific requirements for aspartate and for ATP were masked because glutamate then acted in a triple capacity as a respiratory substrate for the generation of ATP, as a precursor for oxaloacetate formed oxidatively, and as donor of nitrogen to form aspartate by transamination. As far as tissue slices were concerned, permeability barriers toward aspartate and glutamate prevented the demonstration of precursor-product relationships and obscured the adventitious dependence on coupled phosphorylation.[4,5]

Although oxidative ATP generation was thus separated from the ATP requirement for arginine synthesis it was nonetheless equally essential to take cognizance of the new roles for oxaloacetate and α-ketoglutarate in relation to the oxidative operations of the Krebs citric acid cycle and in relation to the transfer of nitrogen for urea formation. We proposed in 1949 that under physiological conditions the ornithine cycle must be linked to the citric acid cycle through glutamic–aspartic transamination, glutamic dehydrogenase, and phosphorylation coupled to respiration.[4,5] These metabolic interrelations are shown in Fig. 2 as originally formulated. The scheme also explained the experimental observations discussed above and assigned a central metabolic function to aspartic-glutamic transaminase, not fully understood at that time. The bicycle has been updated frequently through further elaboration of the reaction sequence of the urea cycle pathway, the discovery of the mitochondrion and elucidation of mitchondrial functions, and related transport processes, and by full appreciation of the role of glutamine in nitrogen transfers and NH_3 uptake, the outcome of many studies by A. Meister.

The observation that fumaric acid was a product of the interaction of citrulline with aspartate, and the fact that the rate curves showed an induction period, suggesting accumulation of an intermediate, implied that two enzymatic steps were involved in the conversion. One enzyme catalyzes the ATP-utilizing condensation of citrulline with aspartate to form

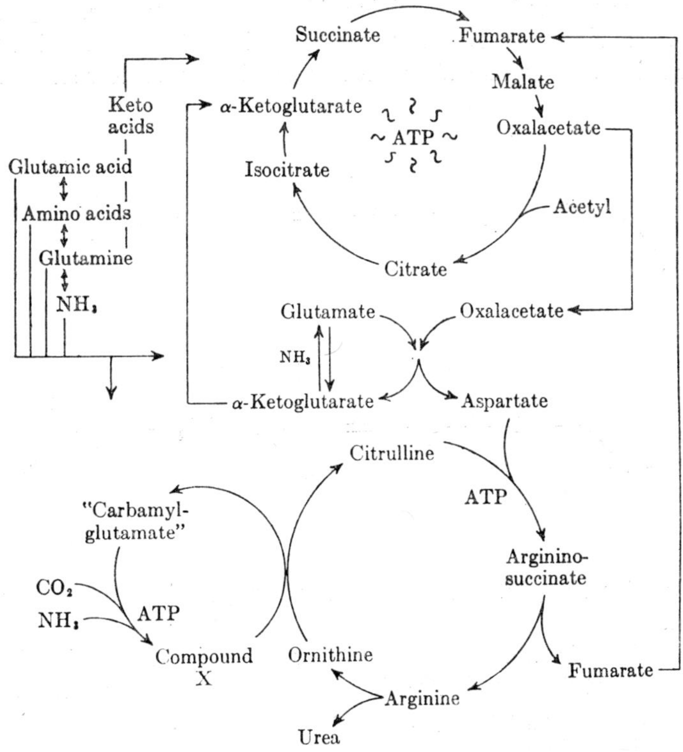

FIG. 2. The relationship of the urea cycle to the citric acid cycle and to transamination. Pathway of amino nitrogen transfer to form citrulline, arginine and urea. (From refs. 4, 5.)

an intermediary condensation product, which we called argininosuccinate, and a second enzyme catalyzes its cleavage to arginine and fumarate.

$$\text{Citrulline} + \text{aspartate} + \text{ATP} \overset{Mg^{2+}}{\rightleftharpoons} \text{argininosuccinate} + \text{AMP} + \text{PP}_i \quad (1)$$

$$\text{Argininosuccinate} \rightleftharpoons \text{arginine} + \text{fumarate} \quad (2)$$

It was only after we separated the two enzymes, thus permitting isolation of the new amino acid,[5-8] that all the pieces fell into place to my satisfaction. I knew that in the structure of this new compound lay the proof I had been seeking and I felt then the sense of discovery that the scientist experiences.

GENERAL FEATURES OF NITROGEN TRANSFER FROM ASPARTATE IN FORMATION OF C—N BONDS

To form the guanidino group of arginine the transfer of nitrogen from aspartate occurs so as to circumvent the participation of free NH_3. In step (1a) condensation of the acceptor carbon, in this case the isoureido carbon of citrulline, with the aspartate nitrogen gives rise to an intermediate which retains the aspartate carbon chain. In step (2a) chain detachment occurs so as to eliminate fumaric acid. Within a few years it was shown in other laboratories that the conversion of inosinic acid to adenylic acid, supposedly through the uptake of NH_3, actually proceeds through the formation of adenylosuccinate by an analogous pair of reactions. Soon thereafter Buchanan and collaborators showed that nitrogen

$$\begin{array}{l} \text{H} \\ \text{HN} \\ \quad\diagdown \\ \quad\quad \text{C—OH} \\ \quad\diagup \\ \text{HN} \\ \quad\diagdown \\ \quad\quad \text{R} \end{array} + \text{H}_2\text{N—}\underset{\text{H}}{\text{C}}\text{—}\underset{\text{H}}{\text{C}}\text{—} \underset{}{\overset{\text{ATP}}{\rightleftharpoons}} \begin{array}{l} \text{HN} \\ \quad\diagdown \\ \quad\quad \text{C—NH—}\underset{\text{H}}{\text{C}}\text{—}\underset{\text{H}}{\text{C}}\text{—} \\ \quad\diagup \\ \text{HN} \\ \quad\diagdown \\ \quad\quad \text{R} \end{array} \qquad (1a)$$

$$\begin{array}{l} \text{HN} \\ \quad\diagdown \\ \quad\quad \text{C—NH—}\underset{\text{H}}{\text{C}}\text{—}\underset{\text{H}}{\text{C}}\text{—} \\ \quad\diagup \\ \text{HN} \\ \quad\diagdown \\ \quad\quad \text{R} \end{array} \rightleftharpoons \begin{array}{l} \text{HN} \\ \quad\diagdown\!\!\diagdown \\ \quad\quad \text{C—NH}_2 \\ \quad\diagup \\ \text{HN} \\ \quad\diagdown \\ \quad\quad \text{R} \end{array} + \begin{array}{c} \diagdown \quad\; \text{H} \\ \text{C=C} \\ \text{H} \quad\; \diagdown \end{array} \qquad (2a)$$

atom 1 of the purine ring, introduced as amide nitrogen, is derived from aspartate by analogous steps.

For many species uric acid rather than urea represents the form in which nitrogen is excreted. It is pertinent that the nitrogen in positions 3 and 9 of the purine ring, previously thought to come from NH_3, actually come from glutamine. One-half of the pyrimidine ring nitrogens originates in aspartate.

MECHANISM OF GUANIDINO GROUP SYNTHESIS

We have pursued an in-depth inquiry into the biosynthetic mechanism of formation of this type of C—N bond because it presented features that were unexpected, unique and obscure. Our enzymological studies have clarified many aspects of the mechanism of action and structure–function relationships. A few, most crucial to metabolic function, can be briefly touched on here.

The biosynthesis of argininosuccinate represents the primary synthesis of the guanidino group. Experiments with ^{18}O suggested that the mechanism of citrulline activation involves transadenylation from ATP. To obtain direct proof for this has required some coaxing. The synthesis is catalyzed by a single enzyme but proceeds on the enzyme surface by a step-wise mechanism consisting of two partial reactions (1b) and (1c). In partial reaction (1b) the adenyl group of ATP is transferred to the isoureido group of citrulline, and if aspartate is absent, the citrulline adenylate that is formed remains bound to the enzyme, as does also PP_i. The very fact that they remain bound to the enzyme in partial reaction (1b) posed difficulties in the way of proof. Eventually a citrulline-dependent cleavage of ATP gave direct proof for reaction (1b) and further proof was obtained by pulse-labeling experiments and by isotope exchange in the presence of an aspartate analogue.

On condensation of the activated citrulline with aspartate, argininosuccinate is liberated. At the same time the affinity for PP_i is lowered, presumably through a conformational

$$\mathrm{E + citrulline + MgATP^{2-} \rightleftharpoons E{<}^{\text{AMP-citrulline}}_{\text{MgPP}_i}} \qquad (1b)$$

$$\mathrm{E{<}^{\text{AMP-citrulline}}_{\text{MgPP}_i} + aspartate \rightleftharpoons E + argininosuccinate + MgPP_i^{2-} + AMP^{2-}} \qquad (1c)$$

change, and dissociation of PP_i becomes detectable. Even then the presence of inorganic pyrophosphatase is necessary for optimum catalysis in order to relieve product inhibition by PP_i.[6]

The mechanism of the reversible cleavage of argininosuccinate to arginine and fumarate, reaction (2), involves a stereospecific β-elimination reaction such that the elimination of arginine or addition of fumarate is *trans*. The stereospecificity coincides with that previously found by others for aspartase and fumarase thus revealing similarities in substrate–enzyme orientations for this group of enzymes. The results have bearing on hydrogen donor specificity in the hydrogen transfer of the citric acid cycle.

The ease of reversibility of reaction (2) conveniently lends itself to the preparation of argininosuccinate. Argininosuccinate is an unstable amino acid primarily because it becomes cyclyzed quite easily through anhydride formation. Neither one of the two anhydrides formed in this way are active as substrate. The unstable properties[7–9] explains to some extent why this amino acid has been overlooked.

HOW DOES THE UREA CYCLE OPERATE—WHAT MAKES IT RUN?

The substrate structures and chemical reactions participating in the conversion of citrulline to arginine are given in Fig. 3. The intermediate formed in the conversion of ornithine to citrulline was found by Grisolia and Cohen in 1952.[10] The structure of this compound was shown by chemical synthesis to be carbamylphosphate by Jones, Spector and Lipmann in 1955.[11] Subsequent advances on carbamyl phosphate synthetase and ornithine transcarbamylase in the laboratories of P.P. Cohen, S. Grisolia, P. Reichard, M. E. Jones and M. Marshall have contributed to the enzymology and elucidation of the reaction mechanisms (cf. Fig. 3). The two enzymes are localized within the liver mitochondria, the last three enzymes of the cycle are in the cytosol.

The energy that drives the urea cycle is derived from ATP and the utilization of this energy is confined to the biosynthesis of the carbamyl group and the guanidino group. Considerable attention has been given to the mechanisms of utilization of phosphate bond energy in biosynthetic reactions and to the evaluation of the energy of the bonds thus formed. F. Lipmann had stressed the usefulness of expressing the value for the energy-rich pyrophosphate bond of ATP ($\sim$P) as $\Delta F°$ of hydrolysis for comparison with bonds of lower or higher transfer potential. A number of $\Delta F°$ values had been gathered for phosphate ester bonds among the glycolytic intermediates, and for acetyl and acyl phosphate bonds, thio ester bonds, and peptide bonds. The values varied considerably and each carried significant metabolic implications with respect to the reaction potential of the bond.

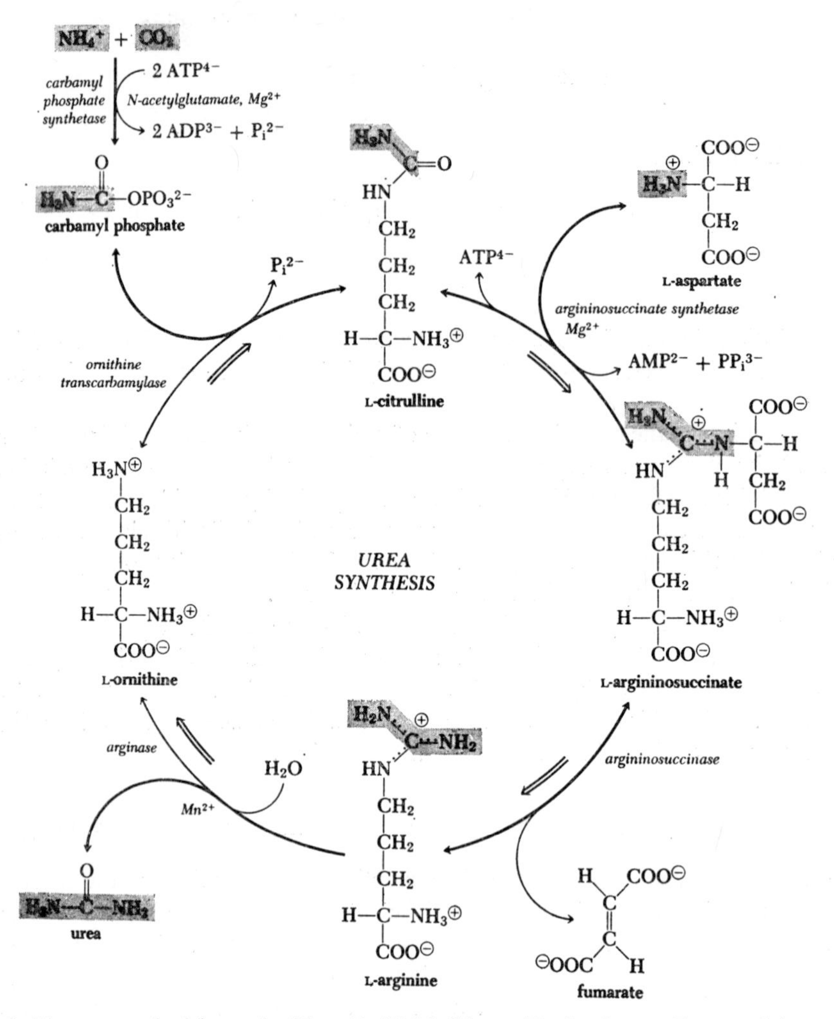

FIG. 3. The urea synthesizing cycle. (From R. W. McGilvery, *Biochemistry: a Functional Approach*, W. B. Saunders Co., Philadelphia 1970.)

Here was an opportunity to evaluate the bond energy in a quite different bond, the C—N bond synthesized in reaction (1) and destined for hydrolytic cleavage to urea. By way of added interest this bond is incorporated in the resonating guanidino group structure of arginine and creatine, compounds that are converted to "high-energy" phosphate esters in muscle to serve as reservoirs of phosphate bond energy.

The $\Delta F^{\circ\prime}$ at pH 7.5 for reaction (1), based on equilibrium measurements, was found to be close to the value for the hydrolytic cleavage of ATP. By taking the magnesium complexed species of ATP, AMP and PP_i into consideration, our calculations showed that $\Delta F^{\circ\prime}$ for the hydrolytic cleavage of ATP at the α,β pyrophosphate bond is more negative by several kcal than cleavage to ADP and P_i.[12]

Thus a mechanism involving transadenylation from ATP is energetically more favorable than one involving transphosphorylation. Since PP_i is the product of reaction (1), and inorganic pyrophosphatase is highly active in most tissues, it can be assumed that the synthetase reaction is driven by two high-energy pyrophosphate bonds. This additional "built-in" energy source marks a second thermodynamic advantage of the transadenylation mechanism. The same advantages accrue to the aminoacyl-tRNA and acetylCoA synthetase reactions; they involve transadenylation.

The synthesis of carbamyl phosphate probably involves at least three partial reactions and here also two moles of ATP are utilized. The equilibrium position for ornithine transcarbamylase highly favors the transfer of the carbamyl group.

To total up the energy cost, phosphate bond energy inherent in four pyrophosphate bonds is consumed to promote the synthesis of urea from NH_3, CO_2 and aspartate (cf. Fig. 3). System operation is further insured through the relative proportions found *in vivo* among the five participating enzymes. The so-called coupling enzymes, ornithine transcarbamylase and arginase, are normally present in very large excess ranging from 20–40-fold and 200–1000-fold, respectively, relative to the two synthetases. Judging from the very low intracellular stead-state concentrations of the intermediates ornithine, citrulline, argininosuccinate and arginine[9] the enzymatic design of the system further insures efficient operation.

MOLECULAR BEHAVIOR IN RELATION TO BIOLOGICAL PROCESSES

The availability of the necessary substrates and methods for estimating the *in vivo* activities of the individual enzymes in tissue homogenates have prompted studies by R. T. Schimke[13] on the regulation of this pathway in livers of rats in response to increases in dietary protein load. His results show adaptive coordinated increases in net amounts of each of the urea cycle enzymes gained over a 4-day period. Somewhat earlier, P. P. Cohen and his colleagues[14] showed that the increases in each of the urea cycle enzymes were coordinated during the development of ureotelism in tadpole metamorphosis. P. P. Cohen and also N. C. R. Räihä demonstrated progressive coordinated increases in the livers of rats during prenatal development. Insights into these relationships have indeed succeeded in linking molecular behavior to biological processes.

A large body of literature has developed dealing with the genetic regulation of the enzymes of arginine biosynthesis from *E. coli* and *Neurospora crassa*, and a body of comparative literature relates the evolutionary development of ureotelism in invertebrate and vertebrate species.

HUMAN MUTATIONS AND NEW QUESTIONS

Familiarity with the intermediates and enzymes participating in urea synthesis has contributed to the recognition of human metabolic abnormalities of genetic origin associated with a deficiency of enzymatic activity in the liver. Inborn errors affecting each of the urea cycle enzymes have been described.[15] Among the earliest to be recognized were partial losses in activity of argininosuccinase and argininosuccinate synthetase since these are associated respectively with elevated plasma levels of argininosuccinate or citrulline, accumulating behind the partial block, and with their excretion in the urine. With early diagnosis the number of reported cases of non-lethal and lethal mutational variants for each enzyme deficiency has increased rapidly. In addition to characteristic biochemical changes, afflicted infants and children with partial deficiencies of any one of the enzymes exhibit certain features in common: mental deficiency, neurological changes, abnormal growth and development to some degree, protein intolerance and hyperammonemia, particularly in response to a protein load.

Possibilities for prenatal diagnosis by enzymatic assay are promising. Dietary therapy varies with the defect and two forms hold promise: low protein diet supplemented with arginine or with "essential" α-keto acids.

The normal level of NH_3 in the plasma is extremely low, about 0.03 mM; neurological manifestations of toxicity begin to appear when the concentration is increased 3- or 4-fold. It is now generally accepted that the convulsions, coma, and other neurological changes are the result of elevated ammonia levels in the blood and brain. A few biochemical studies show that despite a marked degree of enzymatic deficiency (10–20% of normal), the activity is sufficient to support appreciable urea synthesis, amounting to as much as 80% of normal. This is not altogether surprising to enzymologists but the presence of hyperammonemia as a common finding is difficult to understand.

The synchronization of urea production with gluconeogenesis has been repeatedly observed. How does the operation of the urea cycle adjust to this flux and what share is born by metabolic systems in muscle? Assuming appreciable sequestering of NH_3 in the form of glutamine, how does the regulation of glutamine synthetase, glutamine transaminase and the urea cycle interact with gluconeogenesis? What flexibility in kinetic adjustment is available when one rate-limiting step or component is substituted for another. The nature of the metabolic abnormalities of urea cycle enzymes and their biochemical expression has generated new questions and new imperatives.

REFERENCES

1. KREBS, H. A. and HENSELEIT, K. (1932) Untersuchungen über die Harnstoffbildung in Tierkörpes. *Z. Physiol. Chem.* **210**, 33–66.
2. COHEN, P. P. and HAYANO, M. (1946) The conversion of citrulline to arginine (transimination) by tissue slices and homogenates. Urea synthesis by liver homogenates. *J. Biol. Chem.* **166**, 239–250; 251–259.
3. RATNER, S. (1947) The enzymatic mechanism of arginine formation from citrulline. *J. Biol. Chem.* **170**, 761–762.
4. RATNER, S. and PAPPAS, A. (1949) I. Enzymatic mechanism of arginine synthesis from citrulline. II. Arginine synthesis from citrulline in liver homogenates. *J. Biol. Chem.* **179**, 1183–1198; 1199–1212.
5. RATNER, S. (1954) Urea synthesis and metabolism of arginine and citrulline. In *Advances in Enzymology* (F. F. NORD, ed.), **15**, 319–387, Interscience, New York.
6. ROCHOVANSKY, O. and RATNER, S. (1967) XII. Further studies on argininosuccinate synthetase: substrate affinity and mechanism of action. *J. Biol. Chem.* **242**, 3839–3849.
7. RATNER, S., PETRACK, B. and ROCHOVANSKY, O. (1953) V. Isolation and properties of argininosuccinate. *J. Biol. Chem.* **204**, 95–114.
8. RATNER, S., ANSLOW, W. P. and PETRACK, B. (1953) VI. Enzymatic cleavage of argininosuccinic acid to arginine and fumaric acid *J. Biol. Chem.* **204**, 115–125.
9. RATNER, S. (1973) Enzymes of arginine and urea synthesis. In *Advances in Enzymology* (A. MEISTER, ed.), **39**, 1–90, John Wiley & Sons, New York.
10. GRISOLIA, S. and COHEN, P. P. (1952) The catalytic role of carbamyl glutamate in citrulline synthesis. *J. Biol. Chem.* **198**, 561–571.
11. JONES, M. E., SPECTOR, L. and LIPMANN, F. (1955) Carbamyl phosphate, the carbamyl donor in enzymatic citrulline synthesis. *J. Am. Chem. Soc.* **77**, 819–820.
12. SCHUEGRAF, A., RATNER, S. and WARNER, R. C. (1960) Free energy changes of the argininosuccinate synthetase reaction and of the hydrolysis of the inner pyrophosphate bond of adenosine triphosphate. *J. Biol. Chem.* **235**, 3597–3602.
13. SCHIMKE, R. T. (1962) Adaptive characteristics of urea cycle enzymes in the rat. *J. Biol. Chem.* **237**, 459–468.
14. COHEN, P. P. (1970) Biochemical differentiation during amphibian metamorphosis *Science*, **168**, 533–543.
15. SHIH, V. E. and EFRON, M. L. (1972) Urea cycle disorders. In *The Metabolic Basis of Inherited Disease*, 3rd ed. (J. B. STANBURY, J. B. WYNGAARDEN and D. S. FREDRICKSON, eds.), pp. 370–392. McGraw-Hill, New York, New York.

MOLECULAR ECOLOGY

CARLOS ASENSIO
Instituto de Enzimología del CSIC, Facultad de Medicina
de la Universidad Autónoma, Madrid-34, Spain

"I am I and my circumstance".
J. Ortega y Gasset, 1914

A GLIMPSE AT THE ENVIRONMENT

A few years ago I passed through the pretty village of Luarca, in northern Spain. It was summer time and I guessed that its illustrious son Severo Ochoa and his wife Carmen were there. To find out I asked the busy policeman at the central plaza. "No, sir. They left yesterday", was his calm answer while commanding the thick traffic.

The folk of that region are a jocular people with a delightful sense of humor, I was told the blame for this stems from the usual foggy and drizzly weather. They have an old tradition of whale fishing and can also be proud of setting the fastest crossing of the Atlantic by sail. They did it with a brigantine in the last century. As a matter of fact, the men first look to the sea horizons every morning. A mighty sea, indeed. It is still a ritual as in Hemingway's tale. But they also glance backwards, where the earthy fields yield other harvests. Besides, these are in a beautiful evergreen scenery, with an imposing mountainous background. Alas, I must apologize for writing about things apparently far from a biochemical reflexion. Let me give two motives for doing so. First, I am from the same province and, after all, the Editors' message was to be ourselves. Second, modern psychology emphasizes the lasting impact of the early environment. I have tried to describe Ochoa's. He later moved to others not less exciting for different reasons, such as Madrid, Heidelberg, Oxford, and finally New York, his much beloved City. I think all of them contributed somehow to shape him and the work he has done. The man and his circumstance! Actually it points to the principal concerns of this reflective essay.

It is most evident that a new attitude towards our surrounding Nature is emerging. We perceive everywhere a growing pressure to stop, and if ever possible to reverse, the galloping degradation of the biosphere, on which man, a minimal part, is absolutely dependent. This movement, apparently vast and profound, will demand the generalization of novel cultural patterns as well as a considerable amount of new knowledge.

It seems to me that biochemistry should play an important role in these coming events. For this to happen, however, it might also be necessary to change the focus of the biochemist's outlook on nature; a change of mood and style.

Biochemical research has traditionally proceeded as in other biological disciplines by way of simplification. From man, we moved to rats and then to *E. coli* and viruses. And in every case we used their extracts. Meanwhile, somebody from time to time undertakes the task of making some sense out of it, so that we are gaining a beautiful picture of coherence and unity. However, new and pending problems accumulate, some of them as challenging as ever.

In any event, we have today a much better understanding than in the past decade about the organizational levels of living beings and of how they are inextricably linked. Among these levels the upper stage is the ecological, to which we pay the least attention. It seems reasonable to say that advances in this area should be feasible and fruitful at the molecular level as they were in other recent biochemical endeavours. But it is not an easy task to dig and pick novelties on ecosystems by applying our usual tools and attitudes. Human beings find it difficult to study systems that by their very nature resist dissection. Szent-Györgyi has related that when he tried to get into the core of high physics at Princeton, those great scientists suspended the dialogue with him as soon as they learnt that in living systems there are more than two electrons.(1) The same with *E. coli*. We know a lot about the organism but have a meager knowledge of its surroundings, where the cell competes and makes a living in the natural environs. We talk of *E. coli* in the singular but very little in connection with its brothers, cousins and aliens, not to say much bigger cells. And yet, together they make the wholeness where each one is only a part. When I took van Niel's course at Pacific Grove (1960) I remember that somebody asked him which were *really* the carbon and energy sources of *E. coli* in the intestinal system. "I don't know, sir", he answered with his typical candor. Nobody in the audience replied when he passed the simple question onto them. So much we know of our inhabitant and so little about how it interacts with other bacteria and the cells that surround all of them in an ecosystem of obvious interest. Regarding this we are now involved with some observations that, we hope, would be of some value. I shall summarize them in the following.(2)

THERE MUST BE

The human intestinal tract lodges a rather well-defined community of microorganisms that show, however, a dynamic succession of shifts both in the species composition and in the relative abundance of each resident, either under normal or pathological conditions. These successions are not infrequently very fast and specific for some invaders, which are able in some way to displace other very closely related species by mean of mechanisms poorly understood. In the case of enterobacteria, for instance, colicins have been invoked to explain such displacements but the evidence available seems unsuitable.(3) Consequently, one could guess at the existence of other bacterial agents endowed with the needed information to effect specific competition. On this basis, we recently undertook a systematic exploration for compounds of low molecular weight produced by enterobacteria isolated from the intestinal contents of infants, which could act as antibiotics toward other closely related microorganisms. The screening approach was designed to exclude interference by conventional colicins. To our surprise, about 15% of the isolates assayed with several different strains of *E. coli* as sensitivity markers, clearly indicates the occurrence of such compounds. We are now struggling with the isolation and characterization of the most conspicuous substances so far detected, over a dozen, and we suspect there are many more. At present, available data suggest that their chemical nature may involve structures of amino acid derivatives, with molecular weights around 1000 or less. One of them has been in fact identified as l-valine, which was somehow a disappointing finding were it not that it allows a sort of ecological speculation. Though we have not as yet worked out the biochemical basis of the l-valine excretion (several strains were found) it could be presumed as being due to a low affinity for this compound at the allosteric site of the acetohydroxy acid synthetase that initiates the branched-chain amino acid biosynthetic pathway, as has been shown with artificial mutants.[4] The same mechanism is atypical in *E. coli* K12 as studied by Leavitt and Umbarger.[5] Here, however, the resulting behavior is just the opposite: the K12 strains are uniquely sensitive to l-valine because that synthetase shows an unusually high affinity for this amino acid, which thus blocks the pathway at a very low concentration. The corollary of this hypothesis is that allosteric mechanisms, which are considered intracellular regulatory devices, could be used through subtle changes for more dramatic *inter*cellular events that might be properly described at the ecological, and eventually evolutionary, levels.

Returning to the other low molecular weight antibiotics we have detected, they are thermoresistant, soluble in methanol–water (5:1) and insensitive (except one) to proteases. Besides, they show a remarkable specificity, for they act on closely related species, although in a way less restricted than typical colicins, and this makes us think they constitute a novel family of antibiotics, that might be somehow related to the bacterial successions of the intestinal ecosystem.

The above description and comments might constitute an instance, however limited, of the possibilities open to the ecological approach, which could be, as in our case, rather unsophisticated. A clear advantage for research in this field is that we now have available a cohesive frame of biological thought that allows us to jump, to make short cuts in order to pick new concepts and facts for which real evidence is patently absent. This productive frame is being fitly called "la logique du vivant". And the jumps it permits us to dare could be referred to as a sort of "there-must-be approach" which I shall discuss briefly later on.

A BRIEF ECOLOGICAL EXCURSION

As a matter of fact, the ecological perspective at the molecular level is not restricted to the microbial world. Its richness covers all the chemical communication that links all creatures across their external spaces, and it is even more than that. However, microbial systems here again may supply simplifying models of interaction. Biochemical studies of the microbe–microbe type under an ecological view are scarce, to my knowledge, and this is particularly surprising if one considers the success achieved by molecular biologists by the binomial virus-bacterium. Similar considerations apply to cell–cell interactions in eucaryotic systems, which are only recently receiving considerable attention as a consequence of the rising interest in the membrane inhibition phenomenon. In fact, these latter systems hint at the vastness of the ecological scope. Gerard once said that "physiology could be called the ecology of cells in the body, as sociology could be called the physiology of society".[6] Among procaryotic organisms the surrounding habitat is usually little organized and intemperate. In pluricellular organisms, however, cells interact in habitats that are homeostatic and endowed with much information. When did the evolution to a closed milieu occur? One can imagine such a transition as a process that involved several evolutive steps, for which the cellular slime molds might represent a plausible model.

Bacteria are able to excrete cAMP, for missions unknown. We know, however, that this compound attracts the ameboid cells of the mold, which thus feed on the bacteria. When the cells of the cellular slime mold aggregate to form the differentiated stalk, cAMP produced by them now serves as a chemical mediator for cellular aggregation.[7] In this way, the same compound is able to act at different organizational levels, as an external chemical agent, as a primary hormone, or in higher organisms as a second messenger. Through this brief ecological excursion it can be perceived that ecology, developmental biology and physiology are closely linked, the difference being where we draw the line between organisms and environment. As stated by John T. Bonner, the difference between ecological patterns and developmental patterns is simply that in the latter the external chemical processes, with size increase, have become progressively supplemented by self-produced, internal chemical processes. The same is true for the physiological functioning of a multicellular organism; again it is achieved by complex internal chemical systems.[8]

We are beginning to explore, besides, other fields in molecular ecology. These include the territorial chemical communication between plants and between them and their inanimate surroundings, the signaling amidst animal species, the ecology of terpenoids and esteroids as non-hormonal linkers, and the hormonal interactions between plants and insects. The recent developments in this latter area resemble science fiction and I will comment briefly on them.[9]

Since Butenandt's discovery of ecdysone, the insect hormone for metamorphosis, more than two decades have elapsed. The clarification of its chemical structure was a hard job, particularly because of the difficulty in obtaining the product from its natural source, the silkworm. Ten years later its formula was finally elucidated and to everyone's surprise it was a sterol derivative, because unlike most organisms, insects are unable to synthesize the sterol ring structure. Research in this field was greatly facilitated through an unexpected discovery made in 1967 by Japanese workers, who found huge amounts of ecdysone-like substances, including β-ecdysone, in certain plants. Their observations were soon confirmed and extended. To make clear the point, suffice it to say that the Butenandt group needed a ton of silkworms to obtain 25 mg of the hormone, whereas an equivalent amount could be obtained from 2.5 g of dried rhizomes of the common fern *Polypodium vulgare*. The other side of this history is even more striking and relates to the juvenile hormone, which functions as a "brake" for ecdysone to allow completion of the metamorphosis stages. Here again difficulties were encountered in work with this elusive compound. Meanwhile, biologist Karel Slama found that the European bug he brought from Prague, failed to undergo normal development in Williams' laboratory at Harvard. This had never happened to him before. This was a puzzle that mystified everybody for a while until it became apparent that the Harvard cultures had access to some unknown source of juvenile hormone: it was present in the paper in the bottom of the dishes that harbored the bugs! When the Scott Brand paper was substituted for Whatman's filter paper, as in Prague, the entire phenomenon vanished. It was finally found out that some of the typical American sources of paper pulp (balsam fir, eastern hemlock, Pacific yew, and others), unlike their European counterparts, contain enormous amounts of juvenile phytohormone that selectively act on the insects of the family Pyrrhocoridae, to which that bug belongs. This almost incredible series of events brings forward a new vista for the Darwinian doctrine if we come to agree that those plants make exorbitant amounts of phytoecdysones and juvenile phytohormones for something that makes sense, like a self-defense against insect predation. If so, it would represent a parallel evolution between very heterologous species for the synthesis of analogue compounds which, however, accomplish antagonistic missions for ecological reasons.

ON THINGS AND WAYS TO GO

The new ecological attitude is already demanding a revision of our scale of values in some respect, for instance with regard to Western Hippocratic and Judeo-Christian traditionally anthropocentric views,[10] which are often at odds with Darwinian premises. Accordingly, it could be questioned whether Fleming's petri dish came too early, and whether or not we are transgressing the natural law when we treat a case of juvenile diabetes or a phenylketonuric infant with the new science without further eugenic measures, thus favoring individuals at the cost of degrading the species. Likewise, it is interesting that these diseases are called "inborn errors of metabolism", since it is evident that the healthy individual is affected by many such "errors" which are undramatic because of ecological interactions, and thus have given rise to the euphemistic names "vitamin" and "essential amino acid". Furthermore, it seems appropriate to ask whether the human genetic load abounds in conditional lethal mutations, which are in fact ignored in Stanbury, Wyngaarden and Fredrickson's comprehensive treatise.[11] Genetic polymorphism and the rest of our refined homeostatic mechanisms would take care of masking and preserving them through generations. Has anybody considered how many human deaths concomitant with high fever could be due to the expression of underlying thermosensible mutations?

In the preceding, I have tried to make a sort of integrative survey by using more or less obvious ecological knowledge. New developments in this field would probably require, as I said above, newer patterns in the biochemical arts. In these, the so-called there-must-be approach could be eventually a profitable strategy.

We all learned too well how Louis Pasteur, after a series of remarkable experiments, smashed the spontaneous generation theory. He put forth, however, an authoritative "ne pas possible" dictum that brought no good for a long time. In fact, it took many decades to overcome his paralyzing testimonial, until Haldane, Oparin and, years later, Stanley Miller (1953) started to probe a fascinating subject anew.

This example might typify, admittedly as an extreme case, the Baconian style that held powerful reins on experimental science for several centuries. Indeed, biochemists did not try to escape the rules, being on the contrary reliable servants. Here again a giant, Otto Warburg, set a model to be followed. Fritz Lipmann wrote recently: "Warburg's puritanical, pragmatic approach which rejected compromises and speculation filtered through us . . . his stern insistence of letting experiments speak and keeping interpretation to a minimum has dominated our generation".[12] This behavior spread to virtually the entire biochemical community in the 1930s and 1940s, much helped for reasons everybody knows. Warburg's style particularly impregnated the German scholars, which includes people like Lipmann, Krebs, Ochoa, Lynen and others. Well, they made up the core of present biochemistry, so it is hard to argue against such accomplishments.

We contemplate, however, signs of cracking of that orthodoxy and these are protagonized by some of the leading pioneers of molecular biology. The keyword is an old one, teleology, now rechristened as teleonomy.[13] Jacques Monod has forcefully expressed its potential for biological research.[14] Any assumption of "finality" when studying biological systems means a clear transgression of the objectivity principle that provides the foundation for the scientific method. And yet it works. We only have to think of the brilliant *hypo-thetico*–deductive attack on biological regulation made in the 1950s by the Pasteur group. for the scientific method. And yet it works. We only have to think of the brilliant in-ductive–deductive attack on biological regulation made in the 1950s by the Pasteur group. It led these people to anticipate the existence of the repressor and mRNA, without ever "touching" them. In a similar fashion Francis Crick advanced the existence of transfer RNA. The development of the allosteric concept keeps in line with all that. Here, however, some people made the main discoveries (Umbarger, Pardee) and others put forth the doctrine (Monod, Davis, Koshland). The latter, to be fair, opened the way for a general exploration of this phenomenon through the biochemical map, because they provided a teleonomic license for the hunting, a kind of there-must-be incitation that yielded very good results. In this regard, it could be of interest to bring up here a personal recollection of the memorable Cold Spring Harbor Symposium of 1961. Monod proposed there the allosteric concept, apparently for the first time, although under another name: the "NSU-effect" (after Novick and Szilard who made a seminal contribution, and Umbarger). Umbarger, indeed fairly, objected to a generic name. He favored adhering to actual mechanisms instead of generalizing from (then) very few cases, which would otherwise imply risky teleonomy. With our present perspective, Monod's proposition was, in any event, of a highly heuristic value.

It is at least to be hoped that ecology, being placed somewhere at the apex of molecular logic for living organisms, will benefit from teleonomic and maybe other types of un-orthodox excursions. In turn, it will surely pay off in the enrichment of other territories, as is usually the case within the supreme Gestalt of Biology.

REFERENCES

1. Szent-Györgyi, A. (1964) Teaching and the expansion of knowledge. (1976); *Science*, **146,** 1278–1279.
2. Asensio, C., Pérez-Díaz, J. C., Martínez, M. C. and Baquero, F. A new family of low molecular weight antibiotics from enterobacteria. *Biochem. Biophys. Res. Commun. 69*, 7–14.
3. Ikari, N. S., Kenton, D. M. and Young, V. M. (1969) Interaction in the germfree mouse intestine of colicinogenic and colicin sensitive microorganisms. *Proc. Soc. Exp. Biol. Med.* **130,** 1280–1285.
4. De Felice, M., Guardiola, J., Marloni, M. C., Klopotowski, T. and Iacarrino, M. (1974) Regulation of the pool size of valine in *Escherichia coli* K12. *J. Bacteriol.* **120,** 1058–1067.
5. Leavitt, R. I. and Umbarger, H. E. (1962) Isoleucine and valine metabolism in *Escherichia coli.* XI. Valine inhibition of the growth of *E. coli*, strain K12. *J. Bacteriol.* **83,** 624–630.
6. Gerard, R. W. (1952) The organization of science (Prefactory chapter). *Ann. Rev. Physiol.* **14.**
7. Markman, R. S. and Sutherland, E. W. (1965) Adenosine-3′,5′ phosphate in *Escherichia coli. J. Biol. Chem.* **240,** 1309–1314.
8. Bonner, J. T. (1970) The chemical ecology of cells in the soil. In *Chemical Ecology* (E. Sondheimer and J. B. Simeone, eds.), Academic Press, New York.
9. For more information on this subject see: "Hormonal interaction between plants and insects" by C. M. Williams, published in the book cited in ref. 8.
10. White, L., Jr. (1967) The historical roots of our ecological crisis. *Science*, **155,** 1203–1207.
11. Stanbury, J. B., Wyngaarden, J. B. and Fredrickson, D. S. (1972) *The Metabolic Basis of Inherited Diseases*, McGraw-Hill, New York.
12. Lipmann, F. (1971) *Wanderings of a Biochemist*, Wiley, New York.
13. Davis, B. D. (1961) Opening Address: The teleonomic significance of biosynthetic control mechanisms. In *Symp. Quant. Biol.*, vol. XXVI, pp. 1–10, The Biological Laboratory, Cold Spring Harbor, L.I., New York.
14. Monod, J. (1970) *Le Hazard et la Nécessité*, Seuil, Paris.

IV. NUCLEIC ACIDS AND THE GENETIC CODE

FOR THE LOVE OF ENZYMES

ARTHUR KORNBERG
Department of Biochemistry, Stanford University School of Medicine, Stanford, California 94305

I have never known a dull enzyme. Succinic dehydrogenase, aconitase, isocitric dehydrogenase and malic enzyme; these were brief encounters during the first year I met enzymes, but my recollections of them are still vivid and tinged with pleasure.

There is said to be a monogamous relationship with enzymes: "one man, one enzyme". It is even more true that some enzymes support the careers of many men. Despite a happy marriage for 20 years to DNA polymerase, I feel impelled in this age of frank revelation to expose flirtations, affairs and fantasies with other enzymes. I am also impelled by the current infatuation with the sex lives of flies, worms and even DNA molecules to plead for the charms and virtues of enzymes.

I first came to know enzymes through a curious set of circumstances. I was training to be a physician when World War II supervened and so found myself a ship's doctor in an armed force branch of the United States Public Health Service. After a few months of sea duty, a transfer was expedited by a captain exasperated with my inattention to naval etiquette. On the strength of research I had done as a student investigating my own jaundice, I was assigned in 1942 to work on rat nutrition at the National Institute of Health.

Full-time laboratory work was even more engaging than patient care. One could initiate discrete experiments rather than accept an assortment of clinical puzzles. A complex or boring problem could be set aside for the sake of sharpening the focus. Experimental designs were not governed by compassion and chance. Still the duration of a rat nutritional deficiency experiment was about 2 months and so many experiments had to be run concurrently, including some which were redundant or misguided.

After 3 years, I had acquired control of a large colony of rats and had described their dependence on various vitamins and amino acids in some twenty papers. To find out that folic acid was required for producing blood cells was worth-while but not to know why or how was frustrating.

My interest in the vitamin needs of rats paled further when I heard about enzymes. Of course, enzymes were mentioned in my biochemistry course in 1938. The standard textbook by Meyer Bodansky devoted many pages to them. But they did not seem interesting, any more than did the nucleic acids.

One day in 1944, I was enthralled by a seminar on the "one gene–one enzyme" concept by Edward Tatum based on his Neurospora work with George Beadle. An even greater revelation for me were the feats of enzymes described in the work of Otto Warburg, Fritz Lipmann, Herman Kalckar, Carl Cori and Severo Ochoa. Here was a window on a new world of science: enzymes of astonishing specificity and catalytic potency linked the combustion of foodstuffs to the generation of ATP which made the cell grow and the muscle move. What fantastic natural poetry! But there was no one at the NIH doing such enzymology or biochemistry. Well, not quite.

SUCCINIC DEHYDROGENASE

There was Bernard Horecker in the Division of Industrial Hygiene. He had been studying the toxicology of agents used in the War, including action of the insect poison, DDT, on cockroaches. Now he was eager to return to earlier interests in the enzymatic reduction of cytochrome *c*. Late in 1945, I helped him make an extract of rabbit muscle, which after slight manipulation was called succinic dehydrogenase. Its activity was measured by the reduction of cytochrome *c* in the presence of cyanide. He observed the rate of increase in optical density at 550 nm in a spectrophotometer and I recorded the values. We were fascinated by a complex formed between cytochrome *c* and cyanide,[1] and used succinic dehydrogenase simply as a biological reducing agent.

ACONITASE

Severo Ochoa was embarked on the isolation and purification of enzymes of the citric acid cycle when I came to his laboratory in January 1946. He suggested I work on aconitase, the enzyme which catalyzes the equilibrium:

$$\text{citrate} \rightleftarrows \textit{cis}\text{-aconitate} \rightleftarrows \text{isocitrate.}$$

My mission was to resolve heart muscle aconitase into the two enzymes presumed to be responsible for the two component reactions. This was my first stab at enzyme purification and after a few months' work I failed to make any significant progress. I also saw no signs of separation of enzymes responsible for the two reactions; aconitase, as others showed later, is a single enzyme.

Despite initial failures, this introduction to enzymology was exhilarating. Aside from the fascination of seeing an enzyme in action, the momentum of experimental work was breathtaking. By coupling aconitase to isocitric dehydrogenase, spectrophotometric assays, based on TPN reduction, took only a few minutes. Many ideas could be tested and discarded in the course of a day. Late evenings were occupied preparing a series of protocols for the following day. What a contrast with the tedious pace of nutritional experiments on rats.

In directing me in the work on aconitase, Ochoa introduced me to the philosophy and discipline of enzyme purification. The cardinal rule I learned from him and have practiced since is that units of activity and milligrams of proteins must be strictly accounted for in each manipulation and at every stage. In this vein, the notebook record of an enzyme purification should withstand the scrutiny of an auditor or bank examiner. Yet, I have never regarded the enterprise as a banking operation but rather like the ascent of an uncharted mountain: the logistics were like supplying successively higher base camps; protein instabilities and confusing contaminants resembled the adventure of unexpected storms and hardships; and the ultimate reward of a pure enzyme was tantamount to the unobstructed and commanding view from the summit.

I was also influenced by Efraim Racker in the nearby Department of Bacteriology. His enthusiastic pursuit of enzyme purification was marked by success and the dictum: "Don't waste clean thoughts on dirty enzymes."

ISOCITRATE DEHYDROGENASE

Ochoa had partially purified the TPN-dependent isocitrate dehydrogenase from heart muscle:

$$\text{isocitrate} + TPN^+ \rightleftarrows [\text{oxalosuccinate}] + TPNH^+ + H^+ \rightleftarrows \alpha\text{-ketoglutarate} + CO_2$$

I examined the role of metals in the enzymatic catalysis of decarboxylation of the oxidized intermediate, oxalosuccinate. I was impressed that in the conversion of a very unstable metal complex of this β-keto acid[2] the cell did not rely on the spontaneous and rapid decarboxylation. An enzyme was on hand to attain rates, orders of magnitude greater than the spontaneous one, and governed by the amount of enzyme available.

Based on the experiences of others and a few of my own, it seemed to me that enzymes could do anything and did everything. I came to believe that every reaction in the cell is enzyme-catalyzed. Once while giving a seminar in the Chemistry Department of Washington University on the variety of enzymes of pyrimidine biosynthesis I noticed a glazed look overtaking the audience. At this juncture I said bluntly that all cellular chemistry is catalyzed by specific enzymes. Among the people who were awakened and jolted was the late Joseph Kennedy, a nuclear chemist. He said: "Come now, would you go so far as to propose the existence of an enzyme to catalyze the hydration of CO_2?" What a relief it was to be able to tell him about a well-known enzyme, called carbonic anhydrase, which did just that.

"MALIC" ENZYME

Luck is an important ingredient of many scientific successes. I had my share with the malic enzyme. First was the opportunity to join Ochoa and Alan Mehler (then a graduate student of Ochoa) in the study of the malic enzyme. The reversible oxidative decarboxylation (analogous to that of isocitrate) and the CO_2 fixation (in the reverse direction) are catalyzed by this enzyme:

$$\text{malate} + TPN^+ \rightleftarrows \text{pyruvate} + CO_2 + TPNH + H^+$$

At the end of 1946, just before leaving Ochoa's laboratory, we undertook a huge preparation of the "malic" enzyme starting with several hundred pigeons who had donated their livers to Morton Schwartz, Ochoa's talented and devoted assistant. After some weeks he had accumulated enough liver acetone powder for an ethanol fractionation of the enzyme. Rather late one evening, Ochoa and I were dissolving the final ethanol precipitate collected in a number of 250-ml, curved bottom, glass centrifuge bottles. I had just poured the dissolved contents of the last bottle into a cylinder when I brushed against one of the bottles on the crowded bench. A domino effect reached the cylinder; it fell and spilled all of the precious enzyme fraction on the floor. By the time I got home an hour later Ochoa had already called; he was anxious about my state of mind and wanted to console me. The next morning in the laboratory, I noticed that the supernatant fluid from the last fraction, which had been saved and stored at $-15°$, was turbid. I collected the precipitate, dissolved and assayed it. Holy Toledo—the fraction had the bulk of the starting activity and it was several-fold purer than the best of our previous preparations. This step (without the cylinder breakage routine) became part of the procedure.[3]

Today we recognize that the "malic" enzyme serves in fatty acid synthesis by generating

TPNH and in the transport cycle that brings acetyl CoA out of the mitochondria. The contribution of the "malic" enzyme or other similar "malo-lactic" decomposing activities in the secondary fermentation of wine is also appreciated, but less well understood. But it can be said that this fermentation is crucial in the development of fine Burgundy and Bordeaux wines, and of the premium red wines of California, such as Cabernet Sauvignon.

LACTIC DEHYDROGENASE

When I came to Washington University in January 1947, Cori assigned me the purification and crystallization of lactic dehydrogenase from rabbit muscle. Glycogen phosphorylase, aldolase and triose phosphate dehydrogenase had been crystallized from this source in his laboratory in previous years. As a byproduct, Gerty Cori had accumulated a paste of proteins precipitated between 52% and 72% of saturation with ammonium sulfate. It was an excellent starting material for obtaining, in a few weeks, a near homogeneous preparation of lactic dehydrogenase. (It impressed me that in a nine-times recrystallized aldolase sample she gave me to test, there was still a 2% content of lactic dehydrogenase; the relatively high turnover number of the latter enzyme made the preparation almost as active a dehydrogenase as an aldolase.)

AEROBIC PHOSPHORYLATION

After failing in a few attempts to crystallize the enzyme, I rebelled at continuing. I had come to the Cori laboratory to resolve the enzymes of aerobic phosphorylation and was determined to work on that problem. But I had not fully grasped why Cori, Ochoa, Kalckar and Lipmann, each of whom had contributed so much to the recognition of aerobic phosphorylation, were no longer working on it; they were working on glycogen phosphorylase, citric acid cycle dehydrogenases, nucleoside phosphorylase and the coenzyme of acetate activation. I came to realize they had chosen to forge ahead with soluble enzymes rather than struggle with the particulate suspensions which sustain aerobic phosphorylation. They were observing the Warburg dictum that where structure exists, enzymology is excluded; for the next 20 years I stayed clear of insoluble enzymes.

Cori suggested I join a young Swedish biochemist, Olov Lindberg. He was examining an observation made about a half-dozen years earlier by Ochoa when he worked with the Coris. Dialyzed liver dispersions metabolizing glutamate, pyruvate or succinate produced inorganic pyrophosphate.[4] The mechanism of its origin was unknown, and it might have been formed from a more labile compound during isolation. It was funny that I should have ended up with the problem. During the previous year with Ochoa he had mentioned the inorganic pyrophosphate story more than once. The strangeness of the compound and the mystery of its origin made it difficult to retain details of the story and we would badger him to repeat it. Even Ochoa's patience could be tried and finally on one occasion he forbade anyone from mentioning the inorganic pyrophosphate business again.

NUCLEOTIDE PYROPHOSPHATASE

Lindberg and I observed inorganic pyrophosphate production by respiring kidney dispersions but learned nothing about its source. But we did observe that respiration was greatly enhanced by the addition of DPN. Its cleavage furnished adenylate (a phosphate acceptor)[5]:

$$\text{DPN} + H_2O \rightarrow \text{adenylate} + \text{nicotinamide mononucleotide.}$$

Purification of nucleotide pyrophosphatase, the enzyme responsible for this cleavage, was the first problem I tackled when I returned to the NIH in the fall of 1947.

My infatuation with enzymes made purifying them fun. I doubt that any effort I put into purifying an enzyme was ever wasted. The purified enzyme might blaze a new pathway, disclose unanticipated subtleties in a reaction, or, at the very least provide a unique and useful analytical or preparative reagent. I recall a visit from Hans Krebs about 1950 at a time when I was working on the enzymatic synthesis of coenzymes and needed an assay for ATP. I was purifying glucose 6-phosphate dehydrogenase (Zwischenferment) for a spectrophotometric determination of TPN reduction (linked to hexokinase):

$$\text{Glucose} + \text{ATP} \xrightarrow{\text{hexokinase}} \text{glucose 6-P} + \text{ADP}$$

$$\text{Glucose 6-P} + \text{TPN}^+ \xrightarrow{\text{dehydrogenase}} \text{6-P-gluconate} + \text{TPNH} + \text{H}^+$$

I explained to Krebs I was "driven to purifying Zwischenferment for this assay". "Who's driving you?", he asked. I have realized since that I was really attracted to this job rather than "driven" to it. Krebs, on the other hand, has acquired major insights into biochemistry without the labor of purifying enzymes.

Purification of nucleotide pyrophosphatase from potatoes furnished a means for degrading TPN and distinguishing the location of its third phosphate. Cleavage of DPN yielded nicotinamide mononucleotide; FAD yielded flavin mononucleotide. With these cleavage products in hand, I was led to search for enzymes that might use them as building blocks for synthesis of coenzymes. Work on the biosynthesis of these coenzymes, the simplest of the nucleotide condensation products, led me to the biosynthesis of the more complex polynucleotides, RNA and DNA. By strange chance, this work also provided the first insight into an origin of inorganic pyrophosphate.

DPN PYROPHOSPHORYLASE

In the synthesis of DPN, attack by nicotinamide mononucleotide on the α-phosphate of ATP results in the release of inorganic pyrophosphate:

$$\text{Nicotinamide mononucleotide} + \text{ATP} \rightleftharpoons \text{DPN} + \text{inorganic pyrophosphate.}$$

The reaction is freely reversible and the enzyme catalyzing it has been called DPN pyrophosphorylase. This transfer of one 5′-nucleotide to another produces a pyrophosphate bond between them. The reaction is the prototype of a large number of such nucleotidyl transfers to other phosphate compounds. These include transfers to various sugar phosphates to form the nucleoside diphosphate sugar coenzymes, to choline phosphate to form cytidine diphosphate choline, and to phosphatidic acid to form cytidine diphosphate diglyceride. This nucleotidyl transfer is the prototype also for transfers of nucleotides to produce mixed acid anhydrides with fatty acids, amino acids, and sulfates. In each instance inorganic pyrophosphate is produced; this is also true of the nucleotidyl transfers which produce RNA and DNA.

In key steps of virtually all macromolecular syntheses, inorganic pyrophosphate is released. These reactions proceed with relatively small free energy changes, and the action of a potent and ubiquitous inorganic pyrophosphatase may serve to drive them to completion.[6] It seems possible that the respiring liver homogenates, in which the accumulation of inorganic pyrophosphate was first observed by Ochoa and the Coris,[4,7] were relatively deficient in inorganic pyrophosphatase.

POLYNUCLEOTIDE PHOSPHORYLASE

In 1954 we used α-^{32}P-ATP and ^{14}C-thymidine to examine various cell-free extracts for enzymes which might incorporate them into RNA and DNA. Thymidine incorporation by *Escherichia coli* extracts was barely significant but that of ATP was substantial and so we proceeded with purification of what we regarded to be an RNA-synthetic activity.

Some months later, Kalckar, on a visit to St. Louis, brought us the tidings that Marianne Grunberg-Manago and Ochoa had penetrated RNA synthesis in depth. While studying aerobic phosphorylation in extracts of *Azotobacter* they were astute enough to observe the conversion of added ADP to an RNA-like polymer. They purified the responsible enzyme and observed its capacity to polymerize ADP and other ribonucleoside diphosphates, and in the reverse direction, to phosphorolyze the polynucleotide product:

$$\text{nNDP} \rightleftarrows (\text{NMP})_n + \text{nPi}$$

They named the enzyme polynucleotide phosphorylase.[8] On the basis of this intelligence, we switched substrates from ^{32}P-ATP to ^{32}P-ADP. As a result, our activity increased many fold and we succeeded in purifying polynucleotide phosphorylase from *E. coli* instead of discovering RNA polymerase. What a blunder!

DNA POLYMERASE

Upon reexamining the ^{14}C-thymidine incorporation (into an acid-insoluble polymer) about a year later, we were fortunate that the specific radioactivity of the compound, graciously given to us by Morris Friedkin, was nearly ten times greater than that of the thymidine used earlier. The activity now appeared more significant and we proceeded with purification of the responsible enzymes. We learned that preformed DNA and the four commonly occurring deoxyribonucleotides were required. The purified enzyme, called DNA polymerase, used the preformed DNA as a template. The DNA provided directions to DNA polymerase in assembling the DNA chain and it was obvious at once why all four nucleotides were absolutely required. Directing enzyme function with a template was unique and unanticipated by any evidence in enzymology, and was for some biochemists, even after a number of years, very hard to believe. Biologists, on the other hand, were receptive. Although the Watson–Crick proposal for the replication of DNA had not predicted the operation of an enzyme, the properties of DNA polymerase suited the role of accurate chain assembly.

MY FAVORITE ENZYME

Until a few years ago I would have been reluctant to name DNA polymerase my favorite. The sentiment that goes to the first born was a strong attachment to DPN pyrophosphorylase. It opened the door to coenzyme biosynthesis and the mystery of inorganic pyrophosphate. And it led me to the enzymatic synthesis of nucleotides and nucleic acids.

I have also had a favorite spot for the synthetase that makes phosphoribosylpyrophosphate (PRPP):

$$\text{Ribose 5-P} + \text{ATP} \rightarrow \text{PRPP} + \text{AMP}$$

Most of us anticipated that ribosyl activation for nucleotide biosynthesis would use the same device of phosphorylation, so well known for glucose. But the novelty of pyrophosphorylation used by this enzyme (coupled with elimination of inorganic pyrophosphate upon subsequent condensations) established my unalloyed awe for the ingenuity and fitness of an enzyme.

The virtuosity of the DNA polymerase first isolated from *E. coli* still amazes me.[9] Holding one or two turns of one chain of DNA helix as template, it assembles a complementary chain thousands of nucleotides long. It does so by Watson-Crick base pairing with an accuracy which exceeds chemical predictions. The enzyme achieves this fidelity of replication by a proof-reading exonuclease in its active center. This $3' \rightarrow 5'$ exonuclease hydrolyzes the last-added nucleotide when it is seen by the enzyme as an error, an improperly matched base pair. Still another domain within the active center of the enzyme ($5' \rightarrow 3'$ exonuclease) recognizes and excises lesions or aberrant regions at the 5′-chain-end of a gap in the DNA duplex being copied; removed, in this manner, are thymine dimers (introduced by ultraviolet light irradiation) or the vestigial RNA which initiates DNA chains.

What more can this DNA polymerase do? I doubt that we are close to having an adequate catalogue of this enzyme's charms and talents. Why then did DNA polymerase slip from grace about 4 years ago and stand accused as an artful imposter in the replication scene?[10]

OTHER REPLICATION ENZYMES

By 1969 there were several reasons to believe that *E. coli* DNA polymerase could not by itself account for the polymerization events in replication. (a) Genetic analysis of the T4 bacteriophage disclosed that at least six proteins were required for replication of the phage DNA; requirements for the *E. coli* chromosome were at least as complex. (b) Although replication of the DNA duplex was shown to be discontinuous, there was no mechanism known for initiation of the nascent replication fragments; no DNA polymerase was observed to have the capacity to start a DNA chain. (c) An *E. coli* mutant, which appeared to lack DNA polymerase, maintained normal rates and levels of DNA replication but was deficient in repairing DNA lesions.[11] Thus many scientists dismissed the relevancy of this enzyme in replication and relegated it to a subsidiary role in repair.

The replication story has been considerably clarified in the last few years. We now recognize that a multienzyme system is needed for DNA replication;[12] it is at least as complex as the multisubunit RNA polymerase which serves in DNA transcription. Presumably the multienzyme system is organized in the cell as a defined complex, a "replisome", analogous to the ribosome; unlike the ribosome it dissociates readily in cell extracts. One component of the complex is an RNA polymerase whose function is to transcribe a small region of the template in order to produce an oligoribonucleotide primer for the initiation of DNA replication.

Two additional DNA polymerases have been found in *E. coli*, named II and III, in the order of their discovery;[13] the original one is now named I. All operate by essentially identical mechanisms, differing only in their template preferences. The function of DNA polymerase II is still unknown. A multisubunit form of III, called DNA polymerase III*–copolymerase III* (holoenzyme), appears to be chiefly responsible for growth of DNA chains. Since certain mutations in DNA polymerase I, either in its polymerizing or $5' \rightarrow 3'$ exonuclease region, are lethal for the cell, this enzyme appears to be essential for excision of the initiating primer RNA and filling the gaps between nascent replication fragments;[14] the first-discovered DNA polymerase I mutant[11] survived because it was leaky.

MULTIPLE FACES OF ENZYMES

Enzymes have at least three aspects. Most familiar is their classical *catalytic face*. About 20 years ago the *control or regulatory or allosteric* face was recognized and has since come under intensive study. Recently there has been an awareness of a third face, one which recognizes neighboring and related proteins, or lipids in a membranous location, or some other cellular constituent. We might call this the *social* face to emphasize its importance for being in the right place and oriented in the right direction in the cellular community. Understanding the structure and function of all three faces of an enzyme might be all that we could reasonably wish to know. Even so I doubt that the sum of all such knowledge would be sufficient to explain the enzyme's contribution to metabolism and structure of the cell, its physiological role.

The limitations and virtues of studying isolated enzymes were stated clearly by F. G. Hopkins in 1931.[15] He pointed out that while such studies gain their full significance in relation to operations of enzymes in the cell, they are nonetheless essential for describing the molecular events in a cell. The question has been and remains how far one can go with isolated enzymes in analyzing and reconstructing cellular events.

At one pessimistic extreme was the Pasteurian view that alcoholic fermentation of sucrose requires a living yeast cell. Enzyme studies were discouraged for 30 years until the accidental discovery of the cell-free process by Eduard Buchner in 1897. A more modern version of this extreme pessimism was expressed in 1931 by the eminent microbiologist, A. J. Kluyver:[16]

> Attempts to prepare from microbial cells enzymatic agents capable of bringing about more complicated metabolic processes will be fruitless, because either the harmony required will be disturbed by the methods applied or, if this pitfall is avoided by a very mild treatment of the cell, most other metabolic processes will be maintained as well, with the result that the cells will continue their normal development.
> ... We have to resign ourselves to the fact that we will never succeed in proving experimentally the enzymatic character of more complicated metabolic processes.

Buoyed by the successful resolution of the enzymes responsible for the synthesis of fat from sugar, which Kluyver thought impossible, we can broach the optimistic extreme. It is the confidence that there is no limit at all in applying enzymology to explaining physiologic events. It is the point of view I would like to express in the concluding paragraphs.

PROMISE OF ENZYMES

If the guiding theme of earlier enzymologists was: "Don't waste clean thoughts on dirty enzymes", then the motto of the future should be: "don't entertain dirty (pessimistic) thoughts about clean enzymes". We must assume that complex metabolic pathways will be reconstituted from purified enzymes just as the simpler ones have been in the past.

With an understanding of how to fashion membranous vesicles and to orient their surfaces correctly, it should become possible to reconstitute organelles of considerable complexity.

As a long-range proposal I suggest the reconstitution of a successful viral infection—the production, in a few minutes, of several hundred virus particles from a single one—in a molecularly defined and dispersed (cell-free) mixture.

The ϕX174 life cycle can be divided into four stages:[17] (a) Opening of the phage and conversion of the single-stranded viral circle to a duplex form, (b) transcription of the DNA duplex and translation to produce the ten phage-coded proteins, (c) multiplication of the DNA duplex, and (d) production of viral circles and their final assembly into phage particles. Stages (a), (b) and (c) have been or on the way toward being accomplished; stage (d) would appear to offer no formidable obstacles.

What if one were successful in reconstituting each of these stages in the ϕX174 life cycle with characterized molecular components? It is still doubtful that the nice spatial coordination of phage and host components and their temporal juxtaposition, which characterize the living cell operations, would be achieved. We can anticipate missing or superfluous proteins, undisclosed inhibitors and effectors, suboptimal concentrations of cofactors and substrates, and the like. But this is exactly the kind of rigorous test that will reveal the gaps and inconsistencies in our knowledge based on *in vitro* studies. This crucible of cellular events is precisely the kind of challenge we must seek and welcome.

REFERENCES

1. Horecker, B. L. and Kornberg, A. (1946) The cytochrome *c*-cyanide complex. *J. Biol. Chem.* **165,** 11.
2. Kornberg, A., Ochoa, S. and Mehler, A. H. (1948) Spectrophotometric studies of the decarboxylation of β-keto acids. *J. Biol. Chem.* **174,** 159.
3. Ochoa, S., Mehler, A. H. and Kornberg, A. (1948) Biosynthesis of dicarboxylic acids by CO_2 fixation. I. Isolation and properties of an enzyme from pigeon liver catalyzing the reversible oxidative decarboxylation of 1-malic acid. *J. Biol. Chem.* **174,** 979.
4. Cori, C. F. (1942) Phosphorylation of carbohydrates. In *A Symposium on Respiratory Enzymes*, p. 175, Univ. of Wisconsin Press, Madison.
5. Kornberg, A. and Lindberg, O. (1948) Diphosphopyridine nucleotide pyrophosphatase. *J. Biol. Chem.* **176,** 665.
6. Kornberg, A. (1962) On the metabolic significance of phosphorolytic and pyrophosphorolytic reactions. In *Horizons in Biochemistry* (M. Kasha and B. Pullman, eds.), p. 251, Academic Press, N.Y.
7. Cori, G. T., Ochoa, S., Slein, M. W. and Cori, C. F. (1951) The metabolism of fructose in liver. Isolation of fructose-1-phosphate and inorganic pyrophosphate. *Biochim. Biophys. Acta* **7,** 304.
8. Grunberg-Manago, M. and Ochoa, S. (1955) Enzymatic synthesis and breakdown of polynucleotides; polynucleotide phosphorylase. *J. Am. Chem. Soc.* **77,** 3165.
9. Kornberg, A. (1969) Active center of DNA polymerase. *Science*, **163,** 1410.
10. Editorial (unsigned) (1971) Lifting replication out of the rut. *Nature New Biology* **233,** 97.
11. DeLucia, P. and Cairns, J. (1969) Isolation of an *E. coli* strain with a mutation affecting DNA polymerase. *Nature* **224,** 1164.
12. Schekman, R., Weiner, A. and Kornberg, A. (1974) Multienzyme systems of DNA replication. *Science* **186,** 987.
13. Kornberg, T. and Gefter, M. L. (1972) DNA synthesis in cell-free extracts. IV. Purification and catalytic properties of DNA polymerase III. *J. Biol. Chem.* **247,** 5369.
14. Konrad, E. B. and Lehman, I. R. (1974) A conditional lethal mutant of *E. coli* K12 defective in the $5' \rightarrow 3'$ exonuclease associated with DNA polymerase I. *Proc. Nat. Acad. Sci. U.S.A.* **71,** 2048.
15. Hopkins, F. G. (1931) *Problems of Specificity in Biochemical Catalysis*, Oxford Univ. Press, London.
16 Kluyver, A. J. (1931) *The Chemical Activities of Microorganisms*, p. 95, Univ. of London Press Ltd.
17. Kornberg, A. (1974) *DNA Synthesis.* W. H. Freeman, San Francisco.

FROM ENZYME CHEMISTRY TO GENETIC MANIPULATION

PAUL BERG

Department of Biochemistry, Stanford University Medical Center,
Stanford, California 94305

My upbringing in science was amongst people who love enzymes. Harland Wood's department introduced me to the enzymes that drive the cycles and pathways of intermediary metabolism; Herman Kalckar taught me the awe of wondering why and how they do; and Arthur Kornberg persuaded me to be skeptical of experiments (mine and others) done with impure enzymes.

Not surprisingly my first live heroes in science were the pioneers in enzymology. Amongst the handful I worshiped, Severo Ochoa was literally and figuratively a giant. Studying and learning his work was, therefore, a labor of love; but getting to know him personally was even more rewarding. His encouragement and interest in what I was doing made doing it more fun. So, it is a very real pleasure and privilege to share with Severo, his friends and colleagues, my current excitement with enzymes: their use in genetic analysis.

My involvement with enzymes, from the beginning, focused on what they do and how they do it. Such concerns led Bill Joklik and me, as postdoctoral fellows in Herman Kalckar's laboratory, to the discovery of nucleoside diphosphokinase(1) and the recognition of its essential role in using ATP for the formation of the then newly discovered other nucleoside triphosphates.

I arrived in Arthur Kornberg's new laboratory at Washington University full of enthusiasm to isolate and study the enzyme-AMP compound that Lynen and Lipmann had suggested as an intermediate in acetyl-CoA formation.(2) Kornberg's skepticism about the existence of such a compound was sobering but not persuasive; nevertheless, I accepted his advice to obtain a purer enzyme and during that search I realized that a new enzyme-bound intermediate, an acetyl adenylate, is the precursor of acetyl-CoA.(3) Here I must rectify glaring omissions from the historical record: the advice and guidance of Dave Lipkin in the first chemical synthesis of an acyl adenylate and the technical assistance Jerry Hurwitz gave me in carrying out several crucial experiments in the hectic period just prior to our departure for the San Francisco Federation meetings in 1956 to report these findings.

Acyl adenylates helped explain another curious observation: the exchange of inorganic pyrophosphate with the β, γ-phosphoryl groups of ATP promoted by amino acids.(4) A summer (1957) spent in Gobind Khorana's laboratory in Vancouver taught me how dicyclohexylcarbodiimide could be used to synthesize amino acyl adenylates(5) and made it possible to establish that these compounds were indeed formed enzymatically from ATP and amino acids.(6) That summer was even more rewarding because it began a friendship with the Khorana family and provided a preview of the advances in nucleic acid chemistry that were to come from his laboratory. Having available synthetic amino acyl adenylates allowed Jim Ofengand, Fred Bergmann, Jack Preiss, Marianne Dieckmann and me to prove that amino acyl groups were transferred to the 3′-hydroxyl terminus of specific tRNA molecules, by specific enzymes we named amino acyl tRNA synthetases.[7–11] Ann Norris Baldwin and I then isolated the sought-after enzyme-amino acyl adenylate complex and showed that the adenyl moiety could be utilized for ATP synthesis and that the amino acyl portion was transferable to tRNA for protein synthesis.[12]

An enzyme that can distinguish a single amino acid among many, and one or a few tRNAs from 50 to 100 structurally similar molecules, is fascinating indeed. So much so that studies of the structure, catalytic properties and specificity of various amino acyl tRNA synthetases occupied a number of us for several years.[13–17] To measure "recognition" between enzyme and tRNA, Mike Yarus and I developed a rapid and simple method involving the binding of cognate complexes to nitrocellulose filters,[18] a method that was to prove useful for studying other protein–nucleic acid interactions,[19] particularly repressor–operator recognition.[20]

My close friendship with Charles Yanofsky and the times we spent together revealed to me the power of genetics for probing structure–function relationships of enzymes. Consequently, Bill Folk, Maurizio Iaccarino and I examined mutational alterations affecting amino acyl tRNA synthetases[21,22] and John Carbon, Charles Hill, Larry Soll, Folk and Moshe Yaniv studied genetically altered tRNA;[23–28] the latter investigations proved to be the most productive because they showed that changes in nucleotide sequences of tRNA could affect the specificity of amino acylation and the translation of codons in mRNA.[28]

During this period another enzyme captured our attention. Mike Chamberlin, as a graduate student, detected and purified an enzyme which synthesized RNA using DNA as a template; the DNA-directed RNA polymerase of *E. coli*.[29] Thereupon followed a succession of studies concerning the mechanism of RNA synthesis[30,31] which Chamberlin and his students have since greatly enlarged upon.[32] The hypothesis that RNA, transcribed from DNA, is the "messenger" for polypeptide assembly prompted another student, Bill Wood, to test whether RNA synthezised *in vitro* with RNA polymerase and T4 phage DNA, could direct protein synthesis in a fractionated cell-free system. I still recall our excitement at the outcome of the first positive experiments[33] and the later chill upon learning of Marshall Nirenberg and Heinrich Matthaei's startling finding that poly U could direct the assembly of the polypeptide, polyphenylalanine, *in vitro*.[34] Only more recently has the coupling of DNA transcription to protein synthesis *in vitro* come into its own in studies of the mechanism of regulation of gene expression.[35,36]

So much for the excursion through my past and the preoccupation with what enzymes do and how they do it. Many lifetimes could be devoted to further study of the enzymes I came to know intimately; perhaps there will be time to renew acquaintances at some future time. For now there is another opportunity: to exploit the unique catalytic capability and exquisite specificity of enzymes as reagents to solve novel genetic problems.

Enzymes have often been used for analytic and synthetic purposes; the pages of this collection are filled with examples. Who can forget the role of polynucleotide phosphorylase in breaking the genetic code;[37] and Khorana's[38] and Kornberg's[39] success in harnessing enzymes to synthesize genes and chromosomes? Now enzymes provide the geneticist with surgical tools, scalpels and sutures, to dissect and reconstruct genetic assemblies! Though primitive, the promise these genetic manipulations reveal is breathtaking. Let me elaborate with some examples from our work with the tumor virus SV40.

The SV40 viral genome is a bihelical DNA molecule of about 5100 base pairs in a covalently closed circle; it codes for no more than five, and possibly as few as three, genes. It is an ideal molecule for studying molecular anatomy and practicing genetic surgery.

A restriction endonuclease cleaves both strands of DNA at a specific nucleotide sequence. A variety of such endonucleases have been obtained from different bacterial species but none so far from higher organisms; the number of such purified restriction enzymes already exceeds twenty-five. Each enzyme produces a characteristic fragment pattern from any DNA it digests; the number and size distribution of the fragments reflects the frequency and spacing of the restriction sites.

Four of the restriction enzymes cleave SV40 DNA only once. Thus each one converts the circular chromosome to a full-length linear molecule (see Fig. 1). *Eco*RI opens the

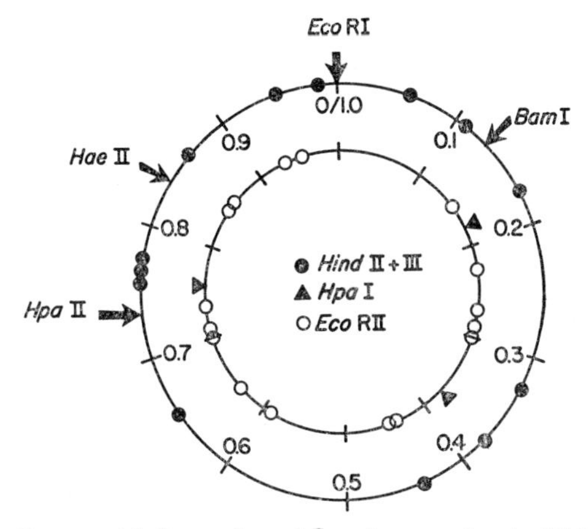

FIG. 1: Some restriction endonuclease cleavage sites in SV40 DNA.

circle at a point arbitrarily defined as the map coordinate, 0/1.0. The cleavage sites for enzymes from *Hemophilus parainfluenzae* (*Hpa*II), *H. aegyptius* (*Hae*II) and *Bacillus amyloliquifaciens* (*Bam*I) are assigned coordinates in terms of fractional molecular length; 0.735, 0.83 and 0.135, respectively. Other restriction endonucleases cleave the DNA at several sites; the keyed circles arrayed on the circular map show the sites for four other such enzymes.

Restriction sites provide chemical rather than genetic markers; just as mileage markers or signposts on a freeway, they show us precisely where we are on the DNA molecule. When the DNA of naturally-arising deletion mutants of SV40 is resistant to cleavage by *Eco*RI or *Hpa*II, we know what part of the chromosome has been affected. The extent of the deletion can be deduced from which fragments are missing or shortened in *Hind*II + III or *Eco*RII endonuclease digests.

Some restriction endonucleases cleave DNA in a special way. Instead of hydrolyzing the diester bonds of a directly opposed base pair, they introduce staggered cleavages (see next page), thereby generating single-strand termini that are complementary and identical, the so-called "cohesive" or "sticky" ends. Consequently, DNA segments or whole molecules generated by cleavages with any of these enzymes, can be annealed to one another and then covalently sealed with DNA ligase.

	↓
*Eco*RI	—GAATTC—
	—CTTAAG—
	↑
	↓
*Eco*RII	—CCAGG—
	—GGTCC—
	↑
	↓
*Hin*dIII	—AAGCTT—
	—TTCGAA—
	↑

The ability to join together segments of any two DNAs, irrespective of their genetic origins, is the basis for the recent excitement in genetic manipulation. I shall comment on that aspect later but now I want to illustrate another use which this type of endonuclease cleavage permits.

Cleavage and creation of cohesive ends enables us to perform microsurgery on the viral chromosome (Fig. 2). After limited digestion of SV40 DNA with *Eco*RII endonuclease (on the average, two hits) and subsequent ligation, Max Herzberg and Janet Mertz isolated SV40 mutants lacking regions bounded by contiguous *Eco*RII restriction sites. The extent and location of the deletions were deduced from the fragments which were missing in *Eco*RII digests of the mutant DNA and by visualization, in the electron microscope, of the part of the parental DNA missing from the mutant's DNA.[40]

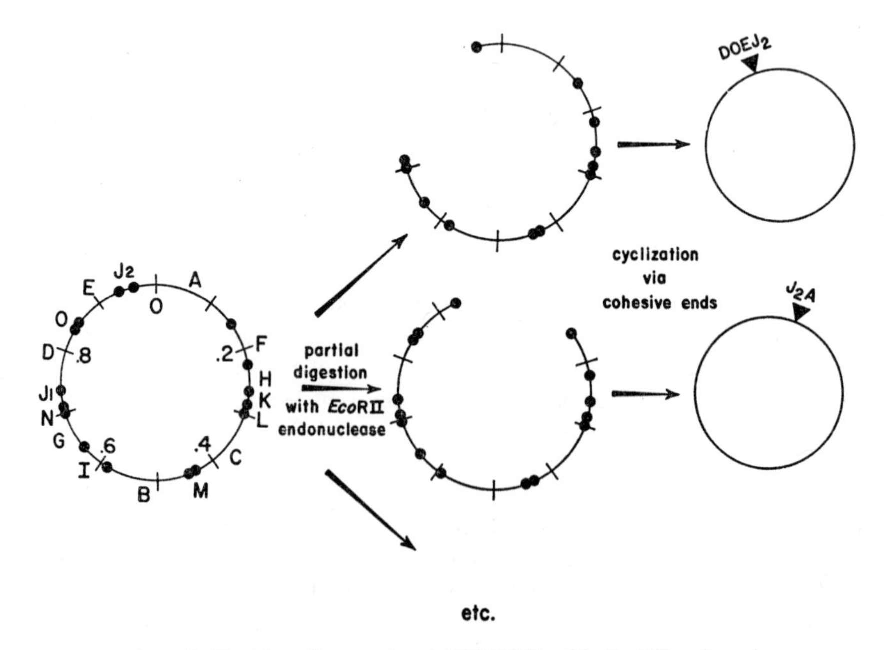

FIG. 2. Excision of segments of SV40 DNA with *Eco*RII endonuclease.

There are other ways that enzymes can be used to produce deletions in the SV40 chromosome. After digestion with any enzyme that cuts SV40 DNA once, the 5′-ends can be trimmed away with phage λ-exonuclease; then, the new ends can be rejoined to produce a chromosome lacking from 10 to 200 nucleotides around the cleavage site. Such small deletions, as well as the larger ones described above, permit us to define the location and boundaries of the structural genes and their controlling elements in the SV40 chromosome.

Enzyme specificity has also served us well for mapping mutational changes that are not susceptible to conventional genetic analysis. When identical linear DNA molecules are denatured and renatured, the original fully duplex structure is regenerated. If two DNAs should differ in a portion of their molecular sequence, heteroduplexes are produced by the denaturation and renaturation procedure. These are molecules that are base paired in their homologous regions and single-stranded only where they differ (Fig. 3). For example, when wild-type DNA and a deletion mutant DNA are cleaved with *Eco*RI or *Bam*I endonuclease to produce linear molecules and these are denatured and renatured, heteroduplexes which contain a single-stranded loop at the site of the deletion are formed.

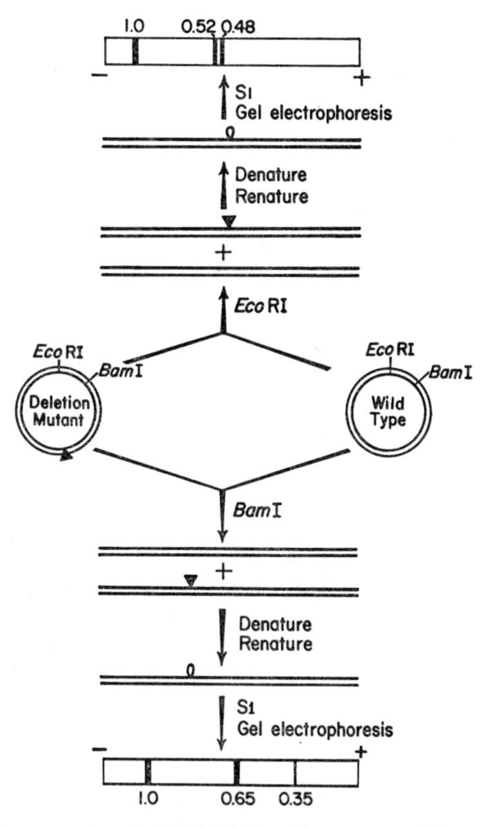

FIG. 3. Mapping deletion mutations in SV40 DNA with enzymes. The small triangle represents the site of the deletion in the mutant molecules, and the loop the unpaired region in the heteroduplexes. The size of the fragments produced by S1 nuclease of the heteroduplexes is deduced from their electrophoretic mobility in agarose gels (diagram of gels at top and bottom).

With deletions of 200 nucleotides or more, the loop can be visualized in the electron microscope[40] but small loops of 10 to 150 nucleotides are difficult or impossible to see.

An enzyme from *Aspergillus*, which is specific for degrading single-stranded DNA, the S1 nuclease, can digest away a small loop and cleave the exposed portion of the opposite DNA strand to produce two fragments whose length define the position of the deletion. In this simple way, a deletion of only twenty base pairs in SV40 DNA can be located readily with respect to the *Eco*RI and *Bam*I cleavage sites. Figure 3 illustrates how a deletion at map coordinate 0.48 was mapped. S1 nuclease is remarkable in detecting non-homologous regions in a duplex structure. We believe that it can detect even a single base mismatch.[41] Thus enzymes are useful for mapping mutational changes where genetic means are not yet available.

The growing sophistication in nucleic acid enzymology has made genetic engineering a reality. Three years ago, in an attempt to use SV40 DNA as a vector to transport new genes into animal cells, Dave Jackson, Bob Symons and I introduced a DNA segment, that codes for the enzymes of galactose utilization in *E. coli*, into the SV40 viral chromosome.[42] This was accomplished by relatively straightforward enzymologic procedures using readily available purified enzymes (Fig. 4). The circular SV40 DNA and a λ-bacteriophage-derived

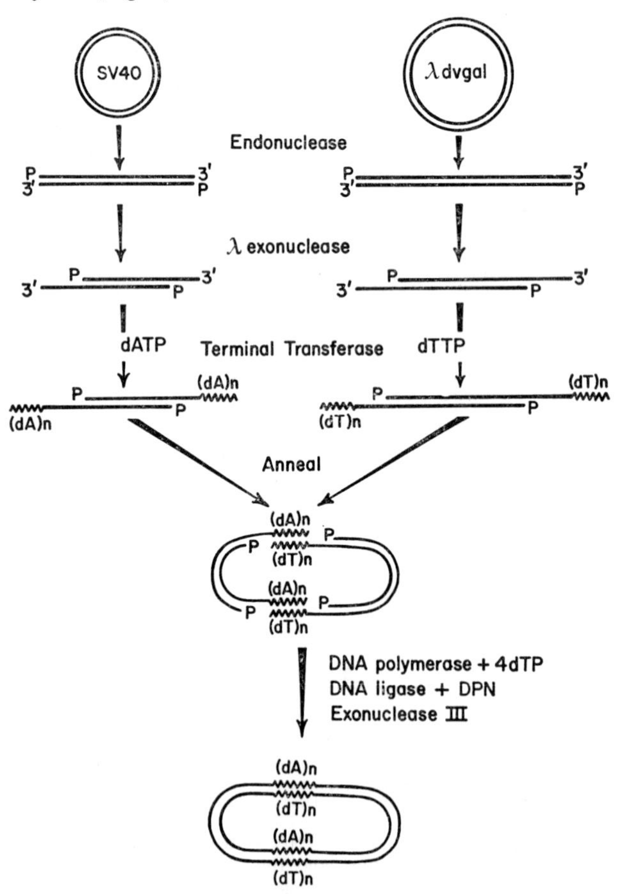

FIG. 4. Construction of a hybrid genome containing genes of SV40, bacteriophage λ and *E. coli*.

plasmid containing the galactose genes were first cleaved with a specific endonuclease. And then, after brief digestion with pure λ-exonuclease to remove about fifty nucleotides from the 5′-termini, short "tails" of either deoxyadenylate or deoxythymidylate residues were added to the 3′-termini with purified deoxynucleotidylterminaltransferase. The two DNAs, with their complementary "tails", were joined and cyclized simply by mixing and annealing them under appropriate conditions. The gaps, occurring at regions where the two parental molecules are joined, can be filled in with DNA polymerase I and the resulting molecules covalently sealed with DNA ligase; exonuclease III repairs nicks or gaps created during the manipulations. This method is general and can be used to join together any two DNA molecules. Dave Hogness has used this method to introduce segments of *Drosophila melanogaster* DNA into autonomously replicating bacterial plasmids, and thereby clone and propagate discrete regions of the fly's chromosomes.[43]

There is now a simpler, though less readily controlled, way of joining DNA molecules together. DNA segments, with cohesive ends generated by restriction endonuclease cleavage, e.g. with *Eco*RI endonuclease, can be joined to similarly cleaved DNA of a bacterial plasmid or phage and then propagated in an appropriate bacterial host.[44] Cloned segments of DNA from *Xenopus laevis*, sea urchins, human mitochondria and plant viruses, are now growing in *E. coli*!

Do such experiments pose a potential biohazard for man and his environment?[45] The ability to join together genetic material from any two sources and to propagate these recombined elements in bacterial and animal cells has produced a qualitative change in the field of genetics. This is not simply another, or easier, way of doing what has been done for a long time. For the first time there is a way to cross very large evolutionary boundaries and to move genes between organisms that have never before exchanged genetic material.

We are now in an area of biology with many unknowns. Such ignorance persuades me that considerable caution must be used in performing this research. Appropriate safeguards, principally biological and physical containment, should become an integral part of the experimental protocol to restrict the spread of newly created organisms.[46] The scientific community can hardly do less to earn the freedom to pursue this challenging and promising line of research.

REFERENCES

1. Berg, P. and Joklik, W. K. (1954) Enzymatic phosphorylation of nucleoside diphosphates. *J. Biol. Chem.* **210,** 657.
2. Jones, M. E., Lipmann, F., Hilz, H., and Lynen, F. (1953) On the enzymatic mechanism of Coenzyme A acetylation with adenosine triphosphate and acetate. *J. Am. Chem. Soc.* **75,** 3285.
3. Berg, P. (1956) Acyl adenylates: an enzymatic mechanism of acetate activation. *J. Biol. Chem.* **222,** 991; the synthesis and properties of adenyl acetate. *Ibid.*, p. 1015.
4. Berg, P. (1956) Acyl adenylates: the interaction of adenosine triphosphate and L-methionine. *J. Biol. Chem.* **222,** 1025.
5. Berg, P. (1958) The chemical synthesis of amino acyl adenylates. *J. Biol. Chem.* **233,** 608.
6. Berg, P. (1958) Studies on the enzymatic utilization of amino acyl adenylates: the formation of adenosine triphosphate. *J. Biol. Chem.* **233,** 601.
7. Berg, P. and Ofengand, E. J. (1958) An enzymatic mechanism for linking amino acids to RNA. *Proc. Nat. Acad. Sci.* **44,** 78.
8. Preiss, J., Berg, P., Ofengand, E. J., Bergmann, F. H. and Dieckmann, M. (1959) The chemical nature of the RNA-amino acid compound formed by amino acid-activating enzymes. *Proc. Nat. Acad. Sci.* **45,** 319.

9. Berg, P., Bergmann, F. H., Ofengand, E. J. and Dieckmann, M. (1961) The enzymic synthesis of amino acyl derivatives of ribonucleic acid. I. The mechanism of leucyl-, valyl-, isoleucyl-, and methionyl ribonuleic acid formation. *J. Biol. Chem.* **236**, 1726.
10. Bergmann, F. H., Berg, P. and Dieckmann, M. (1961) The enzymic synthesis of amino acyl derivatives of ribonucleic acid. II. The preparation of leucyl-, valyl-, isoleucyl-, and methionyl ribonucleic acid synthetases from *Escherichia coli. J. Biol. Chem.* **236**, 1735.
11. Ofengand, E. J., Dieckmann, M. and Berg, P. (1961) The enzymic synthesis of amino acyl derivatives of ribonucleic acid. III. Isolation of amino acid acceptor ribonucleic acids from *Escherichia coli. J. Biol. Chem.* **236**, 1741.
12. Norris, A. T. and Berg, P. (1964) Mechanism of aminoacyl RNA synthesis: studies with isolated aminoacyl adenylate complexes of isoleucyl RNA synthetase. *Proc. Nat. Acad. Sci.* **52**, 330.
13. Baldwin, A. N. and Berg, P. (1966) Purification and properties of isoleucyl ribonucleic acid synthetase from *Escherichia coli. J. Biol. Chem.* **241**, 831.
14. Calendar, R. and Berg, P. (1966) Purification and physical characterization of tyrosyl ribonucleic acid synthetases from *Escherichia coli* and *Bacillus subtilis. Biochemistry*, **5**, 1681; The catalytic properties of tyrosyl ribonucleic acid synthetases from *Escherichia coli* and *Bacillus subtilis. Ibid.*, p. 1690.
15. Arndt, D. J. and Berg, P. (1970) Isoleucyl-tRNA synthetase is a single polypeptide chain. *J. Biol. Chem.* **245**, 665.
16. Ostrem, D. L. and Berg, P. (1970) Glycyl-tRNA synthetase: an oligomeric protein containing dissimilar subunits. *Proc. Nat. Acad. Sci.* **67**, 1967.
17. Ostrem, D. L. and Berg, P. (1974) Glycyl-tRNA synthetase from *Escherichia coli:* purification, properties, and substrate binding. *Biochemistry*, **13**, 1338.
18. Yarus, M. and Berg, P. (1967) Recognition of tRNA by amino-acyl tRNA synthetases. *J. Molec. Biol.* **28**, 479; (1969) II. Effect of substrates on the dynamics of tRNA-enzyme interaction. *Ibid.* **42**, 171.
19. Jones, O. W. and Berg, P. (1966) Studies on the binding of RNA polymerase to polynucleotides. *J. Molec. Biol.* **22**, 199.
20. Riggs, A. D., Suzuki, H. and Bourgeois, S. (1970) Lac repressor-operator interaction. *J. Molec. Biol.* **48**, 67.
21. Iaccarino, M. and Berg, P. (1969) The requirement of sulfhydryl groups for the catalytic and tRNA recognition functions of isoleucyl tRNA synthetase. *J. Molec. Biol.* **42**, 151.
22. Folk, W. R. and Berg, P. (1970) Isolation and partial characterization of *Escherichia coli* mutants with altered glycyl tRNA synthetases. *J. Bact.* **102**, 193.
23. Carbon, J., Berg, P. and Yanofsky, C. (1966) Studies of missense suppression of the tryptophan synthetase A-protein mutant, A36. *Proc. Nat. Acad. Sci.* **56**, 764.
24. Hill, C. W., Foulds, J., Soll, L. and Berg, P. (1969) Instability of a missense suppressor resulting from a duplication of genetic material. *J. Molec. Biol.* **39**, 563.
25. Soll, L. and Berg, P. (1969) Recessive lethals, a new class of nonsense suppressors in *Escherichia coli. Proc. Nat. Acad. Sci.* **63**, 392.
26. Soll, L. and Berg, P. (1969) A recessive lethal nonsense suppressor in *Escherichia coli* which inserts glutamine. *Nature*, **233**, 1340.
27. Yaniv, M., Folk, W. R., Berg, P. and Soll, L. (1974) A single mutational modification of a tryptophan-specific transfer RNA permits aminoacylation by glutamine and translation of UAG. *J. Molec. Biol.* **86**, 245.
28. Berg. P. (1972) Suppression: a subversion of genetic decoding. *The Harvey Lectures*, pp. 247–272, Academic Press, New York.
29. Chamberlin, M. and Berg P. (1962) Deoxyribonucleic acid-directed synthesis of ribonucleic acid by an enzyme from *Escherichia coli. Proc. Nat. Acad. Sci.* **48**, 81.
30. Chamberlin, M., Baldwin R. L. and Berg, P. (1963) An enzymically synthesized RNA of alternating base sequence: physical and chemical characterization. *J. Molec. Biol.* **7**, 334.
31. Chamberlin, M. and Berg, P. (1964) Mechanism of RNA polymerase action: formation of DNA–RNA hybrids with single-stranded templates. *J. Molec. Biol.* **8**, 297; characterization of the DNA-dependent synthesis of polyadenylic acid. *Ibid.*, p. 708.
32. Chamberlin, M. (1970) Transcription 1970: summary. *Cold Spring Harbor Symp. Quant. Biol.* **35**, 851.
33. Wood, W. B. and Berg, P. (1962) The effect of enzymatically synthesized ribonucleic acid on amino acid incorporation by a soluble protein-ribosome system from *Escherichia coli. Proc. Nat. Acad. Sci.* **48**, 94.
34. Nirenberg, M. W. and Matthaei, J. H. (1961) Dependence of cell-free protein synthesis in *E. coli* upon naturally occurring or synthetic polynucleotides. *Proc. Nat. Acad. Sci.* **47**, 1588.
35. Zubay, G., Chambers, D. A. and Cheong, L. C. (1970) Cell-free studies on the regulation of the *lac* operon in the lactose operon. (J. R. Beckwith and D. Zipser, eds), pp. 375–391. Cold Spring Harbor Laboratory, N.Y.

36. ZALKIN, H., YANOFSKY C. and SQUIRES, C. L. (1974) Regulated *in vitro* synthesis of *Escherichia coli* tryptophan operon messenger ribonucleic acid and enzymes. *J. Biol. Chem.* **249**, 465.
37. SPEYER, J. F., LENGYEL, P., BASILIO, C., WAHBA, A. J., GARDNER, R. S. and OCHOA, S. (1963) Synthetic polynucleotides and the amino acid code. *Cold Spring Harbor Symp. Quant. Biol.* **28**, 559.
38. KHORANA, H. G. *et al.* (1972) Total synthesis of the structural gene for an alanine transfer RNA from yeast. *J. Molec. Biol.* **72**, 209 and continuing papers.
39. GOULIAN, M., KORNBERG, A. and SINSHEIMER, R. L. (1967) Enzymatic synthesis of DNA: synthesis of infectious phage ΦX174 DNA. *Proc. Nat. Acad. Sci.* **58**, 3231.
40. MERTZ, J. E., CARBON, J. HERZBERG, M., DAVIS, R., and BERG P. (1974) Isolation and characterization of individual clones of SV40 mutants containing deletions, duplications and insertions in their DNA. *Cold Spring Harbor Symp. Quant. Biol.* **39**, 69.
41. SHENK, T. E., RHODES, C., RIGBY, P. W. J. and BERG, P. (1975) Biochemical method for mapping mutational alterations in DNA with S1 nuclease: the location of deletions and temperature-sensitive mutations in SV40. *Proc. Nat. Acad. Sci.* **72**, 989.
42. JACKSON, D. A., SYMONS, R. H. and BERG, P. (1972) A biochemical method for inserting new genetic information into SV40 DNA: circular SV40 DNA molecules containing lambda phage genes and the galactose operon of *E. coli. Proc. Nat. Acad. Sci.* **69**, 2904.
43. WENSINK, P. C., FINNEGAN, D. J., DONELSON, J. E. and HOGNESS, D. S. (1974) A system for mapping DNA sequences in the chromosomes of *Drosophila melanogaster*. *Cell.* **3**, 315.
44. COHEN, S. N. (1975) The manipulation of genes. *Sci. Am.* **233**, 24.
45. BERG, P. *et al.* (1974) Potential biohazards of recombinant DNA molecules. *Science*, **185**, 303.
46. BERG, P., BALTIMORE, D., BRENNER, S., ROBLIN, R. O., III, SINGER, M. F. (1975) Asilomar Conference on Recombinant DNA Molecules. *Science* **188**, 991.

THE SEARCH FOR THE STRUCTURAL RELATIONSHIP BETWEEN GENE AND ENZYME

CHARLES YANOFSKY

Department of Biological Sciences, Stanford University, Stanford, California 94305

I welcome this opportunity to describe some interesting experiences I have had in my career. These include, of course, my recollections of developments in the 1950s and 1960s when I participated in investigations on what was then a central problem in genetics and biochemistry—how gene structure is related to protein structure. I was introduced to this problem as a graduate student at Yale University where I worked with an extremely warm and colorful advisor, David M. Bonner.

When I arrived at Yale in 1948 I was assigned the task of identifying intermediates in niacin synthesis accumulated by two mutants of *Neurospora crassa*. I succeeded in isolating two compounds, quinolinic acid and α-*N*-acetylkynurenine, and then went on to investigate their role in niacin biosynthesis.[1,2] While performing these studies, the research projects of my fellow students, Gabriel Lester, Otto Landman, Sigmund Suskind, Naomi Franklin and Howard Rickenberg, began to seem more challenging conceptually. Each of them was analyzing some aspect of genetic determination of protein structure or genetic control of enzyme formation. During my last year as a student Dave Bonner encouraged me to work on a gene–enzyme problem and so I started a project with the enzyme tryptophan synthetase. At the time, tryptophan synthetase was one of the two partially characterized biosynthetic enzymes of *N. crassa* for which mutant strains had been isolated.[3,4] My initial objective, as I recall, was to test a then current hypothesis that proteins were synthesized stepwise by sequential reactions, much like amino acids and vitamins. We considered this hypothesis seriously because it had been found that extracts of tryptophan synthetase and pantothenate synthetase mutants, when appropriately treated, exhibited the enzyme activity presumed to be lacking.[4,5] In addition, when extracts of some lactose non-utilizing mutants of *E. coli* and *N. crassa* were carefully examined, β-galactosidase activity was detected.[6] These findings could indicate that enzyme precursors were accumulating in mutant strains.

My approach in studying tryptophan synthetase was greatly influenced by the rigorous training in enzymology I had at Yale in tutorial and laboratory courses with the biochemist Joseph Fruton. Because of this experience, I was determined to purify the protein and characterize it further before I attempted an analysis of the available mutant strains. This was a wise decision, I feel, because there was much to learn about the enzyme.[7] From my studies I concluded that the previously reported activation of a possible tryptophan synthetase precursor[4] was in error.[8] The initial investigators also failed to repeat the activation.[9]

With these studies I was committed to working with tryptophan synthetase and I moved on eagerly to further analyses of the enzyme changes resulting from mutations. By this time the techniques of mutant isolation in *Neurospora* had improved appreciably so that Bonner and I were able to obtain a large set of non-identical tryptophan synthetase mutants.

With these we showed convincingly and for the first time that many independent mutations eliminating detectable amounts of a single biosynthetic enzyme occurred at a single genetic locus.[11]

My next step on the trail of the gene–protein relationship grew out of a term paper I wrote for a course given at Yale in 1950 by a guest of the department, the geneticist Louis Stadler. The course was devoted almost entirely to Stadler's work on the R locus of maize, and dealt with the question: Are mutational changes at a single locus genetically separable? The only assignment in the course was to write a term paper on any poorly understood genetic phenomenon and to propose experiments which might distinguish between alternative explanations. I selected the topic "Genetic Suppression" and, being molecularly oriented, offered explanations involving the repair of damaged, inactive mutant proteins. Little did I realize then that the following year I would be engaged in full-time research on this phenomenon. This is how it happened. While Stadler's course was under way I was performing heterocaryon (complementation) tests with two tryptophan synthetase mutants. The results were negative, as expected, but in one experiment I noticed that there was growth in a control flask, i.e. a flask which had been inoculated with only one of the mutants. This could have been due to contamination, or to back mutation—in either case uninteresting. However, since genetic suppression was also a possibility I decided to examine the growing mycelium further.

I was surprised to find that the growth in the control flask was in fact due to a suppressor mutation at a locus unlinked to the tryptophan synthetase locus.[8] To my satisfaction, extracts of the suppressed mutant contained low but detectable levels of tryptophan synthetase activity.[8] Further analyses showed that the restored enzyme was indistinguishable from the wild type enzyme.[8] Encouraged by these results I put other studies aside and, with the aid of Miriam Bonner and my wife, Carol, examined many tryptophan synthetase mutants for "reversion" by unlinked suppressor mutations. Our hopes were realized for we readily detected several suppressor mutations. These, for the most part, were allele specific, i.e. each suppressor gene suppressed only one or two of the set of tryptophan synthetase mutants tested.[10] In addition, the suppressed mutants invariably contained tryptophan synthetase activity.[10]

Since only some of our mutants were suppressible, Bonner and I considered the possibility that only these mutants contained inactive tryptophan synthetase in some modifiable form. The test of this hypothesis was based on Sig Suskind's interest in immunology. He proposed to obtain antibodies in rabbits to partially purified *Neurospora* tryptophan synthetase and then examine extracts of suppressible and non-suppressible mutants for cross-reacting material (hence the now generally used abbreviation, CRM) related to the enzyme. I vividly remember our first experiment because Suskind and I were so excited by its outcome. Suskind had just begun postdoctoral work with A. M. Pappenheimer at New York University Medical School. I prepared and assayed extracts of all the strains we decided to examine, and, to make the experiment more exciting, coded the extracts. Suskind came to Yale for a weekend and ran the critical tests. The results were beautiful—clear and convincing—it was one of those rare and completely satisfying experiences scientists yearn for. Two groups of mutants were apparent one of which contained tryptophan synthetase CRM while the other did not.[11] With one exception, the suppressible mutants were the CRM producers. Furthermore, extracts of most of the suppressed mutants contained CRM in addition to active enzyme.[11]

In 1955 these findings were presented at the Henry Ford Hospital in Detroit, at a symposium entitled, "Gene Interactions in Enzyme Formation".[12] I concluded from our results that there was substructure in the genetic locus for tryptophan synthetase and that different subunits of the locus were damaged in our distinguishable mutants. Concerning suppression I proposed that "in the suppressed mutant the altered (mutant) template* (the product of the structural gene for tryptophan synthetase) is still formed, but some product of a suppressor gene cooperates with the altered template in the formation of small amounts of tryptophan synthetase". We knew virtually nothing then about protein synthesis so it seemed pointless to speculate on the identity of the suppressor gene product.

Immediately following my talk at the Henry Ford Hospital, I was cornered by a scientist I had not met before, Seymour Benzer. He wanted me to tell him more about the nature of the changes in our tryptophan synthetase mutants and to discuss possible mechanisms of suppression, neither of which I was able to do. Only later did I learn that Benzer was then completing the definitive genetic studies establishing fine structure within a gene. His observations provided convincing support for the hypotheses that a genetic map is a linear representation of DNA structure, and that mutations involve changes of single nucleotide pairs. Somewhat later Benzer and I exchanged observations and strains in our independent studies on suppression. These ultimately established that both nonsense (chain termination) and missense (amino acid replacement) mutations are suppressible.[13,14]

During this period I was frustrated by my inability to obtain tryptophan synthetase in pure form from *Neurospora*. Therefore, I searched for a more suitable enzyme for studies on protein structure alterations resulting from mutations. I characterized D- and L-serine deaminases from *Neurospora*[15,16] with this objective in mind but was unable to develop a convenient isolation procedure for mutants lacking these enzymes and so stopped these investigations.

When I moved to Western Reserve University in 1954, I planned to continue my studies with tryptophan synthetase mutants of *Neurospora*. However, I decided that it would also be wise to identify the intermediate reactions in tryptophan biosynthesis. I felt that this information would be invaluable when the time came, as I knew it eventually would, to explain the various phenotypes of tryptophan-requiring mutants. I switched from *Neurospora* to *E. coli* for these studies, principally because Gabe Lester had convinced me that *E. coli* was a more convenient organism to use for both genetic and biochemical investigations. He was correct, for tryptophan-requiring mutants of *E. coli* were readily isolated and subjected to fine structure genetic analyses using transducing phage Pl[17] much as Demerec and co-workers used phage P22 in their fine structure studies with mutants of *Salmonella*.[18] Another favorable attribute of *E. coli* was the overproduction of tryptophan biosynthetic enzymes and accumulated intermediates when mutants were cultured on growth-limiting levels of indole or tryptophan.[19]

As I recall it, this was an extremely frustrating period in my career because it took close to a year to do little more than isolate and identify one biosynthetic intermediate, indole-3-glycerol phosphate.[20] The continued effort proved to be worthwhile, however, for with the guidance of my colleague in the Microbiology department, Howard Gest, as well as Bob Greenberg and Dave Goldthwait of the Biochemistry department, I established the role of 5-phosphoribosyl-1-pyrophosphate in tryptophan biosynthesis.[20,21] Fortunately for me Greenberg and Goldthwait had just completed studies on the participation of this

* Messenger RNA had not been discovered as yet.

compound in purine biosynthesis. With the aid of a graduate student, Oliver Smith, the remaining intermediate reactions in tryptophan formation were identified,[22] with the exception of the initial step, chorismate → anthranilate. This was elucidated subsequently by Frank Gibson,[23] a scientist I came to know well during the several months he spent in my laboratory soon after I arrived at Stanford.

During my stay at Western Reserve University I also continued my studies of tryptophan synthetase—but now with the enzyme from *E. coli* as well as the one from *Neurospora*.[24] With Jody Stadler and Martin Rachmeler initial observations were made that suggested that the tryptophan synthetase reaction was more complex than we had imagined. The enzyme converted indole-3-glycerol phosphate to indole and indole -3-glycerol phosphate to tryptophan in addition to catalyzing the previously characterized reaction, indole + L-serine → L-tryptophan[25,26,27].

When I arrived at Stanford in 1958, tryptophan synthetase had not yet been obtained in pure form. Yet I knew a pure protein was necessary if we were to examine the relationship between gene structure and protein structure. Irving Crawford joined me in 1958, and, agreeing with me, set out to purify the enzyme from *E. coli*. The first series of experiments was very disappointing because most of the initial tryptophan synthetase activity was lost during chromatographic purification. Chromatography was attempted several times, always with the same puzzling result. Most investigators, I believe, when confronted with such extreme activity losses would ascribe it to enzyme instability and would try gentler purification procedures. Crawford, however, considered an alternative explanation—that during chromatography the enzyme was dissociating into non-identical subunits. This proved to be correct for we showed that tryptophan synthetase of *E. coli* is composed of two non-identical protein subunits which must complex with each other to obtain enzymatic activity.[28] These subunits are now designated α and β_2. Tryptophan synthetase, I believe, was the first enzyme shown to consist of non-identical subunits. In retrospect this fact makes Crawford's persistence even more impressive. When the subunits were examined separately, one, the α subunit, was relatively acid stable. This finding facilitated its purification to homogeneity.[29] Equally important to further developments was the discovery that each of the subunits activated the other in one of the tryptophan synthetase reactions.[28] We later observed that mutant α-chains also could activate the wild-type β_2-subunit.[30] This finding provided us with the capability of employing an enzyme assay to monitor the purification of each mutant α-chain.

Crawford also showed, in other studies, that he could separate two α-proteins from a suppressed *trpA** mutant of *E. coli*. One of these proteins was enzymatically active and resembled the wild-type α-chain, while the other resembled the α-chain of the unsuppressed mutant.[31] Stuart Brody, an imaginative graduate student, extended these studies. He was the first to show that suppression causes mistakes in amino acid incorporation—and that such mistakes are responsible for this phenomenon.[14] As I recall he was also the first to propose that transfer RNA, then called soluble RNA, might be the molecule altered by suppressor mutations. Peggy Lieb and Len Herzenberg expressed virtually the same idea to me somewhat later. This was subsequently shown to be so, first for nonsense suppression,[32] and then for missense suppression in studies with Paul Berg, Gobind Khorana and their co-workers.[33,34] Interaction with Paul Berg was particularly stimulating, enjoyable and

* *trpA* and trpB are, respectively, the structural genes for the α- and β-polypeptide chains of *E. coli* tryptophan synthetase.

rewarding, in fact the idea for the experiment that resulted in the demonstration that transfer RNA alterations were responsible for missense suppression was conceived after a tennis match between Berg and myself. This has been described by Berg.[35]

At this point it appeared that we had experimental material with all the features needed to solve the gene structure–protein structure problem. We could isolate tryptophan synthetase α-chain missense mutants and map their mutationally altered sites by phage mediated transduction.[17] We could purify the wild-type α-chain[29] and the α-chains of missense mutants.[36,37] We thought we should be able to identify the amino acid change or changes characteristic of each missense α-protein by using the screening procedure employed by Ingram[38] to determine the single amino acid change in sickle cell hemoglobin. Then, using conventional procedures, we hoped to determine the amino acid sequence of the wild-type protein and hence the location of each mutant amino acid change. With this information we would be in a position to compare a genetic map with the map relating the positions of the corresponding mutant amino acid changes. We expected that this comparison would reveal the relationship between gene structure and protein structure.

Before these objectives were realized, quite a few years and the efforts of many talented co-workers were required. In particular, Donald Helinski and Ulf Henning were instrumental in applying the Ingram technique to wild-type and mutant α-chains.[36,37,39] They showed that two distinguishable mutants with genetic alterations which were barely recombinable had amino acid changes at the same position in the α-chain.[36,37] These initial observations gave us confidence in the belief that we would find a colinear relationship. Virginia Horn, the super-technician of the group, isolated most of the missense mutants we studied and performed the fine structure genetic analyses. Bruce Carlton, a horticulturist by training, undertook determination of the sequence of the wild-type protein. When he began his studies no sequence was known for any protein as large as the tryptophan synthetase α-chain. Thus this was quite a challenge for someone without prior experience with protein chemistry! Carlton was followed on this aspect of the program by John Guest and then by Gabriel Drapeau.

Proof of colinearity had been sought for so many years by so many groups of investigators working with different proteins[40,41,42] that it was somewhat of a letdown when the expected result was obtained. By the early 1960s the Watson–Crick model of DNA[43] was widely accepted and generally interpreted in terms of linear correspondence of gene structure and protein structure. Nevertheless, it was essential to provide the experimental verification for so fundamental a concept. I remember presenting rather shyly a brief report at the Cold Spring Harbor Symposium of 1963[44] of our findings demonstrating colinearity of gene structure and protein structure. The following year we published the details.[45] The complete sequence of the protein was not established until 1967.[46] The end result of all our genetic and protein structure studies is summarized in Fig. 1. Incidentally we now know much of the nucleotide sequence corresponding to the first 120 amino acid residues of the tryptophan synthetase α-chain (T. Platt, M. Tal and C. Yanofsky, unpublished observations). Colinearity was established independently at about the same time by Sydney Brenner's group.[47] They studied phage head-protein mutants of the nonsense variety and analyzed the fragments of the head protein these mutants produced. Since two such different approaches gave results suggesting colinearity and since these results verified a widely accepted concept, there was every reason to accept the conclusion that the relationship between gene structure and protein structure had been established.

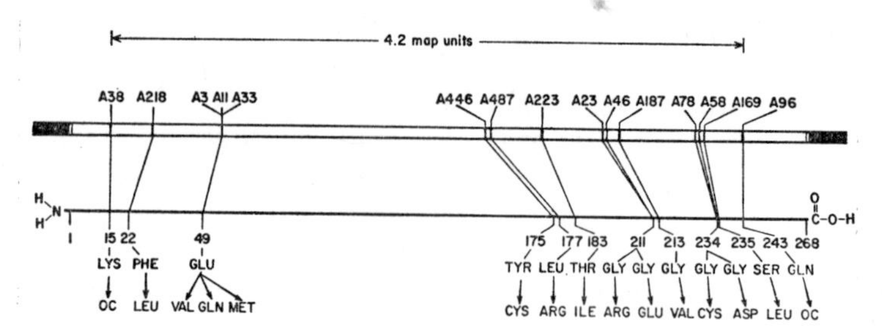

FIG. 1. Colinearity of the genetic map of *trpA* of *E. coli*. and the tryptophan synthetase α-chain.[45,46,53] The genetic map and polypeptide chain are each drawn to scale so one can compare relative distance on the genetic map with relative distance in the polypeptide chain. oc = chain termination codon.

As we were gathering data on the colinearity problem it became evident that our system could potentially be used to probe still further and decipher the genetic code. How did we plan to do this? It was clear from our investigations and those of others in the early 1960s that amino acids could be grouped into families on the basis of amino acid replacement data, i.e. only certain amino acids could replace an existing one in a protein as a consequence of single mutational events. For example, we observed two inactivating "mutant" changes at position 211 of the α-chain, glycine → glutamic acid and glycine → arginine. When the respective mutants reverted and the revertant α-chains were analyzed, we found that several amino acids replaced arginine and glutamic acid. However, with the exception of glycine, the amino acids replacing arginine and those replacing glutamic acid differed (Fig. 2).[48] This only could be explained if mutational changes were generally restricted to single base pair changes and if certain coding sequences (codons) could not be derived from other codons by single base changes. If this interpretation were correct, then we should be able to use mutational data to relate the different codons to one another. In order to identify the base pair change in each mutant we hoped to use base pair–change–specific mutagens as Benzer and Freese did[49,50] in their studies with phage T4rII mutants. For example, since both glutamic acid_{211} and arginine_{211} "reverted" only to glycine in response to 2-amino purine treatment, and since this treatment presumably causes only AT ⇄ GC transitions, we could draw definite conclusions about the relationships between the nucleotide sequences of the codons for glycine, arginine and glutamic acid (Fig. 2).[48] To establish the order of the bases in a codon clearly some other trick was needed. We reasoned that if two mutants had base changes at different positions in a single codon and, if the mutants were crossed with one another, recombinants should be obtained with the corresponding wild-type codon. If an outside genetic marker were employed in the cross, then the relative positions of the mutant base changes in the affected codon could be determined. Analyses of this type were performed,[55,52] but by the time these experiments were well underway, Marshall Nirenberg, Severo Ochoa and their co-workers were reporting newly identified codons monthly and they with Gobind Khorana and his co-workers cracked the code. Thus the results of our studies essentially provided *in vivo* confirmation of codon assignments deduced from the elegant *in vitro* studies.

Despite knowledge of the genetic code and many aspects of protein synthesis, several basic questions relevant to the relationship between gene structure and protein structure

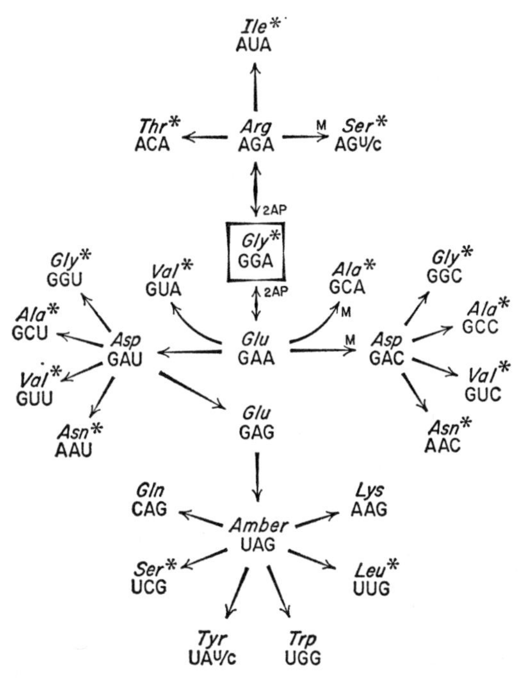

FIG. 2. Amino acid replacements observed at position 211 of the tryptophan synthetase α-chain.[48,54] α-chains with amino acids marked by asterisks are functional, while α-chains with the other amino acids are inactive. Arrowheads labeled 2AP indicate changes from arginine or glutamic acid specifically promoted by 2-amino treatment of mutant bacteria. Similarly M refers to changes favored by the Treffers mutator gene.[48]

remain unanswered. For example, we do not understand how ribosomes recognize the specific translation initiation regions in messenger RNAs. We also do not know the full significance of secondary structure and modification in messenger RNA, where they occur. Nor do we understand the basis of translation reinitiation when it is caused by the introduction of a translation termination codon immediately preceding a potential reinitiation site in messenger RNA. In addition, the explanation for the secondary consequences associated with the introduction of chain termination codons in polycistronic messages—the phenomenon of translational polarity—remains obscure. Along different lines we do not know whether the structural variations responsible for the diversity of antibody specificities reflect amplification of normal processes of gene and chromosome evolution, or the participation of some new feature of genetic determination. Besides these, broader questions of evolutionary importance remain unanswered. Thus, the finding that transfer RNAs can acquire new coding specificities as a consequence of single base changes leads one to wonder why the genetic code is universal. Also, explanations for the wide variations seen in the base pair composition of DNAs of different organisms are not entirely satisfactory. Finally, the significance of synonymous codon changes in messenger RNA and the role of neutral mutations in evolution is largely unknown.

In molecular biology, as you can see, it seems that new findings inevitably raise new questions. It is my guess that this will always be true!

ACKNOWLEDGMENTS

I am indebted to Drs. S. Suskind, I. Crawford, D. Helinski, S. Brody and T. Ocko for allowing me to test their memories and/or benefit from their reading this manuscript. I am also pleased to acknowledge the support of the National Science Foundation and the United States Public Health Service. Their aid made it possible for my co-workers and me to perform the studies described in this article. Regretfully, because of their untimely deaths, neither David Bonner nor Gabriel Lester can share the enjoyment of this review of past personal interactions and pleasant experiences.

REFERENCES

1. BONNER, D. M. and YANOFSKY, C. (1949) Quinolinic acid accumulation in the conversion of 3-hydroxyanthranilic acid to niacin in *Neurospora*. *Proc. Nat. Acad. Sci.* **35**, 576–581.
2. YANOFSKY, C. and BONNER, D. M. (1950) Evidence for the participation of kynurenine as a normal intermediate in the biosynthesis of niacin in *Neurospora*. *Proc. Nat. Acad. Sci.* **36**, 167–176.
3. MITCHELL, H. K. and LEIN, J. (1948) A Neurospora mutant deficient in the enzymatic synthesis of tryptophan. *J. Biol. Chem.* **175**, 481–482.
4. GORDON, M. and MITCHELL, H. K. (1950) Tryptophane desmolase in *Neurospora*. *Genetics*, **35**, 110.
5. WAGNER, R. P. (1949) The *in vitro* synthesis of pantothenic acid by pantothenicless and wild type *Neurospora*. *Proc. Nat. Acad. Sci.* **35**, 185–190.
6. BONNER, D. M. (1951) Gene-enzyme relationships in *Neurospora*. *Cold Spring Harb. Symp. on Quant. Biol.* **16**, 143–157.
7. YANOFSKY, C. (1952) Tryptophan desmolase of *Neurospora*—partial purification and properties. *J. Biol. Chem.* **194**, 279–286.
8. YANOFSKY, C. (1952) The effects of gene change on tryptophan desmolase formation. *Proc. Nat. Acad. Sci.* **38**, 215–226.
9. HOROWITZ, N. H. and MITCHELL, H. K. (1951) Biochemical genetics. *Ann. Rev. Biochem.* **20**, 465–487.
10. YANOFSKY, C. and BONNER, D. M. (1955) Gene interactions in tryptophan synthetase formation. *Genetics*, **40**, 761–769.
11. SUSKIND, S. R., YANOFSKY, C. and BONNER C. M. (1955) Allelic strains of *Neurospora* lacking tryptophan synthetase: A preliminary immunochemical characterization. *Proc. Nat. Acad. Sci.* **41**, 577–582.
12. YANOFSKY, C. (1956) Gene interactions in enzyme synthesis. In *Enzymes: Units of Biological Structure and Function*, Henry Ford Hospital International Symposium, pp. 147–160, Academic Press.
13. BENZER, S. and CHAMPE, S. (1961) Ambivalent rII mutants of phage T_4. *Proc. Nat. Acad. Sci.* **47**, 1025–1039.
14. BRODY, S. and YANOFSKY, C. (1963) Suppressor gene alteration of protein primary structure. *Proc. Nat. Acad. Sci.* **50**, 9–16.
15. YANOFSKY, C. (1952) D-serine dehydrase of *Neurospora*, *J. Biol. Chem.* **198**, 343–352.
16. YANOFSKY, C. and REISSIG, J. (1953) L-serine dehydrase of *Neurospora*. *J. Biol. Chem.* **202**, 567–577.
17. YANOFSKY, C. and LENNOX, E. S. (1959) Transduction and recombination study of linkage relationships among the genes controlling tryptophan synthesis in *Escherichia coli*. *Virology*, **8**, 425–447.
18. DEMEREC, M., HARTMAN, Z., HARTMAN, P. E., YURA, T., GOTS, J. S., OZEKI, H. and GLOVER, S. W. (1956) *Genetic Studies with Bacteria*. Carnegie Institution of Washington Publication 612, Washington, D.C.
19. YANOFSKY, C. (1957) Enzymatic studies with a series of tryptophan auxotrophs of *Escherichia coli*. *J. Biol. Chem.* **224**, 783–792.
20. YANOFSKY, C. (1956) The enzymatic conversion of anthranilic acid to indole. *J. Biol. Chem.* **223**, 171–184.
21. YANOFSKY, C. (1956) Indole-3-glycerol phosphate, an intermediate in the biosynthesis of indole. *Biochim. Biophys. Acta*, **20**, 438–439.
22. SMITH, O. H. and YANOFSKY, C. (1960) 1-(*o*-carboxyphenylamine)-1-deoxyribonulose-5-phosphate, a new intermediate in the biosynthesis of tryptophan. *J. Biol. Chem.* **235**, 2051–2057.
23. GIBSON, M. I. and GIBSON, F. (1964) Preliminary studies on the isolation and metabolism of an intermediate in aromatic biosynthesis: chorismic acid. *Biochem. J.* **90**, 248–252.
24. LERNER, P. and YANOFSKY, C. (1957) An immunological study of mutants of *Escherichia coli* lacking tryptophan synthetase. *J. Bacteriol.* **74**, 494–501.
25. YANOFSKY, C. and RACHMELER, M. (1958) The exclusion of free indole as an intermediate in the biosynthesis of tryptophan in *Neurospora crassa*. *Biochim. Biophys. Acta* **28**, 640–641.

26. STADLER, J. and YANOFSKY, C. (1959) Studies on a series of tryptophan-independent strains derived from a tryptophan requiring mutant of *Escherichia coli*. *Genetics*, **44**, 106–123.
27. YANOFSKY, C. (1959) A second reaction catalyzed by the tryptophan synthetase of *Escherichia coli*. *Biochim. Biophys. Acta*, **31**, 408–416.
28. CRAWFORD, I. P. and YANOFSKY, C. (1958) On the separation of the tryptophan synthetase of *Escherichia coli* into two protein components. *Proc. Nat. Acad. Sci.* **44**, 1161–1170.
29. HENNING, U., HELINSKI, D. R., CHAO, F. C. and YANOFSKY, C. (1962) The A protein of the tryptophan synthetase of *Escherichia coli*. *J. Biol. Chem.* **237**, 1523–1530.
30. YANOFSKY, C. and CRAWFORD, I. P. (1959) The effects of deletions, point mutations, reversions and suppressor mutations on the two components of the tryptophan synthetase of *Escherichia coli*. *Proc. Nat. Acad. Sci.* **45**, 1016–1026.
31. CRAWFORD, I. P. and YANOFSKY, C. (1959) The formation of a new enzymatically active protein as a result of suppression. *Proc. Nat. Acad. Sci.* **45**, 1280–1287.
32. CAPECCHI, M. and GUSSIN, G. N. (1965) Suppression *in vitro*: Identification of a serine tRNA as a "nonsense" suppressor. *Science*, **149**, 417–422.
33. CARBON, J., BERG, P. and YANOFSKY, C. (1966) Studies of missense suppression of the tryptophan synthetase A-protein mutant A36. *Proc. Nat. Acad. Sci.* **56**, 764–771.
34. GUPTA, N. and KHORANA, H. G. (1966) Missense suppression of the tryptophan synthetase A-protein mutant A78. *Proc. Nat. Acad. Sci.* **56**, 772–779.
35. BERG, P. (1973) Suppression: a subversion of genetic decoding. *Harvey Lectures*, **67**, 247–272.
36. HENNING, U. and YANOFSKY, C. (1962) An alteration in the primary structure of A protein predicted on the basis of genetic recombination data. *Proc. Nat. Acad. Sci.* **48**, 183–190.
37. YANOFSKY, C. and HELINSKI, D. (1962) Correspondence between genetic data and the position of amino acid alteration in A protein. *Proc. Nat. Acad. Sci.* **48**, 173–183.
38. INGRAM, V. (1958) Abnormal human haemoglobins I: The comparison of normal human and sickle-cell haemoglobins by "fingerprinting". *Biochem. Biophys. Acta*, **28**, 539–545.
39. HELINSKI, D. R. and YANOFSKY, C. (1962) Peptide pattern studies on the wild-type A protein of the tryptophan synthetase of *Escherichia coli*. *Biochim. Biophys. Acta*, **63**, 10–19.
40. STREISINGER, G., MUKAI, F., DREYER, W. J., MILLER, B. and HORIUCHI, S. (1961) Mutations affecting the lysozyme of phage T_4. *Cold Spring Harb. Symp. Quant. Biol.* **26**, 25–30.
41. GAREN, A., LEVINTHAL, C. and ROTHMAN, F. (1961) Alterations in alkaline phosphatase induced by mutations. *J. Chim. Phys.* **58**, 1068–1071.
42. YANOFSKY, C., HELINSKI, D. and MALING, B. (1961) The effects of mutations on the composition and properties of the A protein of *Escherichia coli*. tryptophan synthetase. *Cold Spring Harb. Symp. Quant. Biol.* **26**, 11–23.
43. WATSON, J. D. and CRICK, F. H. C. (1953) The structure of DNA. *Cold Spring Harb. Symp. Quant. Biol.* **18**, 123–131.
44. YANOFSKY, C. (1963) Discussion following article by W. Gilbert. *Cold Spring Harb. Symp. Quant. Biol.* **28**, 296–297.
45. YANOFSKY, C., CARLTON, B. C. GUEST, J. R., HELINSKI, D. R. and HENNING, U. (1964) On the colinearity of gene structure and protein structure. *Proc. Nat. Acad. Sci.* **51**, 266–272.
46. YANOFSKY, C., DRAPEAU, G. R., GUEST, J. R. and CARLTON, B. C. (1967) The complete amino acid sequence of the tryptophan synthetase A protein (α-subunit) and its colinear relationship with the genetic map of the A gene. *Proc. Nat. Acad. Sci.* **57**, 296–298.
47. SARABHAI, A. S., STRETTON, A. O., BRENNER, S. and BOLLE, A. (1964) Colinearity of the gene with the polypeptide chain. *Nature*, **201**, 13–17.
48. YANOFSKY, C., ITO, J. and HORN, V. (1966) Amino acid replacements and the genetic code. *Cold Spring Harb. Symp. Quant. Biol.* **31**, 151–162.
49. BENZER, S. and FREESE, E. (1958) Induction of specific mutations with 5-bromouracil. *Proc. Nat. Acad. Sci.* **44**, 112–119.
50. BENZER, S. (1961) Genetic fine structure. *Harvey Lectures*, **56**, 1–21.
51. GUEST, J. R. and YANOFSKY, C. (1966) Relative orientation of gene, messenger and polypeptide chain. *Nature*, **210**, 799–802.
52. HENNING, U. and YANOFSKY, C. (1962) Amino acid replacements associated with reversion and recombination within the A gene. *Proc. Nat. Acad. Sci.* **48**, 1497–1504.
53. YANOFSKY, C. and HORN, V. (1972) Tryptophan synthetase α-chain positions affected by mutations near the end of the genetic map of *trpA* of *Escherichia coli*. *J. Biol. Chem.* **247**, 4494–4498.
54. MURGOLA, E. J. and YANOFSKY, C. (1974) Selection for new amino acids at position 211 of the tryptophan synthetase α-chain of *Escherichia coli*. *J. Molec. Biol.* **86**, 775–784.

SYNTHESIS IN THE STUDY OF NUCLEIC ACIDS

H. G. KHORANA

Departments of Biology and Chemistry, Massachusetts Institute of Technology
Cambridge, Massachusetts 02139

I learnt of the discovery of polynucleotide phosphorylase in Severo Ochoa's laboratory in the spring of 1955 in British Columbia where I was then working. Soon after, I met Arthur Kornberg, who told me about his first results on the enzymatic synthesis of DNA in his laboratory. These discoveries immediately proved to be electrifying and they were instrumental in a large measure in bringing about a revolution in the activity and research in different aspects of nucleic acids. To give a single example, the biochemical approach, which eventually led to the elucidation of the genetic code, was a direct consequence of the discovery of polynucleotide phosphorylase.

Working in British Columbia in the early 1950s, I had become interested in the chemistry of phosphate esters of biological interest. New phosphate esters and nucleotide derivatives with new biochemical functions were being discovered in large numbers during that period. A few years earlier, when I was a postdoctoral student in Prelog's laboratory in Zürich, I had been fortunate in coming across accidentally in German literature a class of compounds called carbodiimides. Later, in Cambridge, England, in Todd's laboratory and then in British Columbia, I became very fond of these reagents and found them to be of great utility in tackling synthetic problems in the nucleotide field. Soon, together with my first group of colleagues, John Moffatt, Gordon Tener and Bob Chambers (who shortly after joined Severo Ochoa's department), we began to focus on problems of nucleotide coenzymes and polynucleotide synthesis. The tide set in motion by the work of Severo Ochoa and Arthur Kornberg in the enzymatic synthesis and general chemistry of polynucleotides, further stimulated and enhanced the sense of purpose in our efforts to develop purely chemical methods for the synthesis of oligonucleotides. In dedicating this article to Severo Ochoa's seventieth birthday volume, it need only be added that this is an account principally of one main line of research which my colleagues and I have been pursuing precisely since the time of the discovery of polynucleotide phosphorylase. In this research, the primary objective in the 1950s and early 1960s was to develop chemical methods for the synthesis of ribo and deoxyribo-oligonucleotides of completely defined nucleotide sequences. This work in turn opened doors to two new lines of research. Progress in both these lines of research hinged on the truly miraculous action of two enzymes (DNA polymerase-I and DNA ligase) on our synthetic short-chain oligonucleotides.

METHODOLOGY FOR THE ORGANOCHEMICAL SYNTHESIS OF DEOXY-RIBOPOLYNUCLEOTIDES[1]

The chief problems were: (a) efficient methods had to be developed for the condensation of individual nucleotides, and (b) since the organochemical reactions lack specificity of the enzymatic reactions, selective protection of the vulnerable groups was necessary at all stages. Currently, chemical condensations are carried out between protected 5′-nucleotides or oligonucleotides, and the 3′-hydroxyl groups of suitably protected deoxynucleosides or oligonucleotides. The groups used for protecting the 5′- and the 3′-hydroxyl functions and the amino groups in cytosine, adenine and guanine rings in mononucleotides, are shown in Fig. 1. This figure also shows the condensing agents used for the activation of the phosphate groups.

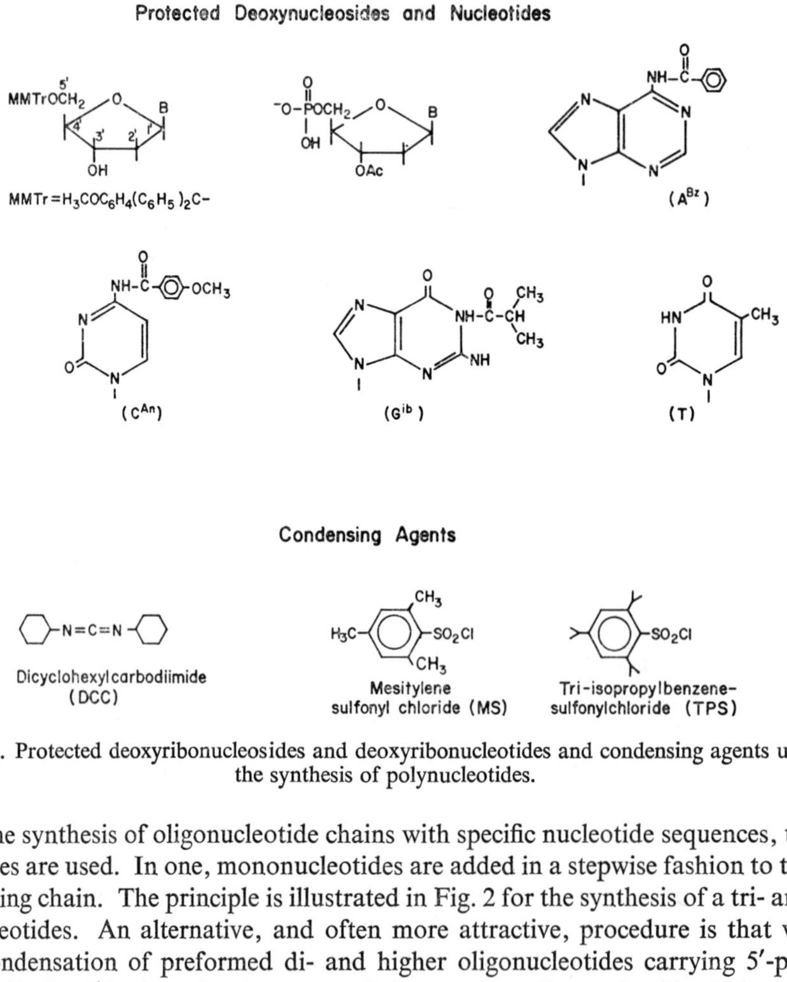

Protected Deoxynucleosides and Nucleotides

MMTr = $H_3COC_6H_4(C_6H_5)_2C-$

(A^{Bz}) (C^{An}) (G^{ib}) (T)

Condensing Agents

Dicyclohexylcarbodiimide (DCC)

Mesitylene sulfonyl chloride (MS)

Tri-isopropylbenzene-sulfonylchloride (TPS)

FIG. 1. Protected deoxyribonucleosides and deoxyribonucleotides and condensing agents used in the synthesis of polynucleotides.

For the synthesis of oligonucleotide chains with specific nucleotide sequences, two basic approaches are used. In one, mononucleotides are added in a stepwise fashion to the 3′-end of a growing chain. The principle is illustrated in Fig. 2 for the synthesis of a tri- and longer oligonucleotides. An alternative, and often more attractive, procedure is that which involves condensation of preformed di- and higher oligonucleotides carrying 5′-phosphate groups with the 3′-hydroxyl end groups of the growing oligonucleotide chains. Thus, it is customary to build oligonucleotide chains by increases of two to four nucleotides at a time.[2,3]

POLYNUCLEOTIDE SYNTHESIS AND THE GENETIC CODE

No attempt is made to introduce the problem of the genetic code or to trace the development of work on it. It need only be mentioned that the approach which revolutionized

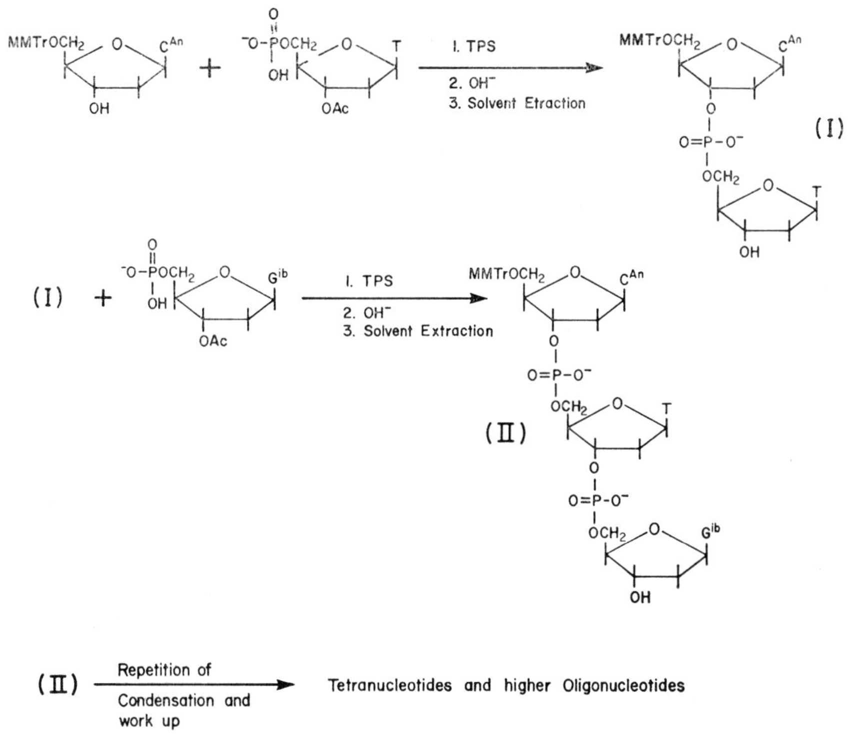

MMTrOCH2 O C^An OH + ⁻O-POCH2 O T OH OAc — 1. TPS, 2. OH⁻, 3. Solvent Etraction → MMTrOCH2 O C^An O O=P-O⁻ OCH2 O T OH (I)

(I) + ⁻O-POCH2 O G^ib OH OAc — 1. TPS, 2. OH⁻, 3. Solvent Extraction → MMTrOCH2 O C^An O O=P-O⁻ OCH2 O T O O=P-O⁻ OCH2 O G^ib OH (II)

(II) — Repetition of Condensation and work up → Tetranucleotides and higher Oligonucleotides

FIG. 2. Stepwise synthesis of a trinucleotide and higher oligonucleotide.

the experimental work on this central problem involved the use of ribopolynucleotides, prepared by the agency of polynucleotide phosphorylase, as messengers in the bacterial cell-free protein synthesizing system.[4] Indeed, a great deal was learned during the years 1961–3, both in the laboratories of Ochoa[5] and of Nirenberg and their coworkers,[6] about the overall nucleotide composition of the coding units.

The hope in my laboratory was to prepare ribopolynucleotide messengers of completely defined nucleotide sequences. However, chemical methodology permitted at this time the synthesis of oligonucleotides containing but a few ribonucleotide units (unambiguous synthesis of the sixty-four ribotrinucleotides derivable from the four mononucleotides was soon accomplished[7] and used in work on the genetic code). In the deoxyribonucleotide field, chemical synthesis of short chains (10–20 nucleotides) of defined sequence was possible. The approach which enabled the synthesis of defined, high molecular weight ribopolynucleotides was based on the earlier observations[8] that both DNA-polymerase-I and DNA-dependent RNA polymerase could use short deoxyribopolynucleotides as templates, especially when the latter contained repeating nucleotide sequences. Most dramatic were the results with the DNA polymerase. Extensive synthesis of high molecular weight DNA-like polymers could be realized by using short deoxyribopolynucleotides in appropriate combination as templates; such as

$$d(TG)_6 + d(AC)_6,\ d(TTC)_4 + d(AAG)_3 \text{ and } d(TATC)_3 + d(TAGA)_2.$$

These reactions as a group are the most satisfying and unique that I have encountered as an organic chemist in my research experience. The specific information contained in the short synthetic polynucleotides could be amplified and multiplied by the polymerase with complete fidelity. Finally, the DNA polymers thus prepared could be, and continue to be, used repeatedly for further production of the same polymers. Therefore, in the years since, it has been unnecessary to go back to the time-consuming chemical synthesis for obtaining the templates again and DNA polymerase has assured their continuity.

A variety of DNA-like polymers containing known repeating sequences were prepared. From this point on, experiments on the genetic code using the cell-free system went remarkably well. In the first step, selective transcription of either of the two strands in all of the synthetic DNA-like polymers was achieved without any difficulty simply by restricting the ribonucleoside triphosphates to those necessary for the transcription of the desired strand. In step 2, the transcription products were used as messengers in the bacterial cell-free amino acid incorporating systems. While detailed accounts of the results of these studies have been given elsewhere (e.g. ref. 9), at least three general features of the results should be mentioned. Polynucleotides with repeating dinucleotide sequences directed, invariably, the synthesis of polypeptides containing two amino acids in alternating sequence; polynucleotides with repeating trinucleotide sequences led to the formation of a series (usually three) of homo-polypeptides containing each a single amino acid. Finally, polynucleotides with repeating tertanucleotide sequences gave polypeptides containing four amino acids in repeating tetrapeptide sequence. These results proved unequivocally the three-letter, non-overlapping properties of the genetic code and provided, for the first time, a direct correlation between a DNA sequence and a polypeptide sequence. Further, together with work in Nirenberg's laboratory on the trinucleotide binding technique,[10]

TABLE 1. THE TRINUCLEOTIDE CODON ASSIGNMENTS FOR THE TWENTY AMINO ACIDS

The abbreviations for the amino acids are standard. The first, second and third nucleotides for each codon are as indicated. The Methionine codon AUG and Valine codon GUG also stand for initiator codons. The codons UAG, UAA and UGA stand for termination signals.

The Genetic Code

1st↓ 2nd→	U	C	A	G	↓3rd
	PHE	SER	TYR	CYS	U
	PHE	SER	TYR	CYS	C
U	LEU	SER	Term.	Term.	A
	LEU	SER	Term.	TRP	G
	LEU	PRO	HIS	ARG	U
	LEU	PRO	HIS	ARG	C
C	LEU	PRO	GLUN	ARG	A
	LEU	PRO	GLUN	ARG	G
	ILEU	THR	ASPN	SER	U
	ILEU	THR	ASPN	SER	C
A	ILEU	THR	LYS	ARG	A
	MET (Init.)	THR	LYS	ARG	G
	VAL	ALA	ASP	GLY	U
	VAL	ALA	ASP	GLY	C
G	VAL	ALA	GLU	GLY	A
	VAL (Init.)	ALA	GLU	GLY	G

it proved possible to derive most of the sequences for the amino acid codons. While space does not permit any further discussion of the structure of any of the aspects of the genetic code, the presently accepted genetic code is shown in Table 1.[11] The codon assignments have been deduced by different workers and often by using more than one method. Although much remains to be done at chemical and biochemical level to obtain an adequate understanding of the very elaborate protein synthesizing system, nevertheless the problem of the genetic code, at least in the restricted one-dimensional sense (the linear correlation of the nucleotide sequence of polynucleotides with that of the amino acid sequence of polypeptides), is now believed to have been solved.

SYNTHESIS OF tRNA GENES AND NUCLEOTIDE SEQUENCES IN THE PROMOTER REGIONS

If one knows the genetic code, then, in principle, one wants to know about the DNA regions which lie between the genes and, presumably, these include the control elements in gene expression. If the assembly of bihelical DNA of any given sequence could be brought within the scope of laboratory synthesis, then there could ultimately be, as perhaps for the genetic code, a new and general approach to the study of DNA expression, such as gene structure–functional relationships and, broadly speaking, DNA–protein interactions including the nature and mode of action of the control elements. With these considerations, my laboratory began to devote attention to the development of methods for the synthesis of bihelical DNA. As a complementary necessity, work on the DNA sequencing in the control regions, which appeared not to be transcribed, was concurrently started. Thus, synthesis and sequencing of DNA have been the major interests of my laboratory in the past 9 years.

As in the work with the genetic code, it was clear at the start of this program that organic synthesis, although the only means for creating specific sequences without the agency of a template, would not be able to take us all the way on the formidably long journey. A new concept would have to be used in conjunction with organic synthesis. The concept we envisaged is the one nature discovered a very long time ago and uses without exception in all the biology of DNA—namely, the ability of polynucleotides to form the two Watson–Crick base pairs. By using short synthetic polynucleotides which would have appropriate overlaps in their partly complementary sequences, it might be possible to form hydrogen-bonded complexes and then to bring about end to end joining to form long polynucleotides. Although, in the early 1960s, experiments using water-soluble carbodiimides had given promising results, the discoveries of the polynucleotide ligase and polynucleotide kinase proved to be boons in the present work. Initial studies with the ligase showed that it required relatively short chains to bring about end to end joining and soon the following general strategy could be formulated for the step-wise construction of any given bihelical DNA. The first and easily the most demanding phase involves the chemical synthesis of polydeoxynucleotide segments of chain length in the range of 8 to 12 units with 3′- and 5′-hydroxyl end-groups free; the segments should correspond to the entire two strands of the intended DNA and those belonging to the complementary strands should have an overlap of four to five nucleotides; (2) the phosphorylation of the 5′-hydroxyl group with ATP carrying a suitable label in the γ-phosphoryl group using the T_4 polynucleotide kinase;

and (3) the head-to-tail joining of the appropriate segments when they are aligned to form bihelical complexes using the T_4 polynucleotide ligase.

From a variety of considerations, the tRNA molecules can be concluded to be one of the most exciting elaborations of evolution. The nucleotide sequences of the tRNAs were being worked on intensively in the early 1960s and that of the major yeast alanine tRNA had just appeared from Holley's laboratory[12] when specific goals for the present work were being considered. Synthesis of the DNA corresponding to the entire sequence (77 nucleotides) of this tRNA was the decision and by using the methodology outlined above it was possible to complete the task of total synthesis in 1970.[13]

While the synthesis of yeast alanine tRNA gene taught us a great deal about laboratory synthesis of DNA, from the biochemical standpoint, limitations were apparent for further work with it. The bacterial tRNA gene, that for tyrosine tRNA suppressor, was selected as the next target. The main arguments, which fortunately became even more compelling as we proceeded with this project, for this choice were as follows: (1) firm knowledge of the primary nucleotide sequence by at least two groups of workers; (2) the marked advantages which this tRNA offered in studying the biochemistry of the amino acid-charging reaction; and (3) the dramatic progress in the biochemistry of *in vitro* protein synthesis in the cell-free *E. coli* system and in study of *in vitro* suppression of amber mutation using suppressor tRNAs; (4) the insertion of this gene into the transducing bacteriophage ϕ80 and the convenience of working with $\phi 80psu^{+}_{III}$ and related derivatives for probing the nucleotide sequences adjoining the tyrosine tRNA structural gene; and (5) the subsequent discovery of the precursor[14] for this tRNA (Fig. 3). This structure containing a 5′-triphosphate end group unambiguously defined the site of initiation of transcription and therefore attention could be focused precisely on the promoter (preinitiation) region and on the process of transcription of this gene.

The total synthetic plan which was formulated in 1968 is shown in Fig. 4. Recently, this job has also been successfully completed: the twenty-six deoxypolynucleotide segments were grouped into parts I–IV as shown. These were joined to form the four duplexes and the latter were then joined with one another to form the entire duplex.[15]

NUCLEOTIDE SEQUENCE IN THE PROMOTER AND TERMINATOR REGIONS

DNA sequencing is now in high momentum and there is available a wealth of techniques and methods. It was not so when the present work was started. The concept we used in our approach, and it is now quite popular, was to use the DNA in a single-stranded form and to focus on the region of interest by specifically hybridizing a complementary deoxyribopolynucleotide in an adjacent site. A primer–template relationship is thus established and the primer may be extended at its 3′-OH end in the direction of the required sequence by using the DNA polymerases. The nucleotide incorporations are dictated by the sequence of the template strand and the sequence of the latter at that site can therefore be deduced by sequencing the extended primer. For sequencing proper a variety of rapid and elegant methods is now available. In the work on the promoter and terminator regions of the tyrosine tRNA gene, all the requirements of the above approach were readily met. Thus, the two strands of the $\phi 80psu^{+}_{III}$ DNA could be conveniently separated and

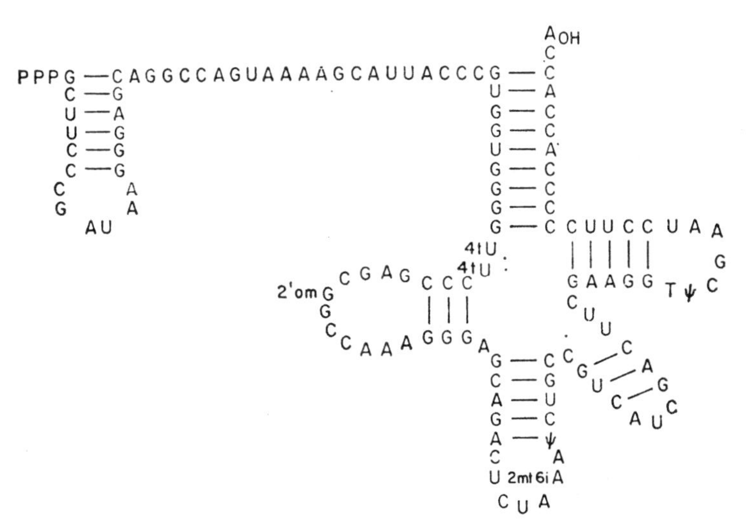

FIG. 3. The primary nucleotide sequence of an *E. coli* tyrosine tRNA precursor.

deoxyribopolynucleotide primers of suitable length were available from the synthetic work. Hybridization of a primer representing the 3′-end of the tRNA to the r-strand of $\phi 80psu^{+}_{III}$ DNA gave a system for sequencing the terminator region, while hybridization of a suitable primer with the l-strand at the 5′-end of the tRNA precursor gave a system for sequencing the promoter region.[16]

Figure 5 summarizes the progress to-date in both the synthetic and sequencing aspects of our work. The illustration has three parts: first at the bottom the base-pairs 1′–59′ contain the sequence in the promoter region (preinitiation region), transcription starting

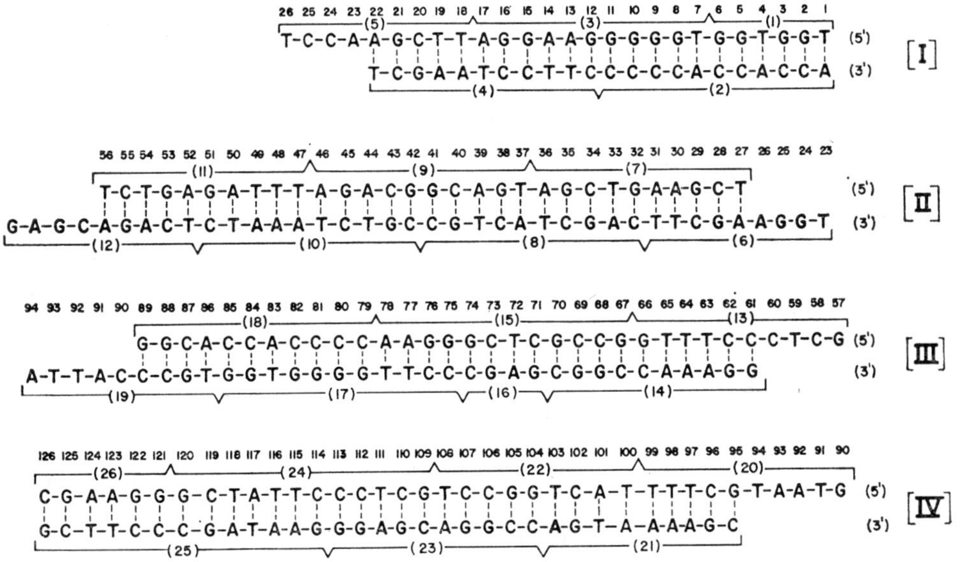

26 25 24 23 22 21 20 19 18 17 16 15 14 13 12 11 10 9 8 7 6 5 4 3 2 1
(5) (3) (1)
T-C-C-A-A-G-C-T-T-A-G-G-A-A-G-G-G-G-G-T-G-G-T-G-G-T (5′)
T-C-G-A-A-T-C-C-T-T-C-C-C-C-C-A-C-C-A-C-C-A (3′)
(4) (2)
[I]

56 55 54 53 52 51 50 49 48 47 46 45 44 43 42 41 40 39 38 37 36 35 34 33 32 31 30 29 28 27 26 25 24 23
(11) (9) (7)
T-C-T-G-A-G-A-T-T-T-A-G-A-C-G-G-C-A-G-T-A-G-C-T-G-A-A-G-C-T (5′)
G-A-G-C-A-G-A-C-T-C-T-A-A-A-T-C-T-G-C-C-G-T-C-A-T-C-G-A-C-T-T-C-G-A-A-G-G-T (3′)
(12) (10) (8) (6)
[II]

94 93 92 91 90 89 88 87 86 85 84 83 82 81 80 79 78 77 76 75 74 73 72 71 70 69 68 67 66 65 64 63 62 61 60 59 58 57
(18) (15) (13)
G-G-C-A-C-C-A-C-C-C-C-A-A-G-G-G-C-T-C-G-C-C-G-G-T-T-T-C-C-C-T-C-G (5′)
A-T-T-A-C-C-C-G-T-G-G-T-G-G-G-G-T-T-C-C-C-G-A-G-C-G-G-C-C-A-A-A-G-G (3′)
(19) (17) (16) (14)
[III]

126 125 124 123 122 121 120 119 118 117 116 115 114 113 112 111 110 109 108 107 106 105 104 103 102 101 100 99 98 97 96 95 94 93 92 91 90
(26) (24) (22) (20)
C-G-A-A-G-G-G-C-T-A-T-T-C-C-C-T-C-G-T-C-C-G-G-T-C-A-T-T-T-T-C-G-T-A-A-T-G (5′)
G-C-T-T-C-C-C-G-A-T-A-A-G-G-G-A-G-C-A-G-G-C-C-A-G-T-A-A-A-A-G-C (3′)
(25) (23) (21)
[IV]

FIG. 4. Plan for the total synthesis of the gene for the precursor of tyrosine tRNA. Subdivision of the total chemically synthesized segments into four parts for the purpose of enzymatic joining.

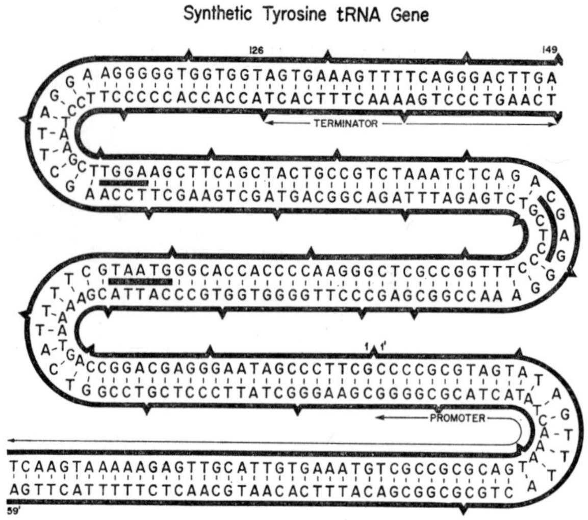

Synthetic Tyrosine tRNA Gene

FIG. 5. Present status of the work on the sequencing of the promoter and terminator region and of the total synthesis of the gene for an *E. coli* tyrosine suppressor tRNA. The nucleotides between the numbers 1′ and 59′ represent the now known sequence in the promoter region. Transcription starts at nucleotide 1. The DNA between nucleotides 1 and 126 (top line) has been synthesized. The nucleotides 126–149 represent the sequence adjoining the C–C–A end of the tRNA gene. This sequence has also been synthesized. The distances between the carets represent the polydeoxynucleotide segments which were synthesized chemically. Carets indicate the sites of ligase-caralysed joinings. The black strips inserted between the complementary sequences show the sites where the synthetic duplexes with protruding single-stranded ends and corresponding to parts of the gene were joined to form the entire duplex.

at left of 1′ with nucleotide marked 1. The synthesis of this duplex is in progress. The DNA between nucleotides 1 and nucleotide 126 (top row), which covers the entire length of the tRNA precursor, has been synthesized. The sequence from nucleotide 126 to 149 contains the region adjoining C–C–A end of the tRNA. This DNA has also been synthesized.

At the present time, we believe that the promoter region is fully included in the sequence shown. Using DNA fragments, which contain the above gene and its control elements and are derived by the action of restriction endonucleases on $\phi 80psu^{+}_{III}$ DNA, efforts are being made to understand the promoter and the process of initiation. Following the total synthesis of the sequence shown in Fig. 5, specific transcription to form the precursor RNA and processing to the mature, functional tRNA would be the next goal. Successful realization of these aims would, it is hoped, pave the way for an extended use of the synthetic approach to the pricise studies of many of the biologically interesting problems, especially in the field of the mechanisms of gene expression and function as well as for the total laboratory syntheses of defined genes.

REFERENCES

1. KHORANA, H. G. (1968) Nucleic acid synthesis. *Pure Appl. Chem.* **17**, 349.
2. WEBER, H. and KHORANA, H. G. (1968) Studies on polynucleotides. CIV. Total synthesis of the structural gene for an alanine transfer ribonucleic acid for yeast. Chemical synthesis of an icosadeoxyribonucleotide corresponding to the nucleotide sequence 21–40. *J. Molec. Biol.* **72**, 219 (and accompanying papers).

3. VAN DE SANDE, J. H., CARUTHERS, M. H., KUMAR, A. and KHORANA, H. G. (1976) Studies on polynucleotides. CXXXII. Total synthesis of the structural gene for the precursor of a tyrosine suppressor transfer RNA from *E. coli* (2). Chemical synthesis of the deoxypolynucleotide segments corresponding to the nucleotide sequence 1–31. *J. Biol. Chem.* **251**, 571.
4. MATTHAEI, J. H. and NIRENBERG, M. H. (1961) Characterization and stabilization of DNAase-sensitive protein synthesis in *E. coli. Proc. Nat. Acad. Sci. U.S.A.* **47**, 1580.
5. OCHOA, S. (1963) Synthetic polynucleotides and the genetic code. *Federation Proc.* **22**, 62.
6. NIRENBERG, M. H., MATTHAEI, J. H., JONES, O. W., MARTIN, R. G. and BARONDES, S. H. (1963) Approximation of genetic code via cell-free protein synthesis directed by template RNA. *Federation Proc.* **22**, 55.
7. LOHRMANN, R., SÖLL, D., HAYATSU, H., OHTSUKA, E. and KHORANA, H. G. (1966) Studies on polynucleotides. CI. Synthesis of the 64 possible ribotrinucleotides derived from the four major ribomononucleotides. *J. Am. Chem. Soc.* **88**, 819.
8. KHORANA, H. G. (1965) Polynucleotide synthesis and the genetic code. *Federation Proc.* **24**, 1473.
9. KHORANA, H. G. (1968) Polynucleotide synthesis and the genetic code. In *The Harvey Lectures*, Series 62, p. 79, Academic Press, New York.
10. NIRENBERG, M. H. and LEDER, P. (1964) RNA codewords and protein synthesis. The effect of trinucleotides upon the binding of sRNA to ribosomes. *Science* **145**, 1399.
11. CRICK, F. H. C. (1966) The genetic code—yesterday, today and tomorrow. *Cold Spring Harbor Symp. Quant. Biol.* **31**, 3.
12. HOLLEY, R. W., APGAR, J., EVERETT, G. A., MADISON, J. T., MARQUISEE, M., MERRILL, S. H., PENSWICK, J. R. and ZAMIR, A. (1965) Structure of a ribonucleic acid. *Science* **147**, 1462.
13. AGARWAL, K. L., BÜCHI, H., CARUTHERS, M. H., GUPTA, N., KHORANA, H. G., KLEPPE, K., KUMAR, A., OHTSUKA, E., RAJBHANDARY, U. L., VAN DE SANDE, J. H., SGARAMELLA, V., WEBER, H. and YAMADA, T. (1970) Total synthesis of the gene for an alanine transfer ribonucleic acid from yeast. *Nature* **227**, 27.
14. ALTMAN, S. and SMITH, J. D. (1971) Tyrosine tRNA precursor molecule polynucleotide sequence. *Nature New Biol.* **233**, 35.
15. KHORANA, H. G., AGARWAL, K. L., BESMER, P., BÜCHI, H., CARUTHERS, M. H., CASHION, P. J., FRIDKIN, M., JAY, E., KLEPPE, K., KLEPPE, R., KUMAR, A., LOEWEN, P. C., MILLER, R. C., MINAMOTO, K., PANET, A., RAJBHANDARY, U. L., RAMAMOORTHY, B., SEKIYA, T., TAKEYA, T. and VAN DE SANDE, J. H. (1976) Studies on polynucleotides. CXXXI. Total synthesis of the structural gene for the precursor of a tyrosine suppressor transfer RNA from *E. coli* (1). General introduction (and accompanying papers). *J. Biol. Chem.* **251**, 565.
16. SEKIYA, T., VAN ORMONDT, H. and KHORANA, H. G. (1975) Studies on polynucleotides. CXXVIII. The nucleotide sequence in the promoter region of the gene for an *E. coli* tyrosine transfer ribonucleic. acid. *J. Biol. Chem.* **250**, 1087.

A PHAGE IN NEW YORK

C. WEISSMANN

Institut für Molekularbiologie I, Universität Zürich, 8049 Zürich, Switzerland

Early in October 1961 I left behind me Switzerland, a short-lived career in medicine, a somewhat longer one in organic chemistry and plunged into the turbulent field of molecular biology. The decision to do so dated back to a seminar by Severo Ochoa which I had heard in Zürich several years earlier. Captivated by what was to me a new and exciting science and attracted by the speaker's enthusiasm, I approached Ochoa after his talk and on an impulse requested to join him after completing my degree. Ochoa—apparently also on an impulse—immediately accepted me.

Ochoa assigned me to work with Z. who, in searching for the enzyme responsible for the replication of Tobacco Mosaic virus RNA, had isolated "RNA synthetase" from spinach leaves, an enzyme he believed synthesized RNA from the four ribonucleoside triphosphates, using RNA as template. I do not remember why spinach had been chosen as an enzyme source, but its ready availability at the local grocery store surely played a role. In any event, little of the starting material went to waste—the tender, inner parts of the leaves were homogenized for enzyme extraction while the tougher outer parts were converted into spinach soup and consumed for lunch by Edith, Z.'s thrifty Swiss technician and myself. I confirmed Z.'s basic observation, namely that RNA synthesis by the partly purified enzyme was dependent on the addition of RNA, however I soon found that this requirement was also satisfied by oligonucleotides, which suggested that the enzyme might require a primer rather than a template. My personal relationship with Z. rapidly deteriorated as I began to follow up this unwelcome finding and reached a low point when Z. asked me to discuss in advance each of my experiments before handing me the labeled ATP I needed. At this point Joe Krakow, who had considerable experience in working with nucleic acid enzymes, took me under his scientific wing. With his quaint habit of celebrating every successful experiment by ordering a small piece of equipment or an expensive chemical, Joe had accumulated a vast supply of radioactive nucleoside triphosphates and these, along with his expert advice, he put at my disposal. I was now able to experiment freely, at least at night and weekends, trying to establish the true nature of "synthetase". Finally Krakow, in whose mind the suspicion had been growing that synthetase might simply be an impure preparation of polynucleotide phosphorylase, induced me to carry out some simple but critical experiments. These showed that the synthetic activity of the spinach enzyme was completely blocked both by low levels of inorganic phosphate and by the combination of pyruvate kinase and phosphoenol pyruvate. Inorganic phosphate was well known to inhibit the synthetic activity of polynucleotide phosphorylase by shifting the equilibrium in the direction of phosphorolysis. The ATP-generating system prevents the phosphorylase from incorporating label from ADP (its true substrate) into RNA by converting the diphosphate into ATP. ADP itself, as later emerged, was generated in the assay mixture from the labeled ATP by phosphatases present in the impure "synthetase" preparation. Within a few days it was clear beyond any doubt that "RNA synthetase" was polynucleotide phosphorylase in disguise, rediscovered in spinach leaves where, some years previously, it had already been found by Littauer and Ochoa. All of this led to quite some upheavals and when the dust had settled I found myself sharing a lab. with Krakow, while Z. sought for and eventually found a job elsewhere.

The scientific aspects of this episode made me very wary of the pitfalls of research in general and of nucleic acid biochemistry in particular, and rendered me empathetic to the aphorism ascribed to Ephraim Racker: "First purify, then think—do not waste clean thoughts on dirty enzymes" which I forthwith inscribed on a wall of my new lab., but did not follow.

I should at this point set my own scientific efforts into the proper context. Ochoa's main interest at that time lay in the elucidation of the genetic code, and the lights in the lab. of Peter Lengyel, Joe Speyer and Carlos Basilio, who were carrying out this work, rarely went out before the early hours of the morning. It was just a few weeks after Nirenberg had announced his discovery that poly(U) elicited the synthesis of polyphenylalanine in a cell-free system. This news must have caused anguish to Lengyel, still a graduate student at the time, and to Speyer, since these two had some months earlier conceived the notion that synthetic homopolynucleotides (easily available by synthesis with Ochoa's polynucleotide phosphorylase!) could be used as messengers for cell-free protein synthesis, a system which Speyer had developed in Ochoa's lab. in the past year or so. The project had been batted back and forth, subjected to extensive critical analyses and finally pursued with only low priority after it was pointed out by a far-sighted colleague that homopolynucleotides should not be expected to function as messenger because they could not contain the specific signals for initiation of protein synthesis which a messenger perforce would have to carry. This thought we now know to be correct in principle, and yet in practice homopolynucleotides work very well because, as it turned out, the unphysiologically high Mg^{2+} concentrations used in the *in vitro* systems at that time allowed the natural initiation mechanism to be bypassed. I learned from this that even the best experiments can be talked to death!

The NYU group rapidly confirmed Nirenberg's findings and ingeniously extended the use of synthetic polynucleotides to deduce the composition of almost all aminoacid codons. Ochoa's personal involvement in this project was very intense and the technical resources of his department were fully mobilized to provide the large number of components required for the decoding work. I learned to appreciate the importance of a well-trained, independent technical staff—as personified by Morton Schneider and Pete Lozina—and a well-equipped pilot plant in the basement, all of which I copied as best I could when I started my own Institute years later.

Two other features impressed me as being characteristic of Ochoa's lab. and important for its success. One was the feeling of belonging to a "family", which accordingly brought duties and privileges with it—the duty of helping and the privilege of being helped (as well as the occasional squabbles characteristic of family life). The other feature was the intense exchange of information, both within the lab. and with the outside world. Ochoa promoted communication between members of his department by bringing up technical problems, results, criticisms and questions at lunch and at the long coffee-and-cake sessions in the library every afternoon. I felt that these sessions were very important for the education of the younger members of the Department, since expertise was available in a variety of domains. Physical chemistry was expounded by Bob Warner, organic chemistry, in particular of nucleic acids, by Bob Chambers, general enzymology was the realm of Yoshito Kaziro and later on of Dan Lane. Al Wahba was up-to-date on all significant new events in biochemistry and mediated information between interested parties. Charlie Gilvarg, one of the brightest minds I had yet encountered, had the capacity of criticizing and suggesting experiments in any field—even in those he was barely familiar with.

Interchange with scientists from outside the lab. was free of any secrecy, and—occasionally to the despair of some of his younger collaborators—Ochoa would expound on the newest and most exciting results from his lab. as though competition did not exist.

It was now early in 1962; I worked under the direct supervision of Ochoa and in close contact with Joe Krakow. Ochoa and I decided to continue the search for an RNA-dependent RNA polymerase, switching to *Escherichia coli* infected by the recently discovered single-stranded RNA-containing phage f2, although this meant that I could no longer use the remains of our starting material to supplement my lunch. Already the first step in this undertaking, that of obtaining phage f2 from its discoverers, proved a major hurdle. While some petitioners—so went the legend—managed to recover infectious phage by shaking out the letter of refusal over a lawn of susceptible bacteria I, like many others, found it easier to obtain an RNA phage named MS2 from A. J. Clark. Working with Lionel Simon, we soon discovered an RNA-synthesizing activity which was not present in uninfected cells but arose within a few minutes after infection with MS2. This activity, which could not be ascribed to any previously known enzyme, inherited the title of RNA synthetase. We purified the enzyme some 80-fold but failed to obtain a preparation which was dependent on exogenously added RNA. Despite this deficiency I was so pleased with our achievement that when Zinder was pointed out to me at a meeting, I approached him, introduced myself and proudly informed him that I had found a virus-specific polymerase in RNA phage-infected cells. He looked up at me from his breakfast, said "who hasn't?" and turned back to his coffee. I was soon to learn what he meant by his dour comment. Meanwhile, by the end of 1962, we had submitted a paper on MS2 RNA synthetase to the *Proceedings of the National Academy of Science*[1] which, due to its conjunction of timeliness and imperfection, was to irritate some of my colleagues in the field.

On discussing the possible mechanism of replication of RNA viruses with me, Charlie Gilvarg suggested that I try to find out whether the single-stranded RNA of the phage, on entering its host, was converted to a double-stranded molecule which could then be replicated by a mechanism recommended by Watson and Crick. I carried out the appropriate experiment early in 1963, infecting *E. coli* with ^{32}P-labeled phage, recovering the ^{32}P-labeled RNA from the infected cell and testing it for resistance to pancreatic RNase, a property characteristic for double-stranded RNA. I found that about 10% of the radioactive RNA was RNase-resistant, but when I tried to denature this material by heating it to 100° (as one would do to denature DNA) the RNase-resistance remained unchanged. I felt that this, along with the low proportion of parental RNA becoming RNase-resistant, was a discouraging result and gave up the approach—prematurely, as I soon found out. Instead, I started out on the 1963 lecture circuit: Federation Meetings, Cold Spring Harbor Symposium, Gordon Conference. While the Federation Meetings brought me a no worse experience than having to present much the same data as Tom August—after him—the Cold Spring Harbor Meeting gave me a taste of what the great wide world of molecular biology was like. After Tom August and then I had told our modest tale to the 200-odd afficionados of nucleic acid biosynthesis, Sol Spiegelman—whom I saw for the first time—took the stage, and, starting out in a deceptively mild voice, launched into a talk which soon made me wish I had stayed in organic chemistry. His first sentences still ring in my ears: "You have heard from Dr. August and Dr. Weiss. . . Weiss. . . Weissberg? [voice from the audience—my own—"Weissmann"] Weissmann, thank you, about their attempts to obtain an RNA-dependent enzyme. *We* have been luckier in our efforts . . . ." He proceeded to show that his enzyme had been purified free of RNA template, that its activity depended on the addition of MS2 RNA but that it was not stimulated by RNA from *E. coli*, and postulated the novel concept of template specificity. Moreover, he suggested the possibility that replication occurred by a direct copying mechanism, since none of his experiments either *in vivo* or *in vitro* had provided evidence for the existence of a complementary strand. These claims met with what Spiegelman later called "well-controlled enthusiasm". Both DNA-dependent RNA polymerase and the Kornberg DNA polymerase were not known to show any template specificity (not even asymmetric transcription had yet been recognized) and it was felt that what was not true for these major polymerases could not be true for a barely discovered, upstart enzyme. The "direct copying" mechanism was rejected as stark heresy against the Watson–Crick base-pairing mechanism.

To my relief Spiegelman did not show up at the Gordon Conference but the fun had gone out of lecturing on MS2 synthetase, despite my decision to disbelieve Spiegelman's data. On returning to the relatively quiet haven of my laboratory, I took up work again, now collaborating with Piet Borst, a mitochondriologist from Amsterdam, who had arrived some weeks earlier. Our association led to an exciting and productive collaboration and to an enduring friendship. With some prodding from Ochoa, who was not very happy with my synthetase purification, Piet worked out a better procedure, the distinctive features of which were the breaking up of the frozen *E. coli* clumps with a wooden mallet (cf. Fig. 1) and precipitation of the enzyme by Mg^{2+} in two successive steps.[2] Nonetheless, he also failed to obtain a template-dependent system. At about this time Montagnier and Sanders[3] reported finding an infectious double-stranded RNA in cells infected with encephalomyocarditis virus, an animal virus which also had single-stranded RNA as genetic material, raising the possibility that this was an intermediate of replication. They also noted that the melting point of double-stranded RNA was almost 20° higher than that of DNA with a similar base composition which explained my earlier difficulties in the characterization of the RNase-resistant MS2 RNA. We immediately examined the *in vitro* product of MS2 synthetase and to our surprise found it to consist of almost 50% RNase-resistant RNA.[4] Within a short time our group, which as a consequence of Ochoa's heightened interests in RNA phages had been expanded by two new arrivals, Martin Billeter from Zürich and Roy Burdon from Glasgow, characterized this RNA by hybridization methods which we developed for this purpose and found that about 80% of the *in vitro* product consisted of viral strands and the rest of complementary strands,[2] for which we coined the designation "plus" and "minus" strands respectively.[4] In the course of these experiments we also noticed that the enzyme preparation itself contained large amounts of endogenous viral minus strands which we isolated as RNase-resistant RNA. Martin Billeter purified this material on large scale and in collaboration with Bob Warner and Bob Langridge characterized it as double-stranded RNA both chemically and by X-ray diffraction. Similar findings were independently made in several other laboratories, in particular those of Dick

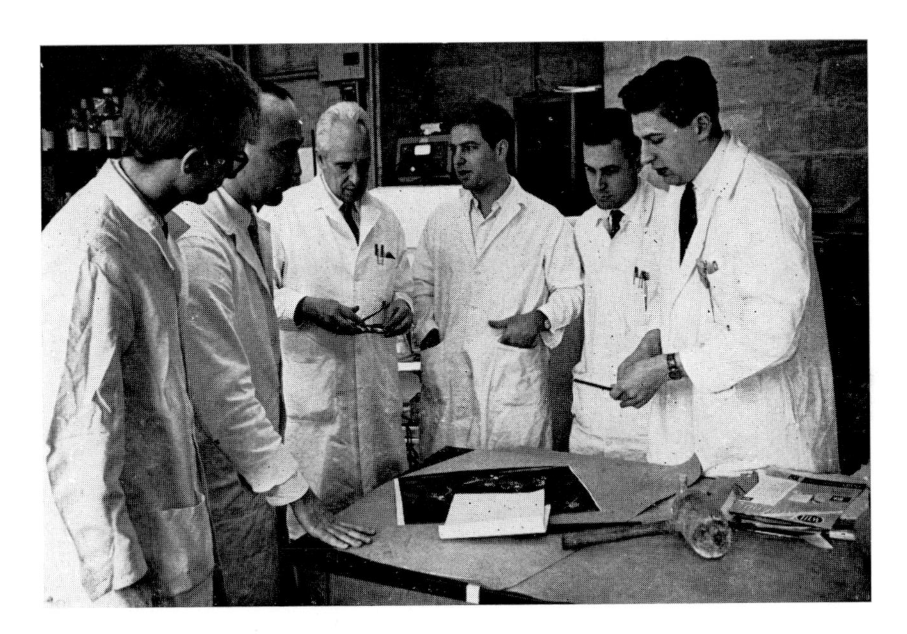

Fig. 1. The RNA synthetase research team at New York University (1964). From left to right: Piet Borst, (Ted Abbot), Severo Ochoa, Charles Weissmann, Martin Billeter and Roy Burdon. On the table: the wooden mallet used in the first step of synthetase purification and the scheme of phage RNA replication (shown in detail in Fig. 2).

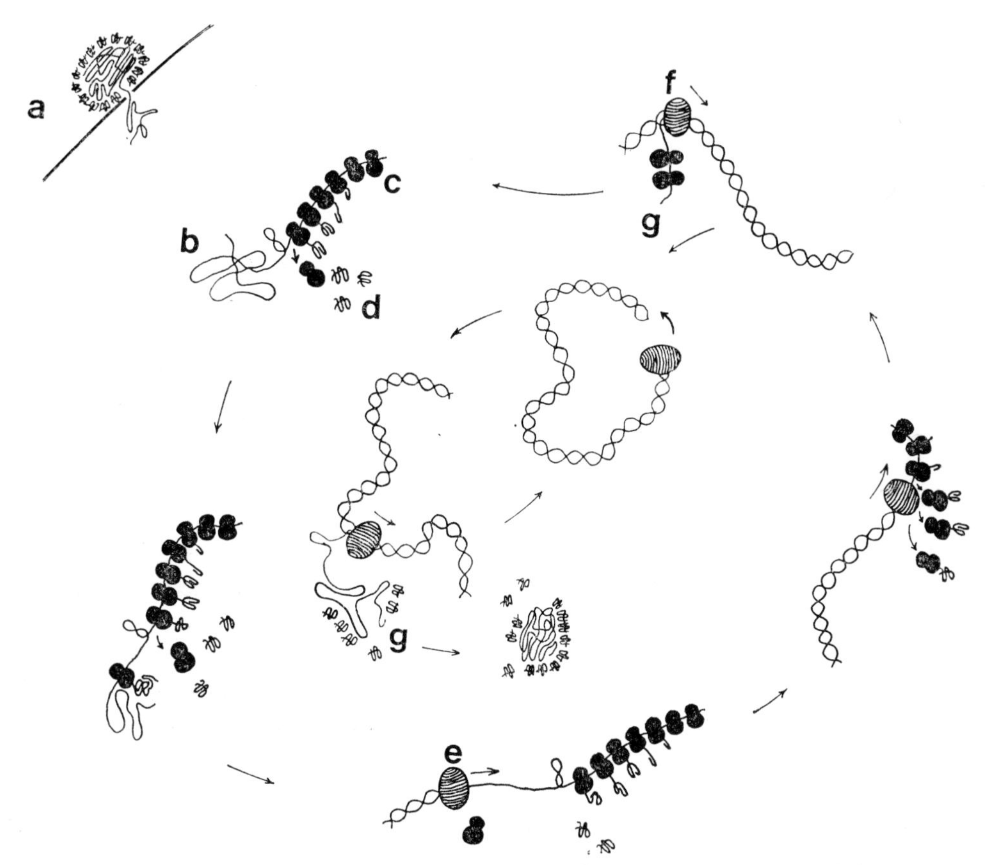

FIG. 2. Scheme of phage replication (1964). (a) Infecting virus; (b) parental RNA; (c) ribosomes; (d) capsomers; (e) enzyme converting single-stranded to double-stranded RNA by synthesis of minus strands; (f) enzyme using double-stranded template for synthesis of parental type viral RNA (plus strands); (g) progeny RNA; (from ref. 5). The major change subsequently introduced into this scheme is the substitution of single-stranded minus strands for double-stranded RNA as template in RNA replication (cf. Fig. 3).

Franklin, Hoffmann-Berling and Peter Hans Hofschneider. Our findings gave rise to a flock of publications in 1964, in which we proposed[5] the mechanism of replication for MS2 RNA shown in Fig. 2.

The paper-writing procedure in Ochoa's department was a process none of his former associates are likely to forget. The prospective authors first submitted their version of the paper to Ochoa, who then ripped it to bits, rejoined some of the more worthy fragments, reordered the sentences, rewrote the discussion and then returned the retyped manuscript with a mild comment such as "Well, it was pretty good, but I have made some small changes". The next few sessions were devoted to haggling about the reinsertion of individual sentences for hours on end. Typically, late on a Saturday afternoon, Ochoa ended a session by inviting us back for Sunday morning at 10 a.m., for a final discussion. The polished output of these collaborations was no longer entrusted to our hands, for fear of last-minute changes of the spotless typescript. Despite his intense participation in the planning of the work and the elaboration of the papers, Ochoa refused to put his name on our RNA phage papers after the first few publications.

Early in 1964 we got a call from Norton Zinder, with whom we were now in frequent contact, complaining that his graduate student Harvey Lodish was unable to detect double-stranded RNA in products made by extracts from f2-infected *E. coli.* Disquieted by their apparent inability to repeat our results, Piet and I rushed to the Rockefeller Institute and in going over Harvey's protocols we noted that in divergence from our procedure, he had not deproteinized his product prior to digestion with RNase. Piet repeated the RNase digestion with native and with deproteinized synthetase products and discovered that RNase-resistance was acquired by the product only after deproteinization! After much further experimentation we concluded that the replicative intermediate consisted not of double-stranded RNA but of a single-stranded template and one or more single-stranded product strands held together by replicase, and that on deproteinization this structure collapsed to the double-stranded structures isolated by us and dozens of other investigators (cf. Fig. 3). The paper published in 1965[6] describing these results, which we considered to be one of the most interesting of the series, found little resonance and it took almost 7 years until the strange character of the replicating intermediate was confirmed by others for RNA phages and found to also apply in the case of at least some animal RNA viruses.

Meanwhile I had successively been promoted to Instructor, Assistant and then Associate Professor, due in part to the proficiency with which I lectured to medical students on

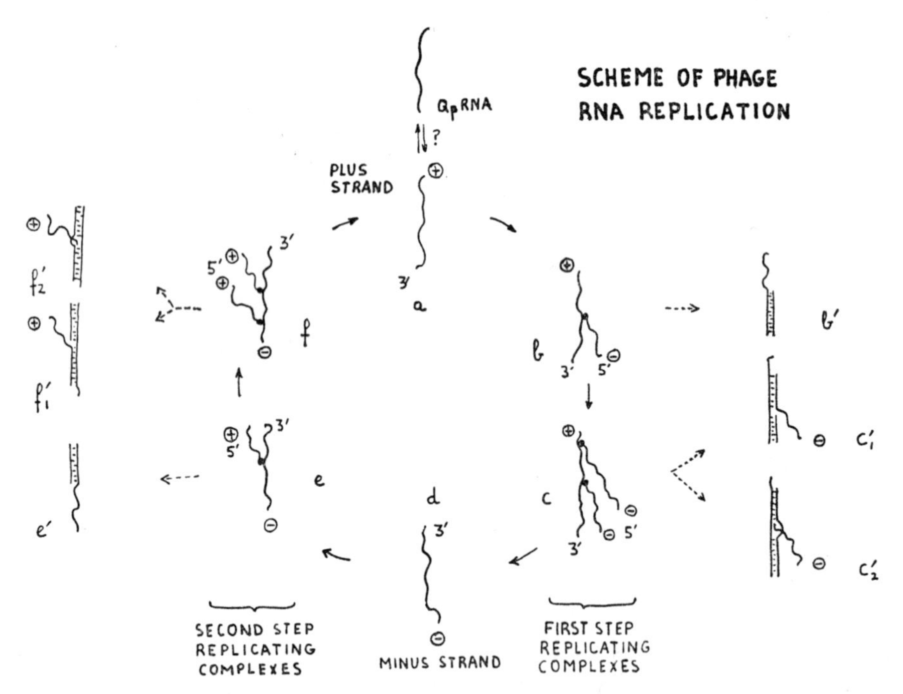

FIG. 3. Scheme of phage RNA replication (1968). In the first stage of replication a single-stranded minus strand is synthesized on a plus strand template; the intermediate, the first-step-replicating complex (b or c), is an "open" structure in which template and product do not form a double helix. In the second step of replication the single-stranded minus strand (d) is used as a template for plus strand synthesis; the intermediate, the second-step-replicating complex (e or f), is analogous to the first, but the full-length template is a minus strand. Replicating complexes may collapse, spontaneously or under the influence of external agents, to yield double-helical structures (b′, c′, e′, f′). (From ref. 16.)

biochemical subjects I had just recently learned myself. Martin Billeter, Massimo Libonati and Eladio Viñuela had obtained convincing evidence that a complex containing minus strands served as an intermediate of viral RNA replication *in vivo*[7] and I was having a good time travelling up and down the country and to Europe, spreading the word of synthetase and the minus strand. We had never succeeded in repeating Spiegelman's purification of template-dependent MS2 "replicase", no further reports had issued from Urbana and we were confident that the enzyme had gone the way of all protein, although Walter Fiers had reported that in his laboratory in Ghent it had on a few occasions proved possible to obtain an RNA-dependent preparation of MS2 replicase. I found it deceptively easy to disregard what I did not believe in, or as Morgenstern[8] put it:

Und er kommt zu dem Ergebnis:
Nur ein Traum war das Erlebnis
Weil, so schliesst er messerscharf,
Nicht sein kann was nicht sein darf.*

However, this blissful state of affairs terminated abruptly in the summer of 1965. During their eclipse period, Haruna and Spiegelman[9,10] had achieved a major breakthrough: working with a newly discovered Japanese RNA phage, Qβ, they had purified its replicase to a degree where it was not only template-dependent but also sufficiently free of RNases to permit net *in vitro* synthesis of infectious Qβ RNA! They also confirmed and extended their observations on the template specificity of the polymerase and, to our chagrin, reasserted their claim that with their purified enzyme neither double-stranded RNA nor minus strands occurred as intermediates of RNA replication.[11] This time the combination of spectacular data and the persuasive power of Spiegelman's rhetoric convinced large segments of the scientific community—especially those not active in nucleic acid biochemistry—of the correctness of his claims. During this depressing state of affairs Ochoa gave me his whole-hearted support and a sense of direction without which I might have felt unable to continue the struggle. On Ochoa's request, Spiegelman, with dismaying promptness, sent us enough replicase to carry out a dozen or so standard incubations. Since the amount appeared insufficient to clarify the central question of replication intermediates, I spent some time miniaturizing the assay and the analytical methods before starting the experiments; by also using substrates of very high specific radioactivity, the range of experimentation was extended some hundred-fold. After confirming the basic observations of Haruna and Spiegelman regarding template specificity, I analyzed the products formed by the replicase using Qβ RNA as template at different times of incubation. The results of the experiment were unambiguous and exhilarating: in the first minutes of incubation exclusively minus strands were synthesized and only subsequently did newly formed plus strands make their appearance. I completed the entire analysis without even using up the whole quota of enzyme. Two publications appeared some months later, back to back in the *Proceedings of the National Academy of Science*, one by ourselves[12] and one by Haruna and Spiegelman,[13] coming to divergent conclusions regarding the synthesis of minus strands and the mechanism of replication. The *denouement*, however, followed rapidly. With the introduction of polyacrylamide gel electrophoresis

* "And he comes to the conclusion:
His mishap was an illusion
For, he reasons pointedly,
That which must not, can not be".

From *The Gallows Song* by C. Morgenstern, translated by Max Knight.[8]

of RNA in Spiegelman's laboratory, the appearance of minus strand-containing replicating intermediates was recognized and the involvement of minus strands in replication acknowledged. There remained the question as to whether the biologically active intermediate was a single minus strand or a partly or totally doubled-stranded RNA structure; the former possibility appeared more likely since both Spiegelman and I had failed to obtain evidence that double-stranded RNA complexes could serve as template for replicase. Günter Feix and I were now preparing our own Qβ replicase, the supply from Urbana having finally run out, and in a *tour-de-force* Bob Pollet, a graduate student at New York University, accomplished the difficult task of purifying single-stranded Qβ minus strands. This RNA, in contrast to viral plus strand RNA, was non-infectious

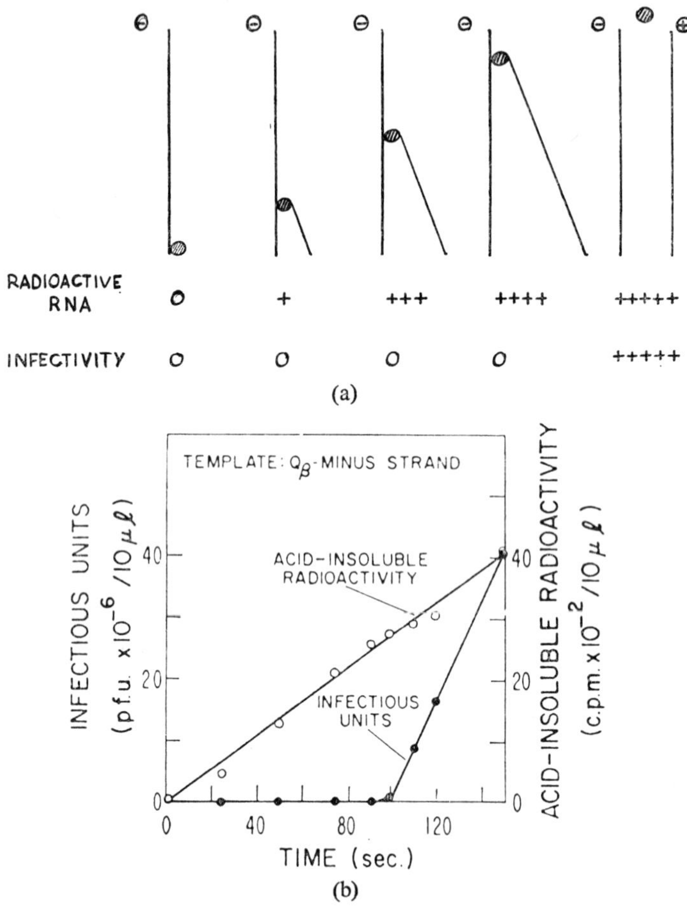

FIG. 4. Synthesis of infectious Qβ RNA using non-infectious minus strands as template. (a) Schematic representation of the experiment, showing that incorporation of radioactive substrate is expected to occur linearly, starting with the beginning of incubation, while formation of infectious RNA starts abruptly after a lag period, after the first plus strands have been completed. (b) Experimental results: Qβ minus strands and Qβ replicase were incubated with ATP, GTP, CTP and [C^{14}] UTP in 0.1 ml of a Mg-Tris buffer. At the times indicated 15 μl-aliquots were removed for the determination of acid-insoluble radioactivity and infectivity. Since infectivity appears after about 100 seconds of incubation and since Qβ RNA contains 4500 nucleotides, the rate of synthesis is about 45 nucleotides per second. (From ref. 14.)

and it proved to be an excellent template for Qβ replicase. Finally, to our immense satisfaction, we found that incubation of non-infectious minus strands with Qβ replicase and the four ribonucleoside triphosphates led to the production of infectious Qβ RNA in high yield within about 2 minutes (Fig. 4).[14] This was the final and formal demonstration that minus strands fulfilled the requirements of an authentic intermediate in Qβ RNA replication. I was very eager to present these results at the 1967 International Congress of Biochemistry in Tokyo, but my abstract was turned down by the organizers and not even Ochoa's intervention could reverse the decision. Ever the optimist, I nevertheless travelled to Tokyo with my slides and on arriving, I learned that a minor miracle had occurred; Spiegelman had telegraphed from Australia cancelling his symposium talk (he was enjoying a boat trip around the Great Barrier Reef) and Ochoa had induced the chairman of the session to use me as a substitute!

With this happy end my years of apprenticeship drew to a close. For several years the University of Zürich had been negotiating for my return and finally, in 1967, following Ochoa's advice, I decided to leave the secure environment of New York University in order to build up an Institute of Molecular Biology in Zürich together with Martin Billeter, my earlier collaborator in New York.

My tale still needs some rounding off. Spiegelman, August and I were pitted against each other once more in a Cold Spring Harbor Symposium in 1968, but on this seventh anniversary of our first meeting, agreement was reached on the general features of Qβ RNA replication.[15–17] All three of us, while continuing to work on Qβ replicase, also entered the RNA tumor virus field, but we have not managed to work up a good fight again. Spiegelman and I became close friends over the years, and I came to realize that our conflict, although unpleasant to me at the time, added to my understanding about the way science is done.

Looking back on the 6 years I spent in Ochoa's department, I appreciate them as the most important and rewarding period of my development. My closest bond remains to Severo Ochoa, teacher and friend, who helped me beyond the call of duty during my stay with him and in the years thereafter.

ACKNOWLEDGEMENTS

I wish to thank many friends for reading this essay, correcting gross historical inaccuracies and suggesting improvements, many of which I did not carry out. Moreover, I wish to apologize to those colleagues whose important contributions to the field were not mentioned in this distorted retrospective; some of them may be more inclined to forgive me if they consider that I have also refrained from mentioning their mistakes.

REFERENCES

1. Weissmann, C., Simon, L. and Ochoa, S. (1963) Induction by an enzyme catalyzing incorporation of ribonucleotides into ribonucleic acid. *Proc. Nat. Acad. Sci. U.S.A.* **49**, 407–414.
2. Weissmann, C., Borst, P., Burdon, R. H., Billeter, M. A. and Ochoa, S. (1964) Replication of viral RNA, IV. Properties of RNA synthetase and enzymatic synthesis of MS2 phage RNA. *Proc. Nat. Acad. Sci. U.S.A.* **51**, 890–897.
3. Montagnier, L. and Sanders, F. K. (1963) Replicative form of encephalomyocarditis virus ribonucleic acid. *Nature*, **199**, 664–667.

4. WEISSMANN, C. and BORST, P. (1963) Double-stranded ribonucleic acid formation *in vitro* by MS2 phage-induced RNA synthetase. *Science*, **142,** 1188–1191.
5. OCHOA, S., WEISSMANN, C., BORST, P., BURDON, R. H. and BILLETER, M. A. (1964) Replication of viral RNA. *Fed. Proc.* **23,** 1285–1296.
6. BORST, P. and WEISSMANN, C. (1965) Replication of viral RNA, VIII. Studies on the enzymatic mechanism of replication of MS2 RNA. *Proc. Nat. Acad. Sci. U.S.A.* **54,** 982–987.
7. BILLETER, M. A., LIBONATI, M., VIÑUELA, E. and WEISSMANN, C. (1966) Replication of viral ribonucleic acid, X. Turnover of virus-specific double-stranded ribonucleic acid during replication of phage MS2 in *Escherichia coli*. *J. Biol. Chem.* **241,** 4750–4757.
8. MORGENSTERN, C. (1964) *The Gallows Song*, translated by M. KNIGHT, University of California Press.
9. HARUNA, I. and SPIEGELMAN, S. (1965) Specific template requirements of RNA replicases. *Proc. Nat. Acad. Sci. U.S.A.* **54,** 579–587.
10. SPIEGELMAN, S., HARUNA, I., HOLLAND, I. B., BEAUDREAU, G. and MILL, D. (1965) The synthesis of a self-propagating and infectious nucleic acid with a purified enzyme. *Proc. Nat. Acad. Sci. U.S.A.* **54,** 919–927.
11. SPIEGELMAN, S. and HARUNA, I. (1966) Problems of an RNA genome operating in a DNA-dominated biological universe. In *Macro-molecular Metabolism*, pp. 263–304, Little, Brown & Co., Boston.
12. WEISSMANN, C. and FEIX, G. (1966) Replication of viral RNA, XI. Synthesis of viral "minus" strands *in vitro*. *Proc. Nat. Acad. Sci. U.S.A.* **55,** 1264–1268.
13. HARUNA, I. and SPIEGELMAN, S. (1966) A search for an intermediate involving a complement during synchronous synthesis by a purified RNA replicase. *Proc. Nat. Acad. Sci. U.S.A.* **55,** 1256–1263.
14. FEIX, G., POLLET, R. and WEISSMANN, C. (1968) Replication of viral RNA, XVI. Enzymatic synthesis of infectious viral RNA with noninfectious Qβ minus strands as template. *Proc. Nat. Acad. Sci. U.S.A.* **59,** 145–152.
15. AUGUST, J. T., BANERJEE, A. K., EOYANG, L., FRANZE DE FERNANDEZ, M. T., HORI, K., KUO, C. H., RENSING, U. and SHAPIRO, L. (1968) Synthesis of bacteriophage Qβ RNA. *Cold Spring Harbor Symp. Quant. Biol.* **33,** 73–81.
16. WEISSMANN, C., FEIX, G. and SLOR, H. (1968) *In vitro* synthesis of phage RNA: the nature of the intermediates. *Cold Spring Harbor Symp. Quant. Biol.* **33,** 83–100.
17. SPIEGELMAN, S., PACE, N. R., MILLS, D. R., LEVISOHN, R., EIKHOM, T. S., TAYLOR, M. M., PETERSON, R. L. and BISHOP, D. H. L. (1968) The mechanism of RNA replication. *Cold. Spring Harbor Symp. Quant. Biol.* **33,** 101–124.

BACTERIOPHAGE ϕ29

ELADIO VIÑUELA and MARGARITA SALAS
Departamento de Biología Molecular, Instituto G. Marañón, C.S.I.C.
Velázquez, 144 Madrid-6, Spain

The study of bacterial viruses has provided many of the most important breakthroughs in modern biology. Thus, the role of nucleic acids in heredity and the basic mechanisms of mutation, replication, transcription and translation of the genetic material were mainly revealed through the use of bacteriophages.

Bacteriophages are also one of the most convenient materials to study macromolecular interactions. Although the host provides most of the machinery needed for the transcription and translation of the invading genome, as well as some factors for morphogenesis, many of the functions required for phage maturation are virus-specific gene products.

The study of the mechanism of assembly of a phage particle was first approached with coliphage T4 making use of the electron microscope and conditional lethal mutants.[1] Since then, the most studied phages, among those that contain double-stranded DNA, have been T4, lambda and P22.

After getting our Ph.D. degree in Madrid, we stayed in Ochoa's laboratory from 1964 to 1967. There, we had our first contact with bacterial viruses (the story of bacteriophage MS2 in the Department of Biochemistry of New York University is told in this volume by Charles Weissmann).

Some time before we left N.Y.U. we had to face the problem of what kind of work to do in Madrid. To continue the work on RNA phages or on protein synthesis, which we had carried out in New York, was quickly dismissed. We had to set up our laboratory starting with an empty room and we knew that it would take some time before we could get things in full operation.

From our work in New York, and a bacterial genetics summer course in Cold Spring Harbor, we had decided to attack some problems on virus assembly. At that time T4 was being intensively explored in Caltech and Geneva and it was clearly a very competitive field. We decided to start a study of *B. subtilis* phage ϕ29, a phage similar to T4 in having a prolate head, and a rather complex morphology, although its DNA was ten times smaller.[2]

Here we describe some of the progress carried out in our laboratory in the study of ϕ29.

STRUCTURAL COMPONENTS

One attractive feature of ϕ29 is the complex morphology of such a small phage, with a double-stranded DNA of only 12×10^6 daltons. This makes the study of its structure interesting and the identification of the viral genes and their role in phage morphogenesis becomes feasible.

Bacteriophage ϕ29 consists of a prolate head with fibers, a neck formed by two collars and twelve radial appendages and a short tail (Figs. 1 and 2). The upper collar is joined

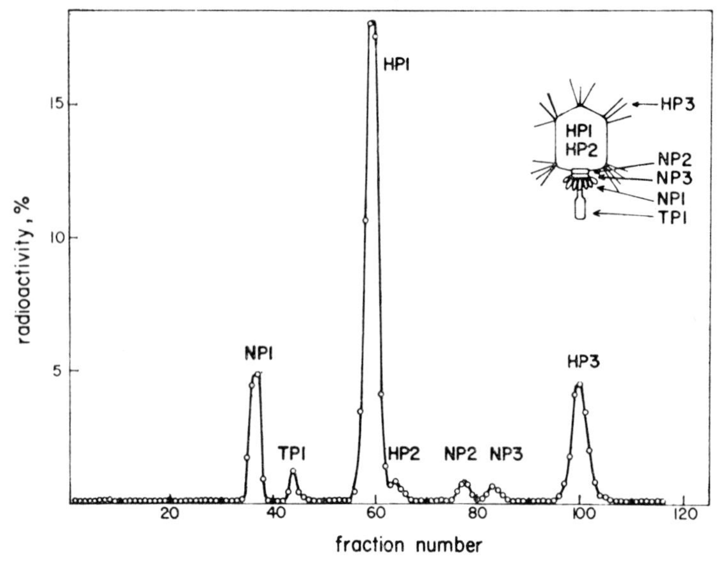

FIG. 2. Gel electrophoresis of the structural proteins of phage $\phi 29$.

to the flattened base of the head, and the lower collar is located just below, having the twelve appendages attached to it.

Structural proteins

The protein moiety of $\phi 29$ consists of seven polypeptides, which can be separated from each other by polyacrylamide gel electrophoresis. Treatment of viral particles with different chemicals provides a variety of viral substructures, which can be purified, analyzed by electron microscopy and their polypeptide composition determined by gel electrophoresis. From experiments of this kind, we could assign each structural polypeptide of $\phi 29$ to specific morphological components, such as it is outlined in Fig. 2.[3]

Viral DNA

The DNA of $\phi 29$ is interesting because a protein is involved in its circularization. Under carefully controlled conditions, a circular complex of one molecule of DNA and, probably, two copies of a DNA-associated protein (DAP) can be isolated.[4] DAP is closely related but not identical to the major head protein, as a comparison of the tryptic peptides of both proteins indicates.

DAP may play a role in the packing of phage DNA inside the capsid. When the tail of $\phi 29$ is removed, the DNA is released as a superhelix which uncoils quickly to produce a relaxed circle of DNA attached to the capsid.[4] Hirokawa showed that a $\phi 29$ DNA-protein complex is highly infectious whereas the DNA without protein is inactive.[5] Recently, we have found that, although protein-free $\phi 29$ DNA is inactive in transfection, it is capable of rescuing mutants in all cistrons found in the viral genome except those which map

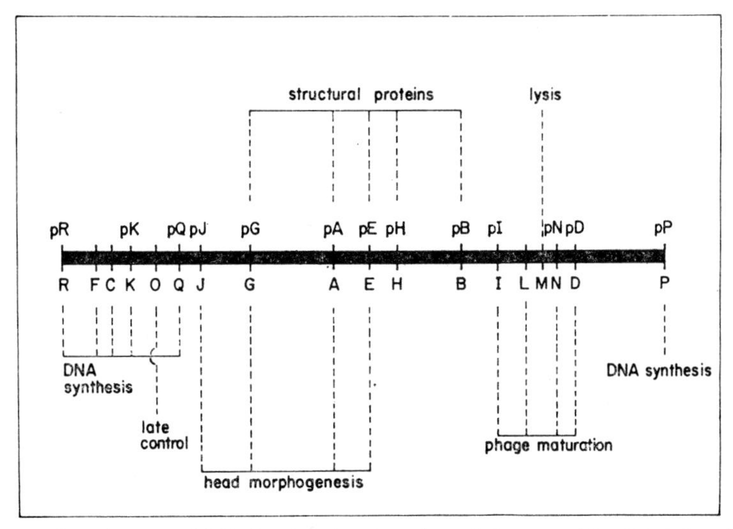

FIG. 3. Genetic map of ϕ29 showing the clustering of cistrons with related functions.

in genes R and P, the two terminal cistrons on each side of the genetic map (Fig. 3). DAP could be one of a new series of proteins, named pilot proteins, involved in the orientation of a particular DNA region towards the cell membrane.[6]

Capsid geometry

The capsid of phage ϕ29, although non-isometric, may be related to the icosahedral symmetry by elongation of a regular icosahedron, as in the case of T2 capsid.[7] That both phages may have a capsid geometry based on similar principles is supported by the finding of similar morphological variants.

Direct observation of surface detail on the negatively stained capsids of ϕ29 has not yet provided information on capsid geometry. A reasonable model for ϕ29 phage structure should account not only for the morphological variants but also for the number of copies of each structural protein per phage particle. These numbers can be determined after gel electrophoresis of viral proteins labeled with a mixture of radioactive amino acids, assuming that the number of counts under each protein peak is proportional to protein mass. From the relative mass contribution of each structural protein, its molecular weight and the total molecular weight of the protein moiety of ϕ29, the number of copies of protein HP1, HP2 and HP3 is about 96, 5 and 66, respectively.

Assuming a basic plane hexagonal lattice p6,[8] the subunits of the major capsid protein HP1 can be arranged in 11 pentamers, formed by $(5 + 1)$ subunits, and 5 equatorial hexamers, formed by $(6 + 1)$ subunits, making a total of 101 subunits. A possible location for protein HP2 could be the center of the hexamers; in that case, the number of copies of protein HP1 per phage would be 96.

Electron micrographs of ϕ29 show that the head fibers are grouped in the upper and lower part of the head and absent in the middle part of the capsid. Since there are eleven 5-fold symmetry vertices and more than one fiber per vertex, the minimum number of fibers would be 55, or 66 if each pentamer, formed by $(5 + 1)$ subunits of protein HP1, contains (5 + 1) copies of protein HP3. The closeness of this value with the number of molecules of protein HP3 per capsid determined experimentally strongly suggests that each fiber is a single molecule of protein HP3.

According to the principle of quasi-equivalence[8] the capsid of ϕ29 would be formed by two 5-fold lattices within each end-cap, corresponding to an icosahedral $T = 1$ symmetry, separated by a band of 5 hexamers (Fig. 4).

GENETIC ANALYSIS

Conventional genetic studies of phage ϕ29 have included the isolation of mutants, the analysis of their patterns of complementation and recombination, and the construction of a genetic map. The collections of temperature-sensitive (ts) and nonsense suppressor-sensitive (sus) mutants isolated in our laboratory and in Minneapolis by Reilly, Anderson and coworkers, have been integrated, resulting a total of eighteen genes, which have been arranged in a linear genetic map by two- and three-factor crosses (Fig. 3).

Gene products

To analyze the protein products of ϕ29 genes a culture of *B. subtilis*, irradiated with UV-light to inhibit host protein synthesis, is infected with ϕ29 in the presence of radioactive amino acids. At the end of the labeling period the cells are harvested, lysed with lysozyme and subjected to polyacrylamide slab gel electrophoresis in the presence of the detergent sodium dodecylsulfate (SDS). After electrophoresis the gel slabs are dried and autoradiographed to determine the positions of the labeled proteins. This system of electrophoresis resolves the protein chains on the basis of size and allows an estimate of the molecular weight of each protein band.

Approximately nineteen ϕ29 proteins can be identified on SDS-polyacrylamide gels. Their molecular weight values range between approximately 4000 and 90,000, and they account for over 90% of the coding capacity of ϕ29 DNA.

When a virus infects a cell, not all the genes are expressed at the same time. If ϕ29-infected cells are pulse-labeled with a radioactive amino acid at different times after infection and the proteins analyzed on acrylamide gels, it is seen that some proteins are synthesized only at early times, other proteins are synthesized late in infection while others are synthesized continuously. The structural proteins of ϕ29 are late proteins; they can be easily identified in the gels by a comparison with the electrophoretic mobilities of the proteins isolated from purified virus (Fig. 5).

To make a correlation of protein bands and genes, nonsense mutants of ϕ29 were utilized. The protein affected by a nonsense mutation should disappear from its normal position in the patterns obtained upon electrophoresis of extracts of mutant-infected cells on SDS-polyacrylamide gels. The proteins specified by thirteen of the eighteen genes known in ϕ29 have been identified. These proteins are named with a p before the letter designating the gene (Fig. 3).

Genes R, K, Q, I and P code for early proteins and genes J, G, A, E, H, B, N and D for late proteins. Since, as will be seen later, cistrons F and C are involved in DNA synthesis, the proteins product of these two cistrons are probably early proteins as it is the case for cistrons R, K, Q and P, also needed for DNA replication. The same must be the

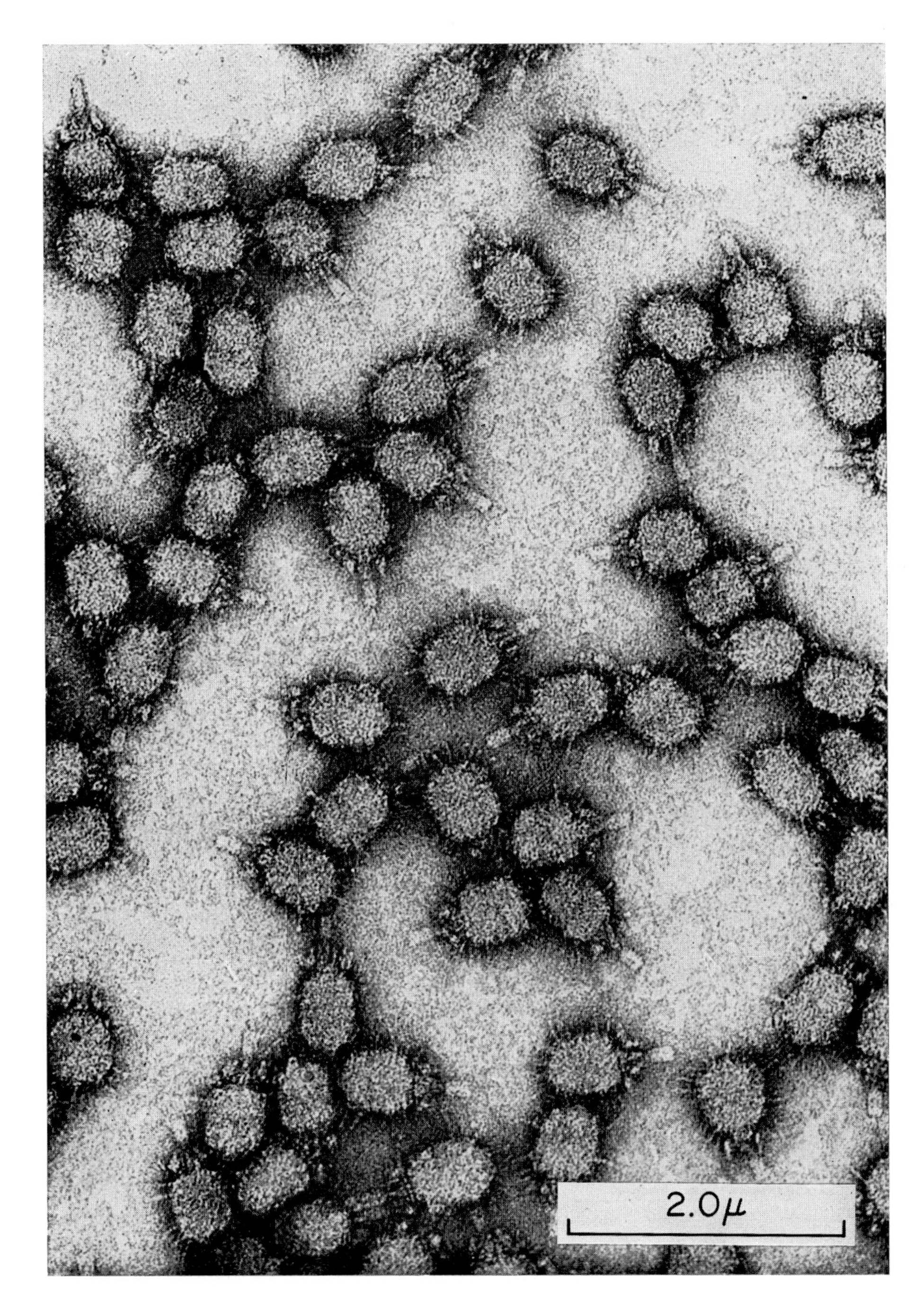

FIG. 1. Electron micrograph of bacteriophage ϕ29 stained with uranyl acetate.

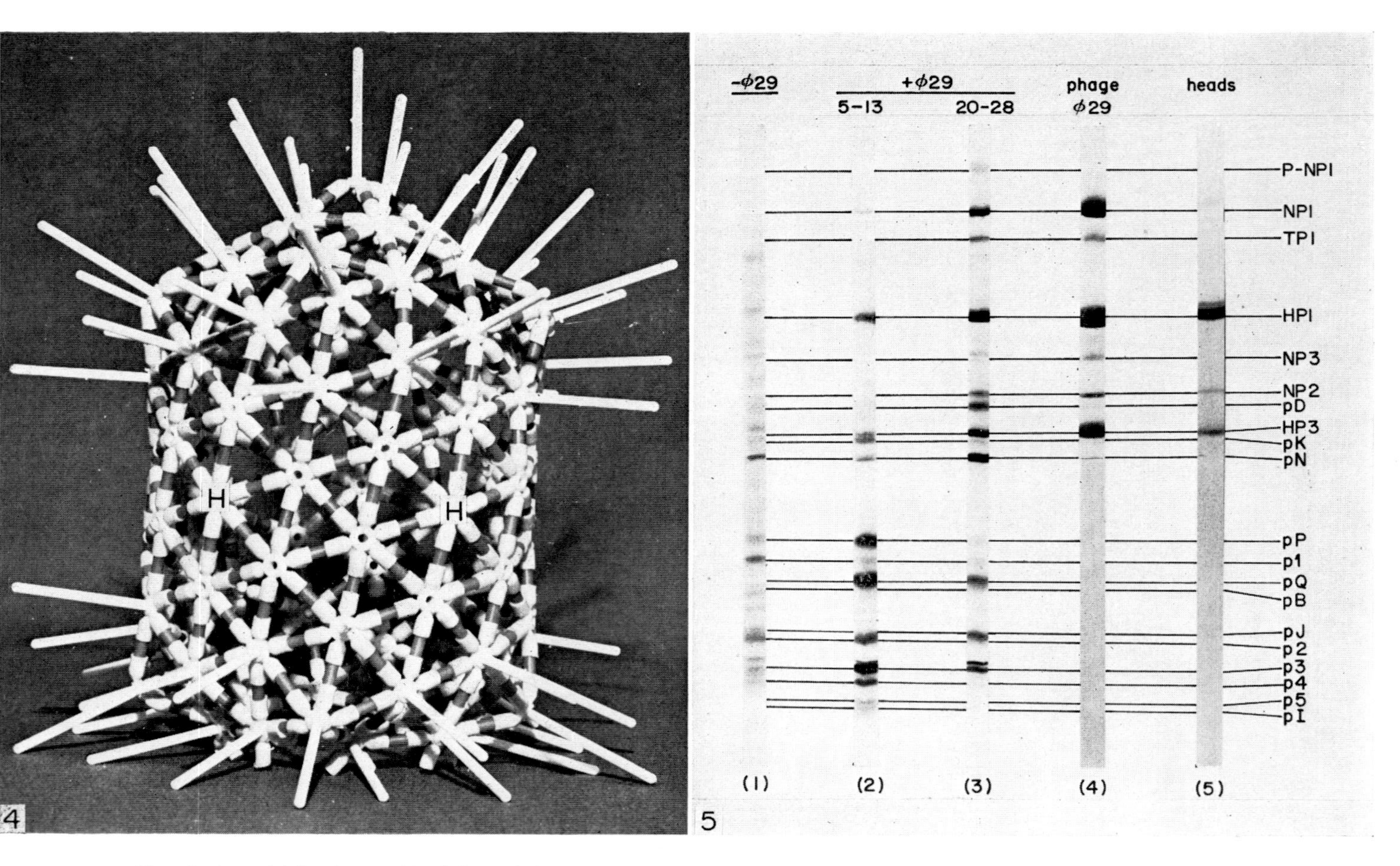

FIG. 4. A model for the capsid of phage $\phi 29$.

FIG. 5. Early and late proteins induced in *B. subtilis* infected with $\phi 29$.

case for cistron O which controls late transcription. Thus, there are three regions in the ϕ29 map coding for early proteins (R to Q, I and P) and two regions coding for late proteins (J to B and N to D). We do not know yet whether cistrons L and M code for early or late proteins. Cistron M is involved in the lysis of the bacteria being probably a late gene.

Fragment map of ϕ29 DNA

To correlate the genetic map of phage ϕ29 and the physical map of the viral DNA molecule we have used the restriction endonuclease EcoRI. This enzyme cleaves ϕ29 DNA into five fragments (A–E). The order of the fragments has been established by analysis of DNA partial digests with the same enzyme. The two terminal EcoRI fragments were identified by treatment of circular ϕ29 DNA with EcoRI. From these experiments the order of the EcoRI fragments of ϕ29 is A–B–E–D–C. The identification of the left and right end fragments relative to the genetic map was carried out by studying their translation in an *E. coli* cell-free system of proteins synthesis. Since fragment A gives place to protein pQ and fragment C to protein pP (Fig. 3), fragments A and C correspond to the left and right ends of the genetic map, respectively.

GENE FUNCTIONS

DNA replication

Since infection of *B. subtilis* with ϕ29 does not stop host DNA synthesis the identification of the genes involved in viral DNA replication is not straightforward. For this kind of screening one can use cells infected with different ϕ29 mutants, incubated in the presence of radioactive thymidine and hydroxyphenylhydrazinopyrimidines. As these drugs inhibit replicative synthesis of host DNA[9] but not the replication of wild-type ϕ29 DNA, uninfected cells or bacteria infected with mutants in genes necessary for viral DNA replication do not incorporate the radioactive precursor. By using this technique it was found that genes R, F, C, K, Q and P are required for the replication of ϕ29 DNA.

DNA transcription

Hybridization-competition experiments have shown the existence in ϕ29-infected cells of at least two classes of viral RNAs, early and late. The use of the two separate strands of ϕ29 DNA has indicated that early and late RNAs are transcribed, respectively, from the light and heavy strands and that most species of early RNA are also transcribed at late times.[10]

Only early RNA is made during infection with gene O conditional lethal mutants. Thus, the gene O product is necessary for transcription of late genes. We have not yet been able to identify protein pO, either because the nonsense fragment is very close to the C-terminal end or because protein pO overlaps with other protein band. In any case, protein pO is not a polymerase similar to the T7-induced RNA polymerase because, in contrast to T7, all ϕ29-specific RNA synthesis is sensitive to rifamycin throughout the development cycle. Since the β subunit of the host polymerase is the site of rifamycin sensitivity, this subunit must be necessary for all phage RNA synthesis. Extensive studies in our laboratory on

RNA polymerase from uninfected and ϕ29 infected *B. subtilis* have shown no difference either in the subunit composition of both enzymes or in their electrophoretic mobilities.

Most of the ϕ29 early RNA species are synthesized *in vivo* in cells where virus-specific protein synthesis does not occur because of the presence of chloramphenicol. This indicates that early RNA is synthesized by the bacterial RNA polymerase. *In vitro*, *B. subtilis* RNA polymerase binds to ϕ29 DNA in four sites, which can be visualized by electron microscopy. Three of these sites correspond or are close to the three initiation sites of transcription of the early genes.

PHAGE ASSEMBLY

The form of the capsid of the simplest viruses, such as the RNA-containing phages, is determined by the properties of the protein components. By contrast, other viruses, including T4, lambda and P22, as well as ϕ29, have several structural parts each composed of a different type of subunit(s). The assembly of these viruses is a complex process involving proteins not all of which are present in the final infective particle.

The morphogenesis of phage particles is a model not only for the assembly of organelles but also for regulation of protein specificity and nucleic acid–protein interactions. The advantage of this over other models is that the study of virus assembly is amenable to a wider variety of chemical, physical and genetic techniques.

The general approach to study virus assembly is to infect the cell under restrictive conditions with sus- or ts-mutants in different genes, in the presence of radioactive amino acids and/or thymidine, and look for the effect of the mutation. For this purpose electron microscopy is a necessary tool. Phage lysates or thin-sections of infected bacteria are examined under the electron microscope and the morphology of the viral particles or substructures present, if any, scored. A further step in the analysis involves the purification of the phage or phage-related structures, usually by gradient centrifugation and the determination of their protein composition by gel electrophoresis.

Head assembly

The morphogenesis of ϕ29 phage capsid is under the control of genes J, G, A and E (Fig. 3). Cistron G codes for the major capsid proteins HP1 and/or HP3 and, probably for the DNA-associated protein, whereas cistrons E and A determine the upper collar protein (NP2), and the tail protein (TP1), respectively.

Protein pJ (M.W. 8000) is a non-structural late protein essential for head assembly because susJ-mutants do not assemble capsids. Furthermore, a role for protein pJ in capsid assembly is indicated by the observation that tsJ-mutants produce membrane-bound isometric heads. Although protein pJ is not present in normal phage, about seventy copies of pJ are present in particles produced after infection with mutants in cistrons I and D.[11]

Sus-mutants in gene G do not induce the synthesis of the major capsid protein (HP1) nor the capsid fiber protein (HP3). As expected, susG-mutants do not produce heads and, what is more interesting, they do not give place to neck–tail complexes; this suggests that the assembly of the neck with the tail requires a preformed head.

Gene A codes for the tail protein TP1. Under restrictive conditions, susA- and tsA-mutants produce DNA-free abortive particles with a core. Morphologically, A-particles, as the normal capsids, are non-isometric, but they are more rounded than wild-type particles. Therefore, the tail protein plays a role in the formation of sharp corners (pentamers). A-particles contain modified head proteins, HP1 and HP3, the upper collar protein (NP2) and a small amount of the non-structural late protein pN.

Gene E codes for the upper collar protein NP2. SusE-mutants produce DNA-free, abortive particles. Morphologically, they are isometric and contain modified head proteins (HP1 and HP3). Therefore, the upper collar protein (NP2) is necessary for capsid elongation.

Phage completion

Besides genes H and B, which code for phage structural proteins, and the genes involved in DNA replication (R, F, C, K, Q and P), the completion of ϕ29 assembly requires genes I, L and D, coding for non-structural proteins.

Gene H codes for the lower collar protein (NP3). Since H-particles contain, besides the normal capsid proteins and DNA, only the upper collar (NP2) and tail (TP1) proteins, these two proteins may interact in the head base to determine the proper assembly of the capsid. An alternative possibility to account for the shape-determining function of the tail protein (TP1) would be that it plays a dual role, catalytic and structural, at two different stages of ϕ29 assembly. In any case, the lower collar protein seems to be assembled after the tail protein.

Gene B codes for protein pB (M.W. 90,000) that is cleaved before assembly to give protein NP1 (M.W. 80,000), the neck appendage protein. Sus-mutants in gene B produce

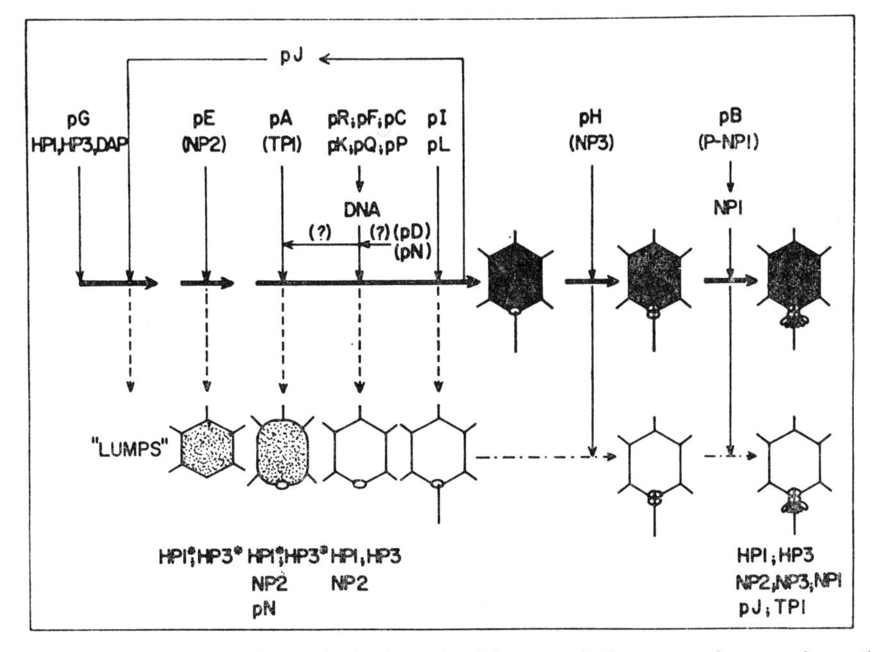

FIG. 6. Assembly of phage ϕ29. The horizontal solid arrows indicate steps in normal morphogenesis. The vertical thin arrows, above the solid arrows, indicate the site of action of the gene products involved in the assembly of ϕ29. The discontinuous arrows, below the solid lines, indicate the structures accumulated as a consequence of mutation in the genes indicated.

particles containing all the normal phage components except proteins NP1 and NP3.

The lack of protein NP3 in B-particles may be due to the instability of the assembled protein in the absence of protein NP1.

Cistron I codes for an early protein, pI, with a molecular weight of about 4000 daltons. Gen L is probably also an early gene, but the protein product has not yet been identified. Under restrictive conditions, sus- and ts-mutants in both cistrons produce DNA-free phage, morphologically similar to DNA-free wild-type particles. I-particles contain, besides the structural protein present in wild-type phage, about seventy copies of protein pJ per particle.

Gene D codes for a late protein, pD, with a molecular weight of about 35,000 daltons. D-particles are DNA-free, morphologically normal capsids, composed by the head proteins, the upper collar protein and the non-structural protein pJ.

Protein pJ could be a scaffolding protein necessary for capsid formation, similar to protein p8 in phage P22.[12] If this is the case, protein pJ would be recycled, as in the case of P22. Figure 6 shows a provisional route of phage ϕ29 morphogenesis.

ACKNOWLEDGMENTS

Different aspects of the work presented here were carried out by: J. Avila, A. Camacho, J. L. Carrascosa, J. Corral, M. Gutierrez, J. M. Hermoso, M. R. Inciarte, F. Jiménez, J. M. Lázaro, R. P. Mellado, E. Méndez, F. Moreno, J. Ortín, G. Ramírez, V. Rubio, J. M. Sogo, A. Talavera, J. de la Torre and C. Vásquez, and supported by grants from the Jane Coffin Childs Memorial Fund for Medical Research and Fondo Nacional para la Investigación Científica.

REFERENCES

1. Epstein, R. H., Bolle, A., Steinberg, C. M., Kellenberger, E., Boy de la Tour, E. Chevalley, R., Edgar, R. S. Susman, M., Denhardt, G. H. and Lielausis, A. (1963) Physiological studies of conditional lethal mutants of bacteriophage T4D. *Cold Spring Harbor Symp. Quant. Biol.* **28**, 375–394.
2. Anderson, D. L., Hickman, D. D. and Reilly, B. E. (1966) Structure of *Bacillus subtilis* bacteriophage ϕ29 and the length of ϕ29 DNA. *J. Bacteriol.* **91**, 2081–2089.
3. Méndez, E., Ramírez, G., Salas, M. and Viñulea, E. (1971) Structural proteins of bacteriophage ϕ29. *Virology*, **45**, 567–576.
4. Ortín, J., Viñuela, E., Salas, M. and Vásquez, C. (1971) DNA-protein complex in circular DNA from phage ϕ29. *Nature New Biol.* **234**, 275–277.
5. Hirokawa, H. (1972) Transfecting deoxyribonucleic acid of *Bacillus* bacteriophage ϕ29 that is protease sensitive. *Proc. Nat. Acad. Sci. U.S.* **69**, 1555–1559.
6. Kornberg, A. (1974) *DNA Synthesis*, W. H. Freeman & Co., San Francisco, California.
7. Branton, D. and Klug, A. (1975) Capsid geometry of bacteriophage T2: a freeze-etching study. *J. Molec. Biol.* **92**, 559–565.
8. Caspar, D. L. D., and Klug, A. (1962) Physical principles in the construction of regular viruses. *Cold Spring Harbor Symp. Quant. Biol.* **27**, 1–24.
9. Brown, N. C. (1971) Inhibition of bacterial DNA replication by 6-(*p*-hydroxyphenylazo)-uracil: differential effect on repair and semi-conservative synthesis in *B. subtilis*. *J. Molec. Biol.* **59**, 1–16.
10. Schachtele, C. F., De Sain, C. F. and Anderson, D. L. (1973) Transcription during the development of bacteriophage ϕ29: definition of "early" and "late" ϕ29 ribonucleic acid. *J. Virol.* **11**, 9–16.
11. Viñuela, E., Camacho, A., Jiménez, F., Carrascosa, J. L., Ramírez, G. and Salas, M. (1976) Structure and assembly of phage ϕ29. *Proc. Roy. Soc. Lond.* B (in press).
12. King, J. and Casjens, S. (1974) Catalytic head assembling protein in virus morphogenesis. *Nature*, **251**, 112–119.

V. PROTEIN BIOSYNTHESIS

PROTEIN SYNTHESIS: EARLY WAVES AND RECENT RIPPLES

PAUL C. ZAMECNIK

Fogarty Scholar, National Institutes of Health; Professor of Medicine, Harvard University; Physician, Massachusetts General Hospital

EARLY WAVES

The pathways for construction of the major classes of macromolecules of the cell and the synthetic enzymes which catalyze their formation were largely unknown until the fourth and fifth decades of this century. In those early years Cori, Lipmann, Krebs and Ochoa in particular awakened the attention of Biochemistry to the role of phosphate bond energy in providing the driving force for the synthesis of complex and polymeric cellular constituents.

At the end of World War II we became aware of the existence of carbon 14, and foresaw the possibility of an amino acid label more sensitive and generally useful than any hitherto available. Our new colleague Robert Loftfield quickly accomplished C^{14}-labeling of amino acids,[1] contemporaneously with several other investigators.[2,3,4,5] It thus became feasible to study the process of protein synthesis in tissue slices, a situation in which the dependence or lack of it on oxygen and on energy requiring reactions could be tested. We were at this moment uncertain whether the synthetic reactions which could be carried out by the proteolytic enzymes of Bergmann,[6] or the phosphorylated intermediates suggested by Lipmann[7] and Kalckar[8] pointed the correct pathway toward protein synthesis. We were in any case impressed that the observations of Caspersson[9] and of Brachet[10] of the relationship between cytoplasmic ribonucleic acid and protein synthesis would have to be taken into account. The tissue slice experiments, using rat liver, were helpful in showing clearly that protein synthesis was dependent on oxygen,[11,12] and was abolished by the energy dissipating compound dinitrophenol.[13] It was further demonstrated that the action of the tissue proteolytic enzymes was not sufficiently specific for protein synthesis.[14] We felt optimistic, therefore, that a new set of enzymes, involving phosphorylated intermediates, perhaps along the lines suggested by Chantrenne,[15] might be involved in protein synthesis. It was clear, however, that an experimental test of this assumption would require a cell-free protein synthesizing system. We set out to try to develop one, a task which also engaged several other laboratories at this time.[2–5] The requirements for success were: (1) that the preparation should be truly cell-free; (2) that the added labeled amino acid should be built into the interior of a long peptide chain in α-peptide linkage; and (3) that a clear-cut dependence on an added energy source should be demonstrated.

Philip Siekevitz, freshly from Greenberg's laboratory, undertook this task. By the time his postdoctoral stay drew to a close, he had developed a cell-free system from rat liver in which amino acid incorporation into a trichloroacetic acid precipitable protein fraction occurred to a small, definite extent, uncomfortably close to the noise level of the system, but clearly dependent on oxidative phosphorylation.[16 17] It had not been demonstrated, however, that the amino acid which was added actually made its way into the interior of a growing peptide chain; and the relationship to ATP as a possible energy donor was unclear. At that time there was also a disturbing observation by Borsook and his colleagues[5,18] that lysine could become bound in apparent covalent linkage to protein in a cell-free system without an added energy source.

In further pursuit of this problem, we were benefited by conditions employed by our colleague Nancy Bucher in the cell-free, cholesterol synthesizing system she worked out at that time.[19] A key feature in her success in preparing active cell-free synthesis of cholesterol, beginning with a C^{14}-acetate precursor, was the use of a "loose" homogenizer—in other words, in gently breaking up liver cells. In retrospect it appears that soluble enzymatic and ribosomal components of the cytoplasm were being liberated by the gentle, incomplete homogenization of minced liver fragments, while lysozomal, degradative enzymes largely remained within the membranous sacs which kept them out of contact with the cell cytosol.

Using this homogenization technique, initially we found no evidence of protein synthesis in the cell-free preparation. Upon addition of ATP and an ATP-generating system, however, there was an agreeable brisk, consistent incorporation of labeled amino acid into protein, considerably greater than that of the earlier system.[16,17] This incorporation was then shown to involve α-peptide bond linkage of the added labeled amino acid into internal peptide bonds of the large peptide chains being formed.[20] At this point the ambivalent term "incorporation"[21] was finally replaced by the firm phrase "protein synthesis" in describing the well-characterized events being followed in the cell-free system.

The first fractionation of this crude, cell-free protein synthesizing system yielded three separable essential parts, in addition to the added labeled amino acid: (1) a soluble, enzyme-containing fraction not sedimentable by high-speed centrifugation; (2) a ribosomal (then called microsomal) pellet; and (3) ATP as an energy source.[22]

At this time, Mahlon Hoagland returned to our laboratories after a period with Lipmann and Linderstrom-Lang, and undertook to find out whether the ATP-requiring step involved enzyme(s) in the soluble protein fraction, or in the ribosomal fraction. His discovery of the first step in protein synthesis followed.[23 24]

RECENT RIPPLES

The aminoacyl-tRNA ligases, of course, provide the mechanism by which an amino acid is activated in the presence of ATP and an activating enzyme specific for that amino acid. The reaction is a reversible one, and can be diagrammed as follows [reaction (I)], bearing in mind that our former colleague Robert Loftfield may be correct in regarding the path from free amino acid to aminoacyl-tRNA in some instances to represent a concerted reaction mechanism, without a definitive intermediate.

$$\text{pppA} + \text{aa}_1 + \text{Eaa}_1 \leftrightharpoons \text{aa}_1\text{-pA}\cdot\text{Eaa}_1 + \text{pp} \qquad \text{(I)}$$

Some years later we discovered[25,26] that in the back reaction of (I) above, a new compound was formed, in an *in vitro* enzymatic system, as shown in reactions (II) and (III).

$$\text{pppA} + \text{lys} + \text{E}_{\text{lys}} \leftrightharpoons \text{lys-pA}\cdot\text{E}_{\text{lys}} + \text{pp} \qquad \text{(II)}$$

$$\text{pppA} + \text{lys-pA}\cdot\text{E}_{\text{lys}} \leftrightharpoons \text{A}_{5'\text{-pppp-}5'}\text{A} + \text{lys} + \text{E}_{\text{lys}} \qquad \text{(III)}$$

This synthesis of Ap_4A (our abbreviation for the compound in reaction (III)) during the course of the usual activation reaction was not restricted to the lysyl synthetase activating enzyme system; but we have found the latter to be a convenient one with which to work, since the enzyme is easily purified, relatively stable over a long period of storage in the frozen state, and inasmuch as its activation does not require the presence of tRNA, although its rate is accelerated by the latter.[27]

In other words, ATP competes with inorganic pyrophosphate in the back reaction. We have indeed found that numerous pyrophosphate compounds will successfully compete with inorganic pyrophosphate in this back reaction, in the *in vitro* system, and an interesting series of mixed-5′5′-nucleoside tri- or tetraphosphate compounds has been found[28] to be formed in this way. These may be designated as A-5′-$(p)_n$-5′-N, where n represents at least three linear phosphoryl moieties, A is adenosine (or deoxyadenosine), and N represents generally a ribo- or deoxyribonucleoside, but certain other moieties as well.

Although these compounds are formed and easily detected *in vitro*, and their formation has been confirmed recently in another laboratory (Hecht, S. M., personal communication), the question arises as to whether they are test-tube curiosities or may be contained also in living cells. It should be mentioned that Finamore and Warner[29] in 1963 discovered the presence of large quantities of the unusual nucleotide P^1,P^4-diguanosine 5′-tetraphosphate in encysted embryos of the brine shrimp, *Artemia salina.* In some as yet obscure way the energy-rich phosphate of this compound is converted to use in the form of ATP, which then reinitiates its life processes when this organism is introduced into water. Bearing in mind our *in vitro* findings and the above example, we searched for the presence of Ap_4A in living *E. coli* and in rat liver. With some difficulty we established that this compound was present in these living forms in a concentration of the order of 1×10^{-8} M.[28]

We found it hard to pursue this problem further at that moment due to lack of clear leads. We watched with great interest, however, the developments which followed the discovery of the "magic spot" compounds I and II by Cashel and colleagues.[30,31] These compounds are pp-5′-G-3′-pp and ppp-5′-G-3′-pp. They appear in prokaryotic cells under stringent growth conditions (and possibly in eukaryotic cells as well) and are related in some way to the cessation of RNA synthesis which follows deprivation of an essential amino acid in stringent forms of *E. coli.* They also participate in place of GTP in certain steps in peptide-chain elongation. Sy and Lipmann[32] and Sy[33] have recently isolated the usually ribosome-bound enzyme which catalyzes the formation of these two compounds, and have demonstrated that the catalysis occurs according to the following pathway:

$$\text{ppp-5'-G} + \text{ppp-5'-A} \leftrightharpoons \text{ppp-5'-G-3'-pp} + \text{pA} \qquad \text{(IV)}$$

Although the reaction is to some extent reversible, it is difficultly so, being inhibited by ATP at 1×10^{-5} M concentration. Upon removal of an essential amino acid in growth of a stringent strain of bacteria, there is prompt formation of pppGpp. Upon addition of the amino acid there is rapid disappearance (in less than a minute) of this compound, and the pathway for its disappearance is still obscure.

Many years ago we discovered the participation of pppG (GTP) in protein synthesis.[34] Over the intervening years we have followed the ramifications of the story, as yet incomplete, of the several roles of GTP in the peptide chain initiation and elongation steps. We recently wondered whether pppGpp and ppGpp would compete with pp in the back reaction of the amino acid activation step, as diagrammed above, and therefore added ppGpp and pppGpp to the lysine activation system. Two unknown new spots appeared in the elution patterns of our PEI-cellulose plates, one when ppGpp was added, and another when pppGpp was added. Because of the position of migration of these ultraviolet absorbing spots, on PEI-cellulose plates in 1 M lithium chloride, 0.75 M K phosphate, or 1.5 M K phosphate, pH 3.4, and in view of our previous experience with pyrophosphate-containing compounds in this reaction, we suspected that these new compounds might be A-5′-ppp-5′-G-3′-pp and A-5-′pppp-5′-G-3′-pp, and the reaction mechanisms involved in their formation to be as shown below. Based on this assumption, the following reactions were carried out to prove or disprove these presumed structures:

(a) Use of ^{3}H-ring-labeled pppA* as a precursor: the new spots contained the tritium label, indicating that the adenine moiety was part of these compounds.

(b) Use of ^{3}H-ring labeled guanine in the ppG*pp employed as precursors in the enzymatic reaction (these compounds kindly furnished by Dr. Jose Sy): the new spots contained the tritium label, indicating that the guanine moiety was part of these compounds.

(c) Use of p^{32}pGpp and pp^{32}pGpp as precursors (kindly furnished by Dr. Sy as formed in crude mixes, and purified for our use in our laboratories): the new spots contained ^{32}P radioactivity.

(d) Use of phosphomonoesterase on the new compounds: treatment of labeled presumed ApppGpp or AppppGpp with phosphomonoesterase yielded labeled ApppG and AppppG respectively. These compounds were identified by their chromatographic properties, which were identical with unlabeled ApppG and AppppG. The two unlabeled markers were in turn prepared by reacting GDP and GTP in the lysine activating system. Furthermore, in this type of experiment the 5′α-^{32}P-label was preserved in the derived compound(s), thus demonstrating that the 5′-pp rather than the 3′pp moiety of the 5′ppGpp-3′ was involved in the linkage with the adenylyl portion.

(e) Addition of snake venom diesterase to the new compound(s): with ^{3}H-ATP as a precursor in the formation of the compound, addition of snake venom diesterase resulted in the liberation of ^{3}H-AMP (pA*) and of ppGpp.

The above reactions and manipulations permit the following formulations to be given:

$$\text{pppA} + \text{lys} + \text{E}_{\text{lys}} \leftrightharpoons \text{lys-pA}\cdot\text{E}_{\text{lys}} + \text{pp} \quad \text{(II)}$$

$$\text{lys-pA}\cdot\text{E}_{\text{lys}} + \text{ppGpp} \leftrightharpoons \text{ApppGpp} + \text{lys} + \text{E}_{\text{lys}} \quad \text{(V)}$$

$$\text{lys-pA}\cdot\text{E}_{\text{lys}} + \text{pppGpp} \leftrightharpoons \text{AppppGpp} + \text{lys} + \text{E}_{\text{lys}} \quad \text{(VI)}$$

The new compounds are thus identified as A-5′-ppp-5′-G-3′-pp and A-5′-pppp-5′-G-3′-pp.[35] The question of reversibility of reactions (V) and (VI) remains to be studied.

Whether they are also formed in living cells remains a problem for the future. An attractive test object for this pursuit is a temperature-sensitive *E. coli* stringent strain, in which the formation of ppGpp and pppGpp can be observed at 42°C, and its disappearance can be followed by lowering the temperature to the permissive range (below 37°C).

In the case of a number of normal cell messenger RNAs and some viral high molecular weight RNAs as well, the 5′-end terminates as ppRNA or pppRNA. We have therefore wondered whether these pyrophosphoryl moieties might also compete with inorganic pyrophosphate, and participate in the back reaction of the amino acid activating step in protein synthesis, as pictured in reaction (I) above. We had recently found that the high molecular weight 35S RNA of the Avian Myeloblastosis Virus terminates in part as 5′-OH-RNA, 5′-ppRNA, or as 5′-pppRNA.[36] We therefore added 35S AMV RNA to the lysine amino acid activating enzyme system depicted in reaction (II) above, in the presence of ^{3}H-pppA. Results indicate the formation of A*pppA-RNA. This result is especially interesting in view of the recent finding of Shatkin and colleagues[37] of the presence of me7-G-5′-ppp-5′-G-2′Ome-RNA at the 5′-terminus of reovirus and other RNAs, and of the essentiality of this "capping" reaction for their function as messenger RNAs.

ACKNOWLEDGEMENTS

The work reported herein was supported by Contracts NOI-CP-33-36 of the National Cancer Institute, E-(1101)-2403 of the Energy Resources Development Agency, and Grant NP-2Q of the American Cancer Society.

On the occasion of this seventieth birthday celebration for Severo Ochoa, we take pleasure in thanking him for important contributions to carbohydrate, nucleic acid, and protein biochemistry; and for the continuous catalytic effect he has had on us, his colleagues, who have benefited by his stimulating presence in these widely varied fields of science.

REFERENCES

1. Loftfield, R. B. (1947) Preparation of C^{14}-labeled hydrogen cyanide, alanine and glycine. *Nucleonics*, **1**, 54–57.
2. Melchior, J. B. and Tarver, H. (1947) Studies in protein synthesis *in vitro*. I. On the synthesis of labeled cystine (S^{35}) and its attempted use as a tool in the study of protein synthesis. *Arch. Biochem.* **12**, 301–308.
3. Anfinsen, C. B., Beloff, A., Hastings, A. B. and Solomon, A. K. (1947) The *in vitro* turnover of dicarboxylic amino acids in liver slice proteins. *J. Biol. Chem.* **168**, 771–772.
4. Winnick, T., Friedberg, F. and Greenberg, D. M. (1947) Incorporation of C^{14}-labeled glycine into intestinal tissue and its inhibition by azide. *Arch. Biochem.* **15**, 160–161.
5. Borsook, H., Deasy, C. L., Haagen-Smit, A. J., Keighly, G. and Lowy, P. H. (1949) The incorporation of labeled lysine into the proteins of guinea pig liver homogenate. *J. Biol. Chem.* **179**, 689–704.
6. Bergmann, M. (1942) A classification of proteolytic enzymes. In *Advances in Enzymology* (Nord, F. F. and Werkman, C. H., eds.) **2**, 49–68, Interscience, N.Y.
7. Lipmann, F. (1941) Metabolic generation and utilization of phosphate bond energy. In *Advances in Enzymology* (Nord, F. F. and Werkman, C. H., eds.), **1**, 99–162, Interscience, N.Y.
8. Kalckar, H. M. (1941) The nature of energetic coupling in biological synthesis. *Chem. Rev.* **28**, 71–178.
9. Caspersson, T. (1941) Studien uber den Eiweissumsatz der Zelle. *Naturwissenschaften.* **29**, 33–43.
10. Brachet, J. (1941) La detection histochimique et le microdosage des acides pentosenucleiques. *Enzymologia*, **10**, 87–96.
11. Frantz, I. D., Jr., Loftfield, R. B. and Miller, W. W. (1947) Incorporation of C^{14} from carboxyl labeled dl-alanine into the proteins of liver slices. *Science*, **106**, 544–545.
12. Zamecnik, P. C., Frantz, I. D., Jr., Loftfield, R. B. and Stephenson, M. L. (1948) Incorporation *in vitro* of radioactive carbon from carboxyl-labeled dl-alanine and glycine into proteins of normal and malignant rat livers. *J. Biol. Chem.* **175**, 299–314.
13. Frantz, I. D., Jr., Zamecnik, P. C., Reese, J. W. and Stephenson, M. L. (1948) The effect of dinitrophenol on the incorporation of alanine labeled with radioactive carbon into the proteins of slices of normal and malignant rat liver. *J. Biol. Chem.* **174**, 773–774.
14. Loftfield, R. B., Grover, J. W. and Stephenson, M. L. (1953) Possible role of proteolytic enzymes in protein synthesis. *Nature*, **171**, 1024–1025.
15. Chantrenne, H. (1948) Un modele de synthese peptidique. Proprietes du benzoylphosphate de phenyle. *Biochim. Biophys. Acta*, **2**, 286–293.
16. Siekevitz, P. and Zamecnik, P. C. (1951) In vitro incorporation of 1-^{14}C-dl-alanine into proteins of rat liver granular fractions. *Fed. Proc.* **10**, 246–247.

17. SIEKEVITZ, P. (1952) Uptake of radioactive alanine *in vitro* into the proteins of rat liver fractions. *J. Biol. Chem.* **195**, 549–565.
18. SCHWEET, R. and BORSOOK, H. (1953) Incorporation of radioactive lysine into protein. *Fed. Proc.* **12**, 266.
19. BUCHER, N. L. R. (1953) The formation of radioactive cholesterol and fatty acids form ^{14}C-labeled acetate by rat liver homogenates. *J. Am. Chem. Soc.* **75**, 498.
20. ZAMECNIK, P. C., KELLER, E. B., LITTLEFIELD, J. W., HOAGLAND, M. B. and LOFTFIELD, R. B. (1956) Mechanism of incorporation of labeled amino acids into protein. *J. Cell. Comp. Physiol.* **47**, Suppl. I, 81–101.
21. ZAMECNIK, P. C. and FRANTZ, I. D., Jr. (1950) Peptide bond synthesis in normal and malignant tissue. *Cold Spring Harbor Symp. Quant. Biol.* **14**, 199–208.
22. ZAMECNIK, P. C. and KELLER, E. B. (1954) Relation between phosphate energy donors and incorporation of labeled amino acids into proteins. *J. Biol. Chem.* **209**, 337–354.
23. HOAGLAND, M. B. (1955) An enzymatic mechanism for amino acid activation in animal tissues. *Biochim. Biophys. Acta*, **16**, 288–289.
24. HOAGLAND, M. B., KELLER, E. B. and ZAMECNIK, P. C. (1956) Enzymatic carboxyl activation of amino acids. *J. Biol. Chem.* **218**, 345–358.
25. ZAMECNIK, P. C., STEPHENSON, M. L., JANEWAY, C. M. and RANDERATH, K. (1966) Enzymatic synthesis of diadenosine tetraphosphate and diadenosine triphosphate with a purified lysyl-SRNA synthetase. *Biochem. Biophys. Res. Commun.* **24**, 91–97.
26. RANDERATH, K., JANEWAY, C. M., STEPHENSON, M. L., and ZAMECNIK, P. C. (1966) Isolation and characterization of dinucleoside tetra- and triphosphates formed in the presence of lysyl-SRNA synthetase. *Biochem. Biophys. Res. Commun.* **24**, 98–105.
27. MARSHALL, R. D. and ZAMECNIK, P. C. (1969) Some physical properties of lysyl and arginyl-transfer RNA synthetases of *Escherichia coli* B. *Biochem. Biophys. Acta*, **181**, 454–464.
28. ZAMECNIK, P. C. and STEPHENSON, M. L. (1969) The role of nucleotides for the function and conformation of enzymes. *In Alfred Benzon Symposium* I (H. M. KALCKAR, H. KLENOW, G. MUNCH-PETERSON, M. OTTESEN and J. H. Thuysen, eds.), pp. 276–291; Munksgaard, Copenhagen.
29. FINAMORE, F. J. and WARNER, A. H. (1963) The occurrence of P',P^4-diguanosine 5′-tetraphosphate in brine shrimp eggs. *J. Biol. Chem.* **238**, 344–348.
30. CASHEL, M. and GALLANT, J. (1969) Two compounds implicated in the function of the RC gene. *Nature*, **221**, 838–841.
31. CASHEL, M. and KALBACHER, B. (1970) The control of ribonucleic acid synthesis in *Escherichia coli* V. Characterization of a nucleotide associated with the stringent response. *J. Biol. Chem.* **245**, 2309–2318.
32. SY, J. and LIPMANN, F. (1973) Identification of the synthesis of guanosine tetraphosphate (MSI) as insertion of a pyrophosphoryl-group into the 3′-position in guanosine-5′-diphosphate. *Proc. Nat. Acad. Sci. U.S.A.* **70**, 306–309.
33. SY, J. (1974) Reversibility of the pyrophosphoryl transfer from ATP to GTP by *Escherichia coli* stringent factor. *Proc. Nat. Acad. Sci. U.S.A.* **71**, 3470–3473.
34. KELLER, E. B. and ZAMECNIK, P. C. (1956) The effect of guanosine diphosphate and triphosphate on the incorporation of labeled amino acids into proteins. *J. Biol. Chem.* **221**, 45–59.
35. RAPAPORT, E., SVIHOVEC, S. and ZAMECNIK, P. C. (1975) Relationship of the first step in protein synthesis to ppGpp: formation of A(5′)ppp(5′)Gpp. *Proc. Nat. Acad. Sci. U.S.A. 72*, 2653 2657.
36. RAPAPORT, E. and ZAMECNIK, P. C. (1975) A new chemical procedure for ^{32}P-labeling of ribonucleic acids at their 5′-ends after isolation. *Proc. Nat. Acad. Sci. U.S.A.* **72**, 314–317.
37. FURUICHI, Y., MORGAN, M., MUTHUKRISHNAN, S. and SHATKIN, A. J. (1975) Reovirus messenger RNA contains a methylated, blocked 5′-terminal structure: $m^7G(5')ppp(5')G^mpCp$. *Proc. Nat. Acad. Sci. U.S.A.* **72**, 362–366.

TEN YEARS IN PROTEIN SYNTHESIS

PETER LENGYEL
Department of Molecular Biophysics and Biochemistry, Yale University
New Haven, Connecticut 06520

STUDIES ON THE GENETIC CODE

In 1961, as a graduate student in Dr. Ochoa's laboratory, I became engaged in studies aiming at deciphering the genetic code. The formulation of a working hypothesis for this endeavor became possible at that time in consequence of the discovery of mRNA as a novel intermediate in information transfer between DNA and protein. It was Jacob and Monod who recognized that the then prevalent assumption of ribosomal RNA as a direct template for ribosomal protein synthesis was wrong. They demonstrated that the template was an RNA species, designated by them as messenger RNA, which became attached to ribosomes and programmed them for protein synthesis.

The working hypothesis for code deciphering was a logical extension of this discovery. Its essence was to provide ribosomes in a cell-free protein-synthesizing system with simple synthetic messengers. The simplest possible messenger is a homopolyribonucleotide. If active, it could direct the formation of nothing but a homopolypeptide. Copolyribonucleotides (of random sequence but defined nucleotide composition), if active, could however direct the synthesis of polypeptides containing more than one amino acid. By using various homopolyribonucleotides and copolyribonucleotides containing two or three types of nucleotide residues in varying ratios, it might be possible to calculate the nucleotide composition (though not the sequence) of codons specifying the various amino acids. (Though it had not yet been proven at that time, we assumed that three adjacent nucleotides in mRNA would specify one amino acid in a polypeptide.)

Polynucleotide phosphorylase, an enzyme discovered by Grunberg-Manago and Ochoa, could be used for synthesizing the required polyribonucleotides. A cell-free protein synthesizing system had been developed following the pioneering investigations of Zamecnik. Joe Speyer, also in Dr. Ochoa's laboratory, had experience with such systems.

These were the foundations on which our work was to be built. Joe Speyer and I were at the preparatory stage of the project (we had received all labeled amino acids ordered for the experiments and were determining the size of our polynucleotides) when the rumor of Nirenberg and Matthaei's momentous discovery reached us in August 1961. While studying the effect of RNAs from different sources on protein synthesis in a cell-free system from *E. coli*, they discovered that the homopolyribonucleotide poly(U) promoted the formation of the homopolypeptide polyphenylalanine. Thereby, the genetic code was broken.

Our obvious disappointment at being scooped was mitigated only slightly by the satisfaction of knowing that at last the first part of our working hypothesis was correct.

Since all our tools had been sharpened in advance, we could confirm the messenger activity of poly(U) in a single day. We proceeded thereafter to test the validity of the second part of our working hypothesis, i.e. to determine if random copolynucleotides did promote the incorporation of several amino acids into polypeptides. Within a month, we were certain that this was the case. The studies were begun with random copolynucleotides containing 5 parts of uridylate residues and 1 part of either adenylate or cytidylate or guanylate residues. Using these, we found right away that the kinds of amino acids incorporated depended on the kinds of nucleotides present in the copolynucleotides. Thus serine was incorporated in response to poly(U_5,C) but not in response to poly(U_5,A) or, poly(U_5,G). We devised a simple calculation for determining the nucleotide composition of codons. We argued that in the random copolynucleotide poly(U_5,C) the ratio of the frequency of UUU triplets (specifying phenylalanine) to UUC or UCU or CUU triplets should be the same as the U to C ratio in the copolynucleotide, i.e. 5:1. Moreover, it followed that the ratio of UUU triplets to UCC (or CUC or CCU) triplets should be 25:1. Since the ratio of phenylalanine to serine incorporation with poly(U_5,C) was found to be 4.4:1 (i.e. close to 5 and far from 25), we proposed that the codon for serine should be either UUC or UCU or CUU.

The use of a large variety of random copolynucleotides along these lines resulted in the determination of the nucleotide composition of close to fifty amino acid-specifying codons in the laboratories of Drs. Ochoa and Nirenberg between the fall of 1961 and the summer of 1963.

In 1962 Speyer, Basilio and I proved the validity of a hypothesis by Spotts and Stanier according to which the difference between streptomycin-sensitive, resistant and dependent *E. coli.* resides in the structure of their ribosomes. We established that streptomycin inhibits poly(U) directed polyphenylalanine synthesis in a fractionated cell-free protein synthesizing system only if the ribosomes in the system are taken from streptomycin-sensitive cells. These results indicated that in sensitive cells streptomycin interferes with ribosome action. As subsequent studies revealed, streptomycin is only one among many antibiotics whose site of action is the ribosome.

My involvement in the studies on the code ended in the summer of 1963. I spent a year at the Pasteur Institute and in 1965 moved to Yale University.

STUDIES ON PEPTIDE CHAIN ELONGATION

At the end of my stay in Paris I became intrigued by the mechanism of the mRNA translating machinery. This was a new interest. The crude mRNA translating system had performed well in the studies on the genetic code, and the mode of its functioning did not concern me at the time. The new endeavor I contemplated was to be based on the pioneering observations of Zamecnik and Lipmann.

In my grant request of 1965 I indicated that my long-range interest was to contribute to the elucidation of the mechanism of action of the peptide-chain elongation factors. Three elongation factors had just been identified by Lucas-Lenard and Lipmann. The main obstacle towards further progress at the time was that one of the three factors from *E. coli.* seemed to be unstable after purification.

I hoped that *Bacillus stearothermophilus*, a thermophilic microorganism growing at temperatures up to 70°, would be a source of factors stable for purification at least at lower temperatures. I became fascinated with this organism in 1961. At the time the only copolynucleotide with a defined sequence available was poly(A–U) (with an alternating base sequence). This could be obtained by transcription from poly(dA–dT), a copolymer with alternating nucleotide sequence, produced by Kornberg's DNA polymerase. I wanted to use the poly(A–U) as mRNA to determine if the coding ratio was odd or even. (We expected the translation of this messenger into one or two homopolypeptides if the coding ratio was even; and into one copolypeptide with an alternating sequence of two types of aminoacyl residues if it was odd.) Since poly(A–U) was double stranded, and thus inactive as a messenger in the temperature range in which the *E. coli* system was active, I hoped to translate it in a cell-free system from the thermophilic microorganism.

In 1965 Algranati and I developed the cell-free system from *B. stearothermophilus*. With gloves to prevent burns on our fingers, *B. stearothermophilus* turned out to be pleasant to work with: it had a generation time of less than 15 minutes, and at the high temperature of growth (70°), there was essentially no need for sterility.

Regrettably, the original aim could not be reached with this system. Though the extract was active up to 70°, poly(A–U) remained double stranded and untranslatable up to this temperature. The efforts had not been wasted, however. *B. stearothermophilus* turned out to be a good source of the three stable elongation factors.

These were isolated in our laboratory by Skoultchi, Ono, Waterson and Beaud. I will designate the factors here in the now accepted way as EF–Ts, EF–G and EF–Tu; at the time they were called S_1, S_2 and S_3, respectively. All three were needed for the poly(U)-promoted formation of polyphenylalanine from the Phe–tRNA. The studies on peptide-chain elongation *in vitro* were greatly facilitated by this model system in which chain elongation takes place without proper chain initiation and termination, and in the absence of initiation and termination factors and initiator tRNA.

To make the studies on elongation with poly(U) more analogous to those *in vivo* (following the example of Lucas-Lenard and Lipmann) we formed *in vitro* a ribosome·poly(U)·acPhe–tRNA complex. The acPhe–tRNA served as an analog of the chain initiator fMet–$tRNA_f$. (Its alpha amino group is blocked, and as fMet–$tRNA_f$, it is bound in the P site of the ribosome.) Using this system, we established that the binding of Phe–tRNA to the above ribosome complex (which, as Ravel has discovered, depended on GTP) was promoted by EF–Tu and EF–Ts, but not by EF–G.

There were indications from several laboratories for the existence of an aminoacyl–tRNA · GTP · EF–Tu complex (ternary complex) serving as an intermediate, carrying the Phe–tRNA to the ribosome. We succeeded in isolating this complex, and determined that in it the molar ratio of GTP to Phe–tRNA was 1:1, and that the same EF–Tu could form a ternary complex with each aminoacyl–tRNA involved in peptide chain elongation.

Further study of the specificity of aminoacyl–tRNA binding of EF–Tu resulted in the solution of the following problem: the chain initiator in *E. coli* is fMet–$tRNA_f$ which is synthesized by formylation of Met–$tRNA_f$. There is a different Met–tRNA species, Met–$tRNA_m$, which provides methionyl residues for internal positions of the peptide chain. Furthermore; as initiator codons in *E. coli*, GUG, as well as AUG, specify fMet–$tRNA_f$, whereas as codons for internal aminoacyl residues GUG stands for Val–tRNA and AUG for Met–$tRNA_m$.

The dual specificity of GUG raises the intriguing problem of how mix-ups between Val–tRNA, fMet–$tRNA_f$ and Met–$tRNA_f$ are avoided during chain elongation.

It is easy to see that fMet–$tRNA_f$ cannot participate in elongation since its alpha amino group is blocked. A more difficult problem arises with Met–$tRNA_f$ which in principle could provide methionyl residues to be incorporated mistakenly in the place of valyl residues in response to GUG codons in chain elongation. Studies with *in vitro* systems in which polynucleotides containing internal GUG codons (e.g. random poly U,G) were used as mRNA proved, however, that this is not the case.

In attempting to solve the above problem, we first established that, as expected, Met–$tRNA_m$ did form a complex with EF–Tu and GTP, whereas fMet–$tRNA_f$ did not. Since a block of the alpha-amino group had been shown to impair the binding of an aminoacyl–tRNA to EF–Tu (Ravel found that acPhe–tRNA was not bound), the lack of activity of fMet–$tRNA_f$ could be attributed to the presence of the formyl residue. We proceeded to check if Met–$tRNA_f$ (unformylated) was bound to EF–Tu or not. The result was clearly negative. This finding provided the solution of the problem. The EF–Tu · GTP · aminoacyl–tRNA complexes are intermediates in binding the aminoacyl–tRNAs to the ribosomes during chain elongation. Consequently, the lack of formation of such a complex with Met–$tRNA_f$ could explain why methionyl residues are not inserted into the polypeptide chain in response to internal GUG codons.

On completing these studies, we examined the fate of the three components of the ternary complex after the transfer of its Phe–tRNA moiety to the poly(U) · ribosome · acPhe–tRNA complex.

We observed (and so did Haenni and Lucas-Lenard, as well as Ravel) that the GTP moiety of the complex was cleaved to GDP and Pi, and established subsequently that one molecule of GTP was cleaved for each Phe–tRNA bound to the ribosome.

That GTP is cleaved during peptide chain elongation in *E. coli* extracts was already known at the time, but it was thought that only one GTP molecule was cleaved for one aminoacyl–tRNA added to the polypeptide, and that the factor involved in the cleavage was EF–G. Consequently, we had to prove that the newly recognized GTP cleavage, in which EF–Tu was involved, and which occurred upon aminoacyl–tRNA binding to the ribosome, was different from the GTP cleavage in which EF–G was involved, and which took place during a different phase of peptide-chain elongation. We accomplished this by showing that the GTP cleavage in which EF–Tu was involved (but not that in which EF–G was involved) affected only GTP moieties in a ternary complex, and only upon the attachment of the ternary complex to a mRNA · ribosome complex. These results indicated to us that at least two GTP molecules were cleaved in the course of the extension of the peptide chain by one aminoacyl residue.

Further studies on the EF–Tu-dependent GTP cleavage process revealed that one of the GTP cleavage products, Pi, was released from the ribosome in a free state, whereas the other, GDP, was released in a complex with EF–Tu. The Phe–tRNA, whose binding to the ribosome preceded the GTP cleavage, formed a dipeptidyl–tRNA (actually, acdiPhe–tRNA) with the acPhe–tRNA attached to the ribosome.

We proceeded by testing if the GTP cleavage is a prerequisite for peptide bond formation. For this purpose, we first performed experiments in which we substituted for GTP an analog, GMPPCP, that has a methylene bridge instead of oxygen between the beta and gamma P atoms, and consequently cannot be cleaved enzymatically into GDP and Pi. When EF–Tu, EF–Ts, Phe–tRNA and GMPPCP were incubated with the ribosome · poly-(U) · acPhe–tRNA complex, we found (in accord with the results of Haenni and Lucas-Lenard) that Phe–tRNA became bound to the ribosomes but no dipeptide was formed. This result was consistent with the need for GTP cleavage prior to peptide bond formation.

It suggested, furthermore, the existence of an intermediate formed after the binding of Phe–tRNA to the ribosomes and persisting until GTP cleavage and dipeptide bond formation take place.

In subsequent experiments, we obtained further data about this intermediate. We established that ribosomes isolated from reaction mixtures, in which a ribosome · poly(U) · acPhe–tRNA complex was incubated with EF–Tu, EF–Ts, GMPPCP and Phe–tRNA, had equimolar amounts of EF–Tu, Phe–tRNA, and GMPPCP (but no acdiPhe–tRNA) bound to ribosomes. However, ribosomes isolated from a reaction mixture identical with the one above, except containing GTP (instead of GMPPCP) had acdiPhe–tRNA bound (but neither EF–Tu, GTP, GDP or Pi). Thus, while we have not isolated a ribosome–poly(U) · acPhe–tRNA · EF–Tu · GTP · Phe–tRNA complex, we nevertheless concluded that the existence of such an intermediate could be inferred from finding EF–Tu and Phe–tRNA attached to ribosomes in the presence of GMPPCP. Consequently, it is probable that EF–Tu and GTP become bound to the ribosome together with Phe–tRNA, but the resulting intermediate has a short life span because of the rapid cleavage of the GTP molecule into Pi and EF–Tu · GDP complex, which are released from the ribosome. The corresponding intermediate formed in the presence of GMPPCP is stable because GMPPCP is not cleaved.

On the basis of these results, we proposed that the binding of Phe–tRNA and EF–Tu depends on intact GTP, whereas GTP cleavage is required for, and has to precede, the release of EF–Tu from the ribosome and peptide bond formation. Thus, GTP may serve as an allosteric effector that is inactivated, or at least whose action is modified by cleavage into GDP and Pi.

Subsequently, other investigators established that the release of some initiation and termination factors from the ribosome also depends on GTP cleavage. Further insight into the role of GTP and its cleavage in protein synthesis has been provided by the studies of Y. Kaziro.

As noted earlier, one of the products of the cleavage of GTP from the ternary complex on the ribosome is an EF–Tu · GDP complex. We wondered how the EF–Tu moiety of this complex is reutilized in chain elongation. The possibility of the involvement of EF–Ts was indicated by the observation that GDP in our EF–Tu · GDP complex was exchangeable with GTP or GDP even at 0°, whereas that in the EF–Tu · GDP preparation of Ravel was not. We suspected that Ravel's complex was essentially free of EF–Ts, whereas ours was not. Since it was difficult to free EF–Tu completely from EF–Ts, and this was necessary for our purposes, we prepared an antiserum against EF–Ts. Using this to free EF–Tu · GDP from EF–Ts activity, we first established that the rate of GDP exchange in EF–Tu · GDP and also the release of GDP from EF–Tu were increased by EF–Ts. Thereafter we determined that in a reaction mixture containing EF–Tu · GDP, as well as Phe–tRNA and GTP, the formation of the EF–Tu · GTP · Phe–tRNA complex was greatly accelerated by EF–Ts. This result, and other studies showing that EF–Ts did not increase the rate of polyphenylalanine synthesis from pre-formed ternary complex, indicated to us that the role of EF–Ts in chain elongation is to promote the conversion of EF–Tu · GDP into the EF–Tu · GTP · aminoacyl–tRNA complex.

The intermediates in this conversion (i.e. EF–Tu · EF–Ts resulting from the reaction of EF–Tu · GDP with EF–Ts, and EF–Tu · GTP resulting from the reaction of EF–Tu · EF–Ts with GTP) were isolated and characterized by Weissbach.

STUDIES ON PEPTIDE CHAIN INITIATION

Another phase of protein synthesis in which I became interested at Yale is peptide-chain initiation. In 1966, J. Eisenstadt and I proved that the chain initiator fMet–tRNA$_f$ is strictly required for the translation of f2 bacteriophage RNA, a natural messenger RNA in a crude extract from *E. coli*. This requirement was, however, only manifested when the magnesium ion concentration in the extract was low. For this demonstration, we had to deplete the pool, and block the formation, of fMet–tRNA$_f$ in the extract. We found two compounds which accomplished this by making formyltetrahydrofolate, the source of the formyl residue in fMet–tRNA$_f$, unavailable: trimethoprim, an inhibitor of dihydrofolate reductase, and hydroxylamine which was shown by Bertino to react with methylenetetrahydrofolate and thereby deplete the pool of formyltetrahydrofolate, produced the result desired.

Next, with M. Kondo we tackled the problem of chain initiation signals in natural mRNAs. As noted earlier, the chain initiator codons AUG and GUG also specify internal aminoacyl residues in the peptide chain. The characteristics which determine if an AUG or GUG codon in a mRNA serves as a signal for initiating a new peptide chain or for the addition of an aminoacyl residue to a chain already started were not known. To facilitate

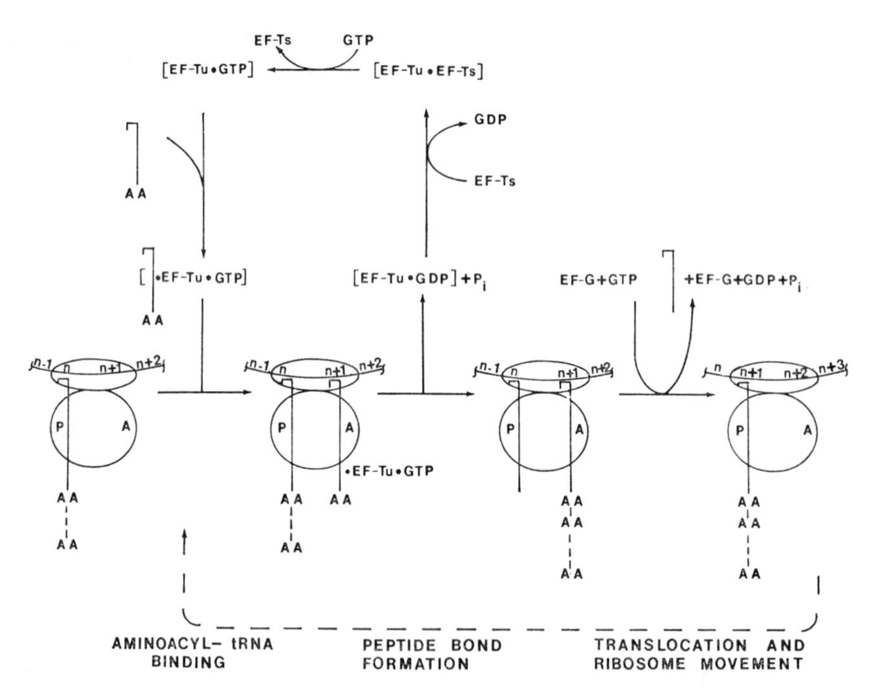

FIG. 1. Scheme of peptide-chain elongation. The small oval shape represents a 30S ribosomal subunit. The large oval shape represents the 50S ribosomal subunit. A and P indicated in the 50S subunit stand for hypothetical tRNA binding sites. The symbols n−1, n, n + 1, n + 2, and n + 3 represent a series of adjacent codons in a segment of mRNA. ┐ represents a tRNA molecule. AA stands for aminoacyl residue. AA | AA stands for peptidyl residue. Further details are discussed in the text.

the study of this problem, we developed a procedure for isolating the peptide-chain initiation signal containing segments of mRNA.

The basis of our procedure was the discovery of Takanami, Yan and Jukes that the region of a natural mRNA which is bound to ribosomes is protected against cleavage by nucleases. To make use of this observation, we needed a way to bind ribosomes selectively to initiation signals. The scheme of Nomura for the steps in chain initiation provided the idea for an approach. According to this scheme, first a 30S ribosomal subunit forms a complex with mRNA and fMet–tRNA$_f$ (in the presence of GTP and initiation factors) and, subsequently, the 50S ribosomal subunit is attached to this complex forming a 70S (initiation) complex. According to this scheme it is expected that the binding of mRNA in a 70S complex should strictly depend on fMet–tRNA$_f$. One would also expect that if fMet–tRNA$_f$ were the only aminoacyl–tRNA present then the ribosomes had to bind only at initiation signals, since in the absence of other aminoacyl–tRNAs they could not move further along the mRNA.

The experiments fulfilled both of these expectations. We developed conditions in which f2 RNA did not become bound in a 70S complex in the absence of fMet–tRNA$_f$. Moreover, in the presence of fMet–tRNA$_f$ only a single ribosome was attached to a f2 RNA molecule. Gupta, Waterson, Weissmann and I isolated by ultracentrifugation such a 70S complex containing f2 RNA, treated it with nuclease, and recovered the 61-nucleotide-long segment of the f2 RNA that was protected against nuclease action by the attached ribosome. The nucleotide sequence of this protected segment indicated that the segment contained an initiation signal, one specifying the initiation of translation of the f2 bacteriophage coat protein. Our procedure for binding ribosomes to initiation signals was used by J. Steitz and others to isolate a variety of initiation signal-containing segments of mRNAs.

STUDIES ON THE MOVEMENT OF THE RIBOSOME ALONG THE mRNA DURING PROTEIN SYNTHESIS

The possibility of attaching a ribosome to a single initiation signal on a mRNA enabled us to study the ribosome movement. These studies were performed with Gupta, Waterson, Sopori and Weissmann.

The problem can be appreciated by looking at a scheme of peptide-chain elongation (Fig. 1). This is a cyclic process which is repeated whenever one aminoacyl residue is added to the peptide chain growing on the ribosome. Various aspects of the first two composite steps of the cycle (i.e. aminoacyl–tRNA binding and peptide bond formation), together with the elongation factors (EF–Tu and EF–Ts) involved, were discussed earlier. The product of these steps is a mRNA–ribosome complex with peptidyl–tRNA bound at the A site of the ribosome and discharged tRNA at the P site. This intermediate is designated as pretranslocation complex.

During the third composite step (translocation), EF–G is bound to the ribosome, and triggers GTP cleavage, the discharged tRNA is released from the P site, and the peptidyl–tRNA (which has been extended by one aminoacyl residue during peptide bond formation) is shifted from the A site to the P site. The product of this step is designated as post-translocation complex. With its formation, the process is ended. The stage is set for attaching another aminoacyl residue to the peptide chain by repeating the cycle.

In the course of one cycle, the ribosome has to move along the mRNA by the length of one codon. We attempted to establish the composite step in which the movement occurs.

The approach was based on our earlier studies in which we isolated and sequenced the f2 RNA segment protected against ribonuclease T_1 cleavage in a 70S initiation complex. The segment in question extended to the codon specifying the seventh aminoacyl residue of the coat protein. We argued that after the ribosome has moved along the f2 RNA in the 5′ to 3′ direction, the 3′ end of the protected segment should also be shifted in the same direction. Since a codon is three nucleotides long, the shift should also be three nucleotides long. To find out if ribosome movement occurs during aminoacyl–tRNA binding and peptide bond formation, or during translocation, the 3′ end of the protected segment in the pretranslocation complex has to be determined.

For these studies we formed an initiation complex on f2 RNA, converted an aliquot into a pretranslocation complex by reacting it with the appropriate EF–Tu · GTP · aminoacyl–tRNA complex. An aliquot of the pretranslocation complex was then converted into a post-translocation complex by treatment with EF–G and GTP. After treating each of the complexes with pancreatic ribonuclease, we sequenced the protected f2 RNA segment in each. The 3′ end of the protected segment was the same in the initiation and pretranslocation complexes; it extended three nucleotides further toward the 3′ end in the post-translocation complex. In accord with these results, the translation of the protected f2 RNA segments from the initiation and pretranslocation complexes resulted in the same pentapeptide, that from the post-translocation complex in the expected hexapeptide. These findings proved that the ribosome movement occurs during translocation and depends on EF–G and GTP.

By 1971, when these experiments were finished, much information had accumulated about the mechanism of protein synthesis. I stopped working in the field and started to follow a new interest. Since that time I have been learning about the interferon system, a part of the first line of defenses of animal and human cells against viral infection.

Writing this essay made me reminisce about the 7 years I was fortunate enough to be working in Dr. Ochoa's laboratory. Looking back I think that what I have admired in him the most are the remarkable clarity of his mind and an uncanny ability to choose the logical next step along the path toward the solution of complicated problems in the forefront of biochemical research.

Abbreviations used

A, C, G and U—adenylate, cytidylate, guanylate and uridylate residues in oligo and polynucleotides; GMPPCP—5′guanylyl-methylene-diphosphonate; acPhe–tRNA—acetyl Phe–tRNA; acdiPhe–tRNA—acetyl PhePhe–tRNA.

REFERENCES

To make the reading smoother no references have been inserted in the text. Interested readers may find the citations of the original publications in the following reviews and articles.

1. LENGYEL, P., SPEYER, J. F., BASILIO, C., WAHBA, A. J., GARDNER, R. S. and OCHOA, S. (1965) Studies on the amino acid code: the ambiguity in translation. In *Nucleic Acids–Structure, Biosynthesis and Function*, pp. 309-332, Council of Scientific and Industrial Research, New Delhi.
2. LENGYEL, P. (1966) Problems in protein biosynthesis. *J. Gen. Physiol.* **49**, 305–330.
3. LENGYEL, P. and SÖLL, D. (1969) Mechanism of protein synthesis. *Bact. Rev.* **33**, 264–301.
4. SKOULTCHI, A., ONO, Y., WATERSON, J. and LENGYEL, P. (1969) Peptide chain elongation. *Cold Spring Harbor Symp. Quant. Biol.* **34**, 437–454.
5. GUPTA, S. L., WATERSON, J., SOPORI, M. L., WEISSMAN, S. M. and LENGYEL P. (1971) Movement of the ribosome along the messenger RNA during protein synthesis. *Biochemistry*, **10**, 4410–4421.
6. LENGYEL, P. (1974) The process of translation: A bird's-eye view. In *Ribosomes* (M. NOMURA, A. TISSIERES and P. LENGYEL, eds.) pp. 13-52, Cold Spring Harbor Laboratory.

BACTERIOPHAGES, COLICINS AND RIBOSOMES: SOME REFLECTIONS ON MY RESEARCH CAREER

MASAYASU NOMURA

Institute for Enzyme Research, Departments of Genetics and Biochemistry, University of Wisconsin, Madison, Wisconsin 53706

It was the spring of 1958 when I saw Severo Ochoa for the first time. I was sitting in a crowded room at the Annual Federation Society meeting in Philadelphia. Severo Ochoa, the chairman of the session, was gallantly introducing Marianne Grunberg-Manago, who had come from Paris to give a talk on polynucleotide phosphorylase ("Ochoa's enzyme", of course!) which Marianne and Severo had discovered only a few years before.(1) Though I do not recall the details of Marianne's talk, I do remember vividly the way Severo introduced her and the enthusiastic way she talked. Even more, I remember the great excitement I felt seeing the hero and the heroine in the story of a great discovery which, together with Arthur Kornberg's discovery of DNA polymerase I in 1956,(2) heralded the beginning of a new era in biochemistry, the enzymology of biosynthesis of informational macromolecules.

After the session, I dared to approach Severo, partly for the opportunity to talk with a famous scientist and partly to ask his opinion on my futile efforts in Sol Spiegelman's laboratory to isolate a "true" RNA polymerase. At that time, despite the glorious discovery of polynucleotide phosphorylase, some people, including myself, thought that the "true" RNA polymerase, which would use DNA as a template, was yet to be discovered. Although I do not remember what Severo told me, I do remember how good I felt when the great biochemist was willing to discuss science with an unknown young scientist from the Orient.

This Federation meeting was the first meeting I attended in the U.S. Just a half year before, in late November of 1957, I had come from Tokyo to Urbana, Illinois, to work in Sol Spiegelman's laboratory. I had been educated in Tokyo during the miserable post-World War II era in Japan. I did not have a good education in basic science. To rectify this weak background and to learn more about the "new" biology, molecular biology as we call it now, I came to the States. At that time, the fundamental concepts of gene replication and gene expression had already been formulated; the former was, of course, the Watson–Crick proposal of semi-conservative replication of DNA molecules, and the latter was Crick's "central dogma" on the information transfer from DNA to protein through an RNA intermediate. Thus, people were very eager to prove (or disprove) these concepts and to study the actual biochemical mechanisms involved in these processes. As mentioned above, I was trying to isolate the "true" enzyme involved in the information transfer, that is, a DNA-dependent RNA polymerase. In retrospect, it was a difficult task for me without the proper training, and I was unsuccessful. Years later, the enzyme was discovered almost simultaneously by several competent biochemists, Sam Weiss, Jerry Hurwitz and Audrey Stevens.

In the summer of 1958 I worked in Jim Watson's laboratory for 3 months. While in the States, I was eager to learn as many different approaches as possible. Somewhat naïvely, I had thought that Jim was a physical chemist. (After all, he had elucidated the molecular structure of DNA! My misconception was perhaps understandable in view of Japan's isolation from the rest of the scientific community at that time.) I wanted to learn physicochemical approaches to the problem of information transfer. Jim and Alfred Tissières were just starting to isolate and characterize ribosomes. Jim was apparently convinced that the RNAs in ribosomes were the information carriers between DNA and proteins. Other people in Jim's laboratory included David Schlessinger and Charles G. Kurland, both of whom are now actively working on ribosomes. They were busy measuring molecular weights of ribosomes and ribosomal RNAs and studying other physicochemical and chemical properties of the ribosomes. Jim suggested that I study the effect of inhibitors of protein synthesis, such as chloramphenicol, on the ribosomes. This was my first formal introduction to ribosome research.

Although I did not learn much physical chemistry from Jim, I did gain awareness of something more important in scientific research. I still remember our discussions and his opinion on the difference between "good" science and "bad" science. I will not elaborate on this here, nor on many other fond memories and events at Harvard; I would merely like to take this opportunity to express my gratitude to him, perhaps not so much for his introducing me to ribosomes, but more for his patience and frankness in discussing with me such serious and important science philosophy.

Young readers may not believe it, but as late as 1959, the central dogma was still a hypothesis, and, in fact, the mechanism of information transfer was in a confused state. The rumor came from Paris that Arthur B. Pardee had observed immediate expression of the β-galactosidase (z^+) gene after its entry from male bacterial cells into z^- female cells in bacterial conjugation (later published in ref. 3). Some people interpreted these experiments to mean that metabolically stable ribosomal RNA is not the intermediate, and hence, DNA might be the direct template for protein synthesis (cf. the discussion in ref. 4). In the summer of 1959, my second summer in the States, I was working again in Jim Watson's laboratory. Apparently, Jim still believed in the central dogma with ribosomal RNA as a possible information carrier. That same summer, I took the phage course at Cold Spring Harbor. I had made arrangements to work with Seymour Benzer, then at Purdue University, in my third year in the States, and Seymour had requested that I take the phage course. Perhaps he did not want to take any postdoctoral fellow who had not had experience in handling phages. While taking the phage course, I was still thinking about ribosomal RNA and information transfer. It seemed to me that the best observation which would still support the central dogma was the synthesis of a small amount of RNA, then called "Volkin–Astrachan's RNA", in *E. coli* after T2 phage infection. Three years earlier, Volkin and Astrachan had discovered this RNA and shown that the base composition of the RNA was similar to that of the T2 phage DNA.[5] However, Volkin and Astrachan were much more concerned with the observed quantitative conversion of this RNA to phage DNA and considered a hypothesis that the RNA might be a precursor for T2 phage DNA.[6]

After the phage meeting, I had 3 months until I would begin work in Seymour Benzer's lab. I decided to isolate the Volkin–Astrachan RNA and to examine the physical properties of the RNA. From discussions with Jim Watson and Sol Spiegelman, I was convinced that the Volkin–Astrachan RNA would be the intermediate information carrier between the phage DNA and phage specific proteins which people such as Seymour S. Cohen had discovered. My main question was whether the Volkin–Astrachan's RNA is identical to the ribosomal RNA, as the hypothesis of ribosomal RNA being the information carrier would predict. The actual experiments were done in Sol Spiegelman's laboratory, and Benjamin D. Hall, then in the Department of Biochemistry at Illinois, joined us in doing the experiments. We soon discovered that the Volkin–Astrachan RNA was associated with ribosomes in the presence of high Mg^{2+}, but was released from ribosomes as free RNA upon decrease of the Mg^{2+} concentration. We also found that "T2-specific RNA" (the name we gave to the Volkin–Astrachan RNA) had sedimentation coefficients of about 8 to 10S and could be physically separated from 16S and 23S ribosomal RNAs.[7] To me, this was the first exciting time in my scientific career, and so I worked very hard. However, at the end of October of the same year, without finishing the experiments I wished to do, I had to leave Spiegelman's laboratory to work with Seymour Benzer at Purdue. Thus, our paper was incomplete; we could not determine whether T2 specific RNA was a new form of RNA functioning together with preexistent ribosomes or if it was "special" ribosomal RNA contained in "special" ribosomes synthesized after phage infection. As is well known, about a year later, brilliant experiments by Sydney Brenner, François Jacob and Matt Meselson, proved the former to be correct.[8] The name "messenger RNA" was coined by Jacob and Monod for this new type of RNA.

The time I worked with Seymour Benzer was rather quiet. At the beginning, I was uneasy because I could not continue the "important experiments" I had started in Sol Spiegelman's laboratory. However, I soon adjusted myself to the new environment. Although Seymour asked me to look for rII gene products (which, like my eminent predecessor, Alan Garen, I failed to identify), I could also start experiments on rII genetics. Since I had never taken any genetics courses, I tried to read many classical phage genetics papers. However, I have to confess that some of the papers were hard for me to understand, and some seemed too abstract. I felt that I was a biochemist. Nonetheless, rII genetics worked beautifully,[9] and I enjoyed doing rather routine experiments, hundreds of phage crosses (in "spot tests") every day in a peaceful and relaxed atmosphere.

Of course, Seymour Benzer was also one of many at that time who were trying to elucidate the mechanism of information transfer from genes to proteins. However, he was very critical of what he called the "rat race" at that time; he was no doubt unique. Later, whenever I was involved in a competitive race in science, I recalled his comments on the "rat race" and avoided being trapped by this danger that so often plagued many competent scientists.

In late 1960, after 3 years in the States as a postdoctoral fellow, I returned to Japan and took an assistant professorship at Osaka University. At the beginning, I thought about continuing the work on messenger RNA which I had started in Spiegelman's laboratory. However, I soon realized that I could not compete in such an isolated and poorly equipped place; many competent molecular biologists in the States and in Europe had started working on the same subjects. The field of messenger RNA and genetic code was most competitive in the early 1960s. Thus, I looked for a more obscure and yet interesting (to me) subject to work on and chose to study the mode of action of colicins. This was late 1961. At that time, no one was seriously working on this subject, except for some geneticists who were studying the transfer of colcinogenic factors between bacterial strains. While I was a graduate student in Tokyo, I had read articles in various areas at random, and one of a few articles which inspired me very much was a review article by André Lwoff on lysogeny, published in 1953.[10] Lwoff described beautifully many challenging problems involved in lysogeny. In the same article, he also mentioned colicins and colicinogeny and cited the paper by Jacob, Siminovitch and Wollmann published in 1952.[11] I found that there had been no major publication since Jacob, Siminovitch and Wollmann studied this subject in 1952. Thus, I started my first project as an independent scientist—studies on the mode of action of colicins.

The colicin project was enjoyable and rewarding. I had several co-workers such as K. Okamoto, M. Nakamura and A. Maeda. When we started working on the colicins, it was thought that all the colicins would kill *E. coli* cells by the same biochemical mechanism. However, by analyzing macromolecular synthesis in *E. coli* treated with various colicins, we quickly noticed differences among the effects caused by the three colicins we had more or less arbitrarily chosen; colicin K inhibited all macromolecular synthesis, E2 caused DNA degradation, and E3 inhibited only protein synthesis.[12] It was exciting for me to find many new facts in a relatively short time, although the detailed mechanisms of colicin action, except for E3, are even now obscure.

Although my main project was the mode of action of colicins, I continued to have interests in other subjects, such as phage genetics, mRNA and ribosomes. Thus, when the crucial paper on mRNA theory by Brenner, Jacob and Meselson appeared in *Nature* in 1961,[8] I was struck by their peculiar observation immediately. When they centrifuged bacterial extracts in CsCl, they observed two bands containing ribosomal particles. The lighter band (the B band), corresponding to a density of 1.61, contained mRNA as well as growing polypeptide chains; the heavier band (the A band), corresponding to the density of 1.65, did not contain either. The presence of the A band was not relevant to the main theme of the paper, but it was suggested that 30S and 50S subunits corresponded to the A band, and 70S ribosomes corresponded to the B band. We knew then that the 30S and 50S subunits and their aggregates, the 70S ribosomes, have the same chemical composition. Why then should their densities be different?

In the summer of 1962 I had an opportunity to visit Matt Meselson's laboratory at Harvard to examine this question. We found that the B band contained undegraded ribosomal subunits, while the A band contained smaller 40S and 23S "core" particles, derived from the normal ribosomal subunits by the splitting off of a part of the proteins during density-gradient centrifugation.[13] This discovery induced my serious return to ribosome work. I have described elsewhere[14] how we discovered reversibility of this partial splitting of ribosomal proteins, how we worked out the partial reconstitution of ribosomes and how eventually we succeeded, in late 1967, in reconstituting 30S ribosomal subunits from 16S RNA and a mixture of 30S ribosomal proteins. Therefore, I will not recount these historical events. Perhaps I should note that I continued to regard my colicin project as the main project and the ribosome projects as side projects perhaps until around 1966. At that time, many molecular biologists and biochemists were busy solving the genetic code. Severo Ochoa's group was one of the three major groups in this game. The other two were Marshall Nirenberg at the National Institutes of Health, Bethesda, Maryland, and Gobind Khorana at the University of Wisconsin—Madison. Seeing and hearing almost every week some new and exciting progress on this subject, it was a temptation for me, as it might have been for many others, to participate in this great historical event. Perhaps as a counter-reaction, I tried to maintain my colicin projects, which were still very quiet and peaceful.

However, I gradually realized that people were so busy solving the genetic code that the ribosomes still remained a black box, a complicated and mysterious machinery required for protein synthesis. In fact, partly because the mRNA discovery negated the possible important role of ribosomal RNA as an intermediate in the information transfer and partly because people gradually found the structure of ribosomes to be almost hopelessly complex, very few people were seriously studying ribosome structure and function. Thus, around 1966, I decided to make the ribosome project my major effort. By then, I had left Japan (in late 1963) and moved to Madison, after making a rather difficult personal decision to leave my own country and commit myself to doing science in the States.

I should also admit that around the time of my decision to switch projects, I started to feel that colicin problems were rather difficult, and it would take a long time to clarify the molecular mechanisms involved in the colicin action. For example, we had already discovered that the ribosomes from colicin E3-treated cells were inactive in *in vitro* protein synthesis. Yet, we could not find any obvious difference between the inactive ribosomes and the control active ribosomes. Only after our successful reconstitution of ribosomes did it become possible to identify the target molecule, 16S ribosomal RNA. We were able to show that a nucleolytic cleavage of the 16S RNA molecule near the 3′-end is responsible for inactivation of ribosomes in E3-treated cells.[15] When I started colicin work, I never anticipated that my colicin project would eventually become so closely related to ribosomes. In fact, my initial effort emphasized colicin E2, which caused DNA degradation. Only later, after realizing the relation of colicin E3 action to the ribosomes, did I gradually concentrate on the studies of E3 action. It was fortunate that the progress in ribosome research eventually helped to clarify many problems, not only those directly involved in the mode of action of colicin E3, but also some other related problems, such as the "immunity" in colicinogenic cells.[16,17]

Early in 1967 Chuck Lowry, then one of my students, and I proposed that the initiation of protein synthesis starts on the 30S subunits, rather than on the 70S ribosomes.[18] I recall that my short talk on this subject at the nucleic acid Gordon Conference left many people skeptical. Only a few people agreed with our proposal. Severo Ochoa, who had discovered the initiation factors and had been working on the initiation of protein synthesis, was one of these few, and I was very pleased to have his support. I believe that this was the start of my close professional interaction with Severo Ochoa. However, the field of the initiation of protein synthesis was also rather competitive. Although Severo Ochoa continued active research, I stopped working on this subject.

I was very fortunate in having several excellent coworkers in my own laboratory. Thus, the complete reconstitution of 30S subunits in late 1967 was largely due to the hard work of Peter Traub,[19] and the complete reconstitution of 50S subunits in 1970 was done in collaboration with Volker Erdmann.[20] Many others, including Christine Guthrie, Shoji Mizushima, Hiroko Nashimoto, William Held, Steve Fahnestock, Larry Kahan and Eberhard Kaltschmidt, contributed to the work done in my laboratory on the structure and function of ribosomes. We have enjoyed doing solid and good science. Thanks also to the efforts made by several other groups, such as Heinz-Günter Wittmann's group in Berlin, Alfred Tissières' in Geneva and Chuck Kurland's in Uppsala, ribosome research has really come of age. I now find many excellent young people, biochemists, biophysical chemists, physiologists, geneticists and others, participating in various aspects of ribosome research. Compared to the time when I was first introduced to the ribosome in 1958, the state of the art is so advanced. The ribosomes are no longer a black box. Almost all the molecular components of the ribosomes have been characterized and their assembly can be studied *in vitro*. People are using various sophisticated biochemical and physico-chemical approaches to study the ribosome structure and function. My own group is still continuing productive experiments.

I suppose I should be satisfied with the present situation. Yet, quite often I hear someone's voice deep within. Perhaps it is Seymour Benzer, who used to criticize the rat race, and who, in one of his essays,[21] quoted Max Delbruck's note addressed to his wife, Dotty Benzer: "... please tell Seymour to stop writing so many papers. If I gave them the attention his paper *used* to deserve, they would take all my time. . . . If he *must* continue, tell him . . . underline what is important". Perhaps, it is Jim Watson who taught me the difference between good and bad science. When I consider the activity in and the quality of the science that Severo Ochoa has maintained, I realize that I still have a few years to test myself as a scientist. Thus, the completion of the present essay is yet to come. Recently, I, along with my collaborators, have started an attempt to identify and isolate all the genes involved in making the ribosomes. Time will tell whether I will still be able to produce work worthy of writing with underlines to my former mentors, Sol Spiegelman, Seymour Benzer and Jim Watson, as well as to people like Severo Ochoa, whom I respect and love.

REFERENCES

1. Grunberg-Manago, M. and Ochoa, S., (1955) Enzymatic synthesis and breakdown of polynucleotides: polynucleotide phosphorylase. *J. Am. Chem. Soc.* **77**, 3165–3166.
2. Kornberg, A., Lehman, I. R., Bessman, M. J. and Simms, E. S. (1956) Enzymic synthesis of deoxyribonucleic acid. *Biochim. Biophys. Acta.* **21**, 197–198.
3. Pardee, A. B., Jacob, F. and Monod, J. (1959) The genetic control and cytoplasmic expression of "Inducibility" in the synthesis of β-galactosidase by *E. coli. J. Molec. Biol.* **1**, 165–178.
4. Pardee, A. B. (1958) Experiments on the transfer of information from DNA to enzymes. *Exp. Cell Res. Suppl.*, **6**, 142–151.
5. Volkin, E. and Astrachan, L. (1956) Phosphorus incorporation in *Escherichia coli* ribonucleic acid after infection with bacteriophage T2. *Virology*, **2**, 149–161.
6. Astrachan, L. and Volkin, E. (1958) Properties of ribonucleic acid turnover in T_2-infected *Escherichia coli. Biochim. Biophys. Acta*, **29**, 536–544.
7. Nomura, M., Hall, B. D. and Spiegelman, S. (1960) Characterization of RNA synthesized in *Escherichia coli* after bacteriophage T2 infection. *J. Mol. Biol.* **2**, 306–326.
8. Brenner, S., Jacob, F. and Meselson, M. (1961) An unstable intermediate carrying information from genes to ribosomes for protein synthesis. *Nature*, **190**, 576–581.
9. Nomura, M. and Benzer, S. (1961) The nature of the "deletion" mutants in the *r*II region of phage T4. *J. Molec. Biol.* **3**, 684–692.
10. Lwoff, A. (1953) Lysogeny. *Bact. Rev.* **17**, 269–337.
11. Jacob, F., Siminovitch, L. and Wollman, E. (1952) Sur la biosynthese d'une colicine et sur son mode d'action. *Annales de l'Institut Pasteur*, **83**, 295–315.
12. Nomura, M. (1963) Mode of action of colicines. *Cold Spring Harbor Symp. Quant. Biol.* **28**, 315–324.
13. Meselson, M., Nomura, M., Brenner, S., Davern, C. and Schlessinger, D. (1964) Conservation of ribosomes during bacterial growth. *J. Molec. Biol.* **9**, 696–711.
14. Nomura, M. (1969) Ribosomes. *Scientific American*, **221**, 28–35.
15. Bowman, C. M., Dahlberg, J. E., Ikemura, T., Konisky, J. and Nomura, M. (1971) Specific inactivation of 16S ribosomal RNA induced by colicin E3 *in vivo*. *Proc. Nat. Acad. Sci. U.S.A.* **68**, 964–968.
16. Jakes, K., Zinder, N. D. and Boon, T. (1974) Purification and properties of colicin E3 immunity protein. *J. Biol. Chem.* **249**, 438–444.

17. SIDIKARO, J. and NOMURA, M. (1974) E3 immunity substance; a protein from E3-colicinogenic cells that accounts for their immunity to colicin E3. *J. Biol. Chem.* **249,** 445–463.
18. NOMURA, M. and LOWRY, C. V. (1967) Phage f2 RNA-directed binding of formyl-methionyl-tRNA to ribosomes and the role of 30S ribosomal subunits in initiation of protein synthesis. *Proc. Nat. Acad. Sci. U.S.A.* **58,** 946–953.
19. TRAUB, P. and NOMURA, M. (1968) Structure and function of *E. coli* ribosomes. V. Reconstitution of functionally active 30S ribosomal particles from RNA and proteins. *Proc. Nat. Acad. Sci. U.S.A.* **59,** 777–784.
20. NOMURA, M. and ERDMANN, V. A. (1970) Reconstitution of 50S ribosomal subunits from dissociated molecular components. *Nature*, **228,** 744–748.
21. BENZER, S. (1966) Adventures in the rII region. In *Phage and the Origins of Molecular Biology* (J. CAIRNS, G. S. STENT and J. D. WATSON, eds.), pp. 157–165, Cold Spring Harbor Laboratory.

RIBOSOMES: STRUCTURE, FUNCTION AND EVOLUTION

H. G. WITTMANN
Max-Planck-Institut für Molekulare Genetik, Berlin-Dahlem, Germany

INTRODUCTION

Protein biosynthesis takes place on ribosomes, which are small organelles present in all organisms. In prokaryotes (bacteria and blue-green algae) the ribosomes are somewhat smaller (sedimentation coefficient 70S) than those in the cytoplasm of eukaryotic cells (80S), whereas ribosomes from chloroplasts and mitochondria have approximately the same size as prokaryotic ribosomes. All ribosomes consist of two subunits of unequal size: those from 80S ribosomes sediment at 40S and 60S and those from 70S at 30S and 50S, respectively. Protein biosynthesis can occur only when the two subunits are associated and combined with messenger RNA.

Since much more is known about the structure and function of bacterial ribosomes, in particular those from *Escherichia coli*, than about those from eukaryotes, the following discussion will mainly deal with *E. coli* ribosomes. In the last chapter a comparison will be made between *E. coli* ribosomes and those of other organisms.

RIBOSOMAL COMPONENTS

RNA

About two-thirds of the mass of both the small and large subunits of bacterial ribosomes consist of three RNA molecules, namely the 16S RNA (with approximately 1600 nucleotides) in the small subunit, in addition to the 23S RNA (with about 3200) and the 5S RNA (with 120 nucleotides) in the large subunit. The complete base sequence of the 5S RNA from *E. coli* has been determined and about 90% of that of the 16S RNA, whereas the present knowledge of the sequence of the 23S RNA is limited to some regions only. So far no identical or very similar regions have been found between 16S and 23S RNA, in contrast to 5S RNA where some duplication of sequence regions occurs. In 16S RNA there are several "palindromes", in which the sequence is the same regardless as to whether it is read from the 5′ or the 3′ end. Almost 70% of the nucleotides are involved in base pairing and the degree of secondary structure of the RNA is approximately the same within the ribosomal particle as in the isolated form.

Proteins

Although the protein moiety makes up only one-third of the mass of the *E. coli* ribosome, it consists of numerous proteins. Analysis by two-dimensional polyacrylamide gel electrophoresis has shown that there are 21 ribosomal proteins (S1–S21) in the small and 34 proteins (L1–L34) in the large subunit. All these proteins have been isolated and their properties studied by chemical, physical and immunological methods. The molecular weights of all the proteins, with one exception, range from 6000–32,000 daltons with an average molecular weight of approximately 17,000; only protein S1 falls outside this range, with a much higher molecular weight of about 65,000 daltons. Many of the proteins are very rich in basic amino acids (up to 35%). This is reflected in their high isoelectric points which are pH 10 or higher for 70% of the proteins, and only three of them are mildly acidic, with isoelectric points of about pH 5.

The isolation of all the proteins from the *E. coli* ribosome has enabled antibodies to be produced against each of them. It is somewhat surprising that there was no detectable cross-reaction between any ribosomal proteins, with the exception of two pairs: proteins L7 and L12 showed complete cross reaction, as did proteins S20 from the small and L26 from the large subunit. Amino acid sequence analysis showed, more precisely, that L7 and L12 have identical primary structures, the only difference being that L7 is acetylated at its *N*-terminus whereas L12 is not. No difference was found between S20 and L26 and it is clear from genetic evidence that both are coded by the same gene. In consequence, the number of different proteins in the 70S ribosome is not the sum of the proteins in the small and large subunits (21 plus 34), but one less, viz. 54.

As mentioned above, it was concluded from immunological studies that there is no extensive homology among *E. coli* ribosomal proteins, and this has also been confirmed by the amino acid sequence analyses which have so far been performed. With the exception of the two pairs of proteins described above, only very short regions of homology (up to five residues) occur. This finding is based on the sequence determination of about 3000 amino acid residues, i.e. 35% of the approximately 8000 amino acids present in the *E. coli* ribosome.

The knowledge of the primary structures which can be expected for most *E. coli* ribosomal proteins within the not too far distant future is an essential prerequisite for a number of further objectives: (a) an understanding at the molecular level of the specific recognition and interaction between ribosomal RNA and proteins; (b) the determination of the spatial structure of ribosomal proteins which is now in its very early stage; (c) the elucidation of amino acid replacements and more drastic differences in ribosomal proteins from mutants with altered phenotypes, e.g. resistance to or dependence on antibiotics; (d) a comparison of the structure of ribosomal proteins from organisms belonging to different bacteria, plants or animals with a view to obtaining information about the evolution of ribosomes.

In conclusion: the *E. coli* ribosome is the first cell organelle for which an understanding at the molecular level of both structure and function can be expected. An essential step towards this very attractive goal is a thorough investigation of the primary structures of all ribosomal components.

TOPOGRAPHY

As mentioned above the *E. coli* ribosome consists of three RNA molecules and fifty-four different proteins. In spite of this great complexity considerable progress has been made in determining the spatial arrangement of the numerous components, especially the ribosomal proteins. This has been achieved by the following methods.

Cross-links

Ribosomal subunits are treated with bifunctional reagents, e.g. bis-methyl suberimidate or other bis-imido esters, which react with two amino acids belonging to the same or different protein chain(s). In the latter case an intermolecular bridge is formed which covalently connects two neighboring proteins. After disruption of the ribosomal structure and separation of the ribosomal components the cross-linked protein pair can be isolated. Identification of the proteins cross-linked by the bifunctional reagent is best achieved by means of specific antibodies against single ribosomal proteins. In this way more than twenty pairs of proteins within the small subunit of *E. coli* ribosomes have so far been identified as neighbors by means of various bifunctional reagents. Similar but so far less extensive studies have been made for the identification of neighboring proteins in the large ribosomal subunit.

Cross-linking has not only been used to identify which proteins neighbor each other within the ribosomal subunit but is now also being used for determining which region of the ribosomal RNA is close to a given ribosomal protein. Cross-linking between RNA and protein can be accomplished by treatment with various reagents or by UV-irradiation of the intact subunits which are then treated with RNase and disrupted. This treatment results in many RNA pieces some of which are covalently bound to proteins. These RNA protein complexes can be isolated by gel electrophoresis and the proteins identified. By sequencing the RNA pieces and comparing their sequences to the known primary structure of ribosomal RNA, e.g. 16S RNA, information about the neighborhood between proteins and RNA regions can be obtained.

RNP fragments

In this approach for the elucidation of RNA–protein and protein–protein neighborhood within the ribosomal subunits, the intact particles are treated mildly with RNase in order to produce a few nicks in the RNA strand at those regions which are accessible to the enzyme. By lowering the magnesium concentration, followed by gel electrophoresis or sucrose gradient centrifugation, fragments of the subunits can be isolated. Each of the fragments consists of several proteins and one (or more) piece(s) of the RNA strand. The proteins can be identified and the sequence of the RNA pieces determined. From this information it can be concluded which ribosomal components are present within a given fragment and are in consequence relatively close to each other.

Assembly map and protein–RNA interaction

An early step in the reconstitution of intact and functionally active 30S and 50S ribosomal subunits is the attachment of the so-called primary binding proteins to specific regions of the RNA strands. Six proteins (S4, S7, S8, S15, S17, S20) specifically bind to the 16S RNA, three proteins (L5, L18, L25) to the 5S RNA and eleven proteins (L1, L2, L3, L4, L6, L13, L16, L17, L20, L23, L24) to the 23S RNA. The binding sites of a number of these proteins on the RNA strands have been localized by the two following methods. (a) Each of the binding proteins is separately bound to the corresponding RNA strand and the unprotected RNA regions are digested away by RNase. The complex consisting of the protected RNA region and the binding protein is isolated and the region of the RNA protected by the protein is determined by sequence analysis. (b) Pieces of RNA strands resulting from very mild RNase digestion are isolated and the ability of the single proteins to rebind to each of them is tested. If the RNA pieces are rather large, further studies in a similar way to that described under (a) can lead to more precise information about the binding sites of the various proteins along the RNA strands.

The specific attachment of the primary binding proteins to the RNA strand enables or facilitates the attachment of other proteins, the secondary binding proteins, either by protein–protein interaction or by new attachment points on the RNA induced by the primary binding proteins. Other proteins can bind only after the attachment of the primary and secondary binding proteins. In this way an "assembly map" can be drawn up which illustrates the sequence of the binding events and probably also reflects the topographical arrangement of the ribosomal proteins. This latter hypothesis is supported by the results from other methods, e.g. cross-links and RNP fragments.

Accessibility

Information about the accessibility of the various proteins at the surface of the ribosomal subunits has become available from the following studies. (a) Chemical modification of ribosomal proteins by lactoperoxidase catalyzed iodination and by carrier-bound reagents, e.g. fluorescein isothiocyanate. In these cases large molecules (enzymes or carriers) are involved in the reaction and therefore the disadvantage inherent in the use of small molecules, e.g. aldehydes or SH-reagents, is avoided; such small molecules might penetrate too quickly into the interior of the particle to allow conclusions as to which proteins are "outside" or "inside". (b) Antibodies directed against individual ribosomal proteins have been extensively used as a probe for surface topography: it was found that all 30S proteins react with their corresponding antibodies, i.e. at least part of each protein carrying one or more antigenic determinant(s) is at the surface of the 30S subunit. This is also true for many of the proteins in the 50S subunit.

Antibodies have further been used for determining which proteins are located at the interface between the subunits and it was concluded that five proteins of the small (S9, S11, S12, S14, S20) and nine proteins of the large subunit (L1, L6, L14, L15, L19, L20, L23, L26, L27) belong to this group. Antibodies against each of these proteins, when attached to their corresponding subunits, inhibit re-association of subunits whereas antibodies against the other proteins do not.

Immuno-electronmicroscopy

The fact that antibodies react with their corresponding ribosomal proteins in the intact particle can be utilized in studies on the localisation of these proteins within the subunits. Due to the bivalancy of the antibody molecule it attaches simultaneously to two subunits, resulting in a "dimer" consisting of the two subunits connected by one antibody molecule (Fig. 1). Since the subunits have characteristic shapes the point of attachment of the antibody onto the subunits can be localized. This is illustrated in Fig. 2 for antibodies against different ribosomal proteins.

The positions of almost all of the proteins in the 30S and 50S subunits have been localized by electron-microscopy. For several of the ribosomal proteins it was observed

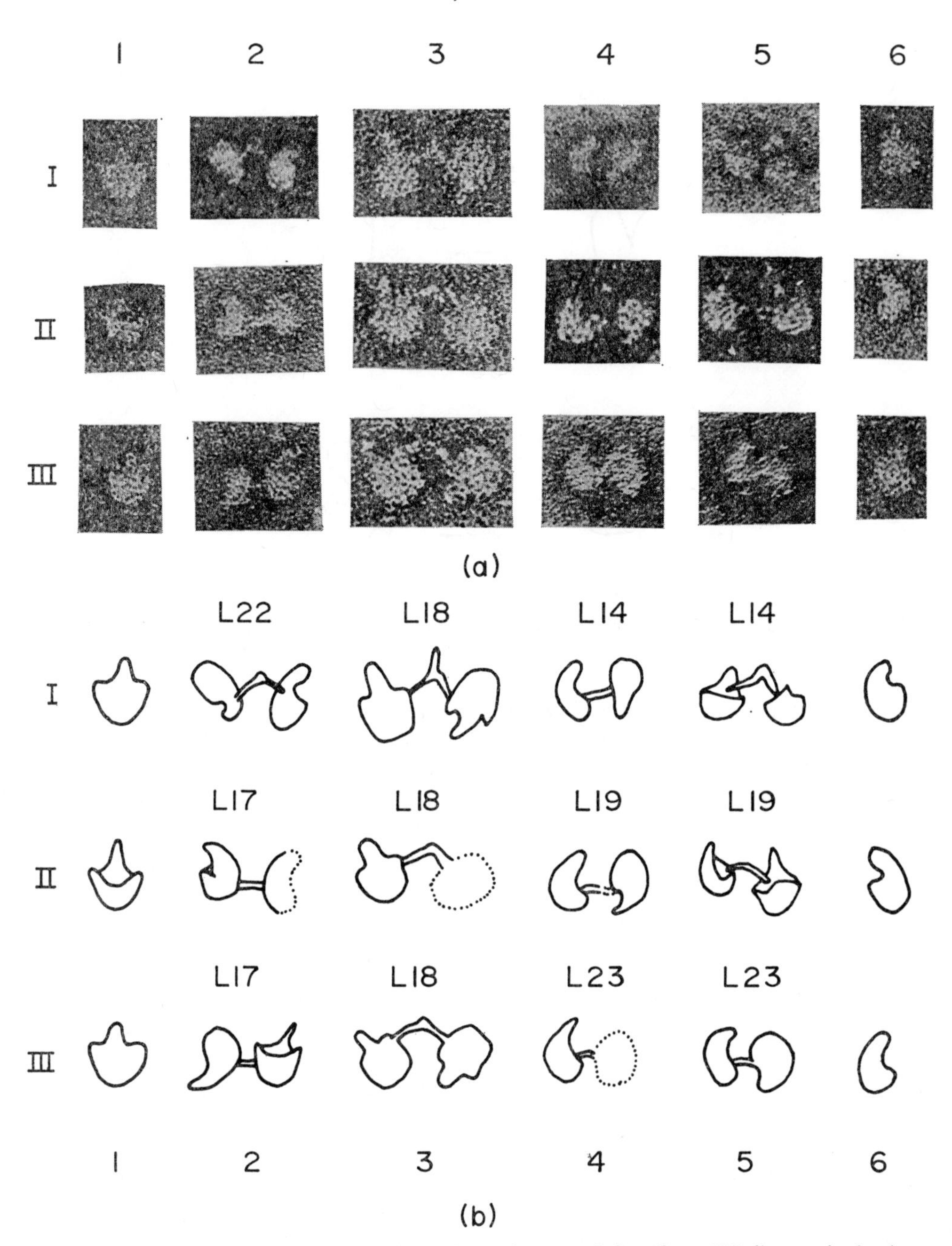

FIG. 1. Electronmicrographs and drawings of complexes consisting of two 50S ribosomal subunits connected by one antibody molecule. Columns 1 and 6 show *E. coli* 50S subunits from different angles and columns 2–5 several antibody–subunit complexes.

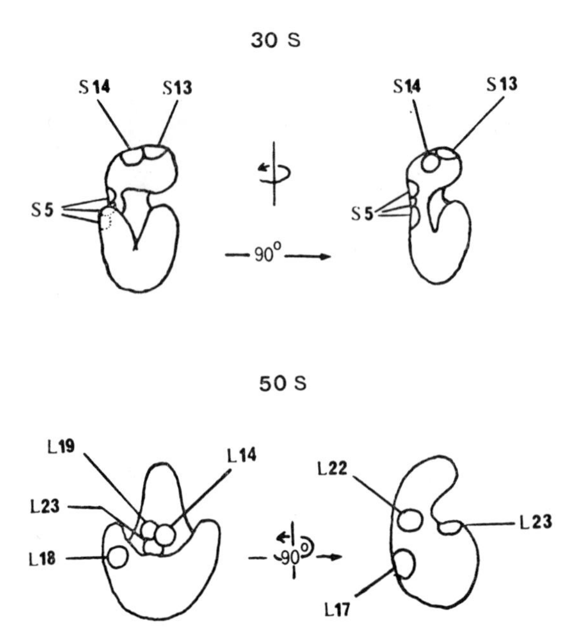

FIG. 2. Shape of *E. coli* ribosomal subunits and location of some proteins on the surface of the subunits by electronmicroscopical visualization of the antibody attachment sites.

that antibodies directed against a given protein attach at different regions of the particle. This is only possible in a protein of moderate size (up to 30,000 daltons) if it is not spherical but elongated. The conclusion that some of the ribosomal proteins have an elongated shape is also supported by hydrodynamic measurements and this finding makes the interpretation of results from cross-link experiments much more complicated. Models of the spatial arrangement of ribosomal proteins based mainly on these data and the assumption that the proteins are of spherical shape have become therefore less plausible.

FUNCTION

The main function of the ribosome is the polymerization of amino acids into a protein chain according to the genetic message encoded in the messenger RNA. This is accomplished by a number of complicated and coordinated events in which ribosomal components and protein factors are involved as illustrated in Fig. 3. The process of protein biosynthesis can be divided into the following main steps: initiation, elongation and termination.

Since other chapters in this book deal in more detail with the steps in protein biosynthesis, in particular elongation, the discussion here will be limited to a short description of the events illustrated in Fig. 3 and of the various methods by which the involvement of ribosomal components in protein biosynthesis have been studied in detail.

Initiation

The first step in the initiation process is the binding of the initiation factor IF-3 to the 30S subunit, thereby inducing a conformational change in the 30S particle. This is followed

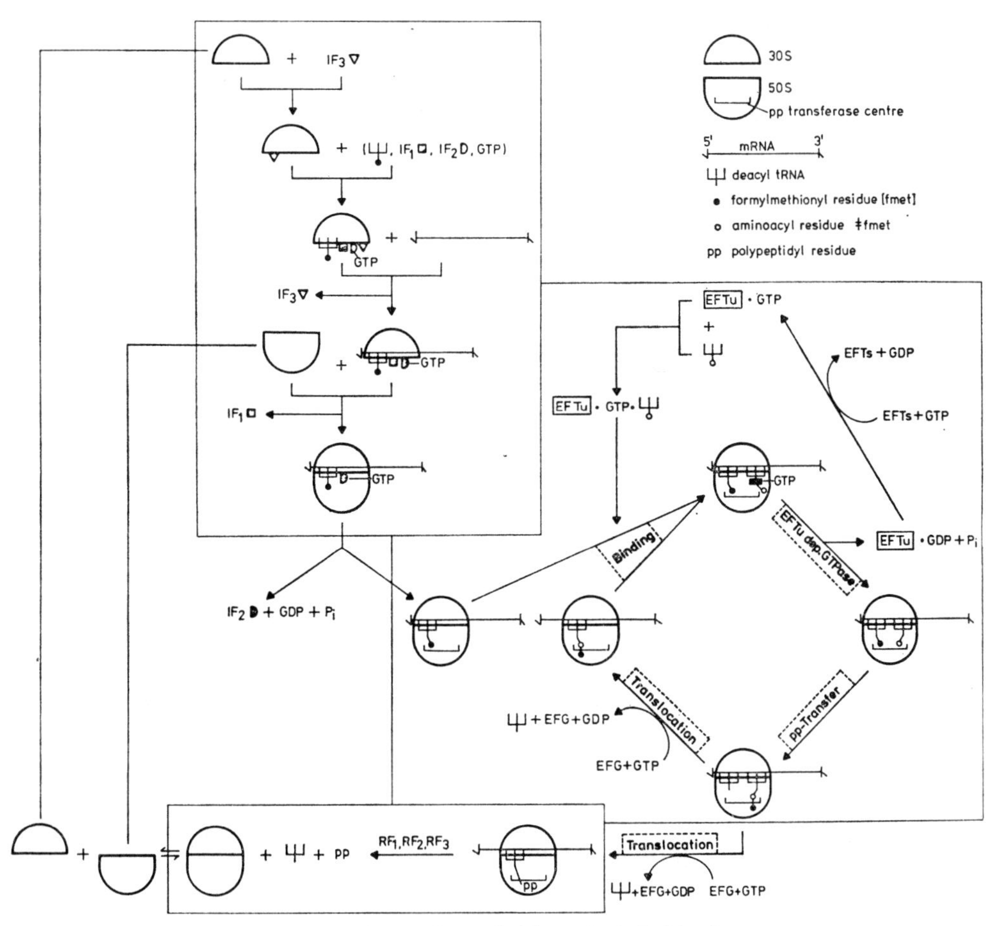

FIG. 3. Scheme of protein biosynthesis in *E. coli.*

by binding of the initiator-tRNA and the messenger-RNA; however, it is still controversial as to which of these two molecules binds first. For the binding of the initiator-tRNA the presence of another initiation factor, namely IF-2, and of GTP is necessary and the tRNA-binding is stimulated by the factors IF-1 and IF-3. These steps lead to the formation of the so-called 30S initiation complex consisting of the 30S subunit, the initiator-tRNA, the messenger-RNA, the initiation factors and GTP. The initiation process is completed by association of the 30S initiation complex and the 50S subunit to the 70S initiation complex and by the dissociation of the factor IF-2 which is accompanied (or caused) by hydrolysis of GTP.

Elongation

After dissociation of the three initiation factors, the 70S initiation complex is ready to bind a ternary complex consisting of aminoacyl-tRNA, elongation factor EF-Tu and GTP. This complex binds to the ribosomal acceptor-site, the A-site, in such a way that the anticodon of the tRNA is in close contact with the complementary codon on the messenger-RNA. The species of tRNA that can bind is determined by the codon in the A-site. After hydrolysis of GTP in the ternary complex, the factor EF-Tu together with GDP dissociate from the aminoacyl-tRNA and leave the ribosome. The amino acid of the aminoacyl-tRNA in the A-site next forms a peptide bond with the amino acid of the initiator-tRNA which is located in the ribosomal peptidyl-site, the so-called P-site. The formation of the peptide bond is catalyzed by the enzyme peptidyltransferase which consists of one or more proteins of the 50S subunit. After peptide bond formation, the fMet-residue of the initiator-tRNA is covalently bound to the amino acid of the tRNA in the A-site. In order to free the A-site for the next aminoacyl-tRNA a translocation step takes place. In this movement the tRNA with the dipeptide attached is translocated from the A- to the P-site, the de-acylated tRNA leaves the ribosome, and the next codon of the messenger-RNA enters the A-site. The translocation step occurs after binding of GTP and of another elongation factor, namely EF-G, which binds to a ribosomal region identical with (or at least very close to) the binding site for EF-Tu. After cleavage of GTP the factor EF-G and GDP leave the ribosome and a second elongation cycle can begin by binding of a new ternary complex to the A-site followed by peptide bond formation and translocation. The number of elongation cycles corresponds to the number of codons in the cistron to be translated.

Termination

Translation ends when a termination codon enters the ribosomal A-site. Release of the newly synthesized protein chain from the ribosome is mediated by the action of the release factor RF-1 specific for the termination codons UAA and UAG or of the factor RF-2 specific for UAA and UGA. A third release factor and GTP stimulate this process. It is likely that the cleavage of tRNA from the completed protein chain is carried out by the same enzyme, namely the peptidyltransferase, which transfers the growing peptide chain from the P-site onto the aminoacyl-tRNA in the A-site. The termination process is completed by release of the newly synthesized protein chain, the de-acylated tRNA and the messenger-RNA from the ribosome. Modification of the protein, e.g. acetylation and methylation, probably occurs when the growing chain is still on the ribosome and as soon as the amino acid to be modified become accessible for the appropriate non-ribosomal enzyme.

Function of ribosomal components

As already mentioned the ribosomal particle consists of numerous components and the question arises as to how many and which of them are involved in each of the various steps briefly described above and illustrated in Fig. 3. Solution of this problem is especially difficult in view of the functional interdependence and cooperativity among the various ribosomal components, and several methods have been applied. The most efficient of them are the following:

(a) *Affinity labeling*. Radioactive components (e.g. aminoacyl-tRNA, messenger RNA or antibiotics) which react with the ribosome are chemically modified in such a way that they can covalently bind to the ribosomal particle. It has to be shown by appropriate controls that the binding of the modified component occurs specifically and at the correct ribosomal site. After disruption of the ribosome, the complex between the radioactively labeled component and a ribosomal protein or RNA is isolated. The identification of those ribosomal components which become covalently bound (e.g. to messenger-RNA or aminoacyl-tRNA) gives direct information about the binding sites of these non-ribosomal components, or about the active sites on the ribosome which are under consideration. The application of the affinity labeling technique has resulted in the elucidation of the involvement of a number of proteins in the various ribosomal functions, e.g. GTP hydrolysis, binding of messenger-RNA, A-site, P-site, etc.

(b) *Partial reconstitution.* Ribosomal subunits are treated with increasing concentrations of salts, e.g. LiCl or CsCl, resulting in a series of protein-deficient "cores" and corresponding "split" protein fractions. The cores are then tested to see which have retained a given functional activity, e.g. peptidyltransferase activity or binding of an antibiotic, and which have lost this activity. By adding back the split proteins to the inactive cores, full activity can be regained. The mixture of the split proteins is then separated into individual proteins, and each protein is added back separately and the reconstituted particle tested for regained activity. In this way it has been possible to elucidate the involvement of individual proteins in various ribosomal activities. A modification of the partial reconstitution method just described is as follows. Some proteins, the so-called fractional proteins, are not present in all ribosomes, and ribosomes lacking these particular proteins are not maximally active. By adding an excess of each of the fractional proteins separately to the ribosomes one can test which of the various ribosomal functions are stimulated and in this way correlate a given function with one or more of the fractional proteins. It was found by this method that proteins S2, S3 and S14 are involved in the binding of aminoacyl-tRNA.

(c) *Inhibition by antibodies.* The antigen binding fragments (Fab) of antibodies directed against individual ribosomal proteins are isolated and added to ribosomes. The treated ribosomes are then tested to see which ribosomal function is inhibited by a particular Fab since the attachment of a Fab molecule to its corresponding protein in the ribosome can strongly impair the function of this protein, e.g. as binding site for the elongation factors EF-Tu or EF-G. It was revealed very early using this method that proteins L7 and L12 are functionally very important.

The three methods described above have been used extensively in recent years for the analysis of the various ribosomal functions, and a great amount of data have accumulated which already allow a good insight into the involvement of ribosomal components, especially ribosomal proteins, in the various steps of protein biosynthesis. It is not possible to discuss these results in detail here and the interested reader is referred to references 2, 4 and 13 for a detailed and up-to-date discussion of this subject.

EVOLUTION

Ribosomes fulfill the same function, namely biosynthesis of proteins, in all organisms. This raises the question as to how far the structure of the ribosome has been conserved during evolution. An answer to this question can be given by the following studies.

(a) *Antibodies* are raised against ribosomes from various classes of organisms, e.g. bacteria, plants and mammals and are tested with the ribosomes from the heterologous species. Relatively strong cross-reaction was found among ribosomes from prokaryotes, i.e. bacteria and blue-green algae, whereas the cross-reaction between prokaryotic and eukaryotic ribosomes was either very weak or non-existent. Stronger cross-reaction was, however, detected between antibodies against individual *E. coli* ribosomal proteins, e.g. L7 and L12, and ribosomes from eukaryotes, e.g. yeast or rat liver. The structure of these particular ribosomal proteins has apparently been conserved relatively strongly during evolution, in contrast to other ribosomal proteins, which have not been conserved to a degree that can be detected by current immunological techniques.

(b) Similar comparisons of ribosomal proteins from various species have been made by *two-dimensional polyacrylamide gel electrophoresis*. The ribosomal protein patterns are rather similar for all species of the family Enterobacteriaceae, to which *E. coli* belongs, whereas, as expected, the similarity between *E. coli* and other bacteria becomes progressively less for more distantly related bacterial species. No similarity is detectable by this technique between ribosomal proteins from bacteria on the one hand and of eukaryotes, e.g. yeast, plants or animals, on the other. However, ribosomes from mammals, from birds and from reptiles are surprisingly similar in their electrophoretic protein patterns. Some similarities still exist between mammals, fishes and amphibians, whereas those between mammals, plants and lower animals (crustaceans and molluscs) are rather low.

(c) The comparison of the *amino acid sequences* of ribosomal proteins from different organisms allows more direct and quantitative conclusions concerning the degree of homology among ribosomal proteins. Unfortunately, information of this kind is so far limited to ribosomal proteins from a very few bacteria. A high degree of sequence homology exists for some proteins but is rather low for others. This is in full agreement with the results from the immunological studies discussed above.

(d) Besides the methods just described (immunology, two-dimensional gel electrophoresis and amino acid sequence analysis) which give information about evolutionary changes in the structure of ribosomes, other methods allow conclusions about *functional* similarities (or differences) during evolution. Hybrid ribosomes, in which one subunit is from *E. coli* and the other from spinach chloroplast ribosomes, are fully active in the poly-U directed poly-Phe synthesis, whereas hybrids between subunits of *E. coli* and of yeast mitochondrial ribosomes are not active in this system. Similarly, it was reported that ribosomal subunits from pea seedlings and mouse liver form functionally active hybrid ribosomes. Furthermore, proteins L7 and L12 from the large *E. coli* ribosomal subunit can replace the homologous proteins from yeast and from rat liver ribosomes with full functional activity. The same is true for hybrid ribosomes in which each of the ribosomal proteins of the *E. coli* 30S subunit is replaced by its homologous protein from 30S subunits of *Bacillus stearothermophilus*. This finding shows that a ribosomal protein can exert its function even in a ribosomal subunit in which all other components are from a rather distantly related bacterial species.

In conclusion

The structure of ribosomes has changed during evolution to such an extent that there is little homology between ribosomes from prokaryotes and eukaryotes. Nevertheless the various steps in protein biosynthesis, the main function of the ribosome, remain the same in all organisms. This raises the interesting question as to how single components in such a complex structure as the ribosome in which numerous proteins and several RNAs interact with each other can change so drastically without impairing the vital function. Too little about the structure of ribosomes other than those of *E. coli* is at present known to answer this question satisfactorily.

RECENT LITERATURE ON RIBOSOMES

(a) *Books*

1. *The Mechanism of Protein Synthesis and its Regulation*, ed. by L. Bosch, North-Holland Publ. Co., Amsterdam, 1972.
2. *Ribosomes*, ed. by M. Nomura, P. Lengyel and A. Tissieres. Monograph Series. Cold Spring Harbor Laboratory, 1974.
3. *Protein Synthesis*, ed. by H. Weissbach and S. Pestka, Academic Press, New York (in press, 1976).

(b) *Review Articles*

4. Brimacombe, R. A., Nierhaus, K. H., Garrett, R. A. and Wittmann, H. G. (1976) The ribosome of *Escherichia coli. Prog. Nucl. Acid Res. Molec. Biol.*, (in press).
5. Caskey, C. T. (1973) Peptide chain termination. *Adv. Prot. Chem.* **27,** 243–276.
6. Erdmann, V. A. (1976) Structure and function of 5S and 5.8S RNA. *Prog. Nucl. Acid. Res. Molec. Biol.* (in press).
7. Garrett, R. A. and Wittmann, H. G. (1973) Structure of bacterial ribosomes. *Adv. Prot. Chem.* **7,** 277–347.
8. Grunberg-Manago, M., Godefroy-Colburn, T., Wolfe, A. D., Dessen, P., Pantaloni, D., Springer, M., Graffe, M., Dondon, J. and Kay, A. (1973) Initiation of protein synthesis in prokaryotes. In *Regulation of Transcription and Translation in Eukaryotes*, pp. 213–249, Springer-Verlag, Heidelberg.
9. Haselkorn, R. and Rothman-Denes, L. B. (1973) Protein synthesis. *Ann. Rev. Biochem.* **42,** 379–438.
10. Kurland, C. G. (1974) The assembly of the 30S ribosomal subunit of *Escherichia coli. J. Supramol. Structure*, **2,** 178–188.
11. Leder, P. (1973) The elongation reactions in protein synthesis. *Adv. Prot. Chem.* **27,** 213–242.
12. Nomura, M. (1973) Assembly of bacterial ribosomes. *Science*, **179,** 864–873.
13. Pongs, O., Nierhaus, K. H., Erdmann, V. A. and Wittmann, H. G. (1974) Active sites in *Escherichia coli* ribosomes. *FEBS Letters*, **40,** 28–37.
14. Spirin, A. S. (1974) Structural transformations of ribosomes. *FEBS Letters*, **40,** S38–S47.
15. Stöffler, G. and Tischendorf, G. W. (1975) Antibiotic receptor sites in *Escherichia coli* ribosomes. In *Topics in Infectious Diseases*, Springer-Verlag, Heidelberg.
16. Vazquez, D. (1974) Inhibitors of protein synthesis. *FEBS Letters*, **40,** 63–84.
17. Wittmann, H. G., Crichton, R. R. and Stöffler, G. (1973) Structure and function of ribosomal proteins of *Escherichia coli. Biochem. Soc. Symp.* **37,** 37–49.

THOUGHTS ON POLYPEPTIDE CHAIN ELONGATION

HERBERT WEISSBACH
Roche Institute of Molecular Biology, Nutley, New Jersey 07110

Once the genetic code was solved, due to the pioneering work of Nirenberg, Ochoa, Khorana and others, it was obvious that elucidation of the mechanism of protein synthesis was within the reach of the biochemist. In looking back at the progress that has been made in understanding the process of protein synthesis, it is apparent that initial success was achieved in bacterial systems, especially in the area of chain elongation. There are several reasons for this, including the large amounts of elongation factors T (EFTu, EFTs) and G (EFG) in cells such as *E. coli*, the ease of purification of these factors and the ability to isolate and characterize intermediate complexes containing these factors.

There have been two recent reviews of protein synthesis that cover the area of peptide chain elongation in great detail.[1,2] I will attempt instead, to look at this process, specifically in the prokaryote system from an historical and personal view point in an attempt to provide the reader with a better understanding of those observations that I feel most influenced the development of this area.*

The steps involved in chain elongation have been most thoroughly investigated in the organism *E. coli*. Once the initiation reaction has occurred, with the formation of an fMet–$tRNA_F$·mRNA·ribosome complex, chain elongation can begin, i.e. the process by which amino acids are added one at a time to the growing polypeptide chain attached to the ribosome in a sequence which is defined by the mRNA. It appears that during elongation the incorporation of all of the amino acids into the growing polypeptide chain occurs by a similar mechanism; therefore, if one can understand how a single amino acid is added during elongation, this can be used as a model for the addition of all the amino acids. A summary of the steps believed to be involved is shown in Fig. 1. The ribosome has two tRNA binding sites named the P (peptidyl) and A (acceptor) sites, respectively. A partially completed peptide containing formylmethionine (fMet) at the aminoterminal end is shown attached to a tRNA on the P site of the ribosome. The codon on the ribosomal A site then determines which aminoacyl–tRNA (AA–tRNA) is to be bound to the ribosome, a reaction that requires two soluble factors in prokaryote cells, EFTu and EFTs, as well as GTP. In the next step the peptide moiety on the P site is transferred to the free amino group of the AA–tRNA on the A site with the formation of a peptidyl–tRNA on the A site containing one additional amino acid, leaving a deacylated tRNA on the P site. The formation of the peptide bond is mediated by a peptidyl transferase complex which is an inherent part of the 50S ribosomal subunit. The final step in the elongation process involves the movement

* I would like to express my deep appreciation to my colleagues who have worked in the area of protein synthesis for the past 8 years, especially Dr. Nathan Brot, Mrs. Betty Redfield and Dr. David Miller. I regret, in an article such as this, that it is essentially impossible to refer to all of the studies that have been done in this area. My attempt to single out specific articles that have *most* influenced *our* thinking will result in an omission of a large number of papers that have contributed to the field.

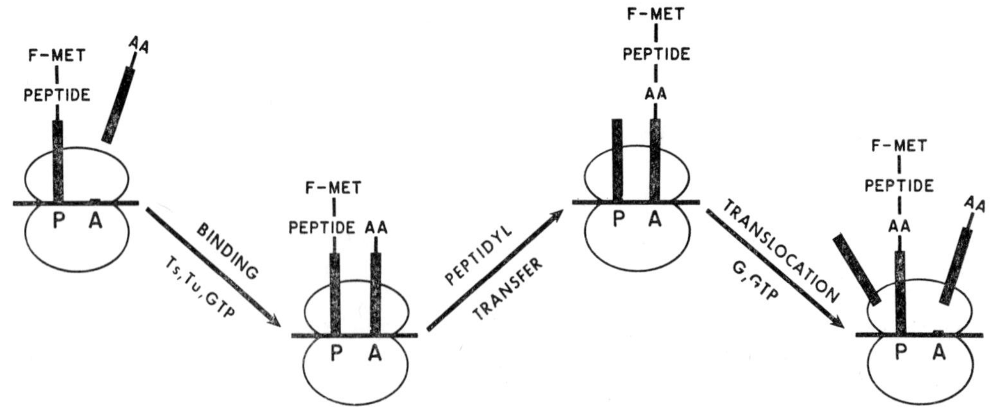

FIG. 1. Polypeptide chain elongation. Description of the partial reactions is described in the text. tRNA is designated as a dark bar. The elongation factors are referred to as Tu, Ts and G.

of the mRNA along the ribosome by three nucleotides (1 codon) which results in the displacement of the deacylated tRNA and the translocation of the peptidyl-tRNA to the P site on the ribosome. The A site is again unoccupied but now contains a new codon. This translocation reaction also requires another soluble factor (EFG) and GTP. Our studies have concentrated primarily on the role of the soluble factors (EFTu, EFTs, EFG) and GTP in this process, and I will begin by reviewing the role of EFT in AA–tRNA binding to the ribosome.

AA–tRNA binding to ribosomes

The initial studies by Lipmann and coworkers[3] led to the separation of two factors, EFT and EFG, that were required for protein synthesis. Subsequently, Lucas-Lenard and Lipmann[4] separated EFT into a heat labile (EFTu) and heat stable factor (EFTs) from *Pseudomonas fluorescens*. A notable advance came from the studies of Ravel[5] who showed that EFT and GTP stimulated the binding of AA–tRNA to the A site on the ribosome, thus pinpointing the site of action of EFT.

Our own involvement in this area of research came about in an indirect way. A major interest in the laboratory had been the role of vitamin B_{12} in methyl group transfer from N^5-methyl-H_4-folate to homocysteine to yield methionine, a problem that was an extension of studies on the isolation of the vitamin B_{12} coenzyme[6] while in Horace A. Barker's laboratory in 1958. When the exciting report by Clark and Marcker[7] appeared showing that *N*-formylmethionine–tRNA (fMet–rRNA) was the initiator or protein synthesis, it was a natural extension of our work to examine the formation of fMet–tRNA, since this reaction also involved both a one-carbon transfer from a reduced folate derivative and the amino acid methionine. Herbert Dickerman, a postdoctoral fellow in the laboratory at that time, was able to purify the transformylase enzyme from *E. coli* extracts[8] that catalyzed eqn. (1):

$$N^{10}\text{-formyl-}H_4\text{-folate} + \text{Met–tRNA}_F \longrightarrow \text{fMet–tRNA}_F + H_4\text{-folate} \quad (1)$$

In retrospect, this problem brought us into the field of protein synthesis. At that time our laboratory at the NIH was adjacent to that of Marshall Nirenberg, and the task of learning the complexities of this problem was made enormously easier by the help we received from many of the talented scientists in Nirenberg's laboratory. In 1967 Jorge Allende was visiting there, and a fruitful collaboration developed. Allende, who had worked with

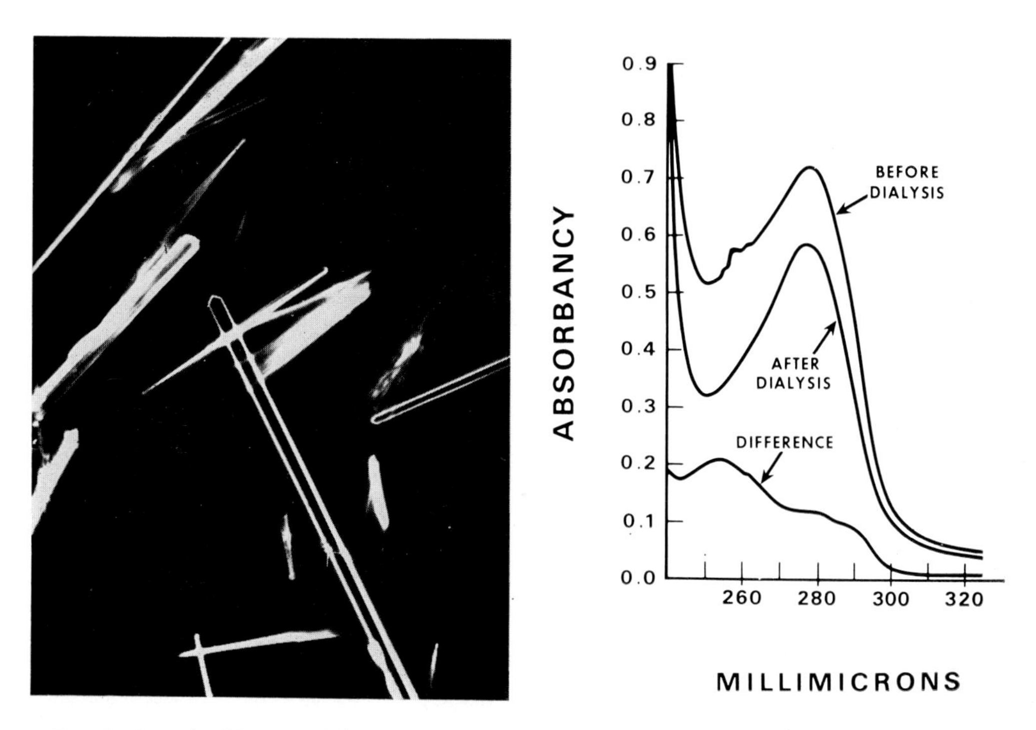

FIG. 2. Crystals of EFTu and the absorption spectrum of the purified protein. Left panel: crystals of EFTu·GDP formed in 0.35 saturated ammonium sulfate. Right panel: absorption spectrum of EFTu. Top curve, spectrum of isolated EFTu; middle curve, spectrum of protein after dialysis in the absence of Mg^{2+}; lower curve, difference spectrum. (From Miller and Weissbach.[12])

Lipmann, had been involved in protein synthesis for several years and was particularly interested in the role of GTP in this process. At that time it had just been shown that GTP was required for chain initiation,[9] and we decided to look for a guanosine nucleotide–ribosome complex that might exist during the binding of fMet–tRNA to the ribosome. The Millipore filter technique, which was already well established as a way to measure the binding of AA–tRNA to ribosomes,[10] was used to assay for the formation of a ribosome. GTP complex. To our surprise, a large binding of GTP to the filter was observed, dependent not on ribosomes but on the crude ribosomal-wash initiation factor preparation. Although originally this binding was thought to be due to an interaction of GTP with one of the initiation factors, this was soon disproved. In fact, the presence of large amounts of a GTP binding factor in the supernatant extracts (S-200), as well as preliminary characteristics of the S-200 material, suggested that the elongation factor T (EFT) might be responsible for the nucleotide binding in the ribosomal wash.

It should be stressed that the Millipore filter procedure[10] has been of enormous help to most investigators in this field. It is convenient, rapid and accurate, and has been used with minor modifications for most of the partial reactions involved in all aspects of protein synthesis. In the case of the binding of guanosine nucleotides to EFT, the assay was made possible because of the quantitative retention of EFT on the filter and the tight binding of GDP, and to a lesser extent GTP, to the protein.

EFT from *E. coli* had not at that time been separated into EFTu and EFTs and this was attempted with the hope of better understanding the nucleotide binding reaction. A purification procedure for the *E. coli* factors was developed by Robert Ertel,[11] based on the original Lucas-Lenard and Lipmann procedure.[4] The two factors were separated and it was shown that both factors were required for GTP binding. Preliminary kinetic studies indicated that EFTs was functioning catalytically,[11] but for some time, further progress was slow and there were several observations that were not readily explained, relating particularly to the effect of Mg^{2+} in this reaction. In the absence of Mg^{2+}, the requirement of EFTs for binding was much less, and EFTu appeared considerably more labile. These two observations proved to be intimately related and once explained, a major step forward became possible.

It was obvious that further progress required substrate amounts of highly purified EFTu and EFTs. At this time David Miller and John Hachmann joined the laboratory, and they soon became involved in the purification of EFTu and EFTs, respectively.[12,13] Both factors were obtained essentially homogeneous, and David Miller was able to crystallize EFTu, which enabled us to begin to resolve the problem. Some of the first crystals of EFTu are shown in the left portion of Fig. 2, and the spectrum of the protein is shown in the right panel. The isolated protein had an A_{280}/A_{260} ratio of 1.1, which suggested that it contained additional A_{260} absorbing material. After dialysis in the absence of Mg^{2+}, the A_{280}/A_{260} value rose to 1.7, and the protein became very labile. The difference spectrum, as seen in Fig. 2, was that of a guanosine-containing derivative, which was easily identified as GDP. Quantitative studies showed that EFTu contained one equivalent of bound GDP. It was immediately apparent that all of the earlier studies with EFTu had, in fact, been done with EFTu·GDP, which in the absence of Mg^{2+} dissociated leaving the unstable free EFTu [eqn. (2)]:

$$\text{EFTu·GDP} \underset{-\text{Mg}^{2+}}{\overset{+\text{Mg}^{2+}}{\rightleftharpoons}} \text{EFTu} + \text{GDP} \qquad (2)$$

This finding explained the increased lability of EFTu seen in the absence of Mg^{2+}. It also became apparent that free EFTu could bind either GTP or GDP in the absence of EFTs; but, when EFTu·GDP was used, EFTs was required for the binding of GTP. EFTs functioned as a catalyst for the displacement of GDP by GTP [eqns. (3)–(5)]:

$$\text{EFTu·GDP} + \text{EFTs} \rightleftharpoons \text{EFTu·EFTs} + \text{GDP} \quad (3)$$

$$\text{EFTu·EFTs} + \text{GTP} \rightleftharpoons \text{EFTu·GTP} + \text{EFTs} \quad (4)$$

$$\text{Sum: EFTu·GDP} + \text{GTP} \overset{\text{EFTs}}{\rightleftharpoons} \text{EFTu·GTP} + \text{GDP} \quad (5)$$

The stability of the complexes shown in eqns (3)–(5) made it possible to isolate each of them, and to study their interconversions and thus confirm the validity of these reactions.[14]

A major development that emphasized the importance of EFTu·GTP was the demonstration[15,16] that a ternary complex could be formed when EFTu·GTP reacted with AA–tRNA. This complex, AA–tRNA·EFTu·GTP, was an intermediate in the transfer of the AA–tRNA to the A site on the ribosome–mRNA complex.[16–18] These reactions, shown below,

$$\text{AA–tRNA} + \text{EFTu·GTP} \longrightarrow \text{AA–tRNA·EFTu·GTP} \quad (6)$$

$$\text{AA–tRNA·EFTu·GTP} \xrightarrow{\text{mRNA–Rib}} \text{AA–tRNA–mRNA–RIB} + \text{EFTu·GDP} + \text{Pi} \quad (7)$$

deserve additional comment. The specificity of eqn. (6) is indeed striking. As Lengyel and coworkers[19] first showed, all AA–tRNAs examined react with EFTu·GTP with the exception of the initiator tRNA, fMet–$tRNA_F$. This specificity insures that the initiator tRNA will not be incorporated into the polypeptide chain during the elongation process. Furthermore, deacylated tRNA[15,16] and N-blocked AA–tRNAs[16] do not react with EFTu·GTP. Of equal significance is the specificity for EFTu·GTP. EFTu·GDP cannot substitute for EFTu·GTP in the reaction with AA–tRNA. David Miller has recently used this observation to look more closely at differences between EFTu·GTP and EFTu·GDP, and there is now clear evidence that the conformation of EFTu is different depending on which nucleotide is bound to EFTu.[20]

The interaction of the ternary complex with the ribosome [eqn. (7)] is extremely complicated, for it involves not only recognition of the AA–tRNA in the complex by the proper codon on the mRNA, but also hydrolysis of GTP. GTP hydrolysis occurs at three different stages in protein synthesis and will be described in further detail below. The hydrolysis of GTP in reaction (7) has been demonstrated by several groups,[21–24] and was of immediate interest to workers in the field, for it seemed ideally suited to study the mechanism of at least one GTP hydrolytic reaction. The substrate, that is the ternary complex, is readily prepared (even with three labels: ^{14}C, ^{3}H, ^{32}P) and relatively stable. The reaction could easily be assayed using a Millipore filter technique, since two of the products, AA–tRNA–mRNA–Rib and EFTu·GDP, were retained on the filter. A most important observation was that the binding of the AA–tRNA to the ribosome and the GTP hydrolysis could be uncoupled. Using the nonhydrolyzable GTP analogue, GDPCP, AA–tRNA binding could still occur, but under these conditions EFTu was not released from the ribosome.[25] This afforded an important clue as to one of the roles of GTP hydrolysis in this reaction, namely that hydrolysis of GTP is required to allow the soluble factor (EFTu) to dissociate from the ribosome and recycle.

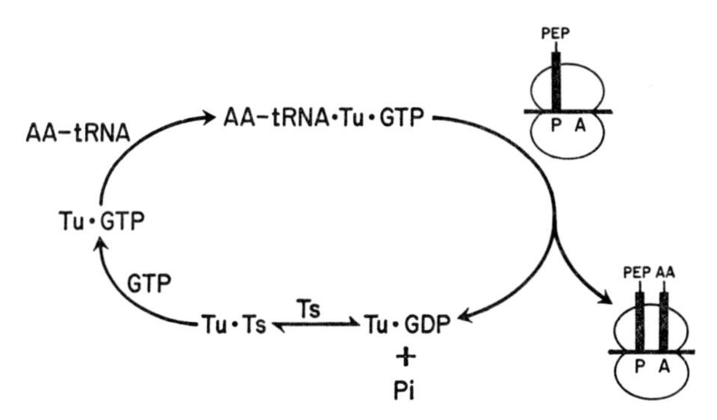

FIG. 3. Summary of the reactions involved in the binding of AA–tRNA to the ribosomes. PEP, peptide; AA, aminoacyl (From Weissbach *et al.*[26])

The ribosome-binding cycle, summarized in Fig. 3, shows the reactions leading to the formation of the ternary complex, the interaction of the complex with the ribosome with the formation of EFTu·GDP, and the reformation of EFTu·GTP via the action of EFTs.[26] I would like to say an additional word about the latter. How did one go about to prove that the only function of EFTs was to catalyze the synthesis of EFTu·GTP by the exchange reaction summarized in eqn. (5)? One approach was based on using the triphosphate-generating system phosphoenolpyruvate (PEP) plus pyruvate kinase (PK), in an attempt to convert EFTu·GDP to EFTu·GTP and thereby eliminate the requirement for EFTs. The results were very clear. For both binding of Phe–tRNA to ribosomes and polymerization of Phe–tRNA to poly-Phe, PEP and PK could replace EFTs.[27] Other studies confirmed that EFTs was required for EFTu to recycle (i.e. act catalytically). This had always been a troublesome point in the *E. coli* system for, in general, the total stimulation by EFTs was only 2-3-fold. However, when the concentration of EFTu was reduced in the incubations, greater than 10-fold stimulation by EFTs was obtained, and EFTu was shown to act catalytically.[27] Thus, there appeared to be direct evidence for the reaction cycle shown in Fig. 3.

Peptidyl transferase and translocation

There are two other steps in the elongation process, i.e. peptidyl transfer and translocation (see Fig. 1). Major progress with the peptidyl transferase system has been difficult because it has not been solubilized. The transferase is an inherent part of the larger ribosomal subunit, and there is evidence that a ribosomal subparticle containing RNA and several proteins (specifically L11) are part of the peptidyl transferase complex (see ref. 2). The peptidyl transfer does not require an exogenous energy source (such as GTP or ATP), and is markedly affected by cations. The reaction is generally assayed using the antibiotic, puromycin, since peptidyl transferase will catalyze the transfer of the peptide on the P site to puromycin to form a peptidyl–puromycin derivative.

As has been apparent from the studies described above, as well as those concerned with the initiation and termination steps, the understanding of the reactions involving soluble factors has outstripped our knowledge of the role of the ribosomal components. Yet, future progress will surely depend on a more intimate understanding of the ribosome, and over the past few years the excellent work from the laboratories of Nomura, Wittmann, Kurland and others has formed the basis for an attack on this difficult biological problem (see ref. 2). One of the most complex steps carried out by the ribosome is translocation. This process requires EFG and GTP, as shown in Fig. 1, and results in the movement of the ribosome along the mRNA by exactly three nucleotides.[28,29] Thus, these chemical reactions are somehow linked to a physical movement of the ribosome relative to the mRNA. This process must be closely regulated, if the ribosomes in a polysome are to move along the mRNA at a constant rate. Are we dealing with conformational changes on the ribosome initiated by EFG and/or GTP hydrolysis, or do some of the ribosomal proteins function in translocation by a mechanism similar to the contractile process seen in muscle? This is not yet clear and may remain one of the last aspects of protein synthesis to be solved. However, despite difficulties, there has been considerable progress in our knowledge of EFG and its interaction with the ribosome.

Nishizuka and Lipmann[3] showed that GTP hydrolysis was associated with the function of EFG (presumably accounting for the naming of the factor). This hydrolysis could occur in the absence of translocation and was dependent on the 50S subunit. The early studies on the interaction of EFG with ribosomes were carried out by several groups: Brot *et al.*,[30] Bodley and coworkers[31] and Kaziro and his colleagues.[32] It became apparent that EFG could form a relatively stable complex with ribosomes (specifically the 50S subunit) in the presence of either GTP or GDP. The data suggested that when GTP was used a 50S·EFG·GTP complex was initially formed. This complex was not isolated due to rapid hydrolysis of GTP yielding a 50S·EFG·GDP complex. Presumably, EFG can dissociate from the latter complex and act catalytically with respect to GTP hydrolysis and translocation. A major advance was made by Bodley and coworkers,[31] when they found that the steroid antibiotic, fusidic acid, which was known to be a potent inhibitor of translocation, resulted in the formation of a very stable complex that contained the 50S subunit·EFG·GDP·fusidic acid. These reactions are shown below.

$$\text{50S} + \text{EFG} + \text{GTP} \longrightarrow [\text{50S·EFG·GTP}] \tag{8}$$

$$[\text{50S·EFG·GTP}] \longrightarrow \text{50S·EFG·GDP·} + \text{Pi}$$

$$\text{50S·EFG·GDP} + \text{Fusidic} \longrightarrow \text{50S·EFG·GDP·Fusidic} \tag{10}$$

Fusidic acid inhibits EFG activity [eqn. (10)] since the stable fusidic acid complex prevents EFG from acting catalytically in translocation. A tight binding of EFG to the ribosome can also be achieved with GDPCP.[32] In this case a 50S subunit·EFG·GDPCP complex is formed. The data indicate that since hydrolysis does not occur with this analog, EFG cannot recycle and translocation is inhibited. In retrospect, the studies with fusidic acid have focused our thinking on the overall role of GTP hydrolysis in protein synthesis.

GTP hydrolysis in protein synthesis

The requirement for GTP hydrolysis in the initiation and elongation process is now well established. In every case, a soluble factor (IF2, EFTu, EFG) and the 50S subunit are required. The available evidence from three lines of investigation suggest that a common site on the larger subunit is involved in the hydrolysis of GTP in these partial reactions (see reviews 1,2). First, it is known that when EFG is bound tightly to the 50S subunit (using

ELONGATION PROCESS

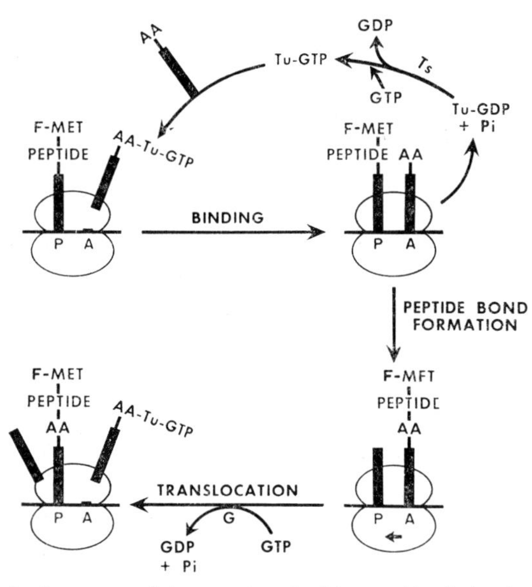

FIG. 4. Summary of the steps involved in peptide chain elongation.

fusidic acid) the hydrolysis of GTP due to EFTu is inhibited. Second, thiostrepton, an antibiotic known to react with the 50S subunit, inhibits the activity of IF2, EFTu and EFG, but is not a general 50S subunit inhibitor since the peptidyl transferase is not affected. Third, removal of ribosomal proteins L7, L12 prevents GTP hydrolysis dependent on either IF2, EFTu and EFG. It is of interest that thiostrepton also inhibits release factor (RF) activity, and RF activity requires ribosomal proteins L7, L12;[33] this suggests that termination may also involve GTP hydrolysis. This has not been shown in the prokaryote system, but termination in eukaryotes does involve GTP hydrolysis.[34] The above findings can be interpreted in the following way: the various soluble factors, present as ternary or binary complexes containing GTP, are recognized by 50S ribosomal proteins, L7, L12. These proteins may not be directly involved in GTP hydrolysis but they may position the factor GTP complex properly on the ribosome so that the GTP hydrolytic site is activated. Thiostrepton appears to prevent the factors from binding to the ribosome and may inhibit the GTPase site. GTP hydrolysis is very likely associated with conformational changes in the factors (whether IF2, EFTu or EFT) and/or ribosome, favoring dissociation of the factor from the ribosome. This is essential for the factors to recycle and function catalytically.

Thus, by the three reactions: AA–tRNA binding, peptidyl transferase and translocation (summarized in Fig. 4), the polypeptide chain is built up one amino acid at a time. During the process of elongation the N-terminal formyl group is lost and in about 50% of the proteins in *E. coli* there is also release of the *N*-terminal methionine. Finally, the protein chain is completed and a terminator codon appears on the A site of the ribosome. In the presence of specific release factors, the completed protein is then released from the ribosome. Subsequently, the ribosomal subunits are released from the mRNA and reenter their respective pools in the cell.

DEDICATION

It is a great personal honor for me to have been asked to contribute an essay in honor of Dr. Severo Ochoa's 70th birthday. His elegant studies on the genetic code and initiation of protein synthesis greatly influenced my own work. Although I had occasional discussions with him over the years because of a common interest in protein synthesis, our closest contact has occurred in the past year since he joined the Roche Institute and became a colleague in the Department of Biochemistry. His contributions to both the Department and the Institute in even this short time have been enormous. The great respect that we at the Institute have for him has only increased with time. My wish on this occasion is that Dr. Ochoa enjoy many more years of fruitful research, and that he and his lovely wife, Carmen, continue to have good health and happiness.

REFERENCES

1. HASELKORN, R. and ROTHMAN-DENES, L. B. (1973) Protein synthesis. *Ann. Rev. Biochem.* **42,** 397–438.
2. COLD SPRING HARBOR LABORATORY (1974) *Ribosomes* (M. NOMURA, A. TISSIERES and P. LENGYEL eds.), Cold Spring Harbor, N.Y.
3. NISHIZUKA, Y. and LIPMANN, F. (1966) Comparison of guanosine triphosphate split and polypeptide synthesis with a purified *E. coli* system. *Proc. Nat. Acad. Sci. U.S.A.* **55,** 212–219.
4. LUCAS-LENARD, J. and LIPMANN, F. (1966) Separation of three microbiol amino acid polymerization factors. *Proc. Nat. Acad. Sci. U.S.A.* **55,** 1562–1566.
5. RAVEL, J. M. (1967) Demonstration of a guanosine triphosphate-dependent enzymatic binding of amino-acyl-ribonucleic acid to *Escherichia coli* ribosomes. *Proc. Nat. Acad. Sci. U.S.A.* **57,** 1811–1816.
6. WEISSBACH, H., TOOHEY, J. and BARKER, H. A. (1959) Isolation and properties of B_{12} coenzymes containing benzimidazole of dimethylbenzimidazole. *Proc. Nat. Acad. Sci. U.S.A.* **45,** 521–525.
7. CLARK, B. F. C. and MARCKER, K. A. (1966) The role of *N*-formyl-methionyl-sRNA in protein biosynthesis. *J. Molec. Biol.* **17,** 394–406.
8. DICKERMAN, H. W., STEERS, E., Jr., REDFIELD, B. and WEISSBACH, H. (1967) Methionyl transmethylase. I. Purification and partial characterization. *J. Biol. Chem.* **242,** 1502–1508.
9. LUCAS-LENARD, J. and LIPMANN, F. (1967) Initiation of polyphenylalanine synthesis by *N*-acetyl-phenylalanyl-sRNA. *Proc. Nat. Acad. Sci. U.S.A.* **57,** 1050–1057.
10. NIRENBERG, M. and LEDER, P. (1964) RNA Codewords and protein synthesis. The effect of trinucleotides upon the binding of sRNA to ribosomes. *Science*, **145,** 1399–1407.
11. ERTEL, R., REDFIELD, B. and WEISSBACH, H. (1968) Role of GTP in protein synthesis: interaction of GTP with soluble transfer factors from *E. coli. Arch. Biochem. Biophys.* **128,** 331–338.
12. MILLER, D. L. and WEISSBACH, H. (1970) Studies on the purification and properties of factor Tu from *E. coli. Arch Biochem. Biophys.* **141,** 26–37.
13. HACHMANN, J., MILLER, D. L. and WEISSBACH, H. (1971) Purification of factor Ts: stability of nucleotide complexes containing transfer factor Tu. *Arch. Biochem. Biophys.* **147,** 457–466.
14. MILLER, D. L. and WEISSBACH, H. (1970) Interactions between the elongation factors: the displacement of GDP from Tu–GDP complex by factor Ts. *Biochem. Biophys. Res. Commun.* **38,** 1016–1022.
15. GORDON, J. (1967) Interaction of guanosine 5′-triphosphate with a supernatant fraction from *E. coli* and aminoacyl-sRNA. *Proc. Nat. Acad. Sci. U.S.A.* **58,** 1574–1578.
16. RAVEL, J. M., SHOREY, R. L. and SHIVE, W. (1967) Evidence for a guanine nucleotide–aminoacyl–RNA complex as an intermediate in the enzymatic transfer of aminoacyl–RNA to ribosomes. *Biochem. Biophys. Res. Commun.* **29,** 68–73.
17. LUCAS-LENARD, J. and HAENNI, A. L. (1968) Requirement of guanosine 5′-triphosphate for ribosomal binding of aminoacyl-sRNA. *Proc. Nat. Acad. Sci. U.S.A.* **59,** 554–560.
18. SKOULTCHI, A., ONO, Y., WATERSON, J. and LENGYEL, P. (1969) Peptide chain elongation. *Cold Spring Harbor Symp. Quant. Biol.* **34,** 437–454.
19. ONO, Y., SKOULTCHI, A., KLEIN, A. and LENGYEL, P. (1968) Peptide chain elongation: discrimination against the initiator transfer RNA by microbiol amino-acid polymerization factors. *Nature*, **220,** 1304–1307.
20. CRANE, L. J. and MILLER, D. L. (1974) Guanosine triphosphate and guanosine diphosphate as conformation—Determining Molecules. Differential interaction of a fluorescent probe with the guanosine nucleotide complexes of bacterial elongation factor Tu. *Biochemistry*, **13,** 933–939.

21. Shorey, R. L., Ravel, J. M., Garner, C. W. and Shive, W. (1969) Formation and properties of the aminoacyl transfer ribonucleic acid–guanosine triphosphate–protein complex. An intermediate in the binding of aminoacyl transfer ribonucleic acid to ribosomes. *J. Biol. Chem.* **244,** 4555–4564.
22. Lucas-Lenard, J., Tao, P. and Haenni, A. L. (1969) Further studies on bacterial polypeptide elongation. *Cold Spring Harbor Symp. Quant. Biol.* **34,** 455–462.
23. Gordon, J. (1969) Hydrolysis of guanosine 5′-triphosphate associated with binding of aminoacyl transfer ribonucleic acid to ribosomes. *J. Biol. Chem.* **244,** 5680–5686.
24. Ono, Y., Skoultchi, A., Waterson, J., and Lengyel, P. (1969) Peptide chain elongation: GTP cleavage catalysed by factors binding aminoacyl-transfer RNA to the ribosome. *Nature*, **222,** 645–648.
25. Haenni, A. L. and Lucas-Lenard, J. (1968) Stepwise synthesis of a tripeptide. *Proc. Nat. Acad. Sci. U.S.A.* **61,** 1363–1369.
26. Weissbach, H., Miller, D. L. and Hachmann, J. (1970) Studies on the role of factor Ts in polypeptide synthesis. *Arch. Biochem. Biophys.* **137,** 262–269.
27. Weissbach, H., Redfield, B. and Brot, N. (1971) Further studies on the role of factor Ts and Tu in protein synthesis. *Arch. Biochem. Biophys.* **144,** 224–229.
28. Erbe, R. W., Nau, M. M. and Leder, P. (1969) Translation and translocation of defined RNA messengers. *J. Molec. Biol.* **39,** 441–460.
29. Gupta, S. L., Waterson, J., Sopori, M. L., Weissman, S. M. and Lengyel, P. (1971) Movement of the ribosome along the messenger ribonucleic acid during protein synthesis. *Biochemistry*, **10,** 4410–4421.
30. Brot, N., Spears, C. and Weissbach, H. (1969) The formation of a complex containing ribosomes, transfer factor G, and a guanosine nucleotide. *Biochem. Biophys. Res. Commun.* **34,** 843–848.
31. Bodley, J. W., Zieve, F. J. and Lin, L. (1970) Studies on translocation. IV. The hydrolysis of a single round of guanosine triphosphate in the presence of fusidic acid. *J. Biol. Chem.* **245,** 5662–5667.
32. Kuriki, Y., Inoue, N. and Kaziro, Y. (1970) Formation for a complex between GTP, G factor, and ribosomes as an intermediate of ribosome-dependent GTPase reaction. *Biochim. Biophys. Acta*, **224,** 487–497.
33. Brot, N., Tate, W. P., Caskey, C. T. and Weissbach, H. (1974) The requirement for ribosomal proteins L7 and L12 in peptide chain termination. *Proc. Nat. Acad. Sci. U.S.A.* **71,** 89–92.
34. Beaudet, A. L. and Caskey, C. T. (1971) Mammalian peptide chain termination, II. Codon specificity and GTPase activity of release factor. *Proc. Nat. Acad. Sci. U.S.A.* **68,** 619–624.

TWENTY FIVE YEARS OF RESEARCH ON INHIBITORS OF PROTEIN SYNTHESIS: PROSPECTS FOR FUTURE DEVELOPMENTS IN THE SUBJECT

D. VAZQUEZ

Instituto de Biología Celular, Velázquez 144, Madrid 6, Spain

INTRODUCTION

Developments concerning selectivity, site and mode of action of translation inhibitors have been reviewed repeatedly since a specific inhibition of bacterial protein synthesis was first described in 1950. This study will be concerned mainly with a general view on the problem over the last 25 years and the prospects for future developments. A complete survey of the literature would not be possible in such a brief contribution as this and a number of reviews will be quoted when required.

SELECTIVE ACTION OF TRANSLATION INHIBITORS

Our studies on [^{14}C] chloramphenicol binding have shown that the antibiotic interacts with all classes of ribosomes of the 70S type tested, but does not bind to any of the 80S type ribosomes.(1) Similar findings were later observed with a number of antibiotics, whereas others have a wider spectrum. Since there are at least two types of systems for protein synthesis (prokaryotic and eukaryotic), their inhibitors can be broadly classified, according to their specificity, into those affecting systems of (a) the prokaryotic type, (b) the eukaryotic type and (c) both the prokaryotic and the eukaryotic types (Table 1) (refs. 2; 3, review). In general, antibiotics affecting prokaryotic-type systems are active in bacteria, blue-green algae, mitochondria and chloroplasts, whereas those acting selectively on eukaryotic-type systems are active in higher cells which are known to have 80S type ribosomes.

EARLY STUDIES WITH INTACT CELLS

It is now 25 years since specific inhibition of protein synthesis by antimicrobial agents was first reported. Specific inhibitory effects on bacterial protein synthesis by chloramphenicol and chlortetracycline, at their minimal growth inhibitory concentrations, were first described by Gale and Paine.(4,5) At those concentrations the antibiotics did not affect respiration, fermentation and amino acid accumulation(5) but caused an immediate cessation of protein synthesis and an increase in the rate of nucleic acid accumulation in bacteria.(6)

The first report concerning inhibitors of translation in eukaryotes did not appear until 1958 when the inhibitory effect of cycloheximide on protein synthesis in *Saccharomyces carlsbergensis* was described. Contrary to what was found in chloramphenicol-treated

TABLE 1. INHIBITORS OF PROTEIN SYNTHESIS
Data taken basically from our previous reviews.[3,19]

Acting on prokaryotic systems	
Althiomycin	Micrococcin
Avilamycin	Multhiomycin
Berninamycin	Negamycin
Chloramphenicol group:	Rubradirin
Chloramphenicol	Spectinomycin
D-AMP-3	Streptogramin A group:
D-Thiomycetin	Ostreogrycin G
D-Win-5094	Streptogramin A
Cloacin DF_{13} (synonym	Streptogramin B group:
bacteriocin DF_{13})	Straphylomycin S
Colicin E_3	Streptogramin B
Kasugamycin	Viridogrisein
Kirromycin	Streptomycin group:
Lincomycin group:	Gentamycin
Celesticetin	Hygromycin B
Clindamycin	Kanamycin
Lincomycin	Neomycin
Macrolide group:	Paromomycin
Angolamycin	Streptomycin
Carbomycin	Streptothricins
Erythromycin	Thermorubin
Forocidin	Thiostrepton group:
Lancamycin	Siomycin
Leucomycin	Sporangiomycin
Methymycin	Thiopeptin
Neospiramycin	Thiostrepton
Oleandomycin	Viomycin
Spiramycin	
Tylosin	

Acting on prokaryotic and eukaryotic systems	
Actinobolin	Guanylyl-methylene-
Adrenochrome	diphosphate
Aflotoxin B_2	Nucleocidin
Amicetin	Pactamycin
Aurintricarboxylic	Polydextran sulphate
acid	Polyvinyl sulphate
Blasticidin S	Puromycin
Bottromycin A_2	Pyrocatechol violet
Chartreusin	Showdomycin
Edeine A_1	Sparsomycin
Fusidic acid	Tetracycline group:
Gougerotin group:	Chlortetracycline
Bamicetin	Doxycycline
Gougerotin	Oxytetracycline
Plicamicetin	Tetracycline
	Toxylphenylalanyl-
	chloromethane

Acting on eukaryotic systems	
Abrin	Pederine
AHR-1911	Phenomycin
Anisomycin	Ricin
Diphtheria toxin	Sodium fluoride
Emetine	Tenuazonic acid
Enomycin	Trichodermin group:
Glutarimide group:	Crotocin
Actiphenol	Crotocol
Cycloheximide	Fusarenon X
Streptimidone	Nivalenol
Streptovitacin A	Toxin T-2
MDMP	Trichodermin
Narciclasine group	Trichodermol
Haemanthamine	Trichothecin
Lycorine	Verrucarin A
Narciclasine	Verrucarol
Pseudolycorine	Tubulosine
	Tylophora alkaloids:
	Cryptopleurine
	Tylocrebrine
	Tylophorine

bacteria, the stringent control of nucleic acid synthesis in yeast was not abolished by cycloheximide and an inhibition of nucleic acid synthesis was also observed in the presence of the antibiotic.[7]

The mechanism of protein synthesis remained obscure during the period 1950–1960, so that it was not possible to clarify the site of action of the known inhibitors of protein synthesis, although it was shown early that chloramphenicol does not inhibit the activation of amino acids or the formation of aminoacyl-tRNA (ref. 8, review).

THE PIONEER CELL-FREE SYSTEMS

Studies on the mode of action of translation inhibitors were facilitated when cell-free systems for translation were developed. Thus, amino acid incorporation by crude cell-free systems, directed either by endogenous mRNA attached to the ribosome preparations or by synthetic polynucleotides acting as artificial messengers, was soon observed to be inhibited by puromycin, streptomycin and chloramphenicol (ref. 8, review). Contributions by Ochoa and his coworkers were very important, showing for the first time the inhibitory effect of tetracycline in these systems,[9,10] the specific action of streptomycin on the ribosome,[11] and the lack of effect of chloramphenicol isomers deprived of antibacterial activity.[10] Furthermore, Ochoa and his coworkers showed that the extent of inhibition by chloramphenicol is highly dependent on the polynucleotide used as a messenger.[12]

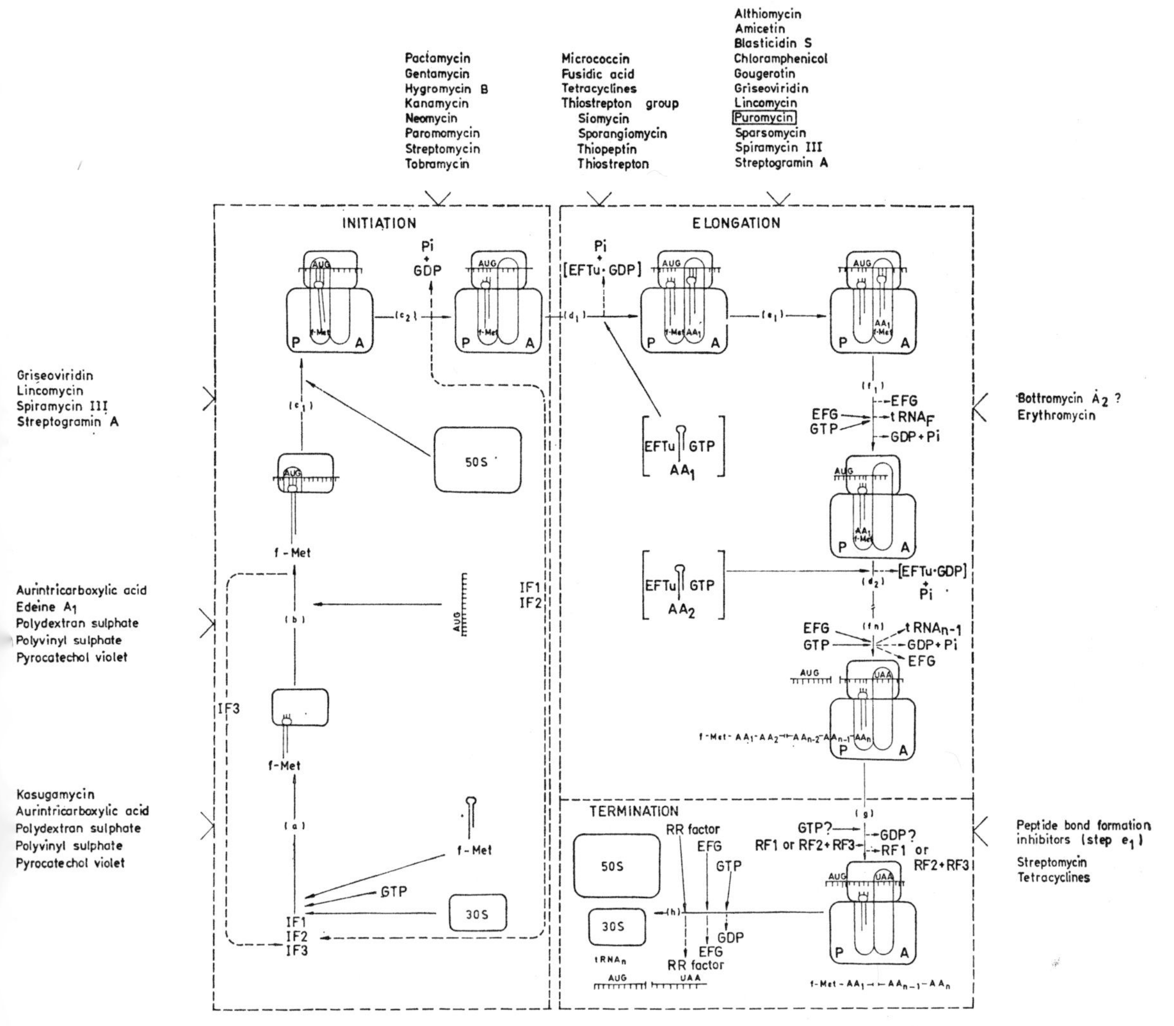

FIG. 1. Schematic representation of protein synthesis by bacterial ribosomes. Taken basically from a previous review.[3]

Abbreviations: IF1, initiation factor 1; IF2, initiation factor 2; IF3, initiation factor 3; EFTu, elongation factor Tu; EFG, elongation factor G; RF1, release factor 1; RF2, release factor 2; RF3, release factor 3; RR, ribosome release factor.

EFFECTS OF INHIBITORS ON THE STEPS OF TRANSLATION

The overall reactions taking place in the process of translation by prokaryotic and eukaryotic ribosomes are shown schematically in Figs. 1 and 2, based on the translocation model with one single entry site on the small subunit (refs. 13–15, reviews; see also contributions in this volume). Following this model, the process of translation can be arbitrarily divided into three phases: initiation, elongation and termination (Figs. 1 and 2). Our knowledge of protein biosynthesis and the specific effects of the various inhibitors has progressed in parallel over the last 10 years. The most recent developments concerning the

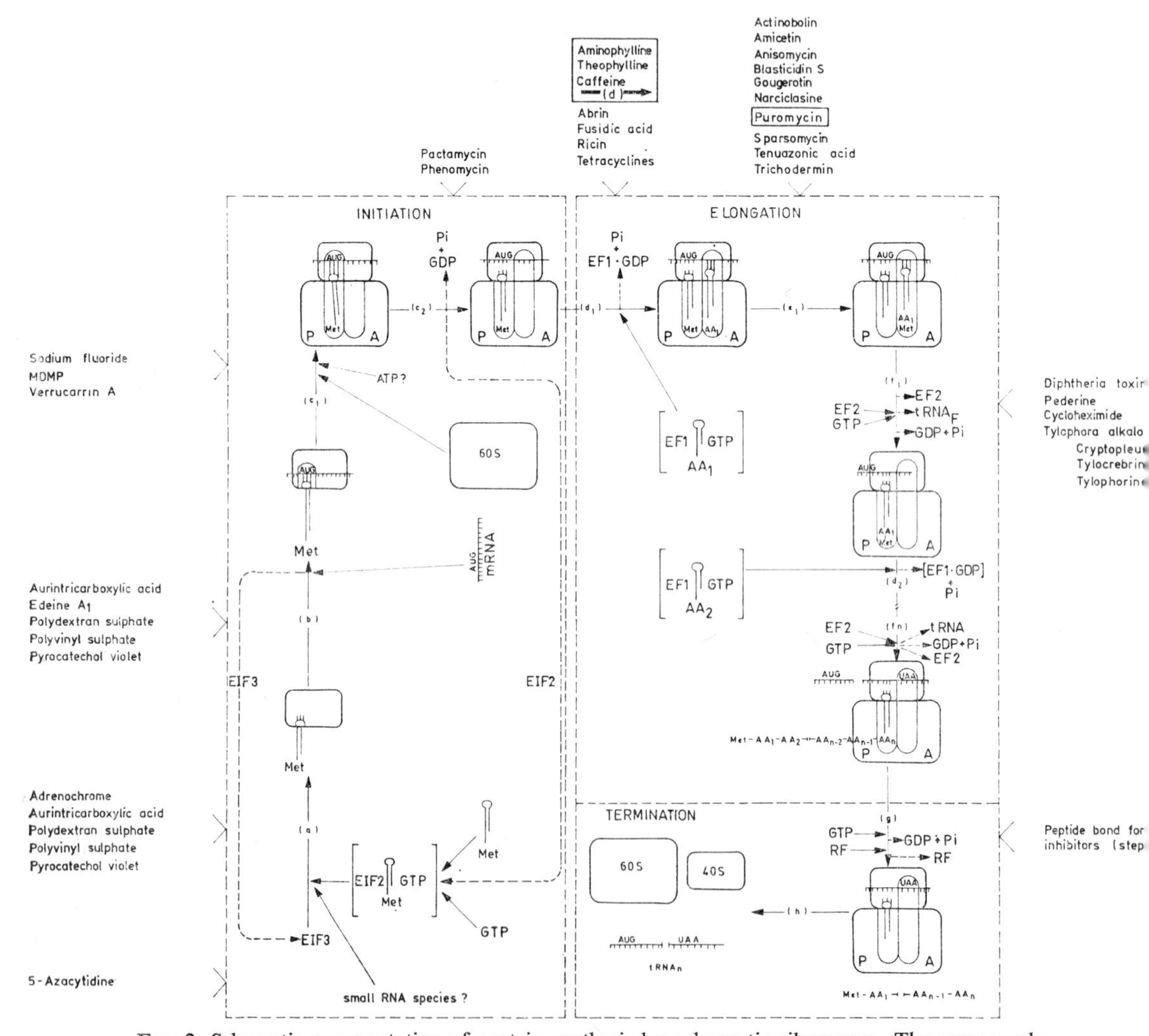

FIG. 2. Schematic representation of protein synthesis by eukaryotic ribosomes. The compounds indicated above the sign—(d) → enhance step (d) in translation.[19]

Abbreviations: EIF2, eukaryotic initiation factor 2; EIF3, eukaryotic initiation factor 3; EF1, elongation factor 1; EF2, elongation factor 2; RF, release factor.

site and mode of action of translation inhibitors have been reviewed recently.[3,16–19] Consequently we will refer only briefly to these topics, shown schematically in Figs. 1 and 2.

Initiation

In the initiation phase we have to distinguish three steps: (a) recognition of initiation factors and binding of the initiator (f-Met–tRNA$_F$ in prokaryotes and Met–tRNA$_F$ in eukaryotes), (b) recognition of the mRNA initiation triplet and (c) joining of the complex to the larger ribosome subunit (Figs. 1 and 2). This step (c) involves the reactions (c_1) in which the larger subunit is attached to the initiation complex and (c_2) in which the 3′ end of the initiator binds to the donor-site of the peptidyl–transferase center and the anticodon interaction with the small subunit is interrupted, leaving the site open for the entry of aminoacyl–tRNA to start the elongation phase. It is widely accepted that the initiation sequence in eukaryotic systems follows the steps (a) → (b) → (c) (Fig. 2), whereas it is still a matter of controversy in bacterial systems whether the step (a) takes place before step (b) or vice versa. For the purpose of clarity we have assumed here that a similar sequence of reactions takes place in the prokaryotic and the eukaryotic systems (Figs. 1 and 2). The initiation phase and role of initiation factors in eukaryotic systems is not well resolved; it is likely that at least two factors EIF2 and EIF3 are required in eukaryotes with similar functions to IF2 and IF3 respectively in bacteria (Figs. 1 and 2). Inhibitors of the initiation phase (steps (a), (b) and (c)) interact with the small ribosome subunit. The sites affected in the initiation phase by aurintricarboxylic acid, streptomycin, edeine A_1, kasugamycin, sodium fluoride, the herbicide MDMP and pactamycin are most probably as indicated in Figs. 1 and 2. A number of inhibitors (streptogramin A, lincomycin, spiramycin III, verrucarin A) appear to block initiation, probably at step (c_1), since they cause polysome breakdown in intact cells, but this mode of action has not been demonstrated in cell-free systems when the individual steps were studied sequentially.

Elongation

The second phase is composed of repeated cycles of elongation. Steps (d) EFTu-(or EFI)-dependent aminoacyl–tRNA binding to the acceptor site (A-site) of the ribosome, (e) peptide bond formation and (f) translocation, can be distinguished in each elongation cycle (Figs. 1 and 2). In the translation process there are as many elongation cycles as there are peptide bonds to be formed in order to synthesize the protein. There is evidence for a single entry site in the small subunit, and therefore the anticodon interaction of either f-Met–tRNA$_F$ or Met–tRNA$_F$ or peptidyl–tRNA in the small subunit probably has to be interrupted for the codon–anticodon interaction of the incoming aminoacyl–tRNA to take place (step (d)). The tetracycline antibiotics were early found to prevent non-enzymic binding of aminoacyl–tRNA to bacterial ribosomes; a greater inhibitory effect of these antibiotics on aminoacyl–tRNA binding was observed when the enzymic assay was developed. A number of the inhibitors of aminoacyl–tRNA binding (step (d)) (abrin, ricin and antibiotics of the thiostrepton group) act on the larger ribosome subunit but it is not clear whether the multiple interactions of the tetracyclines with the smaller or the larger subunit are more relevant to their mode of action. Genetic studies have shown that the EFG factor is altered in bacterial mutants resistant to fusidic acid; however, the inhibitory effect of the antibiotic in translation is due to the formation of a stable EFG–(or EF2)–GDP–ribosome complex on the larger ribosome subunit which does not allow the interaction of EFTu to take place. Recent reports have shown that the antibiotic kirromycin acts directly on the elongation factor EFTu inducing a kirromycin- and EFTu-dependent GTP hydrolysis uncoupled from aminoacyl-tRNA binding. Specific inhibition by abrin and ricin of enzymic binding of aminoacyl–tRNA to 80S type ribosomes has been reported recently.

Contrary to early expectations, the peptide bond formation step (e) does not require any specific supernatant factor and is catalysed by the peptidyl transferase center, which is an integral part of the larger subunit in prokaryotic[20] as well as eukaryotic ribosomes.[21] There is an ample group of compounds which inhibit peptide bond formation (step (e)) in the process of translation. All of them interact with the larger ribosome subunit. Particularly relevant is the antibiotic puromycin which was early postulated as an analog of the 3′ terminal end of aminoacyl–tRNA. Indeed a radioactive puromycin derivative was observed to bind to the NH_2-end of synthesizing peptides. Furthermore, it has been observed that {^3H} puromycin binding to the acceptor site of the ribosome is inhibited by aminoacyl–adenosine which is the 3′ terminal end of aminoacyl–tRNA.[22] However, puromycin lacks that part of the aminoacyl–tRNA molecule responsible for interaction with the template and the small ribosomal subunit. Because of this, the use of puromycin provides a simplified method for the study of peptide bond formation in a reaction where the α-NH_2 group of puromycin becomes linked to the C-terminal end of f-Met or the peptidyl group ("puromycin reaction"). The product of the puromycin reaction is unable to take part in the next step of protein synthesis. However, all the evidence indicates that the formation of a peptide bond between puromycin and the f-Met or peptidyl group takes place by the same mechanism as peptide bond formation in protein synthesis. The terminal fragments CACCA–, AACCA–, ACCA– and CCA–Met-f from f-Met–$tRNA_F$ undergo a ribosome catalysed reaction with puromycin to yield f-Met–puromycin in a simplified system known as the "fragment reaction".[20,21] Studies on the puromycin and fragment reactions have clearly demonstrated that the peptidyl transferase which catalyses peptide bond formation is integrated into the structure of the larger ribosome subunit. Hence the puromycin and fragment reactions are suitable assays for inhibitors of peptide bond formation. Studies in intact cells and cell-free systems have shown the inhibitory effect of actinobolin, althiomycin, amicetin, anisomycin, blasticidin S, chloramphenicol, gougerotin, griseoviridin, lincomycin, sparsomycin, spiramycin, streptogramin A, tenuazonic acid and trichodermin on peptide bond formation. Some of these inhibitors of peptide bond formation (lincomycin, streptogramin A, griseoviridin and spiramycin) interfere with the interaction of the 3′ terminal end of the peptidyl tRNA with the donor site causing polysome breakdown in intact cells and thus might also be considered as inhibitors of step (c_1) in initiation. The alkaloid narciclasine has recently been described as a specific inhibitor of peptide bond formation by eukaryotic ribosomes; it is the only non-antibiotic compound known to block this step of protein synthesis.

The translocation step (f) requires one of the elongation factors (EFG in bacterial and EF2 in eukaryotic systems) and involves movement of the peptidyl–tRNA from the A-site to the P-site, coupled with GTP hydrolysis, and release of the uncharged tRNA and the elongation factor EFG or EF2. Erythromycin blocks the translocation step (f_1) in intact bacteria and some bacterial cell-free systems (Fig. 2). This inhibitory effect on translocation was not observed in some bacterial cell-free systems since erythromycin appears to block specifically the release of the uncharged tRNA bound to the P-site of the ribosome after peptide bond formation. Spectinomycin is also an inhibitor of translocation in bacterial systems, but while erythromycin binds to the 50S subunit, spectinomycin acts on the 30S subunit. There are a number of inhibitors of translocation which specifically act on eukaryotic systems. Of these inhibitors cycloheximide and pederine act on the larger ribosome subunit whereas the *Tylophora* alkaloids (cryptopleurine, tylocrebrine and tylophorine) appear to act on the small ribosome subunit; on the other hand, the diphtheria toxin acts enzymically on the elongation factor EF2 forming, in the presence of NAD, ADP–ribosyl–EF2 which binds to the larger ribosome subunit but is inactive for translocation.

Termination

At least two steps (g) and (h) can be distinguished in the termination phase. Step (g) is initiated with the recognition of the chain-terminating codon (nonsense codon) (UAA, UAG or UGA) and the release factors (RF1 or RF2 + RF3 in bacteria and RF in eukaryotes); the ribosomal peptidyl transferase cleaves the bond between the peptidyl and the tRNA moieties. A further step (h) has recently been resolved in bacterial systems which requires a ribosome release factor (RR factor), the elongation factor EFG and GTP to release the uncharged tRNA from the ribosome (Fig. 1, step (h)).[23] It is likely that a similar step also occurs in eukaryotes (Fig. 2, step (h)) although it has not yet been resolved. Inhibitors of step (d) of elongation also inhibit step (g) of termination by blocking either the release factor(s)-dependent codon recognition or the interaction of the release factor(s) with the ribosome. Peptidyl transferase is widely accepted as being involved not only in peptide bond formation (step (e)) but also in the peptidyl–tRNA hydrolysis reaction required for the termination phase (step (g)). Thus all the peptidyl transferase inhibitors which have been tested block step (g) of termination in bacterial and in mammalian systems with the same selectivity shown in peptide bond formation.

PRESENT AND FUTURE OF RESEARCH ON INHIBITORS OF TRANSLATION

The main objectives in the present and future research on inhibitors of translation are oriented along three lines: (a) to locate the specific molecules of the ribosomes interacting with the inhibitors, (b) to study the changes of conformation of ribosomes and polysomes interacting with the inhibitors and (c) to characterize further the selective action of some inhibitors.

The specific molecules of the ribosome which are the target for the inhibitors have been elucidated only in very few cases (ref. 24, review). This is because (a) most of the inhibitors interact reversibly with the ribosome and (b) the different individual RNA and protein molecules of the ribosome are devoid by themselves of any testable biological activity, unless they are integrated with other ribosomal components in a more complex structure. However, in a number of cases it has been possible to elucidate the ribosome molecules which are (a) involved in resistance to certain antibiotics of (b) covalently linked to the modified antibiotics of (c) required for the inhibitory activity of the inhibitors or (d) irreversibly inactivated by the inhibitors. This was established by genetic evidence or by affinity labeling experiments with modified antibiotics or by experiments on the reconstitution of ribosomes or by studies with ribosomes pretreated with antibodies or chemical reagents. Thus a mutation in protein S12 confers resistance to streptomycin, whereas proteins S3 and S5 are required for dihydrostreptomycin binding and a streptomycin analog binds irreversibly to protein S4. On the other hand, resistance to spectinomycin is due to an alteration of protein S5. Resistance to kasugamycin in *E. coli* is due to a lack of methylation of two adjacent adenine residues in the 16S RNA of the 30S ribosome subunit. Inactivation of the 30S ribosomal subunit in intact bacteria by colicin E_3 is due to a specific cleavage of the 16S RNA near the 3′ terminal. A similar action was proposed for cloacin DF_{13}. As for inhibitors acting on the larger ribosome subunit, reconstitution and affinity labeling experiments suggested that protein L16 is somehow invovled in chloramphenicol binding; however, in other affinity labeling experiments with a chloramphenicol analog the compound was associated with proteins L2 and L27. Some affinity labeling experiments with a puromycin derivative involve protein L16 whereas the 23S RNA is involved when a different puromycin derivative is used. Ribosomal mutants resistant to spiramycin and lincomycin were described in which proteins L4 and L6 are affected respectively. On the other hand, changes in either proteins L4 or L22 confer resistance to erythromycin. Furthermore, induced and constitutive resistance to the antibiotic has been associated with a modification of 23S RNA in the larger ribosome subunit in which formation of dimethyladenine has been observed (refs. 3, 24, reviews). The examples illustrated above clearly show that the results obtained are not always in agreement concerning the specific ribosome molecules interacting with the inhibitors, since the results obtained depend on the experimental approach or even the experimental conditions. Therefore the above studies have to be continued in prokaryotic ribosomes and initiated with eukaryotic ribosomes to advance our knowledge on the specific target of inhibitors on the ribosome.

Not much is known about the conformational changes caused in the ribosome by its interaction with the inhibitors and this topic will obviously be the subject of future research. There are, however, recent data suggesting important conformational changes in ribosomes in the process of translation. Thus erythromycin,[25] streptogramin A,[26] streptogramin B[26] and some other antibiotics interact with washed bacterial ribosomes but not with polysomes. Furthermore, anisomycin shows a differential affinity on interaction with yeast ribosomes depending on their state within the ribosomal cycle; thus the affinity of the antibiotic for ribosomes differs in (a) washed and run-off ribosomes and (b) polysomes with the peptidyl–tRNA bound to the A- or the P-site.[27] Interestingly enough, anisomycin affinity for the polysomes is increased when the peptidyl–tRNA bound to the A-site is translocated to the P-site, which is precisely the position in which the antibiotic should enter to block peptide bond formation.[27]

Despite the broad classification of inhibitors of translation presented in Table 1, there are some cases in which certain antibiotics appear to have narrower spectra of selectivity. Thus the macrolide antibiotics, lincomycin, streptogramin A and streptogramin B are very active on ribosomes from Gram-positive bacteria but have a lower affinity for ribosomes from Gram-negative bacteria. Furthermore, tenuazonic acid inhibits protein synthesis by mammalian but not by yeast ribosomes, cycloheximide is inactive on ribosomes of wild type *Saccharomyces fragilis*, fusidic acid is inactive in *Neurospora crassa* mitochondria and in protein synthesizing systems from sporulating *Bacillus subtilis* (ref. 3, review) and whether or not erythromycin and lincomycin inhibit protein synthesis by rat liver mitochondria is the subject of conflicting reports (ref. 3, review). On the other hand, a wider spectrum of activity than that indicated in Table 1 has been observed in a few cases; thus sensitivity of brain mitochondrial ribosomes to emetine has been reported (ref. 28 and references therein) and yeast mutants having cytoplasmic ribosomes (80S type) sensitive to streptomycin have also been described (ref. 29 and references therein). These few examples illustrate the limited but important differences in sensitivity to inhibitors of ribosomes which are not expressed in Table 1. Furthermore, ribosomes, initiation factors and elongation factors might play an important role in the translational control of protein synthesis in eukaryotic and transformed cells and it is likely that sensitivity to inhibitors of protein synthesis is highly dependent on the state or phase of the cells; therefore inhibitors of protein synthesis might be important tools to study ribosome diversity and control of translation in higher cells. Indeed a preferential inhibition of protein synthesis by diphtheria toxin in malignant cells has been reported.[30] This finding should be confirmed and extended with other inhibitors of translation.

REFERENCES

1. Vazquez, D. (1964) Uptake and binding of chloramphenicol by sensitive and resistant organisms. *Nature*, **203**, 257–260.
2. Vazquez, D. and Monro, R. E. (1967) Effects of some inhibitors of protein synthesis on the binding of aminoacyl-tRNA to ribosomal subunits. *Biochim. Biophys. Acta*, **142**, 155–173.
3. Vazquez, D. (1974) Inhibitors of protein synthesis. *FEBS Letters*, **40**, supplement, 63–84.
4. Gale, E. F. and Paine, T. F. (1950) Glutamic acid accumulation and protein synthesis in *Staphylococcus aureus*. *Biochem. J.* **47**, xxvi.
5. Gale, E. F. and Paine, T. F. (1951) The assimilation of aminoacids by bacteria. 12. The action of inhibitors and antibiotics on the accumulation of free glutamic acid and the formation of combined glutamate in *Staphylococcus aureus*. *Biochem. J.* **48**, 298–301.
6. Gale, E. F. and Folkes, J. P. (1953) The assimilation of amino-acids by bacteria. 15. Actions of antibiotics on nucleic acid and protein synthesis in *Staphylococcus aureus*. *Biochem. J.* **53**, 493–498.
7. Kerridge, D. (1958) The effect of actidione and other antifungal agents on nucleic acid and protein synthesis in *Saccharomyces carlsbergensis*. *J. Gen. Microbiol.* **19**, 497–506.
8. Gale, E. F. (1963) Mechanisms of antibiotic action. *Pharmacol. Rev.* **15**, 481–530.
9. Rendi, R. and Ochoa, S. (1961) Enzyme specificity in activation and transfer of amino acids to ribonucleoprotein particles. *Science*, **133**, 1367.
10. Rendi, R. and Ochoa, S. (1962) Effect of chloramphenicol on protein synthesis in cell-free preparations of *Escherichia coli*. *J. Biol. Chem.* **237**, 3711–3713.
11. Speyer, J. F., Lengyel, P. and Basilio, C. (1962) Ribosomal localization of streptomycin sensitivity. *Proc. Nat. Acad. Sci.* **48**, 684–686.
12. Speyer, J. F., Lengyel, P., Basilio, C., Wahba, A. J., Gardner, R. S. and Ochoa, S. (1963) Synthetic polynucleotides and the amino acid code. *Cold Spring Harb. Symp. Quant. Biol.* **28**, 559–567.
13. Lucas-Lenard, J. and Lipmann, F. (1971) Protein biosynthesis. *Ann. Rev. Biochem.* **40**, 409–448.
14. Haselkorn, R. and Rothman-Denes, L. B. (1973) Protein synthesis. *Ann. Rev. Biochem.* **42**, 397–437.
15. Modolell, J. and Vazquez, D. (1975) In *MTP International Review on Science*. Biochemistry: Vol. 7. *Synthesis of Amino Acids and Proteins*, pp. 137–178 (H. R. V. Arnstein, ed.) Medical and Technical Publishing Co., Oxford.
16. Pestka, S. (1971) Inhibitors of ribosome functions. *Ann. Rev. Microbiol.* **25**, 487–562.
17. Gale, E. F., Cundliffe, E., Reynolds, P. E., Richmond, M. H. and Waring, M. J. (1972) In *The Molecular Basis of Antibiotic Action*, pp. 278–379, John Wiley & Sons, London.
18. Kaji, A. (1973) In *Progress in Molecular and Subcellular Biology* (F. E., Hahn, ed.), vol. 3, pp. 85–158, Springer-Verlag, Berlin–Heidelberg-New York.
19. Carrasco, L., Fernandez-Puentes, C. and Vazquez, D. (1976) Antibiotics and compounds affecting translation by eukaryotic ribosomes. Specific enhancement of aminoacyl-tRNA binding by methylxanthines. *Molec. Cel. Biochem.* **10**, 97–122.
20. Monro, R. E., Staehelin, T., Celma, M. L. and Vazquez, D. (1969) The peptidyl transferase activity of ribosomes. *Cold Spring Harb. Symp. Quant. Biol.* **34**, 357–366.

21. VAZQUEZ, D., BATTANER, E., NETH, R., HELLER, G. and MONRO, R. E. (1969) The function of 80S ribosomal subunits and effects of some antibiotics. *Cold Spring Harb. Symp. Quant. Biol.* **34**, 369–375.
22. FERNANDEZ-MUÑOZ, R. and VAZQUEZ, D. (1973) Binding of puromycin to *E. coli* ribosomes. Effects of puromycin analogues and peptide bond formation inhibitors. *Molec. Biol. Reports*, **1**, 27- 32.
23. HIRASHIMA, A. and KAJI, A. (1972) Factor-dependent release of ribosomes from messenger RNA. Requirements for two heat-stable factors. *J. Molec. Biol.* **65**, 43–58.
24. PONGS, O., NIERHAUS, K. H., ERDMANN, V. A. and WITTMANN, H. G. (1974) Active sites in *Escherichia coli* ribosomes. *FEBS Letters* **40**, supplement, 28–37.
25. PESTKA, S. (1974) Antibiotics as probes of ribosomes structure: Binding of chloramphenicol and erythromycin to polyribosomes; effects of other antibiotics. *Antimicrobial Ag. Chemother.* **5**, 255–267.
26. ENNIS, H. L. (1974) Binding of the antibiotic vernamycin B to *Escherichia coli* ribosomes. *Arch. Biochem. Biophys.* **160**, 394–401.
27. BARBACID, M. and VAZQUEZ, D. (1975) Ribosome changes during translation. *J. Molec. Biol.* **93**, 449–463.
28. IBRAHIM, N. G., BURKE, J. P. and BEATTIE, D. S. (1974) The sensitivity of rat liver and yeast mitochondrial ribosomes to inhibitors of protein synthesis. *J. Biol. Chem.* **249**, 6806–6811.
29. BAYLISS, F. T. and INGRAHAM, J. L. (1974) Mutation in *Saccharomyces cerevisiae* conferring streptomycin and cold sensitivity by affecting ribosome formation and function. *J. Bacteriol.* **118**, 319–328.
30. IGLEWSKI, B. H. and RITTENBERG, M. B. (1974) Selective toxicity of diphtheria toxin for malignant cells. *Proc. Nat. Acad. Sci.* **71**, 2707–2710.

VI. CELL BIOLOGY AND NEUROBIOLOGY

A BIOCHEMIST LOOKS AT THE NERVOUS SYSTEM

J. FOLCH-PI

Department of Biological Chemistry, Harvard Medical School, Boston, Mass 02115 and McLean Hospital, Belmont, Mass 02178

The core of this essay will be a brief discussion of neurochemistry, as a way of explaining its emergence as a semi-independent discipline within the broad field of modern biochemistry. The subject matter will be presented against an autobiographical background, not only because this is the way it was learned, but because it will permit to dwell briefly on the process of becoming a scientist in Spain, a point relevant to the goal of this meeting since my own experiences may well parallel many incidents in Ochoa's life, and their relation may help in the understanding of the process by which Ochoa evolved as he did.

Born in Barcelona in 1911, I followed the regular Spanish curriculum, except for 4 years of attendance at French schools, and finished by obtaining a medical degree in February 1933. Throughout my medical studies I spent part-time as a clinical clerk, and part-time assisting in biochemical research. Upon graduation and a short period of medical practice I returned to biochemical research. This eventually led to my going to the Rockefeller Institute Hospital to work with Dr. Van Slyke in 1935. Once there, the Spanish Civil War, among other factors, transformed an expected stay of 2 years into a life-long career of research which drifted rapidly from biochemistry into neurochemistry. Finally, after 9 years at the Rockefeller Institute, I moved to Harvard Medical School and McLean Hospital, where I was able to pursue neurochemical research with ample means.

It is difficult for me to trace the development of my interest in biomedical research. My family's background was humanistic, not scientific. My medical training was the usual training of physicians in Spain. The curriculum made more demands on memory than on deductive reasoning. Within such a framework, the quality of instruction depended very much on the qualities of the individual instructors. Thus many courses were from bad to mediocre, and a few were outstanding. The basic medical sciences were taught with little exposure to laboratory work. By comparison, the clinical courses were much more challenging especially because during the 6 years of medical studies, students could work as clinical clerks from the first year on. The system was doubly beneficial because it provided free labor to the understaffed teaching hospitals, and it gave the students the opportunity of complementing their course work with autodidactic learning. In an inadequate milieu one must realize the extent to which the interested student can learn, and does learn autodidactically. Autodidactism has a poor press in well-run university environments, but, after all, pioneers have often been autodidactic, to a greater or lesser degree. In autodidactism, the systematic doubt becomes a systematic question. It has the advantage of avoiding learning by rote, and the disadvantage of replacing guidance and constructive criticism from teachers, by the slow and uncertain process of uninformed exhaustive search (as opposed to selective search) and of time-consuming critical examination of unlikely alternatives.

In my case this autodidactism was facilitated by my attendance at the Instituto de Fisiologia of the Medical School. This was an independent Institute supported by provincial funds and directed by Professor August Pi-Sunyer, a physiologist of some international reputation, and a man of foresight who had introduced teaching by modern methods in his course of Physiology. The Institute consisted of some laboratories and a well-run physiological and biochemical library. The staff of the Institute consisted of young men who had been trained abroad and who could distinguish good research from routine laboratory work. I assisted Dr. Cesar Pi-Sunyer, who had worked for 2 years with Carl Neuberg. This experience exposed me to good methodology and to the proper designing of experiments. In addition, the excellent library of the Institute was a tremendous opportunity to learn autodidactically.

At that time, Spain had changed its monarchy into a republic in a bloodless revolution of which we were all very proud. Also, the University of Barcelona had been given complete autonomy, with great improvement in teaching methods, and with the opening of research positions for young people. This raised the possibility of full-time research, a consideration which was cardinal in my decision to embark on a research career.

Many people should be mentioned to whom I am greatly indebted for help in those formative years. Since the list would be too long, and it would be of little interest to most readers, I will only mention here, as a major influence in my decision to pursue research as a career, the name of Francisco Duran-Reynals, who was then on the staff of the Rockefeller Institute, and who arranged the invitation for me to spend some time at the Rockefeller Institute Hospital, with Dr. Donald D. Van Slyke.

At the Rockefeller Institute I assisted in a program of study of the physiopathology of the pituitary hormones on which Irvine Page was engaged. My own contribution was supposed to be the study of changes in blood lipids in various pituitary disorders. In practice the work consisted in setting up quantitative methods for analysis of lipids. It soon became apparent that these methods left much to be desired both quantitatively and qualitatively, because on the one hand what was estimated included non-lipid substances, and on the other, the identification of the various lipid fractions was only approximate. The study of these difficulties required a scrutiny of the basis of the procedure, and the preparation of reliable standards. The standards had to be prepared from brain tissue, and these preparations led eventually to the discovery of phosphatidyl serine and the phosphoinositides. This was my first contribution to brain chemistry, and the beginning of a life-long dedication to neurochemistry. The work on extraction of lipids led eventually to the development of the procedure of extraction which is now in general use.

Once awakened, my interest in brain chemistry grew very fast. From my perusal of the available literature it was apparent that whereas much work on brain chemistry had been carried out during the nineteenth century and up to about 1920, little research had been going on during the 1920s and early 1930s. It was also apparent that there was a rebirth of interest, and that the field of brain chemistry might well be at the beginning of a very active period. The relative inactivity during the period that was coming to an end might well have been the consequence not only of the lack of certain technical requirements but of the special features of brain tissue which eventually set its study as an autonomous discipline.

The special features were that brain exhibited a degree of morphological, functional, and chemical heterogeneity that made each one of its areas different from all other areas.

Hence the findings obtained on one area could not automatically be applied to other areas, and for each area, its chemical study had to be carried out with full consideration of the concomitant morphological and functional characteristics. Brain was a hierarchy of different structures each of which had its own specific role and complicated interrelationships with the others, the result being that the whole functioned as an integrated unit. This unit, the central nervous system, lacked a lymphatic system; it had an abundant blood supply relatively impervious to most vasomotor reactions, a high oxygen consumption, a chemical composition different from other tissues, and a specialized metabolism. Finally there appeared to be some sort of mechanism which was referred to as "the blood–brain barrier" and which regulated the exchange of substances between the brain and the rest of the body thus permitting it to attain a high degree of homeostasis.

The various areas of the brain shared the common features of being constituted by neurons and neuroglia. Neurons were the conducting elements. They belonged to a few basic types but the variations within each type were almost infinite. Neuroglia was referred to as the supportive element of the tissue, a description which was a tacit admission of our little knowledge as to the exact function of these cellules.

In the absence of a lymphatic system the exchange of substances between the central nervous system and the rest of the body had to take place either at the level of the capillary wall, or between the brain parenchyma and the cerebrospinal fluid. The latter route was quantitatively negligible, although in certain cases it might have great qualitative importance. This left the capillary wall as the route through which nearly all material exchange between the nervous system and the rest of the body took place. This exchange seldom obeyed the laws of simple diffusion. Instead, this exchange appeared to be facilitated or hampered for different substances, a state of affairs that was described as being due to the existence of a "blood–brain barrier".

The occurrence of this type of regulated exchange was possibly the main single factor that made it necessary to consider the brain as a unit, in spite of the different functions of its constituent parts. The knowledge of that so-called barrier went back more than 50 years. It was first observed in the study of the behavior of aniline dyes, of bacterial toxins, and of other foreign elements, and was, therefore, for a long time of more importance to the bacteriologist and to the student of permeability than to the biochemist. It was only in recent years that the bearing of the blood–brain barrier on brain biochemistry had been recognized. By the use of isotopic tracers it had been shown that the blood–brain barrier, which was already known to hinder the passage across the capillary walls into the brain of acid aniline dyes and of certain bacterial toxins was also active in preventing the free diffusion of ions and of uncharged small molecule metabolites. It had been found that sodium phosphate, potassium, and chlorine ions, insulin, lactic acid and pyruvic acid—to mention only a few of an ever-increasing number of compounds of all descriptions—diffused into brain tissue from the blood-stream at a much slower rate than that at which they diffused from the blood-stream into other tissues. No satisfactory explanation of the mechanism of action of the blood-brain barrier had yet been offered.

From the rich vascularization of brain and from the high arteriovenous difference in oxygen content, it had been assumed that brain had a high oxygen consumption. Just about that time the elegant method of Kety for the measurement of blood-flow through the brain permitted exact measurements of blood-flow and of oxygen consumption. They showed that in the adult human, about one-fifth of the blood leaving the heart flowed through the brain and that although the brain represented only about $2\frac{1}{2}\%$ of body weight, it accounted for about one-fourth of the oxygen consumed by the whole body under basal conditions. This high level of oxygen consumption appeared to be absolutely necessary to the functioning of the brain, since deprivation of oxygen for a few minutes resulted in death. The consumption of oxygen remained the same in sleep or mental activity and was not changed in mental disease, an observation that should not be interpreted as indicating that brain activity did not increase its oxygen consumption but only that variations in oxygen consumption of discrete areas failed to show changes in the total oxygen consumption of the organ.

The main, if not only, fuel that brain used to subserve its high energy requirements was glucose and the arteriovenous differences in oxygen and glucose absorbed and CO_2 produced by brain corresponded to the formula for complete combustion of glucose. This relatively enormous use of glucose did not require insulin.

Brain tissue had a special chemical composition. In the adult it contained more lipids than proteins and little glycogen. Little was known about the proteins, except for the recent findings of an active metabolism of nucleoproteins, presumably the underlying mechanism of chromatolysis. Lipids belonged to the structural, not the metabolic reserve pool, and an adult animal starved to death showed the same amount of brain lipids as its litter mate that had been fed properly. Much work had been done for several generations on the chemistry of brain lipids since they were often quite different from those found in other tissues. Their function was a puzzle; they obviously acted as insulators in myelin and they provided anionic charges to the acid-base equilibrium of the tissue. Beyond that, they could only be supposed to play an active if as yet undefined role in the functioning of nervous membranes.

The special features found in the adult brain were simplicity itself when compared with the continuous changes that the tissue exhibited during its development. The consideration of these changes was obviously of great importance, because the brain was already known not to reach its adult pattern until several years after birth. It was obvious that it was during these years of developmental changes that the material basis of personality, of individuation, was being laid down.

The briefest way to describe the developmental changes of the brain was to treat separately the histological, chemical and physiological changes. From the histological point of view, the first type of mature cells to appear in large numbers were the neurons which in man already had reached their adult numbers before birth. The non-neuronal elements appeared later, or, rather they did not reach until later the numbers in which they were found in the adult tissue. The astrocytes followed in their appearance an increase in vascularization, and they increased in numbers until the progress of vascularization was complete. Myelination was the dominant feature in the late stages of development. It was accompanied by an increase in oligodendroglia, a fact quite in keeping with the discovery, some years later, that oligodendroglia were responsible for the deposition of myelin.

From the chemical point of view, the blood–brain barrier did not appear until the late stages of development, at the onset of function. Lipids were absent from the neural tube. They appeared fairly early in fetal life and their concentration in brain tissue increased steadily until the composition found in adult tissue was reached. It was already known that the appearance and increase in amounts of lipids followed fairly rigid timetables, which changed from species to species, but were fairly consistent in each species. Various lipids appeared at different times and accumulated at different rates, reflecting not only the lipid composition of different cells but especially the massive deposition of myelin. There were incompletely known changes in enzyme systems. Oxygen consumption per unit weight of fresh tissue increased steadily during fetal and early post-fetal development until a maximum was reached, to be followed by a slight decrease with age. It was already known that during fetal life, the glycolysis was very important and that with growth it lost its predominant role to oxidative metabolism. The gradual increase in dependence on oxidative metabolism corresponded to a dramatic decrease in the resistance of the nervous system to anoxia; a newborn mouse or rat could survive 30 minutes in an atmosphere of nitrogen, whereas the adult animal of either species died from deprivation of oxygen for a few minutes. Similarly, the human fetus during child-birth could be subject to 15 or 20 minutes of complete deprivation of oxygen, whereas the human adult died from complete deprivation of oxygen for a few minutes.

The high oxygen consumption of brain led to a corollary worth some consideration. This was that brain developed ahead of the rest of the body and thus represented in the early stages of development a much larger proportion of the total body weight than it did in the adult. In man, at birth, the brain weighed one-seventh of the total body weight. At the age of 1 year, brain weight still amounted to about one-eleventh of the body weight, and this proportion was maintained for the next 2 or 3 years. In practice this meant that even with the smaller oxygen consumption per unit weight of the immature brain, the oxygen used by this organ represented at birth more than one-half of the total body oxygen consumption, and even at 4 years one-half of the oxygen absorbed by the body was used by the brain. The main point to be made from the foregoing summary is that in the first years of life, when the material basis of individuation was being laid down, the brain occupied a central position in the energy balance of the whole organism. This made imperative a consideration of the consequences of starvation on brain development. At that time, the evidence on this point was confusing. On the one hand, it had been found that curtailment of food intake in rats and mice for periods after birth resulted in an actual decrease in the thickness of the brain cortex. On the other hand, it has already been quoted that two dogs of the same litter, one starved to death and the other one fed *ad lib*, appeared to have the same brain weight, i.e. in the case of these dogs brain development was impervious to starvation, whereas in the rat even partial starvation decreased the thickness of the brain cortex.

The physiological changes in brain tissue in the course of development could be summarized by saying that any given area of the central nervous system remained functionally inactive until the onset of function. The onset of function for the different areas took place according to a rigid timetable. For each area, this onset of function was the culmination of a series of morphological, chemical and physiological events. Morphologically, myelination had reached completion with an enormous increase in the amount of lipids present; there had been an enormous increase in oligodendroglia, in vascularization with a parallel rapid increase in the number of astrocytes. Biochemically the oxygen consumption had increased considerably, and the blood–brain barrier had become operative. Physiologically the onset of function was accompanied by the onset of spontaneous electrical activity. The gradual accretion of electrical activity of the different areas resulted for the whole organ in the appearance of the electroencephalogram (EEG) and its gradual elaboration until the adult EEG was established.

The foregoing summary shows that although our neurochemical knowledge at the time was meager, it pointed to many lines of work, the study of which would be exciting and rewarding. This proved to be the case, and the challenge of neurochemistry attracted an ever-increasing number of young scientists. The handful of biochemists who were interested in the brain in the late 1930s has grown to a large contingent of neurochemists, as evidenced by the creation of neurochemical societies, and by the success of their annual meetings. The appeal and growth of neurochemistry has been paralleled by other neurological disciplines. As a result an ever-expanding field of neurosciences has emerged as one of the most active areas in biomedical research.

Any attempt to depict even the most salient results of the present explosion in neurochemical research would be outside the aims of this short essay. To illustrate the progress made, a few lines of work can be mentioned briefly. The application of electromicroscopy and of X-ray diffraction to brain tissue has greatly increased our knowledge of the ultrastructure of nervous tissue.

We had thought the brain constituents to be very stable biochemically. Instead, it has been shown that brain constituents turn over actively, some of them with unusual rapidity. We thought of neurons mainly as conductive elements. We know now that they are true chemical factories. The discovery and intensive study of axoplasmic flow shows axons to be transport chains that carry substances in both directions and at different speeds. The study of neurotransmitters, their synthesis and catabolism, and their mechanism of action has become a discipline in itself. The study of the effects of starvation has established that brain can use ketone bodies as fuel and that starvation may have serious consequences on the development of the brain and on its eventual performance in the adult. The application of the concepts and techniques of genetics has resulted in the recognition of the occurrence of an increasing number of inborn defects of metabolism; the basic enzyme defect has been established in some cases, and with it the possibility of diagnosis *in utero* and the recognition of the corresponding homozygotes and heterozygotes among the population. It is rewarding to consider how some basic studies have resulted in alleviation of disease.

As a final comment, the evolution and progress of neurochemistry in particular and of the neurosciences in general in the last 30 years is a defense and a justification of the "free initiative" or "open market" way of carrying out scientific research. The decision of many young scientists to go into neuroscience research was a free personal decision. The success of the whole program is obvious. It is doubtful that a directed program in neuroscience research to which young graduates would have been assigned would have been half as successful.

BRAIN DAMAGE AND PERINATAL BIOCHEMISTRY

F. MAYOR

Departamento de Bioquímica y Biología Molecular, Facultad de Ciencias,
Universidad Autónoma de Madrid, Spain

From the time of my first letter to Professor Severo Ochoa in 1956 asking for his advice concerning some biochemical studies, to the inauguration of the Center of Molecular Biology honouring him on the occasion of his seventieth birthday, the lives of Severo and Carmen Ochoa have been an enriching lesson as well as an example, which I among most of the biochemists from Spain will never forget.

I received his answer immediately and I began to work on the metabolism of 4-aminobutyrate. Besides its neurohormonal action, this non-essential amino acid constitutes the fundamental step of a very interesting by-pass of the 2-oxoglutarate dehydrogenase system, a limiting reaction of the tricarboxylic acid cycle, as shown in the scheme of Fig. 1. The 4-aminobutyrate system is widespread in microorganisms, seeds, etc., and it exists also in mammals where it occurs *only in the brain* due to the specific location of the enzyme glutamate decarboxylase. The 4-aminobutyrate system is a good example of the metabolic singularity of the different parts of the body and also of the necessity of always keeping in mind the functioning of *the whole* living organism, with all its complex links between the various pieces of the metabolic machinery. As A. E. Garrod said in 1908: "The conception of metabolism in block is giving place to that of metabolism in compartments . . . ". At present, from the individual step, enzyme, compound or molecule it seems advisable to come back to the "metabolism in block". It happens quite often that we become so used to dividing reality with the aim of a better analysis that in doing so we risk losing the general view of the complete living organism in its integrated context. And, especially, we are prone to forget that man and his diseases constitute the ultimate target and beneficiary of our research plans and activities.

It is true that in biochemistry "no difference exists between a lettuce and a king", and all living organisms have the same fundamental components, the same energetic resources and the same pathways. Therefore, we must look to the *differential* reactions and compounds which can explain the physiological characteristics of a given organ or organism. In this sense if we say physiological, we also mean pathological as well. Looking for and finding such characteristic differences provides the essential key for the understanding of diseases, as well as special situations and events like birth. I always recall the advice of Professor Krebs: "Search for and keep in mind the great profiles". Sir Hans is still working hard for a better understanding of the main problems of physiology. During the year I had the occasion and the honour of working with him in Oxford (1966–7) I learned many things and also I obtained some results, some evidence and even some questions without an answer (at that moment). And one of the things I realized then was the integration of all perinatal physiological and biochemical profiles!

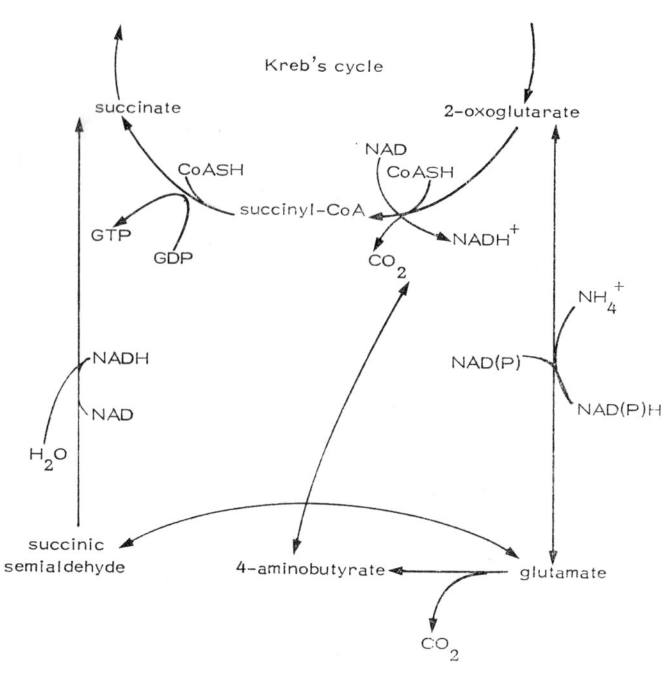

FIG. 1. 4-Aminobutyrate by-pass system.

In the laboratory of Professor Krebs I became aware of the trick which has permitted him—and surely all great minds—to do so much good work: the choice of collaborators. Among them was Derek Williamson, an outstanding person and biochemist from whom I learned many things. I am grateful to have had his help and advice then, as I have now.

Among the review articles on perinatal biochemistry, those of Olga Greengard deserve particular consideration.[1,2] She has discussed in a brilliant manner the main facets of enzymogenesis, providing us with excellent insights about the links between morphological, functional and enzymatic aspects of development in different organs (intestines, liver, kidney, brain). Information concerning perinatal hormonal regulation has been collected in a review I prepared recently.[3]

The work done and reviewed by Greengard is very well known and, of course, does not need repetition. Thus, *I shall confine myself to some aspects of perinatal biochemistry which seem to have a more direct relation with brain damage.*

I will first comment briefly on a vividly important and peculiar aspect of neonatal biochemistry, that of inborn errors of metabolism. The bibliography on inborn errors has been extremely large in the last years[4–6] and I will not attempt to summarize this increasingly interesting material. I should only like to present some brief reflections about mass screening and research on inborn alterations before focusing our attention on the most relevant features of perinatal biochemistry.

INBORN ERRORS OF METABOLISM

The implementation of genetic screening has been controversial in many countries because it represents one of the frontiers of preventive medicine. And it is quite understandable that decisions in this area are taken with reluctance because they do not form a part of the "classic health problems" and need a certain level of specialized knowledge, as well as a good measure of confidence in the specialists. Thus, it has been said, "physicians are far from ready to participate in mass genetic screening". In fact, declares Barton Childs of the John Hopkins University School of Medicine, "they are generally ignorant of its aims and uses".[7] Anyway, there is no doubt that voluntary genetic screening will ultimately be instituted on a routine basis in many countries. Research in this field will provide methods progressively more accurate, simpler, and within the reach of more people. More diseases will be detected and prevented. Even amniocentesis will be performed in pregnant women over the age of 36 whenever necessary. As knowledge in this field increases, programs of genetic screening will be widely applied.

The screening of newborns for phenylketonuria has reduced the incidence of mental retardation caused by the disease. In Spain, we started the mass screening for inborn errors in the province of Granada in 1968. The following substances were determined in the urine collected on a strip of filter paper 15 days after birth and dried at room temperature: glucose, galactose, protein, cysteine-homocystine, histidine, tyrosine and *o*-hydroxyphenylacetic acid (phenylalanine derivative). Since then several cases of PKU have been detected and prevented, and different kinds of aberrations studied. It has not been an easy task to persuade the health authorities, but I think that in 2 years such screening programs will be widespread throughout the country.

The results obtained using the screening techniques in mentally retarded children prompted us to try to study some inborn errors in "animal models". Experimental phenylketonuria, for example, can be simulated by overloading with phenylalanine or by inhibiting the phenylalanine hydroxylase with *p*-chlorophenylalanine. We have used both *p*-chlorophenylalanine and esculin (to block the dihydropteridine reductase reaction), a treatment which leads to a complete inhibition of the enzyme[8,9] (Fig. 2).

An area of child health which has been greatly neglected is that of biochemical profiling for the determination of the physiological status of the so-called normal newborn. Adequate quantitative multiple tests for the normal metabolites must be developed so that dysfunctions in normal metabolic pathways can be detected at birth or during childhood and their consequences prevented.[10]

It is of importance to have analytical data on the population at different ages. In the north of Granada province, for example, the incidence of rickets in about 40% of the children (3 months–1 year) was discovered before any clinical symptoms appeared, by means of the simple chromatographic method used for the screening of inborn errors. Generalized aminoaciduria was completely prevented with vitamin D and a less restricted and uniform diet.

Animal experiments and human studies suggest that a critical period of brain growth exists during which malnutrition even in a mild form and for a short period produces irreversible damage. The critical period is before birth and during early postnatal life.[10]

The cost of mass screening for PKU and some other related diseases makes it possible at present to do its determination in all countries where the mentally retarded citizens receive institutional and medical or social care from the State. In fact, from the economic point of view, it can be considered as "good business". I have been told by pediatricians and gynecologists that "it will be far more useful to avoid fetal brain damage produced by delivery trauma (hypoxia, forceps, etc.) by means of a better obstetric care than to spend money in such complicated methods for diagnosing infrequent diseases". About

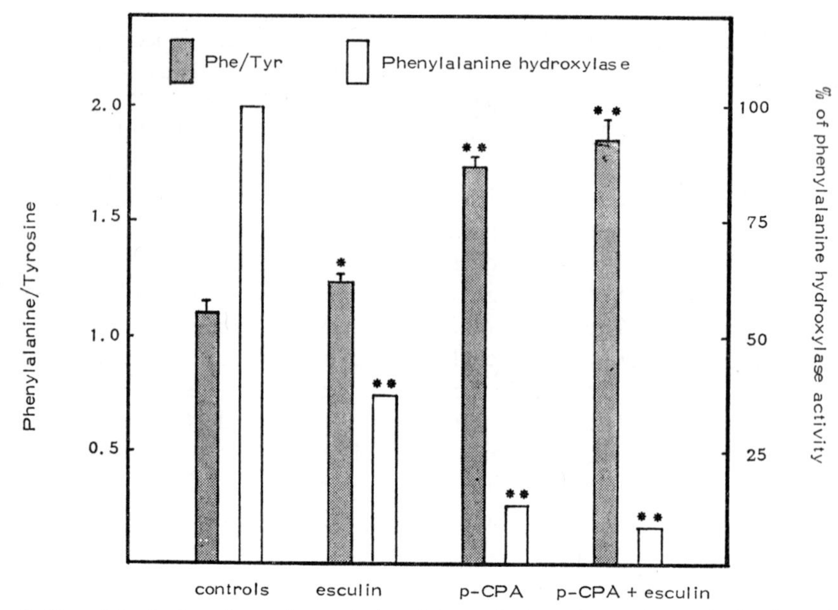

FIG. 2. Phenylalanine/tyrosine ratios and phenylalanine hydroxylase activities in the liver of rats treated with *p*-CPA and/or esculin. (Mean values ± SEM of four to six experiments.) $*p < 0.05$. $**p < 0.001$.

60% of mental retardation in children appears to be due to hypoxia or other trauma at the moment of birth. I am aware of these arguments: what happens is that *both* types of health care must be favored and supported by society. We could say that birth deserves at least as much attention as death (I think that it is quite reasonable because they are undoubtedly the most important moments of life). A considerable number of diseases may be prevented—and a better knowledge of the physiology, biochemistry and pathology of birth will be gained. Such studies will be a very enrichening experience, especially in view of the very poor knowledge we have at present about birth and the even poorer application of this knowledge.

PERINATAL BIOCHEMISTRY: MAIN EVENTS

There is one aspect I should like to emphasize: the irreversibility of the damage produced in the brain *during* the first months of life or the first months of fetal development. It is important to remark that the cause of the damage may not later produce the same kind of permanent lesion or disability caused earlier. Thus, it seems reasonable to assume that during these periods the transport systems between blood–brain and mother–fetus are non-existent or less effective (i.e. the enzymes of the transport systems are still "immature") in regulating the *transport* of substances across the membranes.

Let us consider very briefly the example of phenylketonuria: if during the first months phenylalanine or some of its derivatives enter the brain in higher levels than usual, the damage produced in the brain is not reversible. It may be taken into account that one of the most important characteristics of the brain cells is that *they are not renewed.* The brain of many mammals has at birth over 90% of the adult cell numbers; in the rat the proportion is 94–97%.[11] Environmental influences have their greatest initial effects not on cell numbers but on cell components and cell relationships. Therefore, if the neuronal system is altered at this time the "error" will remain for life.

Together—or alternatively—with the occurrence of chronological differences in transport, resulting in the presence in the brain of substances which are not normal or are found in higher amounts than usual, the existence of specific metabolic pathways in the neonate could also explain its "transient sensitivity". In fact, we know of some reactions, very significant in the neonatal period, which soon become irrelevant or even pathological. For example, ketosis is physiological in the newborn and the ketone bodies produced in the liver constitute an important fuel for energy production, because the brain possesses the enzyme 3-oxoacid CoA transferase. Liver lacks this enzyme. On the other hand adults have some metabolic pathways or enzyme activities not found in the newborn: in the neonatal period, galactose—a very important nutrient because it is a component of the milk sugar lactose and of sphingomyelins, of special significance in the developing brain—is metabolized through pathway A, Fig. 3. In the inborn error known as galactosemia the uridyltranferase is lacking and the accumulation of the phosphorylated derivatives produces a very severe disease. However, galactose can be metabolized to the same end product through pathway B, which appears later (delayed maturation).

It may be emphasized that in perinatal biochemistry there are some events that occur "physiologically" during a very restricted period of time. This is especially true in the moment of birth and during the first days *or even hours*, when the capacity for adaptation to autonomous life is more dramatically required and expressed: for example, the concentration of triglycerides in human blood is 40.02 ± 18.64 mg/100 ml at birth and 62.12 ± 25.63 after 6 hours.[12]

Birth and weaning each represent a very drastic environmental change in the life of a mammal and these two events would be expected to be associated with numerous alterations in enzyme patterns. Significantly, the enzymatic events at birth occur in two clusters,

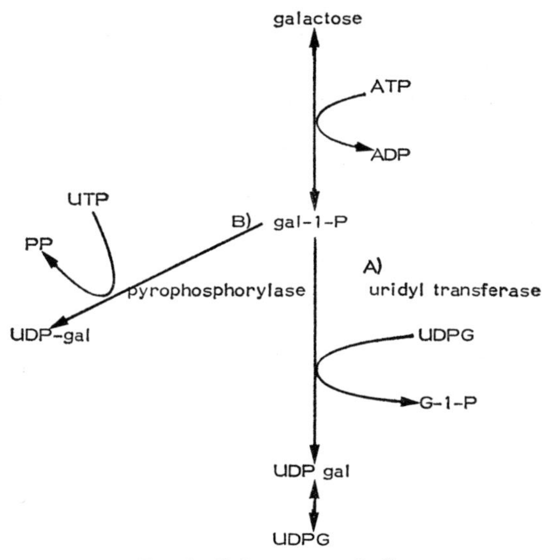

FIG. 3. Galactose metabolism.

the late fetal and the neonatal, suggesting that survival requires the readiness of some functions at the moment of birth while others can develop in the course of the next day.[1]

Enzymatic differentiation is a hormone-regulated process. Some enzyme activities have been expressed prematurely by means of hormone administration. As Greengard has pointed out, it seems reasonable to assume that the hormones that evoke the enzymes experimentally constitute the natural stimuli for the normal developmental formation of these enzymes.[1]

Birth represents a clear turning-point from carbohydrate to lipid metabolism: all the glycogen is dramatically exhausted at birth and in consequence the neonate shows hypoglycemia and lacticacidemia. The continuous supply of maternal glucose is replaced during the suckling period by an intermittent uptake of milk, a diet very rich in fat. Therefore, the transition from mother-dependent uptake to autonomous life involves a sudden metabolic shift.

The most relevant metabolic features of the newborn which I should like to emphasize in this presentation are:

(i) *Increase of gluconeogenesis* in accordance with the low levels of glucose and high amounts of lactate produced by glycolysis in the hypoxic period of birth (from the time of interruption of the mother's oxygen supply to the beginning of breathing). I will briefly discuss below the importance of *hypoxia* during birth, because if the limits of energy production by glycolysis are exceeded, the brain may be seriously and irreversibly damaged.

Concerning gluconeogenesis, we have found[13] that when the livers of well-fed rats are perfused in the presence of low concentrations of glucose, the activity of the phosphoenolpyruvate carboxykinase (PEPCK) increases significantly. This increase is not related to the glycogen content, and seems to be the result of *de novo* synthesis of the enzyme since it is prevented by prior administration of cycloheximide or actinomycin D.

PEPCK activity is not increased in the presence of low concentrations of circulating glucose when imidazole (an activator of phosphodiesterase) is added to the perfusion medium. It is interesting to note that the concentration of cyclic AMP in the liver increases when that of glucose in the medium is low. Therefore, in the absence of hormonal factors, the regulation of PEPCK may be accomplished by glucose itself and the mediator in this regulation, as in hormonal regulation, seems to be cyclic AMP.

As stated by Greengard, prior to the appearance of recognized endocrine glands, specific metabolic products of one area of the embryo may similarly act to influence specific enzyme synthesis in neighbouring areas. The consequence of the interactions of new stimuli with new receptors is the ever-increasing morphological and functional varieties of cell types in the embryo.[1]

As already mentioned, hypoxia may be pronounced at the moment of birth and may result in very serious and permanent pathological alterations. In the absence of oxygen, glycolysis becomes an "emergency" source of energy. The inefficiency of glycolysis, as Krebs[17] has pointed out in his review on the Pasteur effect, is minimized in the animal body because the lactate formed can eventually be utilized, either directly as a fuel, or after resynthesis to glucose in liver or kidney. Striated muscle, intestinal mucosa, renal medulla, erythrocytes, *foetal tissues during parturition*, malignant tumors and retina obtain energy by glycolysis, because the *in vivo* oxygenation of these tissues is more or less limited. Other tissues, notably the heart *and central nervous system*, are strictly dependent on respiration.[15] Two minutes after birth the concentration of lactate in the liver of rats is 5.1 mM (the normal value for adult rats is 1.0 mM). As reported by Krebs[14] the glycogen reserves (10% of the wet weight of the liver!) and the high glycolytic capacity "have an important function in asphyxia when the infant is separated from the parental sources of energy at birth. The fact that human babies can survive anaerobiosis for 30 minutes is no doubt connected with the high glycolytic capacity of the new-born".

The primary site of the Pasteur effect is phosphofructokinase. When this enzyme is inhibited, fructose 6-phosphate and glucose 6-phosphate accumulate. As Alberto Sols points out in his contribution to this volume, the Pasteur effect involves the integration of a sequential series of feed-back mechanisms. However, the isoenzymes present at birth are the less regulated forms. Glycolysis is, therefore, the mechanism which supplies energy during parturition to tide over a period of hypoxia. If this period is exceeded the brain may be seriously damaged, due to a direct effect (i.e. accumulation of intermediates) or indirect effects or both. It should be remembered that the conversion of pyruvate to acetyl-CoA and the subsequent events leading to lipid formation depend on the availability of oxygen.

The formation of lactate represents the primary mechanism for the regeneration of NAD in glycolysis, but there are other reactions leading to NAD from NADH and some of them provide simultaneously intermediates for synthetic processes (Fig. 4).

In yeast grown under hypoxic conditions succinate accumulates[15] and the Krebs cycle is split in two branches, one diverging from oxaloacetate to succinate with simultaneous NAD recovery and the other one from oxaloacetate to glutamate, in which the NADP-linked isocitrate dehydrogenase produces the substrates needed for the NADP-linked glutamate dehydrogenase, namely 2-oxoglutarate and NADPH.[17,18]

(ii) *Ketone body production and utilization.* The concentration of ketone bodies in the blood increases rapidly after birth, remains high during the suckling period and falls after weaning to reach normal adult levels by 30 days of age (Table 1). At 5 days of age the concentration of 3-hydroxybutyrate in the blood is 10 times higher and that of aceto-acetate more than 3 times higher than that in the adult blood. Thus, the neonate may be described as being in a physiological state of ketosis as a direct result of the high rates of utilization and oxidation of fatty acids.[19]

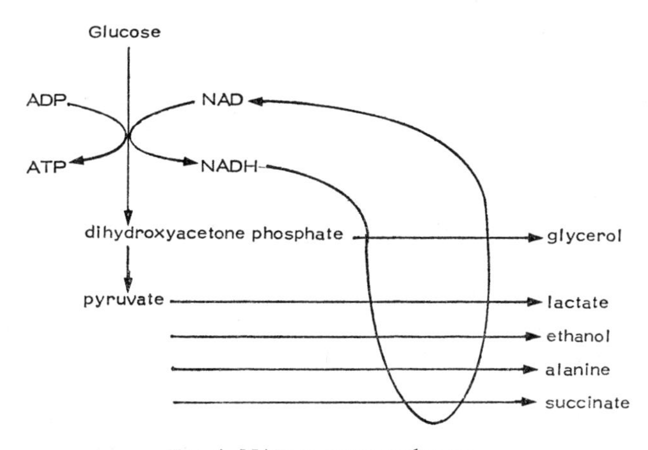

FIG. 4. NAD recovery pathways.

TABLE 1. KETONE BODIES IN RAT BLOOD DURING DEVELOPMENT[10]

Age group	Total ketone bodies
5 day	1.53
10 day	1.10
20 day	0.47
Adult, fed	0.25

The values are expressed as μmol/ml whole blood.

The utilization of ketone bodies by the brain is not limited to the neonatal period: Smith *et al.*[20] reported a six-fold increase of the activity of 3-hydroxybutyrate dehydrogenase in rat brain in starvation. Unlike the hyperketonaemia of starvation, the hyperketonaemia of suckling is accompanied by normal blood glucose concentrations[21] and it is thus possible that in the brain the ketone bodies serve another function (i.e. myelinization) apart from serving as a fuel for respiration and the sparing of glucose.

Hyperketonemia in human neonates and infants is well established and during pregnancy humans often have elevated ketone body concentrations, especially in the third trimester. Thus, ketone bodies are available to human brain during the fetal and neonatal periods. It has been demonstrated that the ketone bodies are transported across the placenta as rapidly as glucose. Accordingly the fetal liver does not have hydroxybutyrate dehydrogenase activity. However, 3-oxoacid CoA transferase, acetoacetyl-CoA thiolase and 3-hydroxybutyrate dehydrogenase activity have been found in fetal brain. These activities can be altered by the nutritional status of the mother. Starvation of mothers for 5 days (16–21 in the pregnant rat) results in a doubling of the 3-hydroxybutyrate dehydrogenase activity on delivery. Similarly, feeding a 45% fat diet to pregnant rats between the 10th and 20th day of gestation causes a 60% increase in the activities of brain acetoacetyl-CoA thiolase and 100% increase in 3-oxoacid CoA transferase at birth.[19,22] It is important to remark that in the adult brain neither starvation nor fat-feeding have any effect on these activities. A case of neonatal ketoacidosis has been described in which there was an absolute deficiency of 3-oxoacid CoA transferase in brain, kidney and muscle.[23] Acetoacetyl-CoA thiolase occupies a central role in hepatic ketogenesis and in extrahepatic acetoacetate oxidation; in the former it sets the acetoacetyl-CoA concentration while in the latter it displaces the unfavorable equilibrium of the transferase reaction (3-oxoacid CoA transferase).[24]

The existence in the brain cytoplasm of acetoacetyl-CoA synthetase and acetoacetyl-CoA thiolase makes unnecessary the existence of a transmitochondrial membrane transport of acetyl CoA for lipid synthesis. We have found[25] a clear inhibitory effect on the activity of the thiolase by several metabolites of phenylalanine. Also the hydroxybutyrate dehydrogenase of the suckling rat brain is inhibited by these metabolites. These effects are interesting in relation to the pathophysiology of mental retardation in phenylketonuria, in which the concentration of such metabolites are considerably increased. It is worth noting that this inhibition can influence the rate of synthesis of fatty acids and cholesterol (precursors of myelin), especially during brain development.

(iii) *Myelinization.* Gangliosides undergo a rather rapid metabolic turnover in the early neonatal period and thereafter this rate is much reduced.[26] The increase of gangliosides coincides with the multiplication of the synaptic connections which takes place 10 to 12

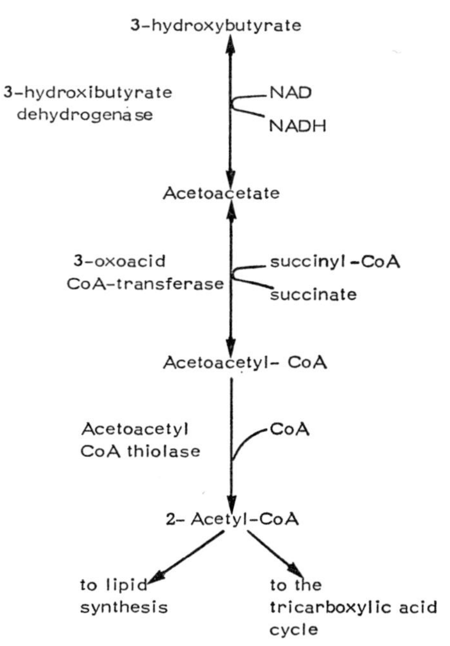

FIG. 5. Utilization of ketone bodies in rat brain mitochondria.

days after birth (in the rat). The main constituents of myelin are galactolipids and proteolipids. Apart from its dependence on the lipid metabolism already mentioned, the formation of myelin is directly related to the metabolism of carbohydrates (α-glycerophosphate) and the supply of amino acids.

Concerning transport of amino acids across the brain–blood barrier, the possible role of the Meister cycle[17] seems very attractive: in the γ-glutamyl cycle, glutathione acts as the "amino acid carrier" and the concentration of glutathione has been found to be significantly higher in the newborn brain than in the adult.

> The knowledge we have is too scanty. Many aspects of perinatal biochemistry remain to be discovered and developed. Though the outlines are clear, the surface has only been scratched. Each enzyme presents its unique problems of chemical control and physiological role and these are not the same in each one of the various tissues. Analysis of the genesis and function of each enzyme in each tissue and each species is a monumental task but one that can be broken into feasible bite-size jobs, some more urgent than others. Integration of these many individual observations can be expected to have medical applications in the very young. However, the general and theoretical implications of these results may even be more important. The study of differentiation, perhaps more than other endeavours, will provide insight into the mechanism of gene control, the metabolic roles of enzymes in particular tissues and perhaps even into that converse process of dedifferentiation of tissues into neoplasia.[1]

As I said at the beginning, I have far more to forecast than to recall: not only the production but the disappearance of some enzymes must be considered as well as the existence of different active forms of the enzymes at different stages of development; attention must be paid to the endocrine regulation of enzyme onset and activity;[28] activation and inactivation, fine and coarse regulation through the differentiation processes; prenatal treatments for the promotion of adequate levels of enzyme activities at birth; scientific bases for a better care of pregnancy and delivery. "There are immediate prospects for enhancing normal human maturation in order to overcome problems arising with prematurity or delayed development during late gestation".[2]

No matter what the type of brain damage produced by an inborn error like phenylketonuria or by hypoxia, it is not reversible and frequently very similar in its symptoms: does this mean that the lesions (i.e. biochemical alteration) produced are also very similar or even unique?

It is necessary to go deeper into perinatal biochemistry and, urgently, in the study of the moment of birth. As I mentioned above, more than 50% of the cases of mentally retarded children could be prevented and this represents a dramatic incentive for a more thorough investigation.

Certainly, perinatal biochemistry is an exciting and recently born subject. A long gestation has been necessary until the development of physiology and biochemistry made the birth possible some years ago. I infer from the aspects pinpointed above that we may now conclude that the newborn is growing well and rapidly, and that it will undoubtedly have a magnificent future.

REFERENCES

1. Greengard, O. (1971) Enzymatic differentiation in mammalian tissues. *Essays in Biochemistry*, **7**, 159.
2. Greengard, O. (1973) Introduction of *Biochemical Bases of the Development of Physiological Functions*, **15**, 1–6.
3. Mayor, F. (1975) *Regulación Hormonal Perinatal*, Simposio Real Academia de Ciencias, Madrid.
4. Bondy, P. K. and Rosenberg, L. E. (Ed.) (1969) *Diseases of Metabolism*, W. B. Saunders & Co., Philadelphia.
5. Stanbury, J. B., Wyngaarden, J. B. and Fredricjson, D. S. (Eds.) (1972) *The Metabolic Bases of Inherited Diseases*, McGraw-Hill, New York.
6. Hommes, F. A. and van den Berg, C. J. (Ed.) (1973) *Inborn Errors of Metabolism*, Academic Press, New York.
7. *Science News*, June 28, 1975, p. 413, Washington, D.C., U.S.A.
8. Gimenez, C., Valdivieso, F. and Mayor, F. (1974) *Biochemical Medicine*, **11**, 1, 81.
9. Valdivieso, F., Gimenez, C. and Mayor, F. (1975) *Biochemical Medicine*, **12**, 72.
10. Ambrose, J. A. (1974) *Child, Health and the Layman*, U.S. Department of Health, Education and Welfare, Center for Disease Control, Atlanta, Georgia, U.S.A.
11. McIlwain, H. (1971) Types of metabolic adaption in the brain. *Essays in Biochemistry*, **7**, 127.
12. Altarriba, O. (1975) Estudios del metabolismo lipídico en el recién nacido. Thesis of the University of Barcelona, Spain.
13. Moreno, F., Sanchez-Urrutia, L., Medina, J. M., Sanchez-Medina, F. and Mayor, F. (1975) *Bioch. J.* **150**, 51.
14. Krebs, H. A. (1972) The Pasteur effect and the relations between respiration and fermentation. *Essays in Biochemistry*, **8**, 1.
15. Hochachka, P. W. and Somero, G. W. (1973) The influence of oxygen availability. In *Strategies of Biochemical Adaptation*, W. B. Saunders & Co. Philadelphia.
16. Lupiañez, J. A., Machado, A., Nuñez de Castro, I. and Mayor, F. (1974) *Molec. Cell. Biochem.* **3**, 2, 113.
17. Nuñez de Castro, I., Arias-Saavedra, J. M., Machado, A. and Mayor, F. (1974) *Molec. Cell. Biochem.* **3** 2, 109.
18. Machado, A., Nuñez de Castro, I. and Mayor, F. (1975) *Molec. Cell. Biochem.* **6**/2, 93.
19. Bailey, E. and Lockwood, E. A. (1973) Some aspects of fatty acid oxydation and ketone body formation and utilization during development of the rat. *Enzyme*, **15**, 1–6, 239.
20. Smith, A. L., Satterthwaite, H. S. and Sokoloff, L. (1969) *Science*, **163**, 79.
21. Page, M. A., Krebs, H. A. and Williamson, D. H. (1971) *Biochem. J.* **121**, 49.
22. Williamson, D. H. and Buckley, B. M. (1973) The role cf ketone bodies in brain development. In *Inborn Errors of Metabolism* (F. A. Hommes and C. I. van den Berg, ed.), Academic Press, London.
23. Tildon, J. T. and Cornblath, M. (1972) *J. Clin. Invest.* **51**, 493.

24. Williamson, D. H. and Hems, R. (1970) Metabolism and function of ketone bodies. In *Essays in Cell Metabolism* (Bartley, Kornberg and Quayle, eds.), John Wiley & Sons, London.
25. Mayor, F., Benavides, J., Gimenez, C. and Valdivieso, F. (1975) Fifth International Meeting of the International Society for Neurochemistry, Barcelona, Spain, 2–6 September 1975, p. 267.
26. Brady, R. O. (1969) in *Diseases of Metabolism* (P. K. Bondy, and L. E. Rosenberg, eds.), W. B. Saunders & Co., Philadelphia.
27. Meister, A. (1973) *Science*, **180**, 4081, 33.
28. Cavanagh, J. B., (Ed.) (1972) *The Brain in Unclassified Mental Retardation*, Churchill Livingstone, London.

MY INTERRUPTED ASSOCIATION WITH POTASSIUM

LEON A. HEPPEL
Section of Biochemistry, Molecular and Cell Biology
Cornell University, Ithaca, N.Y, 14850

This little story is mainly about my associations with potassium and rubidium, two elements in the alkali metal series of the periodic table. My Ph.D. thesis at the University of California in Berkeley (1934–1937) was concerned with potassium metabolism in the rat during pregnancy and lactation. In those days nutrition research was *the* major activity in biochemistry in the United States. In the course of my work I found among other things that the essential element, potassium, could be replaced by rubidium for growth of the rat, although after a time nervous disturbances and other toxic manifestations resulted. Rubidium is, in fact, transported by the same system which tissues use to take up potassium, so that nowadays one often measures potassium uptake activity with the radioactive isotope, ^{86}Rb. The rubidium isotope happens to have a longer half-life and is more convenient to use than ^{42}K.

In 1937, when I finished graduate school, the United States was still in the midst of the Great Depression. A Ph.D. in Biochemistry was of little use for earning a living, so I went to medical school in Rochester, New York. I was lucky, because I was given room and board, and free tuition for 2 years, in exchange for doing part-time research with Professor W. O. Fenn. Wallace Fenn was one of the great scientists of his time—a truly remarkable man. His outstanding contributions to muscle physiology need not be recounted here. I would like to point out, however, that in 1969 at the age of 75, he embarked on a completely new line of research—microbiological studies on the effect of high pressure on streptococcal growth and metabolism. In collaboration with Robert E. Marquis, he showed that barotolerance of cultures growing in a complex medium with ribose as a major catabolite appeared to be determined primarily by the pressure sensitivity of ribose-degrading enzymes. Apparent activation volumes for growth were nearly identical to those for lactate production from ribose.[1] In other words, the effects of high pressure could be quantitatively accounted for in terms of the volume change of a bacterial culture when its ribose is converted into lactate. This was a truly elegant study. Research along these lines is being continued by Dr. Robert E. Marquis.[2]

But to return to the year 1937. In those days the field of electrolyte physiology was dominated by the ideas of two Germans, Mond and Netter.[3] According to their notions, cells are permeable to K^+ but not to Na^+ and are impermeable to anions. I produced evidence in Dr. Fenn's laboratory which contradicted this theory. The experiments were as follows. I took advantage of my experience in working with potassium-deficient rats and I investigated the electrolytes of the muscles of these animals and studied the effect of muscular contraction. To my surprise I found that the muscles of potassium-deprived rats were richer in sodium than in potassium, quite the reverse of the usual situation. Upon stimulation, there occurred the usual loss of potassium so that the K/Na ratio became distorted even more.[4,5] Clearly this excess of sodium could not be accommodated in the extracellular compartment; some of it had to be intracellular. The prevailing notion, that sodium ions were completely excluded from muscle fibers because they could not penetrate the lipoprotein membrane enveloping the sacroplasm was wrong. The diffusion of radioactive sodium into the muscles of potassium-deprived rats was also demonstrated.[6] It had been found earlier that muscles tend to accumulate sodium *in vivo* during fatigue,[7] but the results of experiments with potassium-deficient rats were more striking and aroused a great deal of attention. It should be emphasized that these changes in muscle electrolytes occurred when the rats were still vigorous and active.

During succeeding decades the basic experiment with potassium-deprived rats was confirmed over twenty times in various other laboratories. The work has often been referred in reviews and it is discussed in two recent books.[8] However, after the initial three papers [4–6] I published nothing more on the subject. This confused many people and two theories evolved to account for this fact: (1) that I carried out the work and then died at an early age; (2) that my father carried out the investigation, whereupon *he* died.

The years since 1940 have been pleasant enough and I have had a lot of fun in the laboratory. In 1942 I joined N.I.H. as a commissioned officer. After a period of war-related research, I became a member of a new section of enzymology which was organized by Arthur Kornberg and which also included Bernard Horecker, both of them extremely stimulating people. This ushered in a truly exciting period. We had daily seminars with friends such as Herbert Tabor. A long list of bright young people joined N.I.H. at about this time. And we enjoyed the further advantage of frequent stimulating visits by distinguished scientists such as Severo Ochoa and Fritz Lipmann. Severo always took the time to ask each of us, individually, what we were doing in some detail and this was splendid for the morale of a young investigator. I treasure these early associations.

In 1952 Russell Hilmoe and I became interested in ribonucleases and phosphodiesterases. It was clear that new methods would have to be devised in order to study the mechanism of action of such enzymes. At about this time Markham and Smith published their work on cyclic-terminal nucleotides and the separation of nucleotides by paper electrophoresis. I was excited by their work and arranged to spend the year 1953 in Roy Markham's laboratory in Cambridge, England. This turned out to be a very profitable and enjoyable year. I worked not only with Roy Markham and Paul Whitfeld, but also with Dan Brown, who was in Lord Todd's Chemical Laboratory.

Upon my return to N.I.H. I continued lines of research opened up in Cambridge. Bill Byrne and Henry Kaplan joined me in this effort. Then came the discovery of polynucleotide phosphorylase in *Azotobacter*, by Grunberg-Manago and Ochoa,[9] which marked a notable advance in the history of nucleic acid biochemistry. Severo Ochoa and I arranged to collaborate by mail, in a joint project on structure and properties of various homopolymers and copolymers. I enjoyed this association with Severo Ochoa very much. Maxine Singer and Russell Hilmoe participated in this work. Maxine Singer made a careful study of the phosphorolysis of different kinds of oligonucleotides by polynucleotide phosphorylases from several bacterial sources, and this was the beginning of her long and distinguished study of this enzyme. At about this time we also enjoyed a brief but very pleasant collaboration with Marianne Grunberg-Manago. Later, Marie Lipsett began her very nice studies on the interaction of poly U and adenine oligonucleotides, with Dan Bradley and myself. Still later, Dan Levin and Jack Abrell participated in nucleic acid problems.

At various times, over the years, Gobind Khorana flitted into and out of my laboratory, staying each time for a week, or 10 days at most. Invariably something highly profitable would emerge from such a short visit. I have always been impressed by the highly variable nature of time in biochemical research. The months go by and one's notebooks fill up with useless data and then in just a few weeks, the numbers that appear in the published paper are obtained. Gobind has learned to eliminate all that waste motion. (Note for example, reference 10, which emerged from a short 10-day visit with Ef Racker and Eileen Knowles at Cornell.) The man is a genius!

Another thing that has impressed me is the way that work of a laboratory becomes channeled into new directions by the interest and enthusiasm of a good post-doctoral fellow, without having been guided that way by the Professor. Thus, Audry Stevens started the research which led to her co-discovery of RNA polymerase quite on her own. Harold Neu's coming led to a shift in the research of my laboratory away from nucleic acids and into osmotic shock and periplasmic enzymes, matters that were developed further by Don Harkness, Harold Dvorak, Wallace Brockman and Nancy Nossal (except that Nancy went back into nucleic acids, with much profit). Finally, in my last years at N.I.H., Yasuhiro Anraku completed the shift by initiating work on active transport and binding proteins, subjects which still occupy me today.

Since 1967 I have been at Cornell University in Ithaca, N.Y., where I have been fortunate to have had three excellent graduate students (Joel Weiner, Ed Berger and Paul Sternweis) and a series of excellent post-doctoral fellows, (George Dietz, Clem Furlong, Barry Rosen, Masamitsu Futai, Ilan Friedberg, Janet Wood, James Cowell, Susan Curtis, S. Kang, M. Kasahara and Jeff Smith). Again, I am impressed by how often the work of a laboratory changes direction due to the initiative of one of its junior members. Thus, Ed Berger started the line of investigation which revealed that there are two broad classes of transport systems in *Escherichia coli* which differ in the source of energy which they use to drive concentrative uptake of sugars and amino acids, etc. The first class uses ATP energy to take up substrates against a concentration gradient, uptake systems of the second class use the so-called energized membrane state. I could cite other examples, which lead one to wonder. In research, who really leads and who follows?

In March of 1975 I began a 6 months sabbatical at the Imperial Cancer Research Fund Laboratories in London. Michael Stoker suggested that I might work with Henry Rozengurt because of a common interest in active transport problems. We began a study of potassium uptake (measured by means of the more convenient competitive inhibitor, ^{86}Rb) into cultured fibroblasts. The background for this problem was as follows:

Fibroblasts grown in tissue culture exist in two alternative growth states, one of active proliferation and one of reversible arrest in what is called the G_1 phase of the cell cycle. Addition of serum to resting cells stimulates the reinitiation of DNA synthesis and cell division.[11] In addition to its effect on DNA synthesis, serum also causes changes in the transport rates for inorganic phosphate, nucleosides and glucose,[11] and in the levels of 3′,5′-cyclic adenylic acid and 3′,5′-cyclic guanylic acid, the well-known "second messengers" which almost everyone is working on nowadays. The changes in transport and levels of cyclic nucleotides are, so far, the earliest events detected in this system. The fluctuations in cyclic nucleotides have received particular attention, the thought being that they might play a role as intracellular messengers for events occurring at the cell membrane (called the plasma membrane).

Whether or not the active transport of monovalent cations is *rapidly* increased by serum was not known when our investigation was started. There were reports of a several-fold increase in the activity of the Na^+K^+ ATPase after serum feeding[12] and this is the enzyme responsible for the action of the "sodium pump", which pumps out Na^+ and pumps in K^+.[13,14] However, the ATPase measurements were made 24 hours after serum addition and we were interested in rapid responses. The question of serum effects on cation transport is important because the maintenance of an asymmetric distribution of K^+ and Na^+ ions has a profound influence in the regulation of cell functions such as cell volume,[15] transport of non-electrolytes,[13] membrane potential,[13] activity of glycolytic enzymes[16] and macromolecular synthesis.[17] Thus, alteration in ionic pumping activity is another potential point of control in linking surface and intracellular events.

Henry Rozengurt and I found that serum addition rapidly increases the influx rate of ^{86}Rb into quiescent 3T3 cells. I should explain that 3T3 is a tissue culture cell line developed by George Todaro and Howard Green.[18] Cells grow in a monolayer on the surface of plastic dishes. When these particular cells cover the surface of the plastic dish, growth stops, and the cells are said to enter a "quiescent phase". The protein content of different dishes at this stage is remarkably uniform. This makes possible a very simple assay for transport of an element or compound containing a radioactive label. One simply incubates the cells on the dish with isotope in a suitable medium, sucks off the medium quickly, washes with saline which is also removed by suction and then extracts the accumulated radioactivity with dilute trichloracetic acid. Radioactivity in the acid-soluble pool is measured by conventional methods.

The effect of serum in stimulating the uptake of the rubidium isotope is detectable in only 2 minutes (Fig. 1). The effect is rapidly reversed when serum is removed. It can be

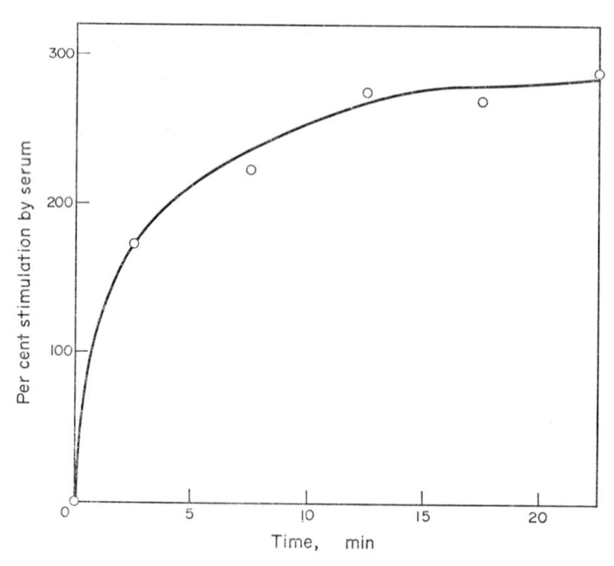

FIG. 1. Stimulation of ^{86}Rb uptake in quiescent 3T3 cells on addition of serum. Serum is added to individual petri dishes and, at the time intervals shown, isotope is added and its rate of uptake is measured over a very brief period. The rates are expressed as per cent stimulation over the basal rate (without serum).

demonstrated whether or not potassium is present in the medium during exposure to serum or during the uptake of ^{86}Rb. The serum effect is most dramatic in quiescent cells but it can also be demonstrated in growing cultures. The maximum stimulation is about 3-5-fold and this requires a concentration of 10% serum in the medium. This concentration, as a matter of fact, is necessary for maximum growth rate of these cells.

This increase in rate of transport after serum addition reaches a maximum in 10 minutes and is readily reversible on removal of serum—the rate of transport returns to the baseline within 15 minutes. Actually, one can repeat the cycle: one can add serum once more and obtain for a second time the rapid increase in rate of uptake of ^{86}Rb. It is as if one were activating a sensitive switch mechanism by addition or removal of serum.

Does serum increase the rate of entry into, or reduce the rate of exit from, 3T3 cells? To answer this question, cells were preloaded with ^{86}Rb then shifted to fresh medium without ^{86}Rb and the rate of exit of the isotope was measured. There was no effect of serum upon the rate of exit. Therefore it must be affecting entry.

We feel that the increase in rate of entry of ^{86}Rb produced by serum is mediated by the sodium pump. What is the "sodium pump"? Briefly stated, it is an enzymatic mechanism in animal tissues for using ATP energy in order for the cell to take up potassium.[13,14] (Remember, animal cells are rich in potassium and very poor in sodium, not because they are impermeable to sodium, but because sodium keeps getting pumped out. That was what my work with Fenn was about.) The following summary of one theory of how the pump works is taken largely from Hokins, review.[14]

In most models for sodium and potassium transport, cation movement is effected by a series of conformational changes in a membrane bound enzyme, called the Na,K-ATPase because it splits ATP and this activity is stimulated by Na^+ and K^+. It has been found that, when ATP is split, phosphate is first transferred from ATP to the dephosphorylated enzyme to make a phospho-enzyme. This step requires Na^+. In the second step, stimulated by K^+, the phosphoenzyme breaks down and liberates inorganic phosphate. According to the theory, the Na,K-ATPase exists in two major conformations. The initial conformation is that of the dephosphorylated enzyme in which Na^+-specific sites face inward. Binding of Na^+ to these sites permits phosphorylation of the enzyme by ATP. This converts the enzyme to a second major conformational form which has K^+-specific sites facing outward. The conversion of the first conformation to the second carries Na^+ to the outside surface of the membrane with simultaneous loss of Na^+ specificity and gaining of K^+ specificity. Binding of K^+ from the outside medium to the second conformational form of the enzyme permits dephosphorylation of the enzyme, converting the enzyme back to its first conformation. This conversion carries K^+ to the inside surface of the membrane with concomitant loss of K^+ specificity and regaining of Na^+ specificity, thus completing the cycle. A very important part of the evidence for all this comes from studies with cardiac glycosides, in particular, ouabain. Ouabain blocks the Na pumps by acting from the outside, where it competes with K for the K binding site.

Henry Rozengurt and I found that the increase in rate of Rb influx produced by serum is markedly sensitive to ouabain. So is the basal rate. This specific inhibitor decreases the rate of influx by 90%. When the concentration of K in the medium is reduced 10-fold, there is a striking increase in sensitivity to ouabain of both the stimulated and the basal rates. This would be expected if the changes in Rb uptake are largely mediated by fluctuations in the activity of the "Na–K pumps", where K competes with ouabain.

When active transport of a substrate is increased by some treatment, one always asks: is this increase caused by a change in V_{max}, the maximum possible rate of reaction, or is it due to a change in affinity of the system for ATP? We observed that the affinity of the system for ATP is not changed after serum treatment but V_{max} is increased up to nearly 4-fold. Is this increase in maximum rate due to the synthesis of new "transporter molecules" of ATPase under the influence of serum? The answer is "no". First of all, the stimulation occurs too rapidly to be accounted for by new protein synthesis. In the second place, cycloheximide, an inhibitor of protein synthesis, does not prevent the stimulation by serum.

Ouabain, which inhibits growth of several lines,[19] prevents the initiation of DNA synthesis in serum-stimulated 3T3 cells. An interesting feature of the inhibition of DNA synthesis by ouabain is its dramatic dependence on K^+ concentration. Thus, we find that the large inhibition of thymidine incorporation obtained at 0.5mM K^+ is completely reversed by raising the level of K in the medium by 5-fold. This experiment indicates that the effect of ouabain on initiation of DNA synthesis is mediated by an inhibition of the sodium pump.

There are other aspects of the story, for example the fact that cyclic AMP also causes an extremely rapid, although smaller increase in rate of ^{86}Rb uptake. However, space does not allow me to go into these matters. Let me summarize, then, by saying that serum has a remarkable effect upon quiescent 3T3 cells. In just a few minutes it causes the rate of entry of potassium (actually measured by using ^{86}Rb) to increase by up to 3.5–4-fold. The effect is on V_{max} and is not due to synthesis of new transporter molecules. The stimulation is rapidly reversible on removal of serum, as if a switch were turned on and off—as if the transporter molecules were being rapidly uncovered and covered again. The effect is sensitive to ouabain, a specific inhibitor of the sodium-K pump. There is also an inhibition of initiation of DNA synthesis by ouabain, which is influenced by the level of K in the medium. It is presumed that a critical ionic environment is necessary for the events concerned in initiation of DNA synthesis, and the Na–K pump is involved in maintenance of this environment.

So it is that, after 37 years, I find myself working once more on potassium and rubidium in animal tissues. I regret this interval of 37 years—it is such a nondescript, random number! It would have been more pleasing to have had a lapse of an even four decades. But life becomes increasingly uncertain with the passage of the years. I couldn't afford to wait!

REFERENCES

1. Marquis, R. E., Brown, W. P. and Fenn, W. O. (1971) *J. Bacteriol.* **105,** 504.
2. Marquis, R. E. and Keller, D. M. (1975) *J. Bacteriol.* **122,** 575.
3. Mond, R. and Netter, H. (1932) *Pfluger's Arch.* **230,** 42.
4. Heppel, L. A. (1939) *Am. J. Physiol.* **127,** 385.
5. Heppel, L. A. (1940) *Am. J. Physiol.* **128,** 440.
6. Heppel, L. A. (1940) *Am. J. Physiol.* **128,** 449.
7. Fenn, W. O. and Cobb, D. M. (1938) *Am. J. Physiol.* **121,** 595.
8. See, for example, *Transport and Accumulation in Biological Systems*, E. J. Harris (1972), Butterworths, London.
9. Grunberg-Manago, M. and Ochoa, S. (1955) *J. Am. Chem. Soc.* **77,** 3165.
10. Knowles, A. F., Kandrach, A., Racker, E. and Khorana, H. G. (1975) *J. Biol. Chem.* **250,** 1809.
11. Pardee, A. B. and Rozengurt, E. (1975) In *Biochemistry of Cell Walls and Membranes* (H. L. Kornberg and D. C. Phillips, eds) pp. 155–185, Butterworths, London.

12. ELLIGSEN, JOANNE D., THOMPSON, J. E., FREY, H. E. and KRUUV, J. (1974) *Exptl. Cell Res.* **87**, 233.
13. BAKER, P. F. (1972) *Metabolic Pathways*, **VI**, 243.
14. DAHL, J. L. and HOKIN, L. E. (1974) *Ann. Rev. Biochem.* **43**.
15. TOSTESON, D. C. (1964) In *The Cellular Functions of Membrane Transport* (J. F. HOFFMAN, ed.), pp. 3–22, Prentice-Hall Inc., Englewood Cliffs, U.S.A.
16. SCHOLNICK, P., LANG, D. and RACKER, E. (1973) *J. Biol. Chem.* **248**, 5175.
17. LUBIN, M. (1967) *Nature*, **213**, 451.
18. TODARO, G. and GREEN, H. (1963) *J. Cell Biol.* **17**, 299.
19. MCDONALD, T. F., SACHS, H. G., ORR, C. W. M. and EBERT, J. D. (1972) *Epxtl. Cell Res.* **74**, 201.

A PERSONAL ACCOUNT OF MY RESEARCH ON COLLAGEN AND FIBROSIS

SIDNEY UDENFRIEND
Roche Institute of Molecular Biology, Nutley, New Jersey 07110

Over the years I have been involved in a number of areas of research. I have chosen to discuss my studies on collagen in this volume for a number of reasons. First I am still active in this area of research. It also represents many approaches to research that I consider typical of my style. Finally, research in collagen is now exciting and productive and is leading to many practical advances in medicine.

This will therefore be a personal account of some of these studies including some details of my own training, that are relevant to research on collagen, with reference to the professors who taught me, the mentors I had as a postdoctoral fellow, my collaborators in research, my students, postdoctoral fellows and colleagues who came as visiting scientists. It will also include some of the personal interactions with other scientists in the field, those with whom I collaborated, those with whom I competed and those with whom I may have had open disagreement.

How did I become interested in collagen? In part, it was because of Severo Ochoa. In 1946, while I was working with him as a graduate student in the Department of Biochemistry at New York University Medical School, he accepted the chair of Pharmacology at that institution. I chose to stay on in the Department of Biochemistry rather than move to Pharmacology with him. It is curious that despite that decision I am now a member of the Pharmacology Society and I believe that I am a respected scientist in the field. This was an ironic turn of events since Severo Ochoa's venture into pharmacology was of short duration. On the other hand, what I learned during the 16 months I spent with him greatly influenced my career.

When Severo Ochoa left the Biochemistry Department at New York University I had to search for another mentor and I was fortunate to be able to continue my graduate work with Albert Keston. While developing the isotope derivative method for amino acid assay[1] we showed for the first time that hydroxyproline was uniquely present in collagen. This observation made a great impression on me. Furthermore, I had the good fortune while at New York University Medical School to meet and to get to know the late Joseph Bunim. It was from him that I learned about the role of collagen in connective tissue and also about collagen diseases.

There followed a long period during which I did not pursue my interest in collagen. In 1948 I left New York University to spend two years in Carl Cori's department at Washington University in St. Louis. There I was fortunate to work with Dr. Sidney Velick, a superb enzymologist. Two years later I moved to the National Heart Institute in Bethesda and began to apply the knowledge of enzymology I had gained from Severo Ochoa and from

Sidney Velick to amino acid metabolism. By 1956 I had become heavily involved in hydroxylation mechanisms, having discovered tryptophan hydroxylase, phenylalanine hydroxylase, and tyrosine hydroxylase and having purified and studied the last two enzymes.

Since hydroxyproline had been shown to be derived from proline I was naturally interested in this hydroxylation as well. It was known from the work of Marjorie Stetten[(2)] that hydroxylation of proline probably occurs not in the free form but at some step during or after protein formation. Of course in 1956 the steps involved in protein synthesis were not known so my laboratory had to limit itself to *in vivo* studies and to the development of analytical methods. By the late 1950s, when the steps involved in protein synthesis had been elucidated, it became possible to design experiments to pinpoint the exact step at which proline was hydroxylated. In 1961 I received some additional help from Severo Ochoa. One of the students in his department, Beverly Peterkofsky, who had been working with Charles Gilvarg, decided to leave New York University to join her husband at the National Institutes of Health. I was asked if I could permit her to continue her graduate work with me. Having Beverly Peterkofsky in my department was one of the best things that ever happened to me. She was able to obtain a cell-free system from chick embryos which incorporated proline as peptide-bound hydroxyproline. This in itself was a major accomplishment. However, Beverly Peterkofsky also noticed that there was a lag between the appearance of proline in the protein and its conversion to hydroxyproline. In pursuing this lead, using data obtained with inhibitors of protein synthesis, with ribonuclease, and with anaerobic conditions, we were able to conclude that peptide chain formation preceded hydroxylation and that hydroxylation must therefore occur on proline already incorporated in peptide linkage. This of course still left open several possibilities: small nascent chains, completed chains still bound to ribosomes, or free chains. Several other laboratories were also pursuing the problem and a number of them had concluded that hydroxylation took place at the tRNA step, i.e. tRNA-Pro was hydroxylated to tRNA-Hyp. These conclusions were based on the reported isolation of radioactive tRNA-Hyp following incubation with radioactive proline, experiments that we could not reproduce. During the height of this controversy in 1964 I attended the Biochemical Congress in New York City. At a symposium there one of the speakers was presenting his evidence for tRNA as the site of proline hydroxylation. I arrived at the large lecture hall a little late, and when the speaker finished, the chairman, Severo Ochoa, who had obviously kept up with the field and wanted to stir up some discussion, pointed to me standing at the back of the room and said, "what do you have to say to that, Dr. Udenfriend?" Unfortunately it was impossible to say very much in those circumstances, except to point out that our results were different. I'm afraid that I didn't help to satisfy the chairman's desire for a lively discussion.

There was another interesting development during this tRNA controversy. One of the proponents of hydroxylation of tRNA-Pro reached the following conclusions in a paper;[(3)] "These findings do not, however, justify the statement of Peterkofsky and Udenfriend that their results rule out s-RNA hydroxyproline as an intermediate in collagen synthesis. The fact that such an intermediate has now been demonstrated by three groups of workers firmly establishes the existence of such an intermediate". Fortunately science does not advance by public opinion polls. Shortly thereafter one of the advocates of the tRNA site for hydroxylation carried out one of the key experiments that finally excluded this mechanism in the biosynthesis of hydroxyproline.

Many other colleagues followed and each made important contributions to the problem. I arbitrarily select another high point in our studies. John Hutton, working in my laboratory in 1965, developed a simple and sensitive method for prolyl hydroxylase and proceeded to purify the enzyme. It was already known that Fe^{2+} and ascorbate were required. However, on dialysis, activity could not be restored by these two agents. An additional factor present in the dialysate was needed. How does one go about identifying a missing factor? John Hutton tried everything that he could find on the laboratory shelf, including nucleotides, coenzymes, amino acids, Krebs cycle intermediates, metals, etc., and found that only α-ketoglutarate restored activity. He then showed that the activity of undialyzed extracts was lost on treatment with glutamate dehydrogenase, ammonia and NADPH, and was restored with added α-ketoglutarate. With these findings a new class of oxygenases requiring α-ketoglutarate was discovered. Soon other investigators were applying the same techniques to a number of Fe^{2+}, ascorbate hydroxylases which had defined purification and finding that they too required α-ketoglutarate.[4]

Elucidation of the role of the keto acid was not possible, even with preparations that had undergone substantial purification, because of the presence of many other α-ketoglutarate requiring enzymes in tissue extracts. In 1967 Robert Rhoads, then a graduate student, succeeded in purifying the enzyme from newborn rat skin essentially to homogeneity. With this preparation he then showed that α-ketoglutarate was oxidatively decarboxylated in the reaction. Furthermore, the oxidation of each Pro–Gly sequence to a Hyp–Gly sequence was shown to be stoichiometrically coupled to the oxidative decarboxylation of a molecule of α-ketoglutarate to succinate and CO_2. This same mechanism has now been shown to be true for all the other α-ketoglutarate hydroxylases. Prompted by studies reported on another α-ketoglutarate hydroxylase,[5] George Cardinale, who is now in my laboratory, carried out an experiment with $^{18}O_2$ and showed that equivalent amounts of $^{18}O_2$ were incorporated into hydroxyproline and into the succinate derived from α-ketoglutarate.[6] Prolyl hydroxylase and the other α-ketoglutarate coupled hydroxylases are therefore dioxygenases. In the case of prolyl hydroxylase the data suggest a peroxy intermediate, perhaps of the type:

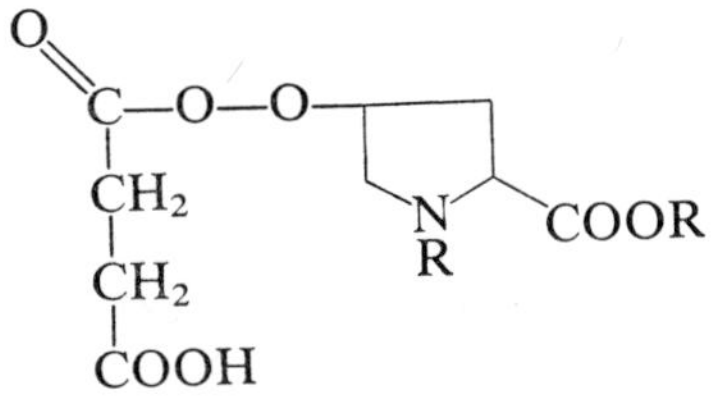

```
O
 \\
  C—O—O—(pyrrolidine ring, 4-position)
  |         N      COOR
  CH2       R
  |
  CH2
  |
  COOH
```

Recently George Cardinale has been able to show that prolyl hydroxylase can catalyze the oxidative decarboxylation of α-ketoglutarate in the absence of a prolyl substrate, at a lower but measurable rate. The same cofactors, Fe^{2+} and ascorbate, are required and optimal conditions are the same for both. This suggests that α-ketoglutarate is the first acceptor of the oxygen and that a peroxy compound, perhaps enzyme bound persuccinate, then attacks the 4-trans position of proline to yield the intermediate shown above. Cleavage of the peroxy compound would then yield equal amounts of succinate and hydroxyproline.

Prolyl hydroxylase has another unique feature. Its substrate is not a simple, small molecular weight metabolite, but a peptide or protein. How did we go about elucidating the sequence recognized by the enzyme and the site of hydroxylation *in vivo*? Soon after the cell-free system was developed by Beverly Peterkofsky, one of my former students, Darwin Prockop, then working independently in Philadelphia, succeeded in solubilizing both enzyme and substrate components of the system. Of course this made possible the enzyme purification discussed above. It also made possible studies on the nature of the substrate. It became obvious that when hydroxylation was made limiting in intact cells one could isolate an unhydroxylated collagen moiety for which Darwin Prockop coined the name "protocollagen". To him it represented an obligatory intermediate. To me it was a useful experimental device. And so I became involved in another controversy, this time with a former student, concerning the nature of the normal intracellular substrate. This controversy continued for some time and in the meantime further information as to the nature of the chemical recognition site came from a different approach. I had become aware of a group at the Weizmann Institute who were synthesizing polymers of the type $(Gly–Pro–Pro)_n$ as collagen models. It immediately struck me that these might serve as synthetic substrates. I wrote to obtain some material from the Weizmann Institute group and then discovered that Darwin Prockop had already written and obtained the material. With it he quickly showed that even the smaller polymers were substrates. Obviously, turning out good graduate students can have its drawbacks, but I still take pride in the fact that I have helped to make them such good competitors. In this case we obtained some of the same polymers and confirmed Darwin Prockop's observations. These polymers were, however, of only limited value in pinpointing the exact recognition sequence. Some time later we learned that the peptide, bradykinin, contained three prolyl residues, one of them a Pro–Gly sequence. Robert Rush, a graduate student, showed that one prolyl residue was hydroxylated specifically, the one in Pro–Gly linkage. Since over the years many bradykinin analogs had been prepared for studies of biological activity it was not long before we were able to obtain enough of these to show that the only absolute requirement for enzyme recognition was the Pro–Gly sequence and that all the other residues in the molecule merely altered the kinetic parameters. The latter studies were carried out by James McGee and Arthur Felix prepared some key bradykinin analogs.

The question remained as to the nature of the peptide which served as substrate in the cell under normal conditions. Darwin Prockop utilized pulse-labeling experiments with ^{14}C-proline to show that hydroxyproline appearance coincided with the estimated time for chain completion. However, a number of laboratories reported data which suggested that hydroxyproline was already present in ribosome associated chains. In 1968 Ronald Miller joined my group as a postdoctoral fellow. His prior experience with Richard Schweet and George Arlinghous made him ideally trained to investigate early stages of biosynthesis of collagen. He showed, quite clearly, that in guinea-pig granuloma cells hydroxyproline appeared in nascent chains even smaller than m.w. 10,000.[7] As much as 50% of newly formed hydroxyproline could be isolated from nascent chains. Subsequently Leonard Lukens[8] and his colleagues utilized cultured fibroblasts and concluded that hydroxylation occurred mainly, if not entirely, on nascent chains. How then, to explain the pulse labeling data? The accepted value for the time required for collagen chain completion was then 1 minute, and this value was used by Darwin Prockop in his calculations. Later, this was corrected to 6 minutes.[9] The longer time made the pulse labeling experiments compatible with hydroxylation occurring during translation. It is now generally accepted that under normal conditions hydroxylation of proline occurs largely on nascent chains. When hydroxylation is made limiting, as by anaerobiosis, ascorbic acid deficiency or by addition of Fe^{2+} chelators, then unhydroxylated chains (protocollagen) accumulate. Apparently, however, the protocollagen which accumulates intracellularly is not secreted until hydroxylation is restored.

I should point out that Darwin Prockop deserves credit for a major contribution in the field, namely, the clarification of the function of hydroxyproline. While we found the studies with prolyl hydroxylase fascinating, others in the collagen field questioned their significance, since no function could be shown for the hydroxyproline in collagen. In the case of hydroxylysine the hydroxyl group served as the site for glycosidic linkage. Not so with hydroxyproline. Furthermore the physical chemists considered that the typical helical structure of collagen required merely imino acid-glycine linkages and also could find no role for the hydroxyl group of hydroxyproline based on their calculations and estimations. We were convinced that nature had not designed such a unique enzyme system to hydroxylate up to 50% of the proline residues in collagen without some physiological purpose. Darwin Prockop[10] and one of his former colleagues, Joel Rosenbloom,[11] independently devised elegant experiments which finally elucidated the significance of hydroxyproline. They synthesized labeled unhydroxylated collagen by limiting hydroxylation conditions, extracted it, and showed that it was thermally much less stable than hydroxylated collagen. Apparently the physical chemists and model builders had missed the stabilizing effect of the hydroxyl group on the collagen helix. Another important conclusion to draw from this is that biochemistry is still largely an *experimental* science.

Our studies on hydroxyproline have taken us into many other interesting areas, including regulation during cell growth, wound healing, and studies of the Clq component of complement. I would like to conclude with some very recent applications of our collagen and hydroxyproline technology as markers in the field of hypertension.

It has been shown by several investigators that damage to blood-vessels by chemical[12] or physical means[13] that leads to atherosclerosis is accompanied, and in fact preceded, by increased biosynthesis and deposition of collagen. It was shown by Russell Ross[12] that collagen is synthesized by smooth muscle cells in the blood-vessels. How I turned our studies on collagen to hypertension is interesting. As a pharmacologist I have long been interested in hypertension through my studies on catecholamines. Because of this interest I participated in a symposium in Japan on the genetically hypertensive rat, where I was scheduled to discuss our work on catecholamines and hypertension. During this symposium there was considerable discussion concerning changes in the walls of blood-vessels during hypertension, because it had been found by those studying cardiovascular hemodynamics in hypertension that the elasticity of blood-vessels (large and small) is altered in a manner indicative of hypertrophy of the blood-vessel wall. When I inquired about this hypertrophy none of the experts knew its cause. When I speculated that it might be collagen I was asked to discard my talk on catecholamines and tell them something about the connective tissue protein. Fortunately I had another set of slides with me for a collagen lecture and I told them about the unique chemistry and biochemistry of collagen. Under normal conditions, collagen is extruded from cells and forms fibers that are the major component of connective tissue. Once deposited these collagen fibers, in most cases, cannot be resorbed. In fact the life of collagen in the skin of the adult rat is about equal to the life of the rat. It is known that when certain tissues are injured they respond by an increased deposition of collagen. This is true in liver cirrhosis, in lung fibrosis and in instances of physical or chemical injury to many other tissues. As noted above, following physical and chemical injury to blood-vessels, the arteriosclerosis is accompanied by collagen deposition. I posed the question: "could the continued insult of an elevated blood-pressure on the blood-vessels lead to increased collagen biosynthesis so that the hypertrophy of the vessel wall was in part composed of collagen fibers?" In other words, does hypertension lead to a generalized vascular fibrosis? Consistent with this is the fact that arteriosclerosis is much more prevalent in hypertensives than in normotensives.

When I returned from Japan I was able to persuade my colleagues to turn their attention to hypertension. Once more we were fortunate to have the right people at the right time. Akira Ooshima, a pathologist from the laboratory in Kyoto that had developed the spontaneously hypertensive rat, was working in our laboratory, and also George Fuller, who had worked on atherosclerosis and collagen. These two were joined by Sydney Spector and George Cardinale. In a fairly short time we showed that hypertension in rats, whether induced by deoxycorticosterone acetate and salt or brought on by inbreeding, leads to increased biosynthesis and net deposition of collagen.[14] We dissected out blood-vessels such as the mesenteric artery and aorta, and measured prolyl hydroxylase activity, incorporation of ^{14}C-proline into protein bound ^{14}C-hydroxyproline and the prolyl hydroxylase antigen as markers of biosynthesis. We showed further that antihypertensive drugs, in doses which prevented hypertension, also prevented the increase in collagen synthesis. Furthermore, when a hypertensive drug was administered to an already hypertensive rat both collagen biosynthesis and blood pressure returned to normal values. Thus at least part of the reported hypertrophy of the blood-vessel wall in hypertension[15] appears to be due to deposition of collagen, in other words, fibrosis.

Most recently we became aware of a simple procedure for isolating the small blood-vessels from the brain.[16] With this procedure we have recently shown that these small vessels in the central nervous system (a mixture of arterioles, capillaries and venules) also produce collagen and respond to hypertension and antihypertensive drugs in the same manner as do the larger blood-vessels in the periphery. We are now beginning to explore the regulatory mechanism which turn on collagen synthesis in hypertension.

Over the years the studies on collagen have encompassed pure chemistry, enzyme isolation, enzyme mechanisms, molecular biology, clinical investigation and pharmacology. Now, 29 years after my initial interest in hydroxyproline and collagen was aroused, I am more deeply involved in their study than ever. As is usual in all research each new finding leads to several new questions so that we are now faced with a far greater number of questions than we had when we started.

Finally, 29 years after Severo Ochoa left me in the Department of Biochemistry at New York University and unwittingly put me in the path of collagen research, we have been reunited in the same Institute. It is a pleasure to have him under the same roof where we can discuss our individual research programs on a day-to-day basis. I value greatly his enthusiasm and the interest he has shown in my research and in the research of my colleagues.

REFERENCES

1. Keston, A. S., Udenfriend, S. and Cannan, R. K. (1949) A method for the determination of organic compounds in the form of isotopic derivatives. I. Estimation of amino acids by the carrier technique. *J. Am. Chem. Soc.* **71**, 249–257.
2. Stetten, M. R. (1949) Some aspects of the metabolism of hydroxyproline, studied with the aid of isotopic nitrogen. *J. Biol. Chem.* **181**, 31–37.

3. JACKSON, D. S., WATKINS, D. and WINKLER, A. (1964) Formation of s-RNA hydroxyproline in chick embryo and wound granulation tissue. *Biochim. Biophys. Acta*, **87**, 152–153.
4. ABBOTT, M. T. and UDENFRIEND, S. (1974) α-Ketoglutarate coupled dioxygenases. *Molecular Mechanisms of Oxygen Activation* (O. HAYAISHI, ed.), pp. 167–214, Academic Press, New York.
5. LINDBLAD, G., LINDSTEDT, G., TOFFT, M. and LINDSTEDT, S. (1969) The mechanism of α-ketoglutarate oxidation in coupled enzymatic oxygenations. *J. Am. Chem. Soc.* **91**, 4604–4606.
6. CARDINALE, G. J. and UDENFRIEND, S. (1974) Prolyl hydroxylase. *Advances in Enzymology*, (A. MEISTER, ed.), vol. 41, pp. 245–300, J. Wiley & Sons, Inc.
7. MILLER, R. L. and UDENFRIEND, S. (1970) Hydroxylation of proline residues in collagen nascent chains. *Arch. Biochem. & Biophys.* **139**, 104–113.
8. LAZARIDES, E. L., LUKENS, L. M. and INFANTE, A. A. (1971) Collagen polysomes: site of hydroxylation of proline residues. *J. Molec. Biol.* **58**, 831–846.
9. VUUST, J. and PIEZ, K. A. (1972) A kinetic study of collagen biosynthesis. *J. Biol. Chem.*, **247**, 856–862.
10. BERG, R. H. and PROCKOP, D. J. (1973) The thermal transition of a non-hydroxylated form of collagen. Evidence for a role for hydroxyproline in stabilizing the triple-helix of collagen. *Biochem. Biophys. Res. Commun.* **52**, 115–120.
11. JIMINEZ, S., HARSCH, M. and ROSENBLOOM, J. (1973) Hydroxyproline stabilizes the triple helix of chick tendon collagen. *Biochem. Biophys. Res. Commun.* **52**, 106–114.
12. ROSS, R. and GLOMSET, J. A. (1973) Atherosclerosis and the arterial smooth muscle cell. *Science*, **18**, 1332–1339.
13. FULLER, G. C., MILLER, E., FABER, T. and VAN LOON, E. (1972) Aortic connective tissue changes in miniature pigs fed a lipid-rich diet. *Connective Tissue Res.* **1**, 217–220.
14. OOSHIMA, A., FULLER, G. C., CARDINALE, G. J., SPECTOR, S. and UDENFRIEND, S. (1974) Increased collagen synthesis in blood vessels in hypertensive rats and its reversal by antihypertensive agents. *Proc. Nat. Acad. Sci.* **71**, 3019–3023.
15. FOLKOW, B., HALLBÄCK, M., LUNDGEN, Y., SIVERTSSON, R. and WEISS, L. (1973) Importance of adaptive changes in vascular design for establishment of primary hypertension, studied in man and in spontaneously hypertensive rats. *Circ. Res.* [Suppl. 1], I2–I13.
16. BRENDEL, K., MEEZAN, E. and AARLSON, E. C. (1974) Isolated brain microvessels: purified, metabolically active preparation from bovine cerebral cortex. *Science*, **185**, 953–955.

VII. SCIENCE IN THE TIME OF OCHOA

REMINISCENCES OF SEVERO OCHOA FROM OUR OXFORD DAYS

ERNST CHAIN
Biochemistry Department, Imperial College of Science and Technology, London, England

I met Severo Ochoa for the first time fleetingly in 1933 when I paid a visit from Cambridge to the National Institute of Medical Research in London where he spent a short time.

I had been lucky to have been accepted at that time as a research worker at the Sir William Dunn School of Biochemistry under Sir Frederic Gowland Hopkins, then PRS, at that time a great world centre of biochemical research.

I had emigrated from Germany to England on 30 January of that year, the day when Hitler and his gang took over political power and destroyed in one day a uniquely brilliant period of culture which it took a hundred years to build up. The cultural activities in Germany, prior to the Hitler period, attained an unparalleled level of excellence in all its manifestations: science, music, literature, art and theatre. Severo experienced personally some of this excellence in the world-famous laboratory of Otto Meyerhof where he worked from 1929 to 1931, during some of its best years. This must have left an indelible impression on him, as it left on most of those privileged to work in this laboratory, many of whom became world leaders in biochemical research.

The first time I saw Severo in London I was immediately attracted by his personality which stood out impressively in its Spanish splendour against the more sober Anglo-Saxon background surrounding him. His characteristic long head and figure gave one the impression of looking at an El Greco picture.

In life very frequently, almost usually the unexpected happens. When we met in London in 1933, Severo seemed in a much more settled and secure position than I was. He seemed to have been destined to a normal academic career in Spain whereas I did not know what fate awaited me. The noise of political rumblings could be heard in the background in Spain, but very few people expected the outbreak of a cruel and bloody civil war, as happened in 1936 and put Severo and many of his colleagues into a very similar condition to my own: that of a political refugee.

I, on the other hand, was offered completely unexpectedly in 1935 a secure job as a lecturer at Oxford in the Sir William Dunn School of Pathology by the newly appointed professor—H. W. (later Lord) Florey—and stayed there until the end of 1948.

After the outbreak of civil war in Spain, Spanish refugees began to appear in Oxford, following the great wave of German refugees a few years earlier. Among them were some outstanding scientists, including the world-famous brain histologist del Rio Ortega, a pupil of Ramon y Cajal and a great artist in his profession, and Severo Ochoa and his charming wife Carmen, a real Spanish beauty, who arrived in 1938. Ortega was given facilities in the Department of Pathology where I worked, and we were privileged to watch his magnificent professional performance which reminded one of a great miniature painter at work. Severo worked in the Biochemistry Department under Professor R. A. (later Sir Rudolph) Peters on problems connected with vitamin B_1 and pyruvate oxidation.

Severo, Carmen and I soon became close friends and this close friendship has lasted over almost four decades. It was primarily based on many similarities of our character, temperament and our attitudes to situations and people. I came to the conclusion that the reason for this was my portion of Russian-Jewish genes inherited from my father which are very dominant. There are many similarities in the character of the Russian and Spanish intellectual.

I lived during most of my 13 years at Oxford in a flat opposite the Dragon School in North Oxford, in one of those quiet and peaceful roads typical of this residential area, lined with cherry, almond and laburnum trees successively blossoming luxuriously in the spring and making North Oxford look like one big garden. I was not yet married at that time but was looked after by my energetic cousin Mrs. Anna Sacharina who became quite a figure in Oxford and is still with us, at 86. We met the Ochoas and their Spanish friends frequently in this flat. We had many evenings of chamber music in which Professor Peters frequently took a leading part as violinist, and to which Severo and Carmen often listened. We discussed for many hours many problems concerning philosophy, human behaviour, politics, art and literature and our views on most subjects of our discussions were very similar. We never talked much about our scientific work, which was partly due to the fact that we were working on totally different problems, partly because we liked to solve our problems in our own way, and partly because it was not the habit in Oxford at that time to talk "shop" outside working hours in the laboratory. The few times I had occasion to talk to Severo on general biochemical subjects, I was very impressed by the perspicacity of his mind and his outstanding capacity of finding relations in complex systems between apparently unrelated experimental findings. It must be remembered that the years 1938–1940 which Severo spent at Oxford were approaching the end of the period of unsophisticated, cheap biochemistry and the beginning of the period of highly sophisticated and very expensive biochemical methodology. We had no isotopes, no chromatography, no mass spectrograph, no nuclear magnetic resonance, only the beginnings of infrared spectroscopy. All we had at our disposal were the classical analytical techniques, manometry, UV spectroscopy, Tiselius cataphoresis with rather complicated optical systems and the beginnings of ultra-centrifugal techniques, at that time limited to molecular weight determinations of proteins. Despite these methodological limitations, however, a number of important basic biochemical discoveries were made during the "cheap" era of biochemical research to which both Severo and I were able to contribute.

In 1940, at the beginning of the war, our Spanish scientists decided to leave Oxford, to our great regret. Del Rio Ortega could not get adjusted to the English climate and particularly the absence of cafés on squares where you could sit in the open, sip strong good coffee, talk and watch the passing population. He emigrated with his friend to Argentina where he died a few years later. Severo went to the Biochemistry Department at St. Louis in the United States which, at that time, under Carl and Gerty Cori, was the most active laboratory in the field of carbohydrate metabolism and the study of its enzymes, with about the same high international standing as was that of Meyerhof's laboratory in the late twenties and early thirties. The stay in both these laboratories had a profound influence on Severo's subsequent career.

For me the departure of Carmen and Severo from Oxford in 1940 signified the loss of very good friends and was a blow, which gave me real pain. Oxford did not seem the same after the Ochoas had left it. However, the war became more menacing soon afterwards and we were kept very busy with our scientific work which appeared to acquire practical importance for the war effort though it was started in 1939, before the outbreak of war, as a purely scientific investigation. We passed through some agonizing years, bringing with it unheard-of destruction, bloodshed and suffering, but the spirit of the population was high and everyone was certain of eventual victory. This came in 1945, and soon afterwards I visited the United States, together with the late Sir Robert Robinson and the late Dr. Arnold King to discuss with our American colleagues the publication of the Anglo-American work on penicillin chemistry and biochemistry. During this visit I saw Carmen and Severo again, and many other friends. Five terrible years which had changed our world and values had passed since we saw each other last, but our friendship was re-established immediately as if there had been no separation. It was a great joy for me to be reunited with them once again, and to be able to ascertain that all of us had come out unscathed, at least physically, from the world catastrophe.

While in New York I was informed that I had been awarded a share of the Nobel prize for the discovery of the therapeutic properties of penicillins; Severo had seen the beginnings of this work while he was in Oxford. This event was duly celebrated by friends among whom were Carmen and Severo. In 1959 Severo was in the same position as I was in 1945; he had been awarded a share in the Nobel prize for his discovery of the enzyme ribonucleotide polymerase which was to play an important part in the unraveling of the genetic code.

It gives me the greatest pleasure to extend to Severo and Carmen my warmest congratulations and best wishes at his 70th birthday and wish him many more years of fruitful creative scientific activity.

HIGHLIGHTS OF A FRIENDSHIP

DAVID NACHMANSOHN
Columbia University, New York, N.Y.

The friendship between Severo and myself started, in 1929, in the Kaiser-Wilhelm-Institutes in Berlin-Dahlem, in Otto Meyerhof's laboratory. Due to the meteoric rise of German science in the preceding half century and in combination with the remarkable spiritual renaissance of Germany during the Weimarer Republik, the Kaiser-Wilhelm-Institutes became in the 1920s one of the greatest scientific centers the world had ever seen. Among the members were Einstein, Max Planck, von Laue, Fritz Haber, James Frank, Otto Hahn and Lise Meitner, to mention just a few of the illustrious names. The "three big" among the biochemists were Otto Warburg, Meyerhof and Carl Neuberg, three leaders whose work was instrumental in laying the foundations of modern biochemistry. There was an additional factor which created an atmosphere hardly encountered in other institutions. The smallness of the campus, formed by about eight buildings two to three stories high, permitted easy contacts between the scientists working there; young post-doctoral fellows had no difficulty meeting the great men and discussing scientific problems. The famous "Haber Colloquia" offered an additional possibility of meeting scientists of different fields. Haber tried deliberately to break down the barriers between physics, chemistry and biology. The vigorous controversies, in which young people were encouraged to participate, offered an exciting experience; they revealed the limitations of science and demonstrated that even the greatest minds were not infallible. Haber, as well as others, was never reluctant to admit mistakes, an attitude which discouraged dogmatic views.

Neuberg was a pioneer in modern concepts of fermentation, Warburg in those of the mechanism of oxidation. Meyerhof was primarily a philosopher with a broad range of other interests, among them literature, history, art and archaeology. His main scientific interest was the transformations of chemical energy into cellular functions. At the end of this essay fundamental notions that he developed will be briefly mentioned.

Meyerhof had only a few collaborators. This factor permitted frequent personal discussions between ourselves and with him. His personality, his critical comments and his penetrating analysis were essential factors in producing an atmosphere which created close links between us and developed longlasting friendships (Fig. 1).

To have had Meyerhof as our teacher, combined with the effects of the unique atmosphere of the Kaiser-Wilhelm-Institutes, would have formed a favorable basis to develop a close friendship between Severo and myself, with our passion and enthusiasm for science. But fate turned this experience to be not the most decisive element in our friendship. The turbulent years of the 1930s: Hitler in Germany, civil war in Spain, Mussolini in Italy, similar upheavals in other European countries, drove many scientists to this country. Between 1939 and 1941, Severo, Meyerhof and I had come to the United States. Meyerhof settled in Philadelphia, we two in New York. Here started a new and perhaps the most prestigious and decisive period of our friendship with Otto and intensified that between ourselves. We three with our wives Hedwig, Carmen and Edith spent the summers from 1941 to 1951 in Woods Hole, a small place in a delightful location dominated by the scientific community working at the Marine Biological Laboratory. The place had an intellectual atmosphere which attracted many Europeans. The Meyerhofs loved this place. Severo and I became his closest and most intimate friends. The three couples spent many delightful hours together. Otto was no longer our teacher but a real friend. The many inspiring discussions with him became an important element in our own relationship and enriched our lives.

At a symposium on "Phosphorus Metabolism" held in 1951 at Johns Hopkins University, Meyerhof was asked to give the opening address. When the book appeared after his death, McElroy and Glass, the organizers of the symposium, wrote in their summary that the symposium was a lasting monument to Meyerhof's greatness and his impact on a whole generation of biochemists: his name will be linked forever with the understanding of the paramount role of phosphate derivatives in the intermediary metabolism.

Meyerhof did not live to read this tribute to his life work. He passed away on 6 October 1951. It was a terrible shock for Severo and myself. We had lost our beloved and inspiring teacher and our personal friend; he was for both of us a kind of fatherly figure. In the next few weeks, Severo, Carmen and I shuttled frequently back and forth between New York and Philadelphia to help Hedwig with innumerable arrangements. These weeks of sharing a great and irreplaceable loss brought us closer together than ever before; they firmly cemented our friendship. Severo and I wrote an obituary, subsequently endorsed by Fritz Lipmann.(1)

Being biochemists and living in New York, Severo and I met frequently at lectures, meetings, symposia, in the famous Enzyme Club, at many official and private parties. We followed closely each other's work and attended, whenever possible, each other's lectures. I greatly admired Severo's originality, imagination and competence. It was for me a great personal joy and a deep satisfaction when he received the Nobel Prize for the enzymic synthesis of RNA. This honor was highly deserved not only by this, but also by his many other outstanding contributions in various fields and by his inspiration of many colleagues, friends and pupils. In the late 1940s our scientific interests, on acetylation mechanisms, came so close that we worked on a problem together and our names appeared on the publication. We travelled frequently together back and forth to Federation Meetings. In the 1950s when Europe began to recover from the havoc of World War II, we attended many international biochemical meetings and had pleasant and stimulating times. Moreover, the two couples met from time to time on a Saturday afternoon, had tea, went to see a good movie and then went to a nice restaurant. These Saturday afternoons became soon a cherished tradition which still continues.

There exists something like a special affinity between personalities which is hard to define. This is probably an important element in our as well as in most friendships. In addition to our passion for science we share many other ones which all have contributed to deepen our friendship. We both love music and going to concerts. We are great opera fans; our regular visits to the Metropolitan Opera mean to us much more than simple entertainment; they are deep emotional experiences, an almost indispensable part of our lives. Whether paintings or sculptures, monuments of the past, cathedrals and monasteries, etc., looking closely and carefully at them, we are enthralled by these expressions of the greatness and beauty of the creative spirit of man. Instead of trying to select some of our many common experiences, I would like to describe briefly four trips which form highlights not only of our friendship but of our lives.

In 1952 we attended a symposium at the Hebrew University in Jerusalem. In addition to participating in the scientific meetings and visiting the Weizmann Institute, we had an opportunity to visit many shrines and monuments, villages and ruins, all familiar monuments of the past. Anybody who knows the Bible and the history of the country knows about Cesarea and Akko, about Haifa and Tiberias, etc. Unfortunately, the Old City of Jerusalem was cut off from the rest. I had seen it in the 1930s. But Carmen, with her determination and will power, managed—with the help of the Spanish Consul—to get a permit for visiting the Old City.

From Tel Aviv we flew to Athens. We had 3 days and had made tentative plans. The first morning we went to the Acropolis. We climbed up the monumental Propylaea, on our left the gigantic North wing, on our right the South wing, and the charming temple of Athene Nike. When we arrived at the top, a view of indescribable beauty opened before our eyes: to the right the breathtaking view of the Parthenon, to the left the elegant form of the Erychtheion with the graceful Karyatides. The beauty of the whole scenery was magnified by the coloring: the azure blue sky above us, the colorful mountains surrounding the Acropolis. We were in a trance. The view surpassed our wildest expectations. In the humanistic gymnasium which I attended, Greek and Latin were the two main languages; knowledge of the details of the history of these two civilizations was of paramount importance for an educated man. What an emotional experience to see all the familiar buildings essentially as they looked at the time of Pericles, the pinnacle of Athen's greatness, the period of Sophocles and Aeschylus, of Phidias and the other artists rebuilding the Parthenon, the Athens of Plato and Socrates. The Splendor surrounding us brought back to us the glory of Greece. We went around and around, we marvelled at the proportions of the Parthenon, we looked at every corner of the Acropolis and were most unhappy when at sunset we were forced to leave. The next morning at breakfast there was no discussion of other plans. We were still so overwhelmed, we must spend the day again at the Acropolis. The same happened on the third day. What more beautiful sight could we expect to see! Three days spent at the Acropolis could hardly satisfy our craving. We were determined to return one day.

From Athens we flew to Rome, expected at the Airport by Ernst and Ann Chain. After two delightful days in Rome we went with the Chains on a tour. We visited Orvieto, Siena, San Giminiano, Florence. Living in Italy at that time, Ernst—an art enthusiast just as great as we are—was a perfect guide and able to select in the short time available the most exquisite monuments, domes, cathedrals, medieval piazzas and streets, etc. He also knew the most delightful restaurants at the Piazza del Campo in Siena, at a terrace in San Giminiano with a beautiful view at an enchanting scenery, and many others. We were happy and gay, always in high spirits, and enjoyed every minute. From Florence, Ann and Ernst returned to Rome. We spent a day in Verona and then drove to Sils Maria in the Engadin, a place which the Meyerhofs had loved and visited regularly. We had two lovely and restful days, walking in the Fextal. Severo and Carmen returned from there to Munich, where Severo worked with Fitzi Lynen, and I went on a lecture tour. A few weeks later we met again at the International Congress of Biochemistry in Paris, also Ernst and Ann Chain, Fitzi and Eva Lynen, and spent again lovely hours together.

The second trip selected for this essay started again in Israel at a meeting on the occasion of the opening of a new Life Science Building in the Weizmann Institute in June 1963. A small but distinguished group of scientists participated at this symposium: "New Perspectives in Biology". Among them were Ernst and Ann Chain, Fitzi and Eva Lynen, Hugo and Margit Theorell, Fritz Lipmann, Jacques Monod. The high level of the presentations and the organization was extraordinary, the warmth of the hospitality unsurpassed. On a well-organized tour we saw again Jerusalem, Cesarea, Haifa, Akko, Safed and Tiberias and spent a night in a kibbutz. Within the short period of a decade the growth and development were incredible. In Jerusalem we had a memorable reception by the President of Israel, Salman Shazar. The meeting was for everybody a great experience, scientifically and emotionally.

From Israel, Severo, Carmen and I flew to Athens and saw again our beloved Acropolis, a stirring reminder of our first visit. But this time we drove the next day through the Peleponnes. We visited Korinth and Mycene. The latter, four millenia ago the site of a highly developed civilization, is just a rather desolate series of ruins. A document of the greatness of this civilization are the stunning gold treasures which we saw in the Museum of Athens. After a delightful rest in Naufila we drove to Olympia, again mostly impressive ruins, reminders of a great past. The highlights of the trip was Delphi, with its charming location and the Parnassus in the background; the relatively large amount of preserved monuments, although not in such good condition as the Acropolis, still provide a good picture of its beauty in the past. The greatest joy, however, was an exciting and unforgettable Sophocles performance ("Bacchus"). We were extremely excited. After a short visit of the monastery in Daphne, with the beautiful Byzantine mosaics of the tenth century, we returned to Athens.

In 1966 we were again at our beloved Acropolis (Fig. 3). The next day we boarded a ship, the S.S. *Romantica*, and visited the Greek Islands. The whole trip was a dream. We visited Delos, Mytilene, Ephesos, Rhodos, with an excursion to the beautiful Temple of Athene in Lindos, Patmos, climbing to the Monastery of Haghios Johannes Theologius. At the end we visited Herakleion at the Island of Crete with a glimpse of the glory of the old Cretan civilization. Thanks to an extremely competent guide, we saw in a relatively short time an amazing amount of magnificent sites, monuments, and works of art.

Our dream has been for 15 years to visit the second Greece, Sicily. After the delightful FEBS Meeting in Madrid in 1969, where again many friends were present (Fig. 2), this dream became reality. We flew from Madrid to Sicily, visited Palermo, Monreale, Segesta, Selinunte, Agrigento and Syracuse, and spent four days in Taormina. From there we drove to Paestum, Ravello, Pompeii, Naples and Rome. The trip again surpassed our wildest expectations. The description of the magnificent sites and monuments and the works of art we had seen, our delight, emotions and elations on any one of these four trips would take manifold the space allotted.

Before closing this all too short essay, another important aspect of our common feelings and attitudes should be mentioned. Science is international in nature and all scientists form a close community. We all share the responsibility to lead the fight against the staggering problems facing mankind. One of the noblest functions of science is building bridges between nations. But scientists are also a product of communities with long and glorious histories. Severo is proud of his Spanish heritage, I of my Jewish heritage, although I am also deeply rooted in German civilization. In addition to our general commitments, we

FIG. 1. Berlin-Dahlem, 1929. Left to right: Fritz Lipmann, David Nachmansohn, Severo Ochoa, Francis O. Schmitt, Ken Iwasaki, Paul Rothschild.

FIG. 2. Madrid, 1969. Left to right: David Nachmansohn, Fritz Lipmann, Hugo Theorell, Severo Ochoa, Ernst Chain, Zacharias Dische, Carl Cori.

Fig. 3. Acropolis, 1966.

both feel a special obligation to the communities of which we are the products. Severo, has greatly contributed to the development of biochemistry in Spain; I devoted much time and effort in my life to help science in Israel.[2] In addition, I tried, during the last two decades, to build bridges between German and Israeli scientists, the most promising way that future generations could overcome a tragic past.

It is difficult in a restricted space to describe even a few highlights of the great assets and the many rewards of our friendship which has lasted almost half a century, has many roots and is founded upon sharing of passions and enthusiasm for values of deep meaning to both of us. These lines may convey an impression of how our friendship enriched our lives in many ways.

Transduction of chemical into electrical energy

In view of the role which Meyerhof played in our friendship, it seems appropriate to indicate briefly the fundamental notions that he developed, since they deeply influenced our scientific thinking. In a lecture, in 1913, in the Philosophical Society in Kiel: "Zur Energetik der Zellvorgaenge", he raised the problem of how the energy of foodstuffs, introduced into the organism, is eventually transformed into the energy required for specific cellular functions. The lecture reveals the depth of his philosophical and scientific thinking.

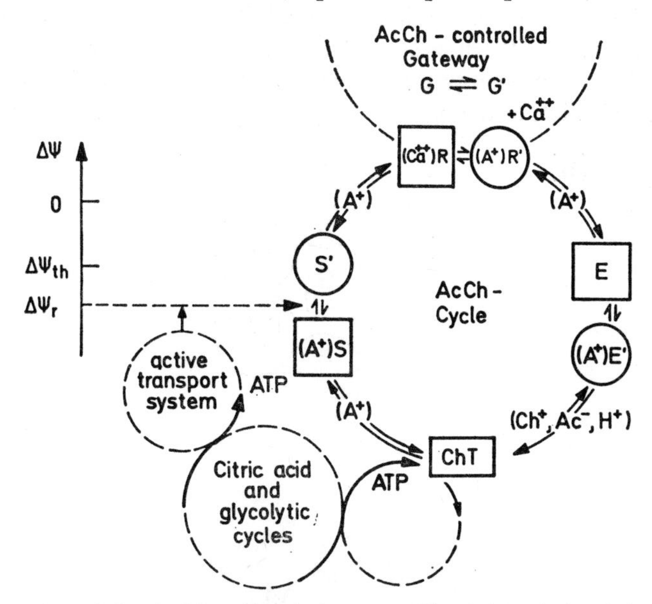

FIG. 4. Transduction of chemical into electrical energy. The citric and glycolytic cycles provide energy (ATP) for two functions required for electrical activity: (1) maintenance of the ionic concentration gradients across the membrane, and (2) of the AcCh-cycle by the acetylation of choline following its hydrolysis by AcCh-esterase, the virtually irreversible step of the AcCh-cycle. When a stimulus reduces the resting potential, Ψ_r, by about 15–20 mV to the threshold value, Ψ_{th}, an action potential is initiated by inducing a conformational change of the storage protein S to S′, thereby releasing AcCh-ions, A^+. A^+ is translocated to the receptor R and induces a conformational change to R′, thereby releasing Ca^{2+} ions into the gateway G, the permeation zone for the rapid Na^+ ion fluxes during the action potential. R has a high Ca-binding capacity. The Ca^{2+} ions perform the changes in the gateway. A^+ is hydrolyzed by AcCh-esterase, E, permitting R′ to return to R and to bind Ca^{2+}; thereby the gateway is closed. A^+ is reformed by choline O-acetyltransferase (ChT).

In 1918 he started his classical work on muscle biochemistry; he selected muscle not because of a particular interest in muscular contraction, but because of its suitability, with the primitive methods available, for his aim of testing the transformation of chemical into mechanical energy. The following notions emerged from this work: (1) The importance of bioenergetics and thermodynamics for understanding reactions in living cells. (2) The necessity of establishing the sequence of energy transformations for determining the reactions directly associated with the elementary process of a specific function. (3) The importance of the analysis of enzyme reactions required for cellular functions. (4) The cyclic character of cell reactions, first demonstrated with the glycolytic cycle in the early 1920s, including the irreversible oxidation of one-sixth of the lactic acid formed that provides the energy for resynthesizing the remaining five-sixths of end product to carbohydrate. (5) The paramount importance of energy-rich phosphate derivatives in intermediary metabolism. He demonstrated, with Suranyi, in 1926 the high enthalpy of phosphocreatine hydrolysis. In rapid sequence he discovered a series of other high-energy phosphate derivatives including ATP. He proposed with Lohmann, in 1934, ATP to be the substance providing the energy for the elementary process of muscular contraction.

As an example of the impact of all these basic notions, my work may be briefly mentioned here: the analysis of the transduction of chemical into electrical energy, the basis of nerve excitability and bioelectricity. One additional factor of crucial importance among several others, was the use, since 1937, of a uniquely favorable material for the analysis of the sequence of energy transformations during electrical activity and the isolation and characterization of the proteins of the acetylcholine (AcCh)-cycle, that controls

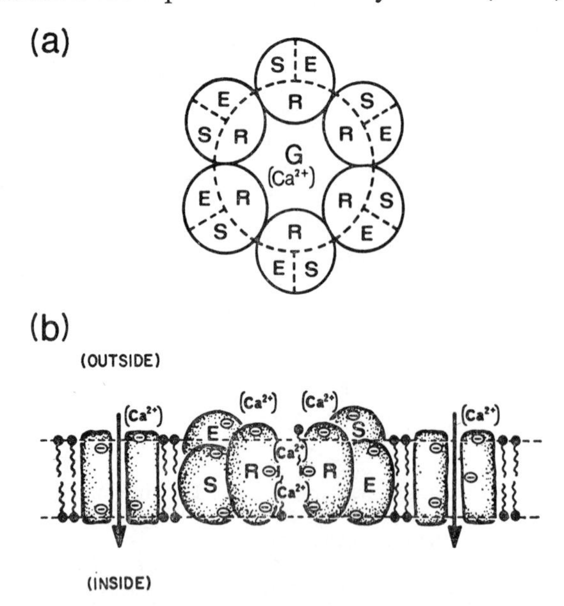

FIG. 5. Basic excitation units (BEU). The gateway G, is surrounded by SRE protein assemblies (in the Fig. by six). When a cooperative number of A^+ is released, activating a critical number of R, the $\Delta(\Delta\psi)$ is strong enough for the initiation of the action potential. (a) BEU viewed perpendicular from the membrane surface; (b) cross-section through a BEU flanked by two units for passages of K^+ ions; the arrows represent the local electric field vectors due to partial permselectivity to K^+ in the resting stationary state. The minus signs, ⊖, symbolize negatively charged groups of membrane components.

the ion fluxes underlying bioelectrical activity: the electric organ of electric fish. The almost unparalleled degree of specialization of this material for bioelectrogenesis renders it as suitable for analyzing the transduction of chemical into electrical energy, as muscle is for the transduction of chemical into mechanical energy. Neumann introduced the notion of Basic Excitation Units and has elaborated an integral model of nerve excitability (bioelectricity) that permits the interpretation of many electrical parameters in molecular terms. The fundamental concepts and the experimental data on which this work is based have been recently summarized in a monograph.[3] Figs. 4 and 5 illustrate some essential features of the present views, which are fully described in ref. 4.

REFERENCES

1. NACHMANSOHN, D., OCHOA, S. and LIPMANN, F. A. (1952) Otto Meyerhof: 1894–1951. *Science*, **115**, 365–369.
2. NACHMANSOHN, D. (1972) Biochemistry as part of my life. Prefatory Chapter, *Ann. Rev. Biochem.* **41**, 1–28.
3. NACHMANSOHN, D. and NEUMANN, E. (1975) *Chemical and Molecular Basis of Nerve Activity*, revised edition. Academic Press, New York.
4. NACHMANSOHN, D. (1976) *Proc. Nat. Acad. Sci. U.S.A.* **73**, 82–85.

DISCOVERY MADE EASY

LUIS F. LELOIR

Instituto Investigaciones Bioquimicas, Buenos Aires, Argentina

As anybody can easily guess, the title of this article is not to be taken seriously. It does give an idea, however, of the subject discussed which is one of the most recurring problems of scientific research: how to organize work in order to make discoveries. Research ability is something one learns by working with experienced investigators and is somewhat similar to the acquisition of any sort of craftsmanship. Undoubtedly, some people have more research ability than others; some make discoveries rather frequently, others once or perhaps never in their lifetime. Which are the factors involved? The subject is of great interest and well worth thinking about. We might not reach any conclusion of value but perhaps somebody reading this article can add new ideas to it and later others might add more until finally we would have a description of the necessary actions and circumstances leading to discovery. We would then be prepared to teach our students how to do research more rationally and perhaps we could program some sort of special computer to make discoveries for us.

These ideas surely sound foolish, but there is no reason why we should not find better ways and new strategies to carry out our research and to make discoveries more frequently. After all, mankind has been using the experimental method for only a few years. We can consider that experimental science started about four centuries ago, a short period compared with the appearance of *Homo sapiens* which is believed to have occurred perhaps more than 400 centuries ago. This is true for rational experimentation because primitive forms of experiments are perhaps as old as man and are practiced also by animals. What else do young animals and humans do during their play? They are learning, experimenting on ways of defending themselves and of dealing with the world around them.

Sir Richard Gregory's description of one of Galileo's experiments[1] gives an idea of the scientific atmosphere that dominated when experimental science was born:

> Members of the University of Pisa, and other onlookers, are assembled in the space at the foot of the wonderful leaning tower of white marble in that city one morning in the year 1591. A young professor climbs the spiral staircase until he reaches the gallery surmounting the seventh tier of arches. The people below watch him as he balances two balls on the edge of the gallery, one weighing a hundred times more than the other. The balls are released at the same instant and are seen to keep together as they fall through the air until they are heard to strike the ground at the same moment. Nature has spoken with no uncertain sound, and has given an immediate answer to a question debated for two thousand years.
>
> "This meddlesome man Galileo must be suppressed" murmured the University fathers as they left the square. "Does he think that by showing us that a heavy and a light ball fall to the ground together he can shake our belief in the philosophy which teaches that a ball weighing one hundred pounds would fall one hundred times faster than one weighing a single pound? Such disregard of authority is dangerous, and we will see that it goes no further!" So they returned to their books to explain away the evidence of their senses; and they hated the man who had disturbed their philosophic serenity. For putting belief to the test of experiment, and founding conclusions upon observation, Galileo's reward in his old age was imprisonment by the Inquisition, and a broken heart. That is how a new scientific method is regarded by guardians of traditional doctrine.

Our mental attitude has changed considerably since Galileo's time. Now many of us trust blindly in the results of experiment. However, probably we have not touched the ceiling of our methodology for discovery and surely there are still some possibilities for improvement.

After a slow start, experimental science developed with an increasing acceleration, the greatest changes having occurred during our lifetime. Therefore, scientists of our generation, who have spent a lifetime in research laboratories and who have met many of the world's leading scientists, should have considerable amounts of information on how many small or large discoveries were made and on the best procedures for discovery.

Types of discovery

The elements involved in uncovering new facts are variable. In some cases discovery is simply an intellectual act. In other cases the mental machinery is supported by observation and in still other cases the mental and observational elements are complemented by experiment. Different disciplines use mainly one of these elements. We might take pure mathematics as an example of a discipline that can advance by purely mental activity. Astronomy, systematic botany and zoology are based mainly on observation, whereas chemistry, physics and biology are experimental, that is, they apply experiment in addition to the mental and observational elements.

The act of creation

Human brain is a machine of rather limited capacity to create and adopt new ideas. The proof of this is how most people stick to their ideas even if abundant evidence shows them that they are wrong. It is not generally known how difficult it is for a human mind to develop a new idea even if it has only a small element of newness. One only has to think on the invention of a simple device such as the wheel, which was discovered by only one or a few groups of people, while fairly advanced civilizations remained ignorant of it.

The usual procedure for progress is to add very small new ideas to previously existing facts. The growth of knowledge occurs by small jumps and thus appears continuous. Large jumps are extremely rare. New ideas and inventions occur to people who are thinking constantly on a problem. This thinking has to produce a great preoccupation and even become painful. Then all of a sudden and perhaps as a consequence of some unconscious thinking the solution emerges. "Eureka", "I have found it". The case of Archimedes in his bath can be taken as an example of discovery by brain work complemented with observation. According to one version Hiero, King of Siracuse, had given a certain amount of gold to his goldsmith. The latter had made a very beautiful crown, but the king suspected that some gold had been replaced by silver. So he asked Archimedes to tell him if the crown was pure gold or not, but without destroying any part of it. Probably Archimedes knew that gold is denser than silver (in today's units the density is 19.3, as compared to 10.5 for silver); but how could he measure the volume? Presumably the problem rolled around his mind for some time. While taking a bath he noticed how his body became lighter on immersion and how the water level increased. There are so many different versions of this episode that it is difficult to guess how his thinking went on, but he may have planned to weigh the crown in air and in water and then do the same with pure gold and pure silver. Anyhow, he solved the problem, and according to the tale he ran in the streets in the nude shouting: "Eureka"!

The tale stresses the fact that Archimedes had been thinking hard, perhaps painfully on his problem and then a mental association with a different set of facts discharged the solution.

Koestler has written a book on the act of creation.[2] According to him the circumstances for creation and those provoking laughter are similar. In both cases what is needed is what he calls a bisociation, that is the meeting of two trains or frames of thought. When these two bisect there may be an act of creation or a hilarious situation. A biochemical example may clarify Koestler's thinking. A lecture hall packed with people is listening to a communication by Marianne Grunberg-Manago. At a certain moment the slides start to be projected. The chairman, Dr. Severo Ochoa, tries to dim the lights and fumbles in a switch-board. He touches several switches and slowly . . . the platform and with it Dr. Grunberg-Manago begin to descend. The public roars with laughter. A descending platform is nothing to laugh about and is a fairly frequent event in theaters. What made the situation laughable was the bisection of the scientific and the theatrical contexts. It would be equally laughable to see a monkey appearing in the scene although we may look at it seriously in a zoo.

One of the examples given by Koestler is less biochemical and goes as follows: "In the happy days of La Ronde, a dashing but penniless young Austrian officer tried to obtain the favors of a fashionable courtesan. To shake off the unwanted suitor, she explained to him that her heart was, alas, no longer free. He replied politely 'Mademoiselle, I never aimed as high as that'." In this story "high" is bisociated with a metaphorical and with a topographical context. What Koestler does not tell us is how to bisociate so that the result is a discovery and not simply a comical episode.

Whatever the mechanism of birth of a new idea, one of the requirements is a situation which stimulates thinking. The story of Archimedes in his bath is one case of a more general fact. People seem to think better, and generate more ideas in the bathroom. One of our coworkers frequently came to the laboratory saying that while he was shaving it had occurred to him that such and such experiments should be tested. Several other colleagues have come various times with bathroom-generated ideas which were often very good. Other scientists report that their best ideas occur while they are in bed and specially in the first hours of the morning. Peace and isolation are probably the main advantages of bath and bed in relation to other parts of the house. There are no reports on new ideas having been generated in the turmoil of a cocktail party. Discussion between four or five people engaged in the solution of the same problem or perhaps working on the same research project can be very fruitful provided the necessary isolation can be obtained. Similar arrangements have been called brainstorm sessions and have had some successes.

Observational sciences

Before experiment became established as a part of the scientific endeavour the mind had to rely on observation only. Darwin's monumental work on evolution relied mainly on observations. Before him several others had held the idea that living things varied in form and that there occurred some kind of evolution. However, it was thanks to Darwin's painstaking and careful set of observations that the theory of evolution gained widespread acceptance. There were many discoveries in his work, but the largest mental "jump" was when he hit on the idea of natural selection. The time must have been ripe for this new idea because as often happens it occurred independently to another person, Alfred Russel Wallace.

It is interesting to point out that Darwin, like many other people engaged in wrestling secrets out of Nature, gave some thought to the mechanism of discovery.

These are his words:[3] "I have been speculating last night what makes a man a discoverer of undiscovered things, and a most perplexing problem it is. Many men who are very clever—much cleverer than the discoverers—never originate anything. As far as I can conjecture, the art consists in habitually searching for the causes and meaning of everything which occurs".

Darwin was well aware of the importance of keeping free from preconceived ideas:[3] "I have steadily endeavored to keep my mind free so as to give up any hypothesis, however much beloved (and I cannot resist forming one on every subject) as soon as the facts are shown to be opposed to it".

Experimental science

The most powerful way devised by man for finding new facts is the experimental method. It consists in the alternative use of hypothesis checked by observation and experiment followed by a new hypothesis which is also checked and so on. This alternating procedure has led to increased rate of discovery and to the explosive development of science. It is difficult to imagine that man can develop a more powerful procedure, but as mentioned before it is less than four centuries old, and surely it can be perfected and superseded.

Although we call it experimental science there is an element of art and perhaps intuition in its practice. The cycle hypothesis–experiment–new hypothesis can be schematized as follows:

.– *Study of the problem*

1.– *Hypothesis.*	1a – imagining a hypothesis
	1b – critical examination
2.– *Experiment.*	2a – design
	2b – doing the experiment
3.– *Observation*	or interpretation of experiment
4.– *New hypothesis*	4a – imagining
	4b – critical examination

For the observational sciences the sequence would go directly from 1 to 3, that is without experiments. The limiting step in the whole process can be any of the different steps. The lack of library facilities or of adequate teachers can slow or stop the study of the problem and inadequate laboratory facilities can make experiment the bottle neck.

However, when the process of discovery is too slow, the step which is usually blamed for is that of finding a good hypothesis. The problem of creativity has been the subject of a lot of writing and discussion. In a conference on "The creative process in science and medicine" the participants, an élite of distinguished scientists, agreed that the major element in scientific success in the ability to distinguish between ideas and good ideas.[4] The important part seems to be the critical examination of the hypothesis so as to discard the bad ones, or those that cannot be tested with the available methods. Critical judgement is required in order to avoid working a lifetime on trivial problems but if it is overstressed it may lead to a paralysis of the research work. So many factors are involved in scientific work that different people work with different styles.

Szent Gyorgi[5] has described two types of scientists: the Dyonisians who work mainly by some sort of intuition or subconscious reasoning, and the Appollonians who rely mostly on the logical–rational elements. The two types complement each other; the Dyonisians can open new horizons or unexpected alleys, and the Apollonians are likely to develop these fields to perfection. According to Szent Gyorgy, most of the Apollonians are very good at writing research projects and reports so that they get the lion's share of the research money. He considers that this is not healthy for the development of original research.

There are different styles in research according to the personality of the researcher and also in the way in which the experiments are designed. Experimenting has some similarities with fishing. The discovery may be compared to the fish; it can be big or small, single or numerous. We can use a hook or a net. The catch depends a lot on our technique. We can bait the hook with worms, fish or artificial imitations, we can use hooks of different sizes and all this determines what kind of fish we are going to get.

The results of fishing with a net are rather different. The probability of a catch is much higher and the size of the fish will be more varied. An example may clarify the idea. Two scientists start to work on the same problem but using different approaches. The basic fact is that an organ extract produces certain effects when injected into animals. One of the scientists, Dr. Hooke, after reading carefully the literature reaches the conclusion that the active substance should have a certain structure. He is a clever chemist and thinks he can synthesize the substance rather easily. He therefore decides to follow this line of work. The other scientist, Dr. Netti, has a more modest plan which consists in purifying the active substance and if possible in determining its structure. He had no idea of what it could be. Dr. Hooke's experiment can only give a single answer, i.e. the substance is or is not what he thinks. In Dr. Netti's project there is no preconceived idea of the structure of the substance; any information on its identity would be useful and furthermore it might turn out to be a very important compound.

Of course Dr. Hooke could be very successful if his synthetic work were very easy but if the work involved were about the same as Dr. Netti's he would need an amount of luck much larger than that of his rival.

The colorful circumstances often attached to discoveries are usually atypical cases, and that is precisely why they make a good story, such as that of Kekulé discovering the structure of benzene while riding on a London bus. One common requirement for getting a new idea seems to be a previous period of obsessive thinking. This seems to have occurred to Archimedes and is evident in Watson's Double Helix.[6] This book contains an entertaining account not only of the scientific aspect but also of the motivations, rivalries and other human characteristics involved in the process of discovery. Here the problem consisted in imagining models for a substance (DNA) of fairly well-known structure and checking if they agreed with the X-ray diffraction data. After many unsuccessful attempts a model which agreed well with the experimental data was found. This was one of the most commented advances in science.

There are parts in Watson's book that reveal how in the course of seeking the solution, his thinking became obsessive. For instance: "Even during good films I found it impossible to forget the bases"; "The fact that we had at last produced a stereochemically reasonable configuration for the backbone was always in the back of my head".

Besides the preoccupation on research there are many other conditions that are necessary for discovery. Most of them are difficult to pinpoint. The reason why some scientists are very productive and others less so is not easy to identify. Usually it is taken as a fact that cannot be changed. I remember that Carl Cori, referring to a case of non-productivity, said "He has no luck". This was a kind way of describing the fact that the person in question did not have the personality nor the ability that make a good scientist. In many cases the failure is due to factors which are difficult to identify. One reason can be that some people never give much thought to their research problems; their mind is occupied with other things and they never get obsessed with their work. Certain people do not have much imagination and have only few ideas while others are so full of new ideas that they may not have the time nor the patience to test them.

Discovery requires many other ingredients besides those already mentioned. Knowledge, curiosity, patience, hard work....

Curiosity is another element lacking in some people; they look at Nature around them but never ask themselves about the mechanism of all the marvellous things that one can see. Curiosity is what made the princes of Serendip look around and discover what went unnoticed to other people. But serendipity, "the faculty of making happy chance finds" requires also careful and wise observation. The creator of the word, Horace Walpole, explains it as follows:[7] "The princes of Serendip were always making discoveries, by accidents and sagacity, of things which they were not in quest of: for instance one of them discovered that a mule blind of the right eye had travelled the same road lately because the grass was eaten only on the side, where it was worse than on the right,—now you understand serendipity".

A trained research worker would surely find the evidence very weak and the behavior of the mule rather odd. He surely would recommend confirmatory work such as observing the eating habits of a mule that has one eye covered.

Finally one should mention one of the less entertaining and less romantic but most important factor: hard work. There does not seem to be any easy way to success other than doing many experiments and doing them well.

Epilogue

Although there are people who blame science and technology for many of the present-day problems and who consider that the rate of discovery should be slowed down, it seems likely that what is required is just the opposite, that is, an increased rate. This will be necessary in order to solve the grave problems brought about by the overpopulation of our planet.

In order to increase the rate of discovery the obvious solution, and the one generally used, is to put more people to work and to allot more resources to research. However, constant consideration should be given to the productivity per worker since there are surely many possibilities for improvement.

REFERENCES

1. GREGORY, R. A. (1923) *Discovery of the Spirit and Service of Science*, p. 2, The Macmillan Company, New York.
2. KOESTLER, ARTHUR (1971) *The Act of Creation*, Pan Books Ltd., London.
3. SHAPLEY, HARLOW (1943) *Samuel Rapport and Helen Wright, A Treasury of Science*, p. 435, Harper and Brothers, New York and London.
4. MAUGH, T. H. (1974) Creativity: Can it be dissected? Can it be taught? *Science*, **184**, 1273.
5. SZENT GYORGI, A. (1973) Dionysians and Apollonians. *Science*, **176**, 966.
6. WATSON, JAMES D. (1970) *The Double Helix*, Penguin Books, England.
7. ROSENAU, M. J. (1935) Serendipity, *J. Bact.* **29**, 91.

COMMENTS ON THE PRODUCTIVITY OF SCIENTISTS

HANS A. KREBS
Metabolic Research Laboratory, Nuffield Department of Clinical Medicine, Radcliffe Infirmary, University of Oxford, Oxford, England

THE PROBLEM

I met Severo Ochoa for the first time in Berlin in 1929—46 years ago, when we worked for about a year under the same roof; Severo in the laboratory of Otto Meyerhof and I, two floors above, in the laboratory of Otto Warburg. At that time we were both engaged on full-time research and we have both continued research throughout our careers, in spite of many distractions and many other obligations, and now in retirement from official responsibilities our passionate interest in research has not flagged. When I compare our lives with those of many other scientists, among them our own contemporaries, I note that many who started keenly, promisingly, and brilliantly, slowed down after a while and showed a sharp drop in productivity, when still rather young. Only a relatively small number continued persistently with undiminished vigour. So I think it is of interest to consider on this occasion some of the factors which affect scientific creativity and productivity in advancing years.

What is true for science—that some cease to be productive while others carry on—may occasionally be true also for other professions involving creative efforts, but it is much less striking in the arts. The majority of creative writers, painters, musicians, continue to produce first-rate work into old age. Verdi completed Falstaff, by some considered his best opera, at the age of 80. Titian created masterpieces when he was well in his 80s and Goethe retained his imaginative creativity and powers of expression until the end of his life at 81. In our own generation P. G. Wodehouse, who died in February 1975 at the age of 95, produced bestsellers until the very end of his life. Robert Browning's last volume of poetry, perhaps his best and most beautiful, was published on the day of his death at the age of 77. An exception is Rossini who had written thirty operas by the time he was 37, but composed next to nothing during the next 33 years of his life.

AGE AND PRODUCTIVITY

Thus, while in the creative arts the quality of productivity does not necessarily change with advancing years the standard of scientific productivity is commonly believed to have an optimal period which is not the same for different subjects. It is taken to end rather early in mathematics. G. H. Hardy,[1] in his autobiographical book *A Mathematician's Apology* commenting on the relation of age to scientific creativity, states that this is especially important for mathematicians. He writes: "No mathematician should ever allow himself to forget that mathematics, more than any other art or science, is a young man's game"... "Newton gave up mathematics at fifty and had lost his enthusiasm long before; he had recognised no doubt by the time he was forty that his creative days were over. His greatest idea of all, fluxions and the law of gravitation came to him in about 1666 when he was 24". Hardy quotes Newton as saying "in those days I was in the prime of my age for invention and minded mathematics and philosophy more than at any time since". "He made big discoveries until he was nearly forty (the 'elliptic orbit' at 37) but after that he did very little but polish and perfect".

Hardy (1887–1947) was recognized by his peers as the leading English mathematician of the first half of the present century.[2] Hardy himself ceased to publish original work at the age of 58—relatively late for mathematicians. In the subsequent twelve years he published several books including the delightful one from which I have quoted.

Another distinguished mathematician (and physicist), P. A. M. Dirac, expressed the same feeling when he wrote,

> Age is, of course, a fever chill
> That every physicist must fear.
> He's better dead than living still
> When once he's past his thirtieth year.[3]

Dirac made his main discovery when he was 26. He was elected Fellow of the Royal Society at 28 and received a Nobel Prize at 31.

What is true for mathematics applies much less to experimental physics and chemistry, and perhaps still less to the biological sciences. But it is certainly true that the pioneering character, i.e. the imaginative opening up of new fields, occurs far more frequently with young scientists. In my context, however, this question of outstanding quality is not of primary importance. What I am concerned with is the maintaining of *efforts* of productivity. Whether such efforts are crowned by outstanding successes is partly a matter of luck and not entirely within the willpower of the scientist. So my main primary problem is the question of why some scientists, after a time, stop making serious efforts to remain active in research, while others continue. I will return later to the question of deteriorating quality.

DIFFERENCES IN THE PREREQUISITES OF PRODUCTIVITY IN ARTS AND SCIENCES

I must first consider some fundamental differences between arts and sciences in the prerequisites of creative work. The roots of productivity in all fields are imagination and what may be called a style of working.

Style in scientific research is made up of two main elements, one practical, the other theoretical. The practical one concerns the methods used in experiments, the theoretical one concerns the kind of questions asked by the experimenter to which the experiments are to find an answer. It is characteristic of science that both experimental methodology and the formulation of questions change very rapidly, nowadays faster than ever. This implies a danger of becoming obsolete in techniques as well as in ideas. The scientist must be prepared to work very hard, just to remain up to date. He must study and learn continuously.

There is, of course, something comparable in the arts, but it is far less marked. If style is defined as the combination of methods of working and the manner of looking at the subject material to be treated, then by these two criteria, style in the arts does not become obsolete anywhere near as quickly as in the sciences. A characteristic style in the arts may remain fresh for a lifetime, and even though styles of painting, sculpture, poetry, drama and prose change gradually, the great achievements in the arts have an eternal quality and never become obsolete. The works of Newton, of Pasteur and even of the great scientists of the turn of the present century—say Emil Fischer, Nernst, Ramon y Cajal—are read today almost exclusively for historical reasons by people interested in historical aspects, although of course the results of their work are incorporated into standard textbooks.

The creative artist, unlike the scientist, does not need continuing rejuvenation of his tools (though he may modify and develop them), nor do his subjects—human nature and man's reactions to the natural world around him—ever change. This is the reason why Horace, Sophocles, Homer, the Bible, Shakespeare and Bernard Shaw, the great painters of the Renaissance and the music of Bach are as alive to us today as they were at the time when these pieces of art were conceived.

This danger of obsolescence, then, is the reason why the productivity of scientists is especially difficult to maintain. As already remarked, a very great effort is needed to prevent falling behind. Science is a country, like the Looking-glass Country of Lewis Carrol, where according to the Red Queen, "it takes all the running you can do, to keep in the same place. If you want to get somewhere else, you must run at least twice as fast as that!" Many a scientist would be in sympathy with Alice's reply "I'd rather not try, please; I'm quite content to stay here—only I am so hot and thirsty!"

DISTRACTION FROM PRODUCTIVITY IN SCIENCE

What I have discussed so far concerns one of the factors that makes continual productiveness difficult for scientists. There are others. The majority of scientists are perpetually tempted away from science—to administration, to positions of power, and to teaching. Many are, of course, primarily paid to administer and teach and their employers expect them to give these activities priority. These responsibilities rob scientists of much of the time they need for research which requires far more time than what is nowadays considered an ordinary working week. Creative artists, as a rule, have all or most of their own time at their disposal. Additional tempting and menacing distractions for the reasonably successful scientist are the invitations to travel, to lecture, to attend conferences and to be feted on these occasions.

A special distraction to which Severo was exposed comes from the fame of a Nobel Prize. Much has been said and written about how a Nobel Prize may affect a scientist's life and interfere with his productive research. Harriet Zuckerman[4,5] came to the conclusion that productivity of U.S. scientists, as measured by the output of papers, falls by 30%, from 6.2 to 4.2 papers per annum in the 5 years following the award. André Lwoff, Tsung-Dao Lee, Brian Josephson, according to a report by Sherwood,[6] commented that their own research had been much hampered by distractions following the award. Others like Geoffrey Wilkinson and Robert Holley[6] recently found the disturbance caused by the Nobel Prize insignificant, as earlier did Otto Warburg, Otto Meyerhof—and Severo Ochoa. These were men who had no difficulty in avoiding many of the distracting intrusions; their insatiable curiosity to find out something new by research prevailed over the attraction of being lionized—a personality characteristic.

Of course Nobel Laureates get rather more invitations than other distinguished scientists. They are also liable to get more requests to sign autographs, to express opinions, to give interviews or to write on many aspects on which they are not experts, to supply testimonials in support of other people, to sit on committees, to make speeches on festive occasions.

A few months after my award I enjoyed the company of James Frank, the physicist, when I stayed for a month at the Faculty Club (Quadrangle Club) of the University of Chicago where he was a resident. From his own experience he warned me that I would get many invitations to dinners and to serve on committees merely as a "table decoration". This prediction proved to be correct and was openly expressed, for example, when I was recently asked to join an Advisory Board in the United States. The organization frankly told me that I was their choice not because of any special competence, but because I had been "sanctified".

Often much more serious are petty wastes of time at other levels which have nothing to do with the Nobel award but with pedantic demands of administrators. In my own case, fortunately, my earlier financial support from the Medical Research Council and the Rockefeller Foundation demanded a minimum amount of time, but at later stages I found the lengthy grant applications a serious and dangerous intrusion into research and study time—of which there is never quite enough.

One distraction mentioned by Macfarlane Burnet[7] is a sense of extended responsibility of a Nobel Laureate. No doubt the general public believes, quite wrongly, that pronouncements by Nobel Laureates even outside their own field are worth listening to. So on occasions Burnet, like others, felt that he had to act accordingly and to lecture or to write on wider issues of science and general academic matters, and even on social problems, emphasizing especially the bearing of the scientific approach to these problems. As it has been said, "Nobel oblige".

AGE AND QUALITY OF RESEARCH

To return to the problem of the loss of outstanding quality of research with advancing age which Hardy emphasized as being of special relevance in mathematics: as I already mentioned this is also true, but less important, in the experimental sciences, as the life history of experimental scientists indicates.

Otto Warburg made outstanding contributions to cancer research between 1922 and 1926 (aged 38 to 43), to the elucidation of the nature of the oxygen activating catalysts of respiration between 1926 and 1928 (aged 43 to 46) and to enzymology between 1931 and 1944 (aged 46 to 61). In these three fields his contributions were judged by the Nobel Committee as worthy of a prize.

The life history of productive scientists also indicates that many start a new line when a "mine" threatens to be exhausted. Sometimes it is impossible to push forward a problem at a particular time. When knowledge in collateral fields and when new techniques have evolved one can return to old problems and tackle them in a new way and from a new angle.

Often lack of imagination is the reason for sterility. The list of publications of many a scientist shows that he only had one idea in his career—and this was often suggested by the Ph.D. supervisor.

In an essay on "Scientific Sterility in Middle Age" Slobodkin[8] expresses the opinion that while motivation of a research scientist is basically curiosity this is in most cases overlain by other things—prestige, money, prospects of security—and these are often found outside research. Productivity in science will thrive only if curiosity is very deep indeed—insatiable.

It may also be of importance that there is one incentive for creative work which is very different in the arts on the one hand and in scientific research on the other. In the arts creative achievements may result in worth-while financial rewards. Creative research in the natural sciences is singularly unrewarding in financial terms, compared with other activities open to a scientist such as lecturing at other institutions, writing books or serving as a consultant on committees and on boards. True, early in the scientist's career successful research leads to promotion to senior posts but once a scientist is established in a top position the financial benefits he may reap from research are very uncertain. The few lucky ones may get one of the big prizes. If they get a full-time established research post their salaries will not depend on the discoveries they make.

Severo is one of those men of science who is continuing with research after he has earned all the honors and all the material benefits that may be expected, and when he has no doubt the opportunity of spending all his time on being celebrated and "enjoying" in this way the fruits of his labours. To be sure Severo and all of us in similar situations can enjoy spending a little time on these sometimes embarrassing pastimes—embarrassing because one may feel that the tributes are overgenerous and unduly exaggerated. I think a fundamentally greater joy is the pleasure of the struggle with Nature to wrench from her some of her secrets, however small. I suppose this sort of feeling was behind the passage in a recent letter (20 November 1974) I had from Severo: "I moved to the Roche Institute on July 1, and I cannot tell you how happy I am here with splendid facilities for work, and with the feeling that I now have time to do what I most like to do".

Some people look forward to retiring from their occupation in order to develop hobbies for which they never quite had enough time. But for the curiosity-motivated research man hobby and work are almost the same thing. As Ernst Henry Starling said at the International Congress of Physiology in Stockholm in 1926: "research is the hardest and the finest game".

It is true that older scientists who still produce research of high quality, do not as a rule reach the very highest level of their earlier work—in contrast to the older artist whose quality is much less age-dependent. Which of the essential qualities required for creative research is liable to deteriorate with advancing years? Not the capacity for hard work. Imagination usually remains but perhaps in a more limited way, not as unprejudiced and not as revolutionary as in youthful days. The accumulated knowledge and experience is no less extensive but may become somewhat stale because retentive memory deteriorates with the years, so that the rapid flow of new information may be less well retained and less effectively assimilated into the general body of knowledge.

Deterioration of memory is a handicap which can be partially compensated by counter-measures. Early in life it is easy to commit matters to memory; later, efforts are needed—efforts to memorize, and more reliance has to be placed on systematic note-taking. It is interesting that there are stage actors whose capacity for memorizing their parts is still very adequate in their eighties. Perhaps continuous practice improves memory.

Thus there may be several reasons why it is more difficult in the sciences to maintain the highest quality of creative activity, but none of them is compelling any more than they are in the humanities. If the majority of scientists lose creativity as they grow older they probably find other activities easier, more attractive and more rewarding.

ORIGINS OF CONTINUED PRODUCTIVITY IN SCIENCE

No doubt the most important single factor in sustaining the research effort is the intensity of motivation. Only those who have a profound love-relation to their work, carry on and on, a love-relation stemming from inherited as well as from environmental factors—from nature and from nurture. They belong to those people in any walk of life to whom action is an inescapable necessity. Goethe has often written about the importance to him of being active, in order to be satisfied. In Wilhelm Meister he says "action is man's first obligation". In Faust (modifying the Gospel according to St. John "In the beginning was the Word") he says "In the beginning was the Deed". In a letter to his friend Knebel he wrote on 3 December 1781: "Even in the smallest village or in the most desolate island I would be just as active merely in order to live. This is an article of my faith." In other words, to the deeply motivated research scientist, and to deeply committed people generally work is a vocation, not only a profession, a vocation being something one lives to do, a profession something one does to live.[9]

And just as most musicians, most painters and most writers are impelled to pursue their skills to their dying days so the researcher in science never stops as long as he is not forced by lack of facilities or by the infirmities of old age.

There are other people to whom creative action is unimportant although they may feel that they must keep busy. But their way of being busy tends to be more passive; they let events come to them rather than creating events. They may enjoy their job, but they enjoy life even more by watching other people doing something, by playing, chatting and pursuing hobbies and sports. Between these extremes there are many intermediate types, for people are not equal, and Nature never meant them to be equal. Indeed, "polymorphism", i.e. individual variation within a species, is a biological necessity.[10] Societies of identical individuals could not survive for long because each hazardous situation requires individuals of different types to overcome the difficulties, and the safest and soundest society is one which is rich in people who are excellent in different ways.

Love alone is, of course, not enough, and the additional constellation of talents required for excellence in creative research—a measure of intelligence, perseverance, a capacity for hard work, some manual dexterity and a pleasure in practising this dexterity, imagination, cool critical objectivity—is not at all common. But then this applies to all spheres of creative skills—literature, fine arts, music. Many try but few get very far.

THE ARCHITECTS AND THE WORKMEN

If excellence in science as exemplified by Severo Ochoa is rare, this should not be a depressing thought for others because apart from architects of the edifice of science—to use a metaphor of C. N. Lewis[11]—there is a need for innumerable workmen. Workmen are no less essential but they require and they are willing to accept direction. Something analogous should also apply to scientific research. A scientist who has become short of ideas and browned off by a sense of frustration may well find continued refreshment if he associates himself, perhaps quite loosely with an inspiring leader like Severo Ochoa who has ample ideas to share. Such a scientist should make a deliberate decision to abandon a little of his "independence" in order to remain productive and happy.†

† The matters discussed in this article have been considered in recent years by sociologists, especially by Merton[12] and his school (Zuckerman,[13] Cole and Cole[14], Roe[5]) and by Ziman.[16] My excuse for raising them is that my approach is somewhat different, stemming as it does from a life-long personal association with the problems of scientific productivity.

REFERENCES

1. Hardy, G. H. (1969) *A Mathematician's Apology*, with a foreword by C. P. Snow, Cambridge University Press.
2. Titchmarsh, E. C. (1949) Godfrey Harold Hardy. *Obituary Notices of Fellows of the Royal Society*, **6**, 447.
3. Dirac, P. A. M. quoted in ref. 13.
4. Zuckerman, H. A. (1967) The sociology of the Nobel Prize. *Scientific American*, **217**, no. 5, 25–33.
5. Zuckerman, H. A. (1968) *Scientific Elite: Nobel Laureates in the United States*, University of Chicago Press, Chicago.
6. Sherwood, M. (1974) Life at the top. *New Scientist*, 3 Oct., pp. 13–17.
7. Burnet, M. (1974) *New Scientist*, 3 Oct., p. 17.
8. Slobodkin, L. B. (1971) Scientific sterility in middle age. *American Scientist*, **59**, 678–679.
9. Parkes, A. S. (1968) The youngest profession. *Perspectives in Biology and Medicine*, **10**, 56–66.
10. Ford, E. B. (1971) *Ecological Genetics*, 3rd ed., London.
11. Lewis, C. N. and Randell, M. (1923) *Thermodynamics and the Free Energy of Chemical Substances*, McGraw-Hill, New York, Toronto, London.
12. Merton, R. K. (1973) *The Sociology of Science*, The University of Chicago Press, Chicago and London.
13. Zuckerman, H. A. and Merton, R. K. (1972) Age, aging and age structure in science. In *Aging and Society*, Vol. 3, *A Theory of Age Stratification* (M. W. Riley, M. Johnson and A. Foner, eds.) Russell Sage Foundation, New York.
14. Cole, J. R. and Cole, S. (1973) *Social Stratification in Science*, The University of Chicago Press, Chicago.
15. Roe, Anne (1963) Changes in scientific activities with age. *Science*, **150**, 313–318.
16. Ziman, J. M. (1969) Some problems of the growth and spread of science in developing countries. *Proc. Roy. Soc.* A, **311**, 349–369.

PREBIOLOGICAL CHEMISTRY AND THE ORIGIN OF LIFE. A PERSONAL ACCOUNT

J. Oró

Departments of Biophysical Sciences and Chemistry, University of Houston,
Houston, Texas,
and
Instituto de Biofísica y Neurobiología, C.S.I.C.,
Barcelona, Spain

Of stardust and life

When one looks at the sky on a clear night in a reflective mood, a few thoughts may come to one's mind: How were the myriads of stars formed? How did life and man come about on this planet? Are there other worlds around the many stars where similar or more advanced forms of life are wondering as we do about their own existence and that of other "civilizations"? Where is the world taking us or where are we going?

These are indeed four of the most fundamental questions that man can ask himself without the hope of receiving any immediate and satisfactory answer. The first and the last questions may never be answered. The second and the third may some day be answered if it proves to be true that they are amenable to the treatment by the scientific method.

A romantic's view

INTRODUCTION

It has been said that the future historian will probably look upon the past few decades as the first golden age of biochemistry.[1] Aside from the knowledge gained on the nature and function of a large number of key enzymes and coenzymes and the elucidation of the most important energy-yielding and biosynthetic reactions, some of the most significant biochemical breakthroughs (e.g. DNA structure, polynucleotide phosphorylase, DNA and RNA synthesis, genetic code, protein biosynthesis) made in the recent past, had to do with compounds involving phosphodiester bonds or high-energy phosphates, key molecules of life. By having the insight and wisdom to follow the thread of the biochemistry of high-energy phosphate, and working with vigor and imagination, Severo Ochoa has been a pioneer or major contributor in many of the above discoveries. These major advances together with the contributions of Mendelian and molecular genetics[2] have provided a clear understanding of the basic principles that govern all living organisms and have demonstrated the fundamental unity of terrestrial life, as required by Darwin's theory of biological evolution.

SEVERO OCHOA, NEW YORK, 1952

It was August of 1952. Upon arriving in New York for the first time on my way to Houston, I visited Severo Ochoa in his laboratory at New York University School of Medicine. This first visit would eventually have a significant impact on the course of my career. I had brought with me the educational and biological background obtained at the State High School of Lerida (1934–1942) and the chemical background at the University of Barcelona (1942–1947), some ideas on chemical evolution and a determination to study the fundamental problem of the origin of life. These ideas were rudimentary in nature and intuitive in their inception, going back to my high school years, but were logically based on a molecular extrapolation of the Darwinian theory of evolution to the predawn of terrestrial life. Although Darwin had his difficulties "in understanding how a simple being can be changed and perfected into a highly developed being", the essential conclusion of this theory was that man and other complex creatures are derived from earlier less complex ones by the processes of variation (requiring genetic change) and natural selection.

In posing the problem, the relatively simple argument I made at that time went somewhat as follows: If in the process of going from complex to simpler living beings, we continue extrapolating backward in time, from organism to organism, we will eventually arrive at a final point where we will find the simplest, most primitive, independently living cell, possibly a prokaryote. At this crucial point one must accept either a continuity or a discontinuity in the overall process of evolution. If we accept a discontinuity (e.g. panspermia theory, special creation, etc.) we refuse to attack the problem frontally and we transfer it to other worlds, places and processes of which we know very little, if anything at all. If we accept the continuity of the overall evolutionary process, then we are logically forced to extrapolate further backward from the cell to lower levels of subcellular and molecular organization in search of simple but progressively more complex molecules, reactions and interactions which eventually can lead to the self-assembly of the first living entity. In other words, if multicellular organisms are derived by association from single eukaryotic cells, and if eukaryotic cells are derived from prokaryotic cells by symbiotic association processes, it is not unreasonable to imagine that a primitive prokaryotic cell could have originated by a similar symbiotic association of independent molecules or molecular complexes, each showing at least one of the essential attributes of life.

I talked about these matters with Severo Ochoa and I discussed with him whether some known chemical synthetic reactions could be considered significant in the context of a very simplified prebiological organic chemistry. For example, in the "formose" reaction studied by A. Butlerow, O. Loew and E. Fischer, a one-carbon compound, formaldehyde, is converted into monosaccharides by a simple base-catalyzed process. In the 1930s similar reactions were still implicated, erroneously of course, in the biosynthesis of sugars by plants following the reduction of CO_2 to bound formaldehyde, as noted in some biochemical texts of that time.

I asked Severo whether it would be worthwhile to study non-enzymic reactions involving formaldehyde, certain one-carbon compounds and other simple organic molecules in the hope of discovering abiotic pathways for the synthesis of the most important blocks of biological macromolecules. His answer was quite encouraging, and although I was not fully prepared to start work in this field immediately, it helped me to plan the program of research which I was to carry out later upon the completion of my doctoral studies.

Fortunately, my graduate research in biochemistry at Baylor College of Medicine, in Houston, involved studies on the mechanism of formic acid oxidation in animal tissues and on the incorporation of this one-carbon compound into nucleic acid components and their precursors. This provided me with experience in the use of isotopic tracers and the background in biochemical research which proved crucial, in later years, for unraveling the intermediates and mechanisms of synthesis of purines and other compounds, when I discovered the prebiotic synthesis of adenine and other building blocks of nucleic acids, and a general pathway or method for the prebiotic formation of oligodeoxynucleotides and peptides.

The following is an informal and personal account of my journey through the field of prebiological organic chemistry during the last 20 years.

PREBIOTIC SYNTHESIS OF BIOCHEMICAL COMPOUNDS—SCOPE OF ESSAY

From a logical point of view, it would have been advisable to present first the analytical and theoretical observations concerning the presence and formation of organic molecules in nature, and then the laboratory experimentation supporting them. However, such an approach would have taken me beyond the scope of this essay. Therefore, I will restrict myself to the laboratory synthesis of biochemical and related organic compounds under presumed primitive earth conditions. These conditions require a non-oxidizing environment, an aqueous system not far from neutrality, moderate temperatures and a variable concentration of organic and inorganic compounds, which can be easily increased by simple evaporation.

In line with the spirit of this book, I will present mainly the highlights of our contributions and those of other laboratories on this subject, with autobiographical comments, following an approximate chronological order. I will leave aside the general aspects of the theory of chemical evolution and will mention briefly some of the more specific observations of organic molecules in interstellar space and comets.

Unfortunately, the interesting findings of organic molecules in meteorites, and of microfossils in the most ancient terrestrial rocks, remain also beyond the scope of this article. Few specific references to our studies and that of others will be given in the text because of the limited space available. For a complete treatment of this subject in its experimental as well as theoretical and observational aspects, the reader is referred to some reviews and books at the end of the essay.

AMINO ACIDS FROM ONE-CARBON COMPOUNDS

As indicated in the Introduction, I had planned to study the reactions of formaldehyde and other one-carbon species with other compounds in an attempt to elucidate the prebiotic synthesis of amino acids, monosaccharides, purines and other key biochemical compounds. I had not yet completed my graduate studies when I became aware of the classic work of S. L. Miller on the synthesis of amino acids under simulated primitive earth conditions.[3] This remarkable synthesis, which opened the field of prebiological organic chemistry, was accomplished by the action of electrical discharges on a mixture of methane, ammonia, hydrogen and water. As pointed out by Miller, the similar experiment performed earlier by W. Loeb was done in an entirely different context in an unsuccessful attempt to provide a chemical model for the mechanisms of nitrogen fixation and amino acid biosynthesis in plants. Besides hydroxy acids, urea and other organic compounds, some of the most abundant biological amino acids obtained by Miller in these experiments were glycine, alanine and aspartic acid.

Upon graduation from Baylor, I went to the University of Houston where I became Instructor in the Department of Chemistry, with a considerable teaching load in Chemistry and Biochemistry. As soon as I was able to establish my own laboratory, I began the experimental research on one-carbon compounds. At that time, I maintained the view that the primitive hydrosphere should have been relatively well supplied from the atmosphere with simple and reactive organic compounds, and that the important prebiotic reactions should have occurred in an aqueous environment either in solution or in a liquid–solid interphase.

First, I tested mixtures of formaldehyde and each of the nitrogen bases which had been suggested as intermediates in nitrogen fixation, namely NH_3, N_2H_2 and NH_2OH, and later with hydrogen cyanide–ammonia mixtures.[4] In both mixtures amino acids were synthesized, in addition to other organic compounds, but what was more surprising is that the most abundant amino acids were again glycine, alanine and aspartic acid, as in the experiments of Miller. I should hasten to say that a common denominator in all these experiments was found to be the participation of hydrogen cyanide (present or formed during the course of the experiment) in simple condensation reactions, such as the Strecker reaction or in more complicated condensations.

We know now that by a variety of experimental methods simulating primitive earth conditions and involving principally electrical discharges, UV light,[5] and/or solution reactions, many amino acids can be synthesized, including most of those found in proteins. Furthermore, an almost identical correlation has been found between the amino acids present in carbonaceous meteorites and those synthesized by electric discharges as shown, not long ago, by S. L. Miller and coworkers.

A PREBIOTIC ROUTE FOR PURINE SYNTHESIS

The prebiological formation of purines on the primitive earth poses *a priori* a more difficult problem because it requires the formation of two fused heterocyclic rings. In principle, there are two possible chemical routes or pathways. In one, the imidazole ring is formed first and is followed by the cyclization of the pyrimidine ring. In the other, this order is reversed.

In the 1950s it was well established, and I also had the fortune to experimentally observe it in the laboratory in the course of my graduate work, that the formation of purines in living organisms occurs by a pathway involving a 4-aminoimidazole-5-carboxamide (AICA) derivative. Also, it was known that the chemical hydrolysis of adenine produces 4-aminoimidazole-5-carboxamidine (AICAI). Thus, on one hand, we have the very mild conditions of enzymic synthesis, and on the other the more drastic conditions of acid hydrolysis, yet in both cases (synthesis and degradation) a 4,5-disubstituted imidazole is the intermediate. Therefore, it appeared reasonable to think that a prebiotic synthesis of purines may proceed through a pathway involving 4,5-disubstituted imidazoles. In fact, this was found to be the case when the prebiotic synthesis of adenine was first accomplished in my laboratory.[6]

ADENINE FROM HCN: FROM POISON TO LIFE GIVER

In spite of the above logical rationalization, the solution to the problem of purine synthesis came in a very roundabout way. You may say by serendipity. The initial spark emerged from a discussion between G. E. Hutchinson, from Yale, P. Abelson and myself (during the 1st International Oceanographic Congress held in New York in 1959) on the nature of the polymer formed by electrical discharges in Miller's experiment. Because of my interest in the above problem, and my familiarity in handling HCN (I had been involved, while in Spain, with my colleague E. Duró, in the commercial synthesis of amigdalin and mandelic acid from benzaldehyde via a Strecker condensation), I did not hesitate to use substantial amounts of HCN in order to elucidate the nature of the products formed in the course of the polymerization of this compound by means of ammonium hydroxide.

The experiment could not be simpler: Hydrogen cyanide was bubbled through an aqueous solution of NH_4OH, the mixture was refluxed overnight and the supernatant was spotted on a filter paper for chromatography.

The first time I treated the paper chromatogram with a specific adenine color reagent (Gerlach-Döring test), but only a barely visible pink spot was seen on the paper. The chromatographic spot was so marginal (I had not concentrated the solution) that I was initially skeptic of its reality and considered it an artifact. Since I was also looking for amino acids, the next day I sprayed another chromatogram with a ninhydrin solution, and a profusion of blue-purple spots appeared on the chromatogram.[4] The excitement of this interesting discovery made me forget, at least temporarily, about the first intriguing observation. However, several months later, upon completion of the work on the prebiotic synthesis of amino acids, I resumed the search for adenine.

This second time, I concentrated the supernatant from the original reaction product several fold and prepared a number of chromatograms. First, I analyzed one of the chromatograms by UV light, and indeed a very strong absorption spot with the R_f of adenine was observed! Immediately, I analyzed a second chromatogram by the Gerlach–Döring test and observed a very large spot giving the specific color reaction for adenine. Furthermore, with the product of R_f corresponding to adenine, I obtained its UV spectrophotometric absorptive curve in acid and in base. Finally, I sprayed another chromatogram with diazotized *p*-aminosulfanylic acid in search of the amino imidazoles. After the identification of adenine by its UV spectrum, I was thrilled to see a rainbow of blue, purple, redish, salmon and yellowish colored spots corresponding to the diazonium derivatives of AICA, AICAI and other diazotizable amines. The results were all positive.[6]

In short, adenine and a host of other organic compounds, including the necessary intermediates for purine synthesis, were found among the reaction products.

If at first it was difficult for me to accept the one-experimental-step transformation of HCN into adenine, after all the above products were identified, such a remarkable synthetic process could no longer be doubted.

In retrospect, if one examines this reaction a little more closely, this unusual synthetic process appears almost inevitable. Indeed, the empirical formula of adenine, $H_5C_5N_5$, corresponds to that of pentameric hydrogen cyanide, that is to say, that the overall reaction can be represented simply as five molecules of hydrogen cyanide producing one molecule of adenine:

$$5HCN \longrightarrow \text{Adenine}$$

The apparent paradox of this reaction is that one of the most toxic substances to the majority of living organisms can be the prebiotic precursor of one of the most important molecules to life.

ADENINE AND OTHER PURINES: MECHANISMS AND YIELDS

Upon completion of the above work, the isolation, derivatization and additional characterization of the most important products formed from HCN (adenine, AICA, AICAI, formamide, formamidine, glycinamide, etc.) was carried out with the collaboration of A. P. Kimball,[7] then a doctoral student, and now a colleague in our Department, doing fundamental research on cancer chemotherapy. Working together, we demonstrated the ready cyclization of AICA and AICAI with formamidine, urea and guanidine into all the functional purines, thus a better understanding of the prebiotic pathways of purine formation emerged.

The mechanism and sequential steps of the overall reaction were presumed to occur by the base-catalyzed oligomerization of HCN, first to a dimer, and then to a trimer (aminomalonodinitrile) followed by the condensation of this trimer with formamidine to 4-aminoimidazole-5-carbonitrile (AICN) or the corresponding amidine (AICAI). The cyclization into adenine is completed by condensation of either AICN or AICAI with formamidine. Guanine and xanthine are formed by condensation of AICA with urea.[8]

The prebiotic synthesis of adenine was confirmed by C. U. Lowe, M. W. Rees and R. Markham,[9] and indirectly by Ponnamperuma and co-workers.[10] The former authors also found the fourth common biological purine, hypoxanthine, in the reaction product from HCN and ammonia mixtures. It may be noted that the above sequence of reactions follows the same direction (first imidazole, then pyrimidine cyclization) as the biochemical synthesis of purines. The kinetics and mechanisms of HCN oligomerization as well as the potential role of the HCN trimer and tetramer for purine synthesis were studied by several investigators, particularly by J. P. Ferris and L. Orgel[11] and coworkers. The yields of purines from HCN are quite low but after isolating the appropriate intermediates, e.g. aminomalonodinitrile (AMN), diaminomaleodinitrile (DAMDN), AICN, AICAI, AICA, and condensing them with formamidine or other one-carbon compounds, substantial yields of adenine and the other purines can be obtained.

Pertinent to this question, an interesting account was related to me by Shiro Akabori on occasion of the International Biochemistry Congress held in Tokyo in 1967. It concerned an industrial company involved in the manufacture of acrylonitrile and other products requiring HCN as reactant. He said that large amounts of black polymers were obtained as side products of the main reaction, and that among these products they found appreciable quantities of an unknown white crystalline powder which they discarded together with the waste material. After the appearance in press of my report on the synthesis of adenine from ammonium cyanide, they analyzed the white powder and found it to be adenine. An industrial method using anhydrous HCN and NH_3 was soon patented by the Japanese company for the commercial production of adenine. Indeed, in the absence of water and at 120°C, yields of adenine from HCN as high as 35% were obtained.

A PREBIOTIC PANDORA'S BOX

It is now quite clear that HCN–aqueous ammonia reaction mixtures have a prominent role in the prebiotic synthesis of biochemical compounds. In addition to a variety of one-carbon compounds, amino acids, amino acid amides, malonic and maleic acid derivatives, imidazole derivatives and purines, these reaction mixtures are also a source of pyrimidine and pteridine derivatives and a number of polymeric molecules, some of which appear to contain amino acid residues in their structure: a real prebiotic Pandora's box. I would like to relate the following additional observations which have a bearing on the chemical exuberance and fecundity of this reaction.

Notwithstanding the highly poisonous nature of these mixtures, the preparation of one of these concoctions is a tremendously fascinating event. When the HCN begins to bubble in the aqueous ammonia solution, many bright colorations follow in quick succession: yellow, orange, red, deep red, and other hues with darker and darker tones until all the solution becomes pitch black. In short, such mixtures are not only rich prebiotic soups but genuine alchemist brews! Not surprisingly, their preparation is not devoid of serious dangers. One should not use a flask with a neck of regular size because even at temperatures below 90°C the ammonium cyanide formed in the reaction sublimes and crystallizes around the lower inside part of the flask's neck, clogging it and eventually leading to a fairly potent explosion. Fortunately, no one was hurt when this happened in one of our early experiments.

The energetic chemical behavior of the above processes is a reflection of the autocatalytic nature of HCN polymerization reactions and suggests that the overall process could take place at much lower temperatures. In fact, the blackening of the solution, corresponding to the formation of hydrogen cyanide polymers, was observed to occur in initially colorless control samples which had been frozen and kept for sometime in a dry-ice chest. Other interesting observations on explosions, and speculations on active processes induced by products closely related to HCN may be found in the old chemical and biological literature.

HCN AND COMETS

Since hydrogen cyanide and ammonia are two of the most important parent compounds present in comets and since these bodies are intensely irradiated when they pass through their closest proximity to the Sun, it is not unreasonable to think that some of the organic compounds described above may also be found in comets. The chemical composition of these bodies is considered to be representative of the composition of the solar nebula from which the primitive earth was formed. Therefore, it offers a solar-system model for studies on prebiological chemistry.[12–14]

I discussed some of these matters with A. I. Oparin and V. Fesenkov during the International Congress of Biochemistry held in Moscow in 1961. The latter scientist had just concluded, after examination of the available records, that the Tunguska explosion, which occurred in Siberia in 1908, had been caused by a small comet. This observation added further support to my suggestion, following the earlier calculations of H. C. Urey, that a substantial amount of carbon and nitrogen compounds could have been added to the primitive earth by comets captured by our planet[12–14] during the anoxygenic period of its first 1 billion years.

In the summer of 1962, in the laboratory of Melvin Calvin at Berkeley, I obtained amino acids, amino acid amides and UV absorbing compounds by electron bombardment of CH_4, NH_3 and H_2O solidly frozen by liquid N_2 in experiments simulating the irradiation of comets.[13] In similar experiments C. Ponnamperuma, M. Calvin and coworkers, using the same components in the vapor or gas phase, demonstrated the formation of adenine and other organic compounds.[10]

ORGANIC COSMOCHEMISTRY

In May of 1962 I was asked to present a paper to the New York Academy of Sciences reviewing some of the work done in my laboratory on the synthesis of organic compounds related to the chemistry of space.

It had been established by H. C. Urey and others that the four most abundant elements in the universe (with the exception of He and possibly Ne) are H, C, N and O, which, as we know, are precisely the four most important elements of organic matter. From the elemental abundance data and the high stability of simple diatomic combinations of these elements in stellar atmospheres, it was logical to conclude as I did then: "There is no doubt that carbon compounds exist widely distributed in the Universe. Whether the more complex biochemical compounds described in this paper are present in cosmic bodies other than the earth, will only be answered with certainty by space probes".[15]

Since then, amino acids and other biochemical compounds have been found in meteorites, phosphorus compounds and dicarbon compounds in Jupiter, and all the other molecules necessary for the synthesis of the most important biochemical monomers have, in fact, been found by radioastronomy in the interstellar clouds of our galaxy.[16] After the initial finding of NH_3 and H_2O, the first organic molecule, formaldehyde, was discovered in 1969. This was followed by the discovery of a large number of organic molecules in interstellar space, including HCN, cyanoacetylene, formamide, acetaldehyde, methyl amine, etc., precisely the same compounds which had been used or involved previously in laboratory experiments on the prebiotic synthesis of biochemical compounds. A number of mechanisms have been suggested for the natural formation of these relatively simple organic molecules.[16,17]

SYNTHESIS OF PYRIMIDINES

Some time ago, I postulated a route involving the condensation of α, β-unsaturated three-carbon compounds (e.g. acrylonitrile) with urea for the synthesis of pyrimidines. Very small yields of uracil were obtained in this manner in our laboratory.[14,18] A better C_3 precursor appears to be cyanoacetylene, which, after HCN, is the major nitrogen-containing product formed from methane and nitrogen by electrical discharges. Cyanoacetylene condenses with urea or cyanate to form cytosine in 20% yield as shown by Sanchez, *et. al.*[19] Cytosine is readily converted into uracil by deamination. Furthermore, by reacting uracil with formaldehyde in the presence of a reducing agent, such as hydrazine, small yields of thymine (probably through hydroxymethyluracil as intermediate) were obtained in our laboratory by E. Stephen-Sherwood and coworkers.[20] More recently, the common biochemical precursor of pyrimidine bases in living organisms, orotic acid, has been found by J. P. Ferris and co-workers to be released by the hydrolysis of HCN oligomers formed from $HCN–NH_4OH$ mixtures.

FORMATION OF RIBOSE AND DEOXYRIBOSE

As indicated in the Introduction, the condensation of formaldehyde into monosaccharides by basic catalysis has been known since the last century. The synthesis starts with the formation of glycolaldehyde which is a slow reaction and is responsible for the induction period observed in the condensation of formaldehyde to sugars. Once sufficient amounts of glycolaldehyde have been formed, an autocatalytic process ensues which transforms glycolaldehyde into glyceraldehyde and dihydroxyacetone, and then into all the possible tetroses, pentoses and hexoses. The principal mechanism is a base catalyzed aldol condensation, somewhat similar to the enzyme catalyzed biochemical transformations of sugars.[14] A common mineral, kaolinite, has been found to be an efficient catalyst for this reaction.[21] Ribose is indeed one of the important monosaccharides formed in this reaction.

As shown by A. C. Cox in my laboratory,[14,22] the 2-deoxypentoses (2-deoxyribose and 2-deoxyxylose) were readily formed by condensation of glyceraldehyde and acetaldehyde in yields of up to 15% as measured by Dische's method. What was interesting about the latter condensation is that it was very smoothly catalyzed by ammonium hydroxide at room temperature and could be followed for a long time without observing any degradation or transformation of 2-deoxyribose into other products. Ribose and 2-deoxyribose can also be produced by the irradiation of CH_2O by UV light.

SYNTHESIS OF NUCLEOSIDES

Since pentoses and phosphates were not included in the HCN–NH_4OH mixtures one could not expect the formation of nucleosides or nucleoside phosphates in such mixtures, as in the biochemical *de novo* synthesis of purine nucleotides, even though an adenine derivative of possible nucleosidic nature was detected in one of our early experiments. Alternate biochemical routes also exist (salvage pathways) for the synthesis of nucleosides from the preformed bases. The latter general approach has been followed in the prebiotic synthesis of purine nucleosides, and the former has been used in the synthesis of pyrimidine nucleosides and nucleotides.

The formation of pyrimidine nucleoside and nucleotides has been found to be somewhat difficult, involving a sequence of complex condensation reactions and rearrangements as shown by Sánchez and Orgel.[23] Thus, when ribose-5-phosphate is treated with cyanamide and then with cyanoacetylene a product is obtained which is mainly α-cytidylic acid. The α-cytidylic acid and the α-cytidine derived from it can be converted (photoanomerized) to the natural biological isomers, β-cytidylic and β-cytidine, respectively, by irradiation with UV light. The latter two compounds are easily transformed into β-uridylic acid and β-uridine by deamination. There is some question on the prebiotic significance of this complex sequence of reactions.

The synthesis of purine nucleosides is easier. Thus, when purine bases are heated with ribose and the salts from sea water (which contain magnesium chloride, a catalyst for this reaction) small yields of the different nucleosides are obtained, approximately 3% of β-guanosine, 2% of β-adenosine and comparable yields of the α-nucleosides.[24]

FORMATION OF NUCLEOTIDES

The phosphorylation of nucleosides to their corresponding monophosphates, has been accomplished in a number of laboratories.

Nucleosides can be phosphorylated directly with orthophosphate by dry-heating, at relatively high temperatures.[25] This phosphorylation is facilitated at lower temperatures (below 100°C) by the presence of urea and ammonium chloride, as shown by R. Lohrmann and L. Orgel.[26] Alternatively, when nucleosides and inorganic pyrophosphate are adsorbed on apatite crystals and dry-heated at temperatures below 100°C, all possible nucleotide isomers are obtained in reasonable yields. In addition to these methods, nucleosides can also be phosphorylated by a system composed of either cyanamide or dicyandiamide plus ammonium oxalate and apatite, as shown by A. W. Schwartz.

In summary, the most successful routes for the phosphorylation of nucleosides involve reactions using moderate temperatures and ambient humidity, simulating the conditions resulting from the drying of primeval lakes. Surface temperatures of up to 80°C have been recorded in desert areas. Reactions carried out at temperatures higher than 100°C (or with polyphosphoric acid or polymetaphosphate esters) are of doubtful significance with respect to prebiological chemistry.

HIGH-ENERGY PHOSPHATES

The synthesis of pyrophosphate under primitive earth conditions has been shown by S. L. Miller and M. Parris.[27] It occurs by reaction of orthophosphate with cyanate on the surface of hydroxylapatite, presumably forming carbamyl phosphate which reacts with another phosphate molecule to generate pyrophosphate. The formation of the pyrophosphate bond has also been accomplished by extension of the above reaction methods. Thus, $NH_4H_2PO_4$ in the presence of urea at low humidity and below 100°C forms inorganic polyphosphate or pyrophosphate. Once inorganic pyrophosphate is available, transphosphorylations can take place. In this way AMP has been further phosphorylated to ADP and ATP.

More recently the synthesis of all deoxyribonucleoside mono-, di- and triphosphates has been accomplished in our laboratory[28] in a single step. This was done by evaporating a solution of the corresponding deoxyribonucleoside, cyanamide and phosphate [e.g. NaH_2PO_4; $(NH_4)_2HPO_4$] at an initial pH between 5 and 8 and heating it at temperatures between 80 and 90°C. Reasonably good yields of the different monophosphates and high-energy phosphates were obtained. For example, in one of the experiments 28% TMP, 15% pTp, 11% TDP and 6% TTP were obtained. Other condensing agents, such as urea, guanidine, etc., were found to be less effective than cyanamide in this reaction and did not produce detectable amounts of triphosphates.[28]

OLIGORIBONUCLEOTIDES

The approaches used for the synthesis of ribonucleotide oligomers follow the general lines of the method used for the synthesis of mononucleotides. Thus, oligonucleotides can be obtained by heating for several days uridine, urea and ammonium dihydrogen phosphate at temperatures between 85 and 100°C under ambient humidity or dry conditions. Some of the synthesized oligomers, e.g. 7% of UpUpU, contained approximately 60% of the natural 3′,5′-phosphodiester bond.

Little can be said with regard to the selection of the natural 3′,5′-linkage, which may have occurred at some later stage of chemical evolution. In this connection it is significant that adenyl-(3′,5′)-adenosine has a greater selectivity for a right-handed stack conformation than adenyl-(2′,5′)-adenosine. This suggests that nucleic acids built with a 3′,5′-linkage may have assumed preeminence during the evolutionary process, because of their superior uniformity of conformation.

The urea–ammonium chloride–phosphate system offers a method for the formation of short-chain oligoribonucleotides on the primitive earth. It does not appear to be applicable to the formation of oligodeoxyribonucleic acids since no oligonucleotide formation was evident after 44 days of heating at 65°C of a mixture of thymidine, urea, ammonium chloride and phosphate. (See also references 29, 30 and 31.)

CYANAMIDE. A PREBIOTIC ANALOG OF DICYCLOHEXYLCARBODIIMIDE

It has been known for a long time that dicyclohexylcarbodiimide (DCC) is one of the best condensing agents for the formation of peptide and phosphodiester bonds. Being concerned with the prebiotic synthesis of biopolymers, while at Baylor, it was a matter of direct generalization to consider that the most simple prebiotic analog of DCC is the decyclohexylated bis-cyclohexylcarbodiimide, that is to say, just plain carbodiimide, which in its tautomeric form is cyanamide. When I wrote my first review on organic cosmochemistry in 1963, I suggested the potential prebiotic role of cyanamide for the formation of polymeric compounds. Aside from the fact that HCN can be readily transformed to cyanamide by UV light, the latter compound has been recently found in interstellar clouds, clearly confirming the cosmic nature of cyanamide, and adding it to the list of important prebiotic molecules (H_2, NH_3, H_2O, CH_2O, HCN, CO, cyanoacetylene, CH_3CHO, H_2S, etc.) associated with the stardust of our galaxy.

OLIGODEOXYRIBONUCLEOTIDES

One of the most elegant applications of DCC is the extensive and now classic synthesis of oligonucleotides by H. G. Khorana and his coworkers. Because of the primordial nature of deoxyribonucleic acids, I undertook with the collaboration of J. Ibañez and other coworkers the prebiotic synthesis of oligodeoxyribonucleotides. For reasons of simplicity, we selected the synthesis of oligothymidylic acid as a model.

We followed the general approach of Khorana's method (DCC in an anhydrous system) and we adapted it to plausible prebiotic conditions, i.e. cyanamide in an aqueous system.[32] Thus an aqueous reaction mixture of TMP and cyanamide was heated at temperatures below 90°C overnight and evaporated to dryness before analysis. Oligomers of thymidylic acid of up to four units in length, in total yields of about 1% were obtained. A large proportion of the products contained 3′,5′-phosphodiester bonds as measured by enzymatic degradation. Imidazole and an imidazole derivative obtained from $HCN–NH_4OH$ (AICA) were also found to promote the condensation of TMP in a similar manner.

In subsequent experiments, with the collaboration of E. Stephen-Sherwood and D. Odom, increased amounts of oligomers were obtained by including TTP and AICA · HCl in the reaction mixture, evaporating the solution and heating the evaporated mixture at temperatures below 90°C at ambient humidity, for about 40 hours. The products were analyzed following Khorana's methods. The total yield of oligodeoxynucleotide product with at least one terminal phosphate was 32%. About 80% of these oligomers contained 3′,5′-phosphodiester bonds. The yield of $(pT)_2$ and $(pT)_3$ was about 9% and 6%, respectively.[33]

The oligomerization is presumed to take place by condensation of thymidine-5-triphosphate on the free 3′OH of TMP with the release of pyrophosphate. It is not known if the mechanism of synthesis occurs through the cyclic trimetaphosphate intermediate postulated in Khorana's chemical synthesis, but the conditions of synthesis (heating after evaporation at temperatures below 90°C) are not incompatible with such a consideration.

Thymidine-5′-phosphate may be considered to act as a "primer" with the more reactive 5′-hydroxyl group blocked with a phosphate. The role of AICA is assumed to be that of a proton donor and the role of cyanamide that of a condensing or transferring agent. Whether there is any structural–functional correlation between the imidazole and guanido groups of the latter two compounds and that of the histidine and arginine group, respectively, which are present in some enzyme active centers, remains an intriguing possibility.

Although the oligomerization of mononucleotides under the above conditions provides plausible models for the prebiotic synthesis of short-chain oligomers, it is doubtful that this type of condensation is applicable to the formation of long polymers.

PREBIOTIC SYNTHESIS OF PEPTIDES

Within the limited scope of this essay it is not possible to examine the extensive work on the thermal synthesis of polymers of α-amino acids carried out by S. W. Fox and other investigators. These polymers have normal α–α bonds, and abnormal α–β, α–γ, and other linkages not commonly found in proteins. The reader is referred to these studies for pertinent information.[34,35] Following the prebiotic conditions indicated at the beginning of this article, I have restricted myself to reactions in aqueous conditions at moderate temperatures, with or without evaporation, which produce simple biochemical monomers or relatively small linear oligomers. Under these conditions peptides can only be synthesized from activated amino acids or by means of water-soluble condensing agents, in the presence or absence of other catalytic agents.

In line with my earlier suggestion on the role of cyanamide, several investigators, particularly G. D. Steinman, R. M. Lemmon and M. Calvin,[36] have shown that this compound and dicyandiamide will condense amino acids under mild aqueous conditions into normal peptides. The HCN tetramer, diaminomaleodinitrile, formed in the HCN–NH_4OH mixtures has also been found to condense amino acids into simple biological peptides as shown by Chang and co-workers.[37] Recent experiments indicate that oligopeptides can also be formed on the surface of minerals or clays by heating, after evaporation, at temperatures below 100°C.

Alternative routes for the prebiotic synthesis of oligo- and polypeptides in aqueous systems, at relatively low temperatures, may well have involved activated monomers like α-amino nitriles, α-amino amides or amino acyl adenylates. Interesting results were obtained earlier with the first two types of compounds in our laboratory and other laboratories. More recently the amino acyl adenylates, in the presence of montmorillonite and at pH

7.9–8.5 have been converted to polypeptides with degrees of polymerization from 9 to 54 units, as reported by M. Paecht-Horowitz, *et. al.*[38] As one may expect from the different reactivity of amino acids with different substituent groups, a certain selectivity or non-randomness, in the condensation of some amino acids, has been observed in several laboratories.

A UNIFIED SYSTEM FOR THE PREBIOTIC SYNTHESIS OF OLIGOMERS

In few, if any, of the studies made so far, has an effort been made to correlate the synthesis of peptides with that of oligonucleotides. A plausible prebiotic system capable of producing both oligonucleotides and peptides would be specially significant. In line with this suggestion, recent studies in our laboratory, with the collaboration of D. W. Nooner and E. Stephen-Sherwood,[39] have shown that the same system that we have used earlier for the synthesis of oligothymidylic acid will also condense free amino acids to oligopeptides.

The fundamental component of the system is again cyanamide, but for optimum yields, the addition of AICA · HCl and high-energy phosphate (in this case ATP) is required. The amino acid solution with the other components is heated after evaporation, at temperatures which may vary from 40 to 90°C, for up to 24 hours. Upon separation by thin-layer chromatography the peptides are derivatized and identified by combined gas chromatography–mass spectrometry. After 1 hour reaction at 88°C the yield of L-phenylalanine dipeptide was 36% and that of the L-isoleucine dipeptide 22%. The yield and number of oligopeptides obtained from L-phenylalanine increased with temperature and time, with about 66% of the amino acid being converted to a mixture of di-, tri- and tetrapeptides in 24 hours at 90°C.[39]

This work is being extended to other amino acids. The significance of these results and their relation to the synthesis of monomeric and oligomeric deoxyribonucleotides has not been completely evaluated but the prominent role of cyanamide by itself, or in combination with other compounds, in all these syntheses is clearly evident.

BASE-PAIR-DIRECTED SYNTHESIS OF OLIGONUCLEOTIDES

The formation of oligonucleotides by evaporation of aqueous reaction mixtures and heating at ambient humidity in the absence of a template can only be expected to produce oligomers with a small number of units. This is so because the elongation occurs by the sequential addition of one mononucleotide at a time to the previously formed oligonucleotide. Indeed, the results show a progressive decrease in yield as the degree of polymerization increases. Only rarely have oligomers with more than 6 units been obtained.

The question arises: What are the conditions required for the replication and elongation of these nucleotidic short chains in an aqueous environment? The information available on this process is rather limited. For instance, using water-soluble carbodiimides, adenylic acid is preferentially condensed on a doubly stranded polyuridylic acid as template and guanylic on a doubly stranded cytidylic. Dinucleotides were isolated in high yield (35%), but the majority of the isomers had the unnatural 2′,5′-phosphodiester bonds. It has also been shown that D-adenosine-5′-phosphorimidazolide reacts much more rapidly with D-adenosine than with L-adenosine on a poly-D-uridylic acid template, suggesting that segregation of D- and L-nucleotides may have occurred at an early stage of molecular evolution.

A problem that has not been resolved so far is that only pyrimidine polynucleotides can serve as templates for base-pair-directed nucleotide synthesis.

In spite of this problem and in spite of the fact that unnatural 2′,5′ bonds are predominantly formed, the above and other results demonstrate that the selectivity with respect to Watson–Crick pairing does exist at the monomer level and that the formation of complementary oligomers would probably have been reasonably accurate with respect to base sequences in a prebiological world. (See also references 29, 30 and 31.)

CYCLIC CHANGES IN PREBIOTIC ENVIRONMENTS

Apart from the major variation in temperature and humidity caused by the seasons, there are more frequent periodic fluctuations that may have played an important role on the repetitive synthesis of biopolymers and facilitated the elongation and replication of polynucleotides. These are, for instance, evaporation, drying and refilling of mainland and coastal pools and lagoons, the similar daily effects of tides, and the fluctuations in temperature between night and day.

In line with this reasoning, and in collaboration with D. Odom we have recently investigated the effects of tidal washing and evaporation on a thymidylic acid–cyanamide–salt system at 60°C and neutral pH, either once or six times, with only replenishment of NH_4Cl-cyanamide each time.[40] The products formed were tri-, tetra- and higher oligomers. The total yields increased from about 18% after one evaporation to about 30% after six evaporations, the principal product being the trinucleotide (13%). With the conditions of relatively low temperature, absence of substituted nucleotides and initial neutral pH, we believe this experiment provides a simple and plausible prebiotic model for the repetitive synthesis of oligodeoxynucleotides in a primeval environment.

If a similar reaction was carried out in the presence of a template with a temperature cycle of about 10°C to 40°C simulating a sharp fluctuation between night and day it is possible that the successive replication and elongation of these polymers could take place.

ELONGATION AND REPLICATION OF OLIGONUCLEOTIDES

Although incomplete, the evidence in the last two sections suggests that the elongation and replication of nucleotides may have been possible in the pools and lagoons of the primitive earth.

From the interesting studies carried out in Arthur Kornberg's laboratory, it is known that a deoxyoligonucleotide of six units, pT $(pApT)_2$pA, can serve as a primer for the synthesis of a high molecular weight dAdT copolymer using *E. coli* DNA polymerase. The replication was found to be profoundly influenced by the temperature, the shorter oligomers requiring lower temperatures in order to maximize hydrogen bonding. For example, $(AT)_4$ replicated best at 10°C and less efficiently at higher temperatures due to breakage of hydrogen bonds. Further studies using DNA-like polymers with repeating tri- and tetranucleotide sequences as primers, indicate that a minimum size of about 9 to 12 nucleotides is necessary for template activity.

A slippage mechanism based on a model originally proposed by M. J. Chamberlin and Paul Berg has been suggested to account for the enzymatic synthesis of higher molecular weight polymers using short-chain primers. Whether a similar slippage model could account for the replication and elongation of short-chain oligomers under conditions simulating primitive pools and lagoons has yet to be tested. The temperature fluctuations, as a result of night and day and other cycles, could account for the forming and breaking of hydrogen bonds and hence promote both elongation and replication.

An alternative possibility, which may have occurred in later stages of molecular evolution, is based on a mechanism similar to the rolling circle model devised to explain the replication of circular DNA in prokaryotic organisms. However, this slippage model would have required the cyclization of oligomers with about 20 units or more, which, so far, has only been accomplished with the help of enzymes. (See also references 29, 30 and 31.)

MAJOR EVOLUTIONARY TRANSITIONS

It is obvious that in trying to relate the advance in our understanding of the non-enzymic formation of biological molecules to the problem of the origin of life, we have only scratched the surface of the real question, which, of course, is the major transition from these synthesized molecules to a living entity. But this is not a total surprise, because as unique as this transition is, it has some common features with other major transitions in the overall process of chemical and biological evolution.

For instance, as I mentioned in the Introduction, among the gravest difficulties which concerned Charles Darwin when he wrote *On the Origin of Species* were precisely "the difficulties of transitions" (from simple to complex organisms). Thus, while his theory beautifully explained the infinite variations found among species, what may be called horizontal evolution or diversification, it did not provide as good an "understanding", in Darwin's own words, of "how a simple being or a simple organ can be changed and perfected into a highly developed being or elaborately constructed organ". This is an example of what may be called ascending evolution or truly progressive evolution.

I believe Fritz Lipmann, in his essay in this book,[41] has also put his finger on a very similar, if not identical, problem when he says: "For some time I have felt unsure whether one can consider the development in bacteria as evolution in the sense applied to multicellular organisms. What we see in bacteria, I feel, is not true evolution but it might be called a diversification".

Indeed, from the preliminary evidence obtained by J. W. Schopf[42] and other micropaleontologists, it is known that it took more than 2000 million years for the prokaryotes to make the major transition into eukaryotes. It took also considerable time for the most primitive eukaryotes to evolve into metazoans at about 700 million years ago.

These are the two major discontinuities in the biological evolutionary record. It should be noted that the emergence of the more complex organisms was not directly predictable from the DNA of the prokaryotic cell (no amount of molecular genetics would have solved the problem), but required the concurrence and interaction of independently living cells and other external factors each initially acting with independence. Or, as F. Lipmann has put it: "Symbiotic fusion of different lines of organisms at a crucial stage of evolution would be an unexpected addition to chance mutation, and to judge from present-day examples, be attributable to choice rather than to chance".

In line with the endosymbiotic theory of Lynn Margulis[43] and others, the eukaryotic cell is the result of the association of several prokaryotic cells, which while still retaining some of their ancestral features, have given up their independence to work for the eukaryotic cell as a whole. In a similar manner, the metazoans and other multicellular organisms are the result of associations of eukaryotic cells each becoming specialized but working in a cooperative way for the common cause of the more complex multicellular organism.

PROTOBIOLOGICAL OR PRECELLULAR EVOLUTION

Whether cooperative associations of non-enzymically synthesized biological macromolecules could provide the basis for the origin of a prokaryotic cell is not known. Some time ago, while participating in a summer biology course at Cambridge, England, I made a very preliminary suggestion in this direction involving four specific types of molecules.[44] After summarizing the major stages of chemical evolution, I discussed briefly precellular evolution somewhat as follows:

The next major event following the formation of biochemical compounds was a series of catalytic and structural self-organizing processes which probably occurred immediately preceding and immediately succeeding the appearance of the first self-replicating molecular system. This so-called precellular evolution period is very little understood but certainly the most significant event of chemical organic evolution since it marks the major apparent discontinuity between the non-living and living worlds.

Of utmost importance within this evolutionary transition are four unique processes of synthesis responsible for the emergence and subsequent cooperative interaction of the following four molecules or molecular complexes:

1. Biocatalytic molecule. Protoenzyme

Small linear peptide of 4 or more units in length, with functional amino acids such as arginine, histidine, serine, etc., and with the ability to form complexes with oligonucleotides.

2. Encoding self-replicating molecule. Proto-DNA or proto-RNA

Linear oligonucleotide of 12 or more units in length, with 4 or fewer different bases and the ability to form complexes with other oligonucleotides (complementary) and oligopeptides.

3. Code-translating molecule. Proto-tRNA

Small linear oligonucleotide combined with a terminal amino acid, with the ability to form intramolecular hairpin structures capable of binding or complexing with the above nucleotides and peptides.

4. Interphasic molecule or system of molecules. Proto-membrane

Amphoteric lipid molecule capable of forming a stable spherical structure with a boundary for the separation of its internal hydrophilic phase from the environment.

In theory, bimolecular and trimolecular complexes could be generated by the self-assembly of the above first two or three types of molecules, showing the most essential attributes, or fundamental processes of life, but without requiring the existence of a discrete living cell in the prebiotic environment. The existence of a more complex living entity, a probable precursor of the primitive prokaryotic cell, embodying the above three molecular systems and respective functions, could have become possible by enclosing these systems within the fourth amphoteric lipid molecule, in a manner reminiscent of the self-assemblying membrane structures of present-day halophilic purple bacteria.

These four molecules, or systems of molecules, are known today in biology as enzyme (which should include the protein–rRNA complexes of ribosomes), DNA (and/or m-RNA), tRNA and membrane. The emergence of the first enzyme (read enzyme–DNA–tRNA-complex) in the words of Sir F. G. Hopkins, may be considered as the most significant single event in the history of the universe.

EXPERIMENTAL AND THEORETICAL MODELS OF PRECELLULAR EVOLUTION

Whereas most of the work discussed so far in this essay has dealt with the synthesis of well-defined biochemical species supporting the theory of chemical evolution as first proposed by A. I. Oparin, one of Oparin's major concerns has been to develop a hypothesis of precellular evolution and to experimentally demonstrate that specific biochemical reactions can occur within simulated precellular entities (coacervates). In an elegant experiment, using polynucleotide phosphorylase in coacervate droplets and the appropriate substrate in the external medium, he showed a continuous uptake of the substrate, a rapid internal synthesis of polynucleotides and a continuous release of phosphate to the external environment. His more recent concepts on evolution of probionts and the origin of cells were presented at the 4th International Conference on the Origin of Life held in Barcelona, Spain, in 1973.[45] Experimental models involving microspheres made of polymers of amino acids have been developed by S. W. Fox and coworkers[34] and other investigators.

A general mathematical treatment of the precellular self-organization of matter and the evolution of macromolecules was presented some time ago by Eigen.[46] A set of selectivity and evolutionary principles were derived for alternative populated states of macromolecules, leading towards states of higher complexity and information content. It would be highly desirable to analyze and correlate the three models presented thus far (molecular, coacervate-microsphere and mathematical), select the realistic and cooperative portions of these models, and integrate them into a theory of precellular evolution which could be experimentally tested.

CONCLUSIONS

In this essay I have not considered explicitly the question of amino acid enantiomer selectivity, the role of various high-molecular weight polymeric products as prebiotic catalysts, the development of the genetic code, and other important questions of prebiological chemistry. In order to limit myself to well-defined molecular species I have attempted to present only some of the experimental evidence, obtained in several laboratories, which demonstrate that most of the building blocks of nucleic acids and proteins, high-energy phosphates, as well as a few oligonucleotides and oligopeptides, can be synthesized in the absence of enzymes under conditions presumed to have existed on the primitive earth.

Also, I have briefly outlined preliminary concepts suggesting that the cooperative and synergistic interaction of four types of linear molecules or complexes could eventually lead to the assembly of a self-replicating and evolving system. These concepts and related hypotheses, admittedly crude as they are, may eventually be modified and developed into a molecular protosymbiotic theory for the origin of the prokaryotic cell. However, more important even than the development of these concepts will be experiments which either support or disprove them.

By proving that the transition from molecules to living systems is possible, the *de novo* synthesis of an independently self-replicating and evolving system would confirm the universality of the principles of life and of life generation (biopoiesis),* whether or not, it solves the problem of the origin of life on the earth, which may be difficult to solve due to the intrinsic limitations of the geological record.

EPILOG (THE CYCLE OF LIFE AND MATTER)

As it must have been obvious from the Introduction, I have to confess that when I first arrived in the United States I was unaware of the studies on chemical evolution being done elsewhere. Upon the publication of the classical experiment of Stanley Miller in *Science* and *J. Amer. Chem. Soc.*, it became common knowledge that credit for the original ideas and first scientific discussions on the problem of the origin of life was due principally to A. I. Oparin, as well as to J. B. S. Haldane, J. D. Bernal, M. Calvin and H. C. Urey, who provided a theoretical basis and preliminary test for these studies.

It is now well known that this chemical theory on the origin of life was first published more than 50 years ago by A. I. Oparin. The essence of this theory is based not on the spontaneous generation of life proposed by earlier authors, but on the progressive chemical evolution and self-organization of organic matter on a reducing, or non-oxidizing, primitive earth environment.

It is interesting to reflect, from the point of view of the evolution of scientific thought, that the origins of this idea go back to a misconception of the origin of petroleum by one of the founders of chemistry, Dm. I. Mendelejeff. As Alexander Oparin kindly related to me, in Tokyo in 1967, Mendelejeff believed that petroleum was formed by the chemical hydrolysis of metallic carbides. Therefore, the following argument could be made: If petroleum, which is related in its chemical composition to that of living organisms, was generated geochemically from natural terrestrial products, analogous processes could have been responsible in the earth's past for the formation of the organic matter from which the organisms evolved.

Apart from any of the scientific conclusions that these studies may eventually bring, I believe of more spiritual significance to us would be the humbling recognition that we may descend from very simple and lowly molecules and the realization that we may not be alone in the Universe. Hopefully, such recognition and realization should instill in us the humility, and love for each other, that we so direly need, if the human race is to survive on this planet. In the long run, however, there is no need to despair because whether we heed this cosmic

* No definition of the terms *prebiological* and *prebiotic* has been given. According to common usage, they apply to conditions, chemicals or processes presumed to have occurred on the primitive earth before the emergence of life. Such an important event has been referred to commonly as *abiogenesis* and less commonly as *biogenesis*. Because of the self-contradiction of the former, and ambiguity of the latter, it would be preferable to adopt the terms *biopoiesis*, as used by N. W. Pirie and myself in previous articles. Furthermore, according to its etymology, the word biopoiesis is of general application and not restricted to any one set of conditions, thus bypassing the problems resulting from not knowing exactly the conditions which prevailed on the primitive earth.

call or not the eternal cycle of matter and life will go on as it was written long ago: "*Pulvus eris et in pulverem reverteris*", or in modern language, "Stardust you were and in stardust you will return".

ACKNOWLEDGMENTS

The ideas, good or bad, and errors in this essay are my own. Some go far back to the time when as a young baker, I could observe, during work breaks, the stars in the night sky. Some were inspired during my high school days reading the writings of C. Darwin and E. Haeckel on biological and organic evolution and of C. Flammarion, the popular French astronomer-poet, on the plurality of the inhabited worlds. My love for biology and organic chemistry came from my Lerida High School teacher J. Sirera, and my Barcelona University professor J. Pascual Vila, respectively.

The vagaries of the Spanish civil war and the Second World War determined that, instead of Germany, I ended in the United States for my graduate work. To this country, my own, and their people, to my universities (Barcelona, Baylor and Houston) and their teachers, and to the encouragement received from Severo Ochoa during our many years of friendship, I owe in large measure the satisfaction of being able to pursue my planned studies and of transforming some of my original ideas into reality.

The research work performed in my laboratory described here, would not have been possible without the valuable collaboration of esteemed graduate students and colleagues, particularly A. H. Bartel, A. C. Cox, C. L. Guidry, J. D. Ibañez, S. S. Kamat, A. P. Kimball, D. W. Nooner, D. Odom, M. O'Rourke, E. Stephen-Sherwood and others who worked on related chemical problems.

I would also like to recognize Drs. R. S. Young, K. Kvenvolden, F. Mayor and J. L. Yuste as well as the National Aeronautics and Space Administration and the Juan March Foundation for support received duiing the course of this work. To all the above colleagues and students and to many others, too numerous to mention, I am greatly indebted.

REFERENCES

I have selected only two or three references for some of the sections of this essay. A few additional books and reviews are included at the end of this list for more detailed information.

1. Bloch, K. On the evolution of a biosynthetic pathway. This Vol., pp. 143–150.
2. See, for instance, the essays by P. Berg, H. G. Khorana, A. Kornberg and C. Yanofsky in this volume.
3. Miller, S. L. (1953) A production of amino acids under possible primitive earth conditions. *Science*, **117**, 528–529.
4. Oró, J. and Kamat, S. S. (1961) Amino acid synthesis from hydrogen cyanide under possible primitive earth conditions. *Nature*, **190**, 442–443.
5. Sagan, C. and Khare, B. N. (1971) Long wavelength ultraviolet photoproduction of amino acids on the primitive earth. *Science*, **173**, 417–420.
6. Oró, J. (1960) Synthesis of adenine from ammonium cyanide. *Biochem. Biophys. Res. Commun.* **2**, 407–412.
7. Oró, J. and Kimball, A. P. (1961) Synthesis of purines under primitive earth conditions. I. Adenine from hydrogen cyanide. *Arch. Biochem. Biophys.* **94**, 217–227.
8. Oró, J. and Kimball, A. P. (1962) Synthesis of purine under possible primitive earth conditions. II. Purine intermediates from hydrogen cyanide. *Arch. Biochem. Biophys.* **96**, 293–313.
9. Lowe, C. U., Rees, M. W. and Markham, R. (1963) Synthesis of complex organic compounds from simple precursors: formation of amino acids, amino acid polymers, fatty acids and purines from ammonium cyanide. *Nature*, **199**, 219–222.
10. Ponnamperuma, C., Lemmon, R. M., Mariner, R. and Calvin, M. (1963) Formation of adenine by electron irradiation of methane, ammonia and water. *Proc. Nat. Acad. Sci. U.S.* **49**, 737–740.

11. FERRIS, J. P. and ORGEL, L. (1965) Aminomalononitrile and 4-amino-5-cyanoimidazole in hydrogen cyanide polymerization and adenine synthesis. *J. Am. Chem Soc.* **87,** 4976–4977.
12. ORÓ, J. (1961) Comets and the formation of biochemical compounds on the primitive earth. *Nature*, **190,** 389–390.
13. ORÓ, J. (1963) Synthesis of organic compounds by high-energy electrons. *Nature*, **197,** 971–974.
14. ORÓ, J. (1965) Stages and mechanisms of prebiological organic synthesis. In *The Origin of Prebiological Systems* (S. W. Fox, ed.), pp. 137–171, Academic Press, New York, New York.
15. ORÓ, J. (1963) Studies in experimental organic cosmochemistry. *Ann. N.Y. Acad. Sci.* **108,** 464–481.
16. ORÓ, J. (1972) Extraterrestrial organic analysis. *Space Life Sci.* **3,** 507–550.
17. ANDERS, E. HAYATSU, R. and STUDIER, M. H. (1974) Catalytic reactions in the solar nebula: implications for interstellar molecules and organic compounds in meteorites. *Origins of Life*, **5,** 57–67.
18. ORÓ, J. (1963) Non-enzymatic formation of purines and pyrimidines. *Federation Proc.* **22,** 681.
19. SÁNCHEZ, R. A., FERRIS, J. P. and ORGEL, L. E. (1966) Cyanoacetylene in prebiotic synthesis. *Science*, **154,** 784–785.
20. STEPHEN-SHERWOOD, E., ORÓ, J. and KIMBALL, A. P. (1971) Thymine: possible prebiotic synthesis. *Science*, **173,** 446–447.
21. GABEL, N. W. and PONNAMPERUMA, C. A. (1967) A model for the primordial origin of monosaccharides. *Nature*, **216,** 453–457.
22. ORÓ, J. and COX, A. C. (1962) Non-enzymatic synthesis of deoxyribose. *Federation Proc.* **21,** 80.
23. SÁNCHEZ, R. A. and ORGEL, L. E. (1970) Synthesis and photoanomerization of pyrimidine nucleosides. *J. Mol. Biol.* **47,** 531–543.
24. FULLER, W. D., SÁNCHEZ R. A. and ORGEL L. E. (1972) Studies in prebiotic synthesis. Solid-state synthesis of purine nucleosides. *J. Molec. Evolution* **1,** 249–257.
25. PONNAMPERUMA C. and MACH R. (1965) Nucleotide synthesis under possible primitive earth conditions. *Science* 148, 1221–1223.
26. LOHRMANN R. and ORGEL L. E. (1971) Urea–inorganic phosphate mixtures as prebiotic phosphorylating agents. *Science* **171,** 490–494.
27. MILLER, S. L. and PARRIS, M. (1964) Synthesis of pyrophosphate under primitive earth conditions. *Nature*, **204,** 1248–1250.
28. ORÓ, J. (1974) Avances recientes en evolución química. In *Simposio sobre evolución molecular y biólogica*, pp. 19–20, Instituto de Biología Fundamental Univ. Autónoma de Barcelona, Barcelona.
29. ORGEL, L. E. and LOHRMANN, R. (1974) Prebiotic chemistry and nucleic acid replication. *Accounts Chem. Res.* **7,** 368–377.
30. ORÓ, J. and STEPHEN-SHERWOOD, E. (1974) The prebiotic synthesis of oligonucleotides. *Origins of Life*, **5,** 159–172.
31. TS'O, P. O. P. (1974) In the beginning. In *Basic Principles in Nucleic Acid Chemistry*, Vol. I (Ts'o, P. O. P. ed.), pp. 1–92, Academic Press, New York.
32. IBAÑEZ, J. D., KIMBALL, A. P. and ORÓ, J. (1971) Possible prebiotic condensation of mononucleotides by cyanamide. *Science*, **173,** 444–445.
33. STEPHEN-SHERWOOD, E., ODOM, D. G. and ORÓ, J. (1974) The prebiotic synthesis of deoxythymidine oligonucleotides. *J. Molec. Evol.* **3,** 323–330.
34. FOX, S. W. and DOSE, K. (1972) *Molecular Evolution and the Origin of Life*, Freeman and Co. San Francisco.
35. HARADA, K. and FOX, S. W. (1975) Characterization of functional groups of acidic thermal polymers of α-amino acids. *Biosystems*, **7,** 213–221.
36. STEINMAN, G. D., LEMMON, R. M. and CALVIN, M. (1965) Dicyandiamide: possible role in peptide synthesis during chemical evolution. *Science*, **147,** 1574–1575.
37. CHANG, S., FLORES, J. and PONNAMPERUMA, C. (1969) Peptide formation mediated by hydrogen cyanide tetramer: a possible prebiotic process. *Proc. Nat. Acad. Sci. U.S.* **64,** 1011–1015.
38. PAECHT-HOROWITZ, M., BERGER, J. and KATCHALSKY, A. (1970) Prebiotic synthesis of polypeptides by heterogenous polycondensation of amino acid adenylates. *Nature*, **228,** 636–639.
39. STEPHEN-SHERWOOD, E., NOONER, D. W. and ORÓ, J. (1974) Condensation of amino acids in a system containing 4-amino-5-imidazole carboxyamide, cyanamide and adenosine-5′-triphosphate. *Fed. Proc.* 33, No. 5, Part II, Biochemistry/Biophysics Meeting, Minneapolis, Minn. June 2–7.
40. ODOM, D. and ORÓ, J. (1976) Prebiotic oligonucleotide synthesis: tidal washing. *Am. Soc. Biol. Chem. Meeting* (Abstract 21251), San Francisco.
41. LIPMANN, F. (1976) Reflections on the evolutionary transition from prokaryotes to eukaryotes. This Vol., pp. 34–39.
42. SCHOPF, J. W. (1974) The development and diversification of Precambrian life. *Origins of Life*, **5,** 119–135.
43. MARGULIS, L. (1970) *Origin of Eukaryotic Cells*, New Haven, Conn., Yale Univ. Press.
44. ORÓ, J. (1968) Synthesis of organic molecules by physical agencies. *J. Brit. Interplanet. Soc.* **21,** 12–25.

45. OPARIN, A. I. (1974) A hypothetic scheme for evolution of probionts. *Origins of Life*, **5**, 222–226.
46. EIGEN, M. (1971) Self-organization of matter and the evolution of macromolecules. *Naturwiss.* **58**, 465–523.

General References—Monographs

OPARIN, A. I. (1957) *The Origin of Life on Earth*, Academic Press, New York.

BERNAL, J. D. (1967) *The Origin of Life*, Weindenfeld and Nicolson, London; World Publishing Co., New York.

CALVIN, M. (1969) *Chemical Evolution*, Oxford University Press, New York.

KENYON, D. H. and STEINMAN, G. (1969) *Biochemical Predestination*, McGraw-Hill, New York.

MILLER, S. L. and ORGEL, L. E. (1973) *The Origins of Life on the Earth*, Prentice-Hall, Englewood Cliffs, New Jersey.

Proceedings (Conferences and Symposia)

OPARIN, *et al.* (eds.) (1959) *The Origin of Life on the Earth*, Pergamon Press, New York.

FOX, S. W. (ed.) (1965) *The Origin of Prebiological Systems*, Academic Press, New York.

BUVET, R. and PONNAMPERUMA, C. (eds.) (1971) *Chemical Evolution and the Origin of Life*, Vols. I and II, North Holland, Amsterdam.

KIMBALL, A. P. and ORÓ, J. (eds.) (1973) *Prebiotic and Biochemical Evolution*, North Holland, Amsterdam.

ORÓ, J., MILLER, S. L., PONNAMPERUMA, C. and YOUNG, R. S. (eds.) (1974) *Cosmochemical Evolution and the Origin of Life*, vols. I and II, D. Reidel Publishing Co., Dordrecht. The Netherlands.

General Reviews (Predominantly Organic Synthesis)

MILLER, S. L. and UREY, H. C. (1959) Organic compound synthesis on the primitive Earth. *Science*, **130**, 245–251.

HOROWITZ, N. H. and MILLER, S. L. (1962) Current theories on the origin of life. *Fortschr. Chem. Org. Naturst.* **20**, 423–459.

PONNAMPERUMA, C. and GABEL, N. W. (1968) Current status of chemical studies on the origin of life. *Space Life Sci.*, **1**, 64–96.

LEMMON, R. M. (1970) Chemical evolution. *Chem. Rev.* **70**, 95–109.

E. STEPHEN-SHERWOOD and ORÓ, J. (1973) Chemical evolution: recent syntheses of bio-organic molecules. *Space Life Sci.*, **4**, 5–31.

General Articles or Monographs (Primeval Conditions and Synthesis)

UREY, H. C. (1952) On the early chemical history of the earth and the origin of life. *Proc. Nat. Acad. Sci.* **38**, 351–363.

WALD, G. (1954) The origin of life. *Scientific American*, **191**, 44–53.

ORÓ, J. (1965) Investigation of organo-chemical evolution. In *Current Aspects of Exobiology* (MAMIKUNIAN, G. and BRIGGS, M. H., eds.), pp. 13–76, Jet Propulsion Laboratory, California Institute of Technology, Pasadena, Calif.

ABELSON, P. H. (1966) Chemical events on the primitive Earth. *Proc. Nat. Acad. Sci. U.S.A.* **55**, 1365–1372.

RUTTEN. M. G. (1971) *The Origin of Life by Natural Causes*, Elsevier, Amsterdam.

MY HOMAGE TO SEVERO OCHOA

"God does not play dice", wrote Albert Einstein long before the discovery of the DNA ladder, on which steps travel the angels, in the Jacob's dream I had the night before drawing that of Severo Ochoa; and they symbolize the genetic code messengers, or molecules of polynucleotides, which were synthesized for the first time in Severo Ochoa's laboratory.

Even though I am not a scientist, I must confess that the scientific events are the only ones that guide constantly my imagination, at the same time that they illustrate the poetic intuitions of traditional philosophers, to the point of the blinding beauty of certain mathematical structures, specially those of the polytopes, and above all those sublime moments of abstraction, which, "seen" through the electron microscope, appear as viruses of regular polyhedric form, confirming what Plato said: "God always makes Geometry".

SALVADOR DALÍ

Barcelona, September 15, 1975

AUTHOR INDEX

SUBJECT INDEX